Pollutants in Acid Mine Drainage

Pollutants in Acid Mine Drainage

Editor

Teresa Valente

MDPI • Basel • Beijing • Wuhan • Barcelona • Belgrade • Manchester • Tokyo • Cluj • Tianjin

Editor
Teresa Valente
Earth Sciences
University of Minho
Braga
Portugal

Editorial Office
MDPI
St. Alban-Anlage 66
4052 Basel, Switzerland

This is a reprint of articles from the Special Issue published online in the open access journal *Minerals* (ISSN 2075-163X) (available at: www.mdpi.com/journal/minerals/special_issues/PAMD).

For citation purposes, cite each article independently as indicated on the article page online and as indicated below:

LastName, A.A.; LastName, B.B.; LastName, C.C. Article Title. *Journal Name* **Year**, *Volume Number*, Page Range.

ISBN 978-3-0365-8329-7 (Hbk)
ISBN 978-3-0365-8328-0 (PDF)

Contents

About the Editor

Teresa Valente

Teresa Valente is an associate professor with habilitation at the University of Minho, Portugal. Her research domains are focused on environmental geochemistry, environmental monitoring, mine waters, and the rehabilitation of contaminated areas. Under these subjects, she coordinates numerous Master's and Ph.D. theses. She has participated in several collaborative projects related to the environmental impact of mining and soil degradation with European partners, as well as in other parts of the world, namely Africa and South America. Her research activity is developed at the Institute of Earth Sciences, which includes three of the leading Portuguese universities. She is the coordinator of the research group "Environmental Monitoring and Remediation for Sustainability". She is an associated editor of *Mine Water and the Environment*, a journal of the International Mine Water Association (IMWA). Her scientific production includes more than 80 ISI/Scopus papers.

minerals

Editorial

Editorial for Special Issue "Pollutants in Acid Mine Drainage"

Teresa Valente

Earth Sciences Department, Institute of Earth Sciences, Pole of the University of Minho, University of Minho, 4710-057 Braga, Portugal; teresav@dct.uminho.pt

Citation: Valente, T. Editorial for Special Issue "Pollutants in Acid Mine Drainage". *Minerals* **2023**, *13*, 931. https://doi.org/10.3390/min13070931

Received: 15 June 2023
Revised: 2 July 2023
Accepted: 11 July 2023
Published: 12 July 2023

Acid mine drainage (AMD) is among the major environmental concerns related to mining activity and often causes the complete degradation of affected ecosystems during and/or after mine closure. AMD is a particular research focus owing to its incorporation of assessments of various pollutants' presence, reactivity, and biogeochemical behavior [1,2]. Additionally, the properties of these polluted mine waters constrain the efficacy of monitoring plans and the environmental remediation approaches.

The systems affected by AMD have high heterogeneity and generate samples with complex matrices [3]. Moreover, AMD frequently creates high amounts of colloids [4], typically in the nanoscale dimensions [5]. Further, this mine drainage represents peculiar ecosystems dominated by acidophilic organisms involved in complex interactions that influence the migration and fate of pollutants [6–9]. In addition, the evolution of AMD promotes the development of new mineral phases that control the concentration of contaminants and seasonal behavior of the affected systems [10,11].

Extensive literature exists about the origin of AMD in relation to the oxidative dissolution of sulfide minerals [12–14]. Nevertheless, the complexity of biogeochemical processes involving the diversity of minerals and cyclic reactions with acidophilic microorganisms continue to be interesting research topics. Also, the state of the art refers to several works dealing with properties, environmental impacts, and treatment processes (e.g., [2,3,15,16]). However, in these peculiar waters, pollutants' diversity, mobility, and geochemical behavior are very site dependent. The paragenesis of each ore deposit, climate, and engineering options of exploitation and waste deposition (among many other factors) controls the nature and degree of pollution. Moreover, the importance of natural attenuation processes and monitoring efficiency depend on the geology (sensu lato), hydrology, and deposition structures, like waste piles and tailing dams. Important research efforts have been undertaken in the laboratory and field to increase knowledge about the behavior of these complex, affected systems [17].

In light of these considerations, this Special Issue addresses a wide variety of topics, such as the source and nature of pollutants, speciation, mobilization/precipitation, and toxicity of trace elements. Additionally, the articles featured in this publication cover the modeling of processes, innovative techniques for removal of hazardous elements, and advanced monitoring techniques, aiming to enlarge the base knowledge about pollutants in AMD. Therefore, the published papers present the latest advances in (bio)geochemistry and mineralogy of AMD and wastes from which AMD develops. The Special Issue contains 19 articles that provide examples of methodological approaches and novel tools and solutions for the monitoring, treatment, and remediation of AMD.

Geochemical modeling is an important research focus in this Special Issue [18–22]. This relevance is emphasized by Nordstrom's work [18], which showed the role of these models in understanding aspects such as speciation and natural attenuation, while exploring treatment processes for the recovery of metals in these complex matrices. In different geological contexts, Drapeau et al. [19] and Skierszkan et al. [20] also presented geochemical modeling to investigate the mobilization of lead and uranium, respectively. Geochemistry and mineralogy were used by Lemos et al. [21] to determine the AMD potential and establish the ability to mobilize potentially toxic elements (PTE), even in alkaline conditions. The authors highlighted the importance of using geochemical experiments to support sustainable tailings management. Combining geochemical modeling with acid–base neutralizing capacity (ANC–BNC) tests, Drapeau et al. [22] proposed an approach for characterizing siliceous and calcareous mining wastes and modeling pH evolution, and advocated element mobility as a function of the added amount of acid or base. However, with regard to the topic of modeling, Raymond et al. [23] focused on the behavior of waste rock cover systems and used a numerical model for understanding metal release as a function of water infiltration. In the domain of previewing the AMD potential, Moyo et al. [24] investigated the uncertainties associated with standard static tests and the importance of sulfur speciation when calculating potential acidity. The impact of the oxidation state of sulfur on the mobility of metals in contaminated sediments was investigated by Langman et al. [25]. The authors concluded that adding algal detritus to Fe- and S-rich sediments in a mining-impacted lake enhances Mn release but may not affect other toxic elements (e.g., As, Cd). Ódri et al. [26] investigated how the degree of isotope fractionation could help to understand the release and transport of pollutants in AMD.

The articles by Pi-Puig et al. [27], Menshikova et al. [28], and Kim [29] focused on the secondary minerals formed in a diversity of geological and climate scenarios. Pi-Puig et al. [27] proposed a genetic model for crusts and efflorescence and highlighted their roles as sources of contamination during the wet season. Menshikova et al. [28] characterized the secondary precipitates, emphasizing the amorphous phases formed under the influence of mine waters. Kim [29] demonstrated the natural attenuation potential of iron and Mn oxyhydroxides, which are able to retain toxicity mobilized from wastes with pyrite and manganese pyroxene.

The papers by Zawierucha et al. [30], Kato et al. [31], Ryskie et al. [32], and Santos et al. [33] were related to diverse processes and tools for the treatment of AMD. Zawierucha et al. [30] showed a novel approach based on the selectivity of elements of environmental concern, especially arsenic. The technique consisted of applying polymer inclusion membranes (PIMs) and revealed an efficiency of 90% in the removal of arsenic. Passive treatment through wetlands was considered by Kato et al. [31]. The authors modeled the mechanisms of pollutant removal by incorporating reactions into GETFLOWS. The results were consistent with adsorption on ferrihydrite observed in analyzed sediment samples. Ryskie et al. [32] presented a review of contaminants of emergent concern (CEC) in mine water and cold climates, e.g., rare earth elements, cyanide, and nitrogen compounds. The authors began by clarifying the definitions of this concept and continued reviewing the treatment strategies based on the best available technologies that are still economically viable.

Santos et al. analyzed microalgae's role in remediation [33], showing the use of algal growth for biosorption processes. The *Scenedesmus* genera revealed the ability to improve AMD quality and reduce pollutants toxicity. Algae were also analyzed in the study of Gomes et al. [34], exposing the stressful ecological conditions of AMD. The association between the algal community and hydrological patterns, obtained from digital surface models and flow maps of toxic elements, was presented as a relevant tool for biomonitoring AMD systems. In the same metallogenetic context (Iberian Pyrite Belt—IPB, SW Europe), there was the paradigmatic Rio Tinto, presented in the research of Olías et al. [35], which also showed the extreme conditions of AMD. The authors analyzed the evolution of pollutants concentrations and loads, highlighting the control by rainfall and concluding with the expected long-term persistence of pollutants mobilization in this river system. Still, in the IPB, the work by Flores et al. [36] presented a methodological approach to monitoring. They applied high-resolution unmanned aerial system (UAS)-based hyperspectral mapping and machine learning tools to detect geochemical and mineralogical variations and then related them with concentrations and physicochemical properties in the river system.

The above-presented works showed AMD to be a global problem worthy of investigation with manifestations worldwide. The diversity of addressed subjects illustrated the magnitude of this environmental problem and the recent evolution of knowledge about pollutants in AMD. Moreover, it contributed to new approaches to and techniques for AMD monitoring, modeling, and remediation.

Funding: This research was funded by FCT—Fundação para a Ciência and Tecnologia through projects UIDB/04683/2020, UIDP/04683/2020, and Nano-MINENV 029259 (PTDC/CTA-AMB/29259/2017).

Acknowledgments: The guest editor thanks the authors, reviewers, Academic Editor, and Minerals' staff for their contributions to this special issue.

Conflicts of Interest: The authors declare no conflict of interest.

References

1. Gray, N.F. Acid mine drainage composition and the implications for its impact on lotic systems. *Water Resour.* **1998**, *32*, 2122–2134. [CrossRef]
2. Nordstrom, D.K.; Blowes, D.W.; Ptacek, C.J. Hydrogeochemistry and microbiology of mine drainage: An update. *Appl. Geochem.* **2015**, *57*, 3–16. [CrossRef]
3. Wolkersdorfer, C.; Nordstrom, D.K.; Beckie, R.D.; Cicerone, D.S.; Elliot, T.; Edraki, M.; Valente, T.; França, S.C.A.; Kumar, P.; Oyarzún, R.; et al. Guidance for the Integrated Use of Hydrological, Geochemical, and Isotopic Tools in Mining Operations. *Mine Water Environ.* **2020**, *39*, 204–228. [CrossRef]
4. Kimball, B.A.; Callender, E.; Axtmann, E.V. Effects of colloids on metal transport in a river receiving acid mine drainage, Upper Arkansas River, Colorado, USA. *Appl. Geochem.* **1995**, *10*, 285–306. [CrossRef]
5. Valente, T.; Barroso, A.; Antunes, I.M.; Gomes, P.; Fonseca, R.; Pinho, C.; Pamplona, J.; Sequeira Braga, M.A.; Sousa, J.P.S. *Acid Mine Drainage Precipitates at the Nanometric Scale—Properties and Environmental Role;* Stanley, P., Wolkersdorfer, C., Wolkersdorfer, K., Eds.; International Mine Water: Cardiff, UK; pp. 625–631.
6. Ehrlich, H.L. *Geomocrobiology;* Marcel Dekker: New York, NY, USA, 1996; 719p.
7. Robbins, E.I. Bacteria and Archaea in acidic environments and a key to morphological identification. *Hydrobiologia* **2000**, *433*, 61–89. [CrossRef]
8. Sabater, S.; Buchaca, T.; Cambra, J.; Catalan, J.; Guasch, H.; Ivorra, N.; Muñoz, I.; Navarro, E.; Real, M.; Romaní, A. Structure and function of benthic algal communities in an extremely acid river 1. *J. Phycol.* **2003**, *39*, 481–489. [CrossRef]
9. Valente, T.M.; Leal Gomes, C. The role of two acidophilic algae as ecological indicators of acid mine drainage sites. *J. Iberian Geol.* **2007**, *33*, 283–294.
10. Bigham, J.; Schwertmann, U.; Traina, S.; Winland, R.; Wolf, M. Schwertmannite and the chemical modeling of iron in acid sulfate waters. *Geochim. Cosmochim. Acta* **1996**, *60*, 2111–2121. [CrossRef]
11. Valente, T.M.; Gomes, C.L. Occurrence, properties and pollution potential of environmental minerals in acid mine drainage. *Sci. Total Environ.* **2009**, *407*, 1135–1152. [CrossRef]
12. McKibben, A.A.; Barnes, H.L. Oxidation of pyrite in low temperature acidic solutions: Rate laws and surface textures. *Geochim. Cosmochim. Acta* **1986**, *50*, 1509–1520. [CrossRef]
13. Evangelou, V.P.; Zhang, Y.L. A review: Pyrite oxidation mechanisms and acid mine drainage prevention. *Crit. Rev. Environ. Sci. Technol.* **1995**, *25*, 141–199. [CrossRef]
14. Rimstidt, J.D.; Vaughan, D.J. Pyrite oxidation: A state-of-the-art assessment of the reaction mechanism. *Geochim. Cosmochim. Acta* **2003**, *67*, 873–880. [CrossRef]
15. Hudson-Edwards, K.; Jamieson, H.E.; Lottermoser, B.G. Mine wastes: Past, present, future. *Elements* **2011**, *7*, 375–380. [CrossRef]
16. Valente, T.; Grande, J.; de la Torre, M.; Santisteban, M.; Cerón, J. Mineralogy and environmental relevance of AMD-precipitates from the Tharsis mines, Iberian Pyrite Belt (SW, Spain). *Appl. Geochem.* **2013**, *39*, 11–25. [CrossRef]
17. Arnold, M.; Kangas, P.; Mäkinen, A.; Lakay, E.; Isomäki, N.; Lavén, G.; Gericke, M.; Pajuniemi, P.; Tommi Kaartinen, T.; Wendling, L. Mine water as a resource: Selective removal and recovery of trace antimony from mine-impacted water. *Mine Water Environ.* **2019**, *38*, 431–446. [CrossRef]
18. Kirk Nordstrom, D. Geochemical Modeling of Iron and Aluminum Precipitation during Mixing and Neutralization of Acid Mine Drainage. *Minerals* **2020**, *10*, 547. [CrossRef]
19. Drapeau, C.; Argane, R.; Delolme, C.; Blanc, D.; Benzaazoua, M.; Hakkou, R.; Baumgartl, T.; Edraki, M.; Lassabatere, L. Lead Mobilization and Speciation in Mining Waste: Experiments and Modeling. *Minerals* **2021**, *11*, 606. [CrossRef]
20. Skierszkan, E.K.; Dockrey, J.W.; Mayer, K.U.; Bondici, V.F.; McBeth, J.M.; Beckie, R.D. Geochemical Controls on Uranium Release from Neutral-pH Rock Drainage Produced by Weathering of Granite, Gneiss, and Schist. *Minerals* **2020**, *10*, 1104. [CrossRef]
21. Lemos, M.; Valente, T.; Reis, P.M.; Fonseca, R.; Delbem, I.; Ventura, J.; Magalhães, M. Mineralogical and Geochemical Characterization of Gold Mining Tailings and Their Potential to Generate Acid Mine Drainage (Minas Gerais, Brazil). *Minerals* **2021**, *11*, 39. [CrossRef]
22. Drapeau, C.; Delolme, C.; Vézin, C.; Blanc, D.; Baumgartl, T.; Edraki, M.; Lassabatere, L. ANC–BNC Titrations and Geochemical Modeling for Characterizing Calcareous and Siliceous Mining Waste. *Minerals* **2021**, *11*, 257. [CrossRef]
23. Raymond, K.E.; Seigneur, N.; Su, D.; Poaty, B.; Plante, B.; Bussière, B.; Mayer, K.U. Numerical Modeling of a Laboratory-Scale Waste Rock Pile Featuring an Engineered Cover System. *Minerals* **2020**, *10*, 652. [CrossRef]

24. Moyo, A.; Amaral Filho, J.R.D.; Harrison, S.T.L.; Broadhurst, J.L. Implications of Sulfur Speciation on the Assessment of Acid Rock Drainage Generating Potential: A Study of South African Coal Processing Wastes. *Minerals* **2019**, *9*, 776. [CrossRef]
25. Langman, J.B.; Ali, J.D.; Child, A.W.; Wilhelm, F.M.; Moberly, J.G. Sulfur Species, Bonding Environment, and Metal Mobilization in Mining-Impacted Lake Sediments: Column Experiments Replicating Seasonal Anoxia and Deposition of Algal Detritus. *Minerals* **2020**, *10*, 849. [CrossRef]
26. Ódri, Á.; Becker, M.; Broadhurst, J.; Harrison, S.T.L.; Edraki, M. Stable Isotope Imprints during Pyrite Leaching: Implications for Acid Rock Drainage Characterization. *Minerals* **2020**, *10*, 982. [CrossRef]
27. Pi-Puig, T.; Solé, J.; Gómez Cruz, A. Mineralogical Study and Genetic Model of Efflorescent Salts and Crusts from Two Abandoned Tailings in the Taxco Mining District, Guerrero (Mexico). *Minerals* **2020**, *10*, 871. [CrossRef]
28. Menshikova, E.; Osovetsky, B.; Blinov, S.; Belkin, P. Mineral Formation under the Influence of Mine Waters (The Kizel Coal Basin, Russia). *Minerals* **2020**, *10*, 364. [CrossRef]
29. Kim, Y. Geochemical Behavior of Potentially Toxic Elements in Riverbank-Deposited Weathered Tailings and Their Environmental Effects: Weathering of Pyrite and Manganese Pyroxene. *Minerals* **2020**, *10*, 413. [CrossRef]
30. Zawierucha, I.; Nowik-Zajac, A.; Malina, G. Selective Removal of As(V) Ions from Acid Mine Drainage Using Polymer Inclusion Membranes. *Minerals* **2020**, *10*, 909. [CrossRef]
31. Kato, T.; Kawasaki, Y.; Kadokura, M.; Suzuki, K.; Tawara, Y.; Ohara, Y.; Tokoro, C. Application of GETFLOWS Coupled with Chemical Reactions to Arsenic Removal through Ferrihydrite Coprecipitation in an Artificial Wetland of a Japanese Closed Mine. *Minerals* **2020**, *10*, 475. [CrossRef]
32. Ryskie, S.; Neculita, C.M.; Rosa, E.; Coudert, L.; Couture, P. Active Treatment of Contaminants of Emerging Concern in Cold Mine Water Using Advanced Oxidation and Membrane-Related Processes: A Review. *Minerals* **2021**, *11*, 259. [CrossRef]
33. Santos, K.B.D.; Almeida, V.O.D.; Weiler, J.; Schneider, I.A.H. Removal of Pollutants from an AMD from a Coal Mine by Neutralization/Precipitation Followed by "In Vivo" Biosorption Step with the *Microalgae Scenedesmus* sp. *Minerals* **2020**, *10*, 711. [CrossRef]
34. Gomes, P.; Valente, T.; Albuquerque, T.; Henriques, R.; Flor-Arnau, N.; Pamplona, J.; Macías, F. Algae in Acid Mine Drainage and Relationships with Pollutants in a Degraded Mining Ecosystem. *Minerals* **2021**, *11*, 110. [CrossRef]
35. Olías, M.; Cánovas, C.R.; Macías, F.; Basallote, M.D.; Nieto, J.M. The Evolution of Pollutant Concentrations in a River Severely Affected by Acid Mine Drainage: Río Tinto (SW Spain). *Minerals* **2020**, *10*, 598. [CrossRef]
36. Flores, H.; Lorenz, S.; Jackisch, R.; Tusa, L.; Contreras, I.C.; Zimmermann, R.; Gloaguen, R. UAS-Based Hyperspectral Environmental Monitoring of Acid Mine Drainage Affected Waters. *Minerals* **2021**, *11*, 182. [CrossRef]

minerals

MDPI

Article

Geochemical Modeling of Iron and Aluminum Precipitation during Mixing and Neutralization of Acid Mine Drainage

Darrell Kirk Nordstrom

U.S. Geological Survey, Boulder, CO 80303, USA; dkn@usgs.gov

Received: 21 May 2020; Accepted: 14 June 2020; Published: 17 June 2020

Abstract: Geochemical modeling of precipitation reactions in the complex matrix of acid mine drainage is fundamental to understanding natural attenuation, lime treatment, and treatment procedures that separate constituents for potential reuse or recycling. The three main dissolved constituents in acid mine drainage are iron, aluminum, and sulfate. During the neutralization of acid mine drainage (AMD) by mixing with clean tributaries or by titration with a base such as sodium hydroxide or slaked lime, $Ca(OH)_2$, iron precipitates at pH values of 2–3 if oxidized and aluminum precipitates at pH values of 4–5 and both processes buffer the pH during precipitation. Mixing processes were simulated using the ion-association model in the PHREEQC code. The results are sensitive to the solubility product constant (K_{sp}) used for the precipitating phases. A field example with data on discharge and water composition of AMD before and after mixing along with massive precipitation of an aluminum phase is simulated and shows that there is an optimal K_{sp} to give the best fit to the measured data. Best fit is defined when the predicted water composition after mixing and precipitation matches most closely the measured water chemistry. Slight adjustment to the proportion of stream discharges does not give a better fit.

Keywords: geochemical modeling; acid mine drainage; iron and aluminum precipitation; schwertmannite; basaluminite

---

1. Introduction

Geochemical modeling is often limited by lack of sufficient mineralogic and hydrologic data, lack of sufficient temporal and spatial data, and lack of sufficient understanding of complex processes that dictate water compositions [1]. Substantial progress has been made over the last century and sophisticated codes use models to simulate mineral precipitation/dissolution, sorption, gas exchange, ionic strength effects, thermal effects, and transport [2]. Even with these advances, numerous assumptions must be made when applying these simulations to field conditions. Continual testing of models against field data is needed to build confidence that models are reasonable for a variety of water-rock interactions.

This paper applies a geochemical model to the process of neutralization of acid mine drainage (AMD) with particular reference to the effects of the precipitation of insoluble Fe and Al phases during mixing. Certain assumptions are examined such as the specific phase of the precipitate, its thermodynamic properties, and the sensitivity to mixing parameters. For example, schwertmannite commonly precipitates from acid mine drainage over the pH range of 2–4 [1], at slightly higher pH values ferrihydrite dominates precipitation, and at pH values near or below 2.0, jarosite precipitates (e.g., [3]). Of course, the presence or absence of specific anions play a role in determining which phases precipitate [4]. Ferrihydrite, akagenéite, lepidocrocite, and goethite may also co-precipitate during oxidation of dissolved ferrous iron. Indeed, mixtures of these minerals have been found associated with

AMD and their identification is complicated by their fine-grained particle size and poor crystallinity (e.g., [3,5,6]). Another complication is that the first phase to precipitate is typically the least stable and it tends to age to a more stable phase of lower solubility. Schwertmannite converts to microcrystalline goethite and/or jarosite [7]. The transformation rate is pH dependent [8]. Basaluminite converts to alunite [9]. Techniques such as differential X-ray diffraction, electron diffraction, transmission and scanning electron photomicroscopy, synchrotron-based X-ray absorption spectroscopy, and X-ray diffraction using advanced light sources have all been important in the identification of these phases and understanding their transformations [10,11].

Similar concerns are found with the precipitation of aluminum hydroxide and hydroxysulfate phases. Basaluminite, hydrobasaluminite, aluminite, meta-aluminite, zaherite, gibbsite, microcrystalline gibbsite, boehmite, diaspore, and alunite have all been found to precipitate in lab solutions and as mixtures of phases (e.g., [12]). A summary review of these aluminous phases is also given in [1].

This paper begins with simulations of acid-base titrations with two examples using actual acid mine water analyses and using $Ca(OH)_2$ as the titrant. Both NaOH and $Ca(OH)_2$ have been used to neutralize AMD in treatment plants although $Ca(OH)_2$ is less expensive and easier to handle. These simulations are important because they demonstrate how geochemical modeling is helpful in determining the minimum amount of base necessary to run a treatment plant, the inflection points that reflect buffering mechanisms, and the change in concentration of constituents during the titration. The final simulation is a field example where a near neutral tributary mixes with AMD to bring the pH up from 3.25 to 4.90 and precipitates large quantities of an aluminum phase at the mixing point. Data for discharges as well as full chemistry of the waters above and below mixing are available so that the simulated chemistry with mass fluxes after mixing can be compared with the actual measured values to evaluate the assumed thermodynamic properties of the precipitating phases. A close match between simulated vs. measured properties depends on the solubility product constants for Fe and Al precipitates as well as the relative proportions of stream discharges in the mixing zone.

2. Materials and Methods

Simulations were obtained using the PHREEQCI code v.3.4.0 [13] using the graphical user interface and the wateq4f.dat database [14]. Of the several available databases in PHREEQC, only the wateq4f.dat database has the thermodynamic properties for colloidal minerals that typically precipitate from AMD.

In the results that follow, two different acid mine water samples were chosen for titration simulations and a field site where full chemistry and discharge measurements were obtained for the confluence of a clean carbonate-buffered stream mixing with AMD. The two AMD sample compositions were obtained from the Leviathan Mine data set [15] and the Berkeley Pit data set [16], and a spreadsheet provided by the Montana Bureau of Mines and Geology. Leviathan Mine is a USEPA Superfund site in eastern California (Figure 1). The Berkeley Pit is a large open pit filled with AMD and is a USEPA Superfund site near Butte, Montana. The Leviathan sample was identified as 82WA118 collected on 6 June 1982 near the portal of the main tunnel. The Berkeley Pit sample was collected on 4 June 2012 from a depth of 15.2 m. The field samples used for simulation of the mixing of streams were AMD sample 82WA110 from Leviathan Creek, clean tributary Mountaineer Creek sample 82WA108, and the confluent water sample 82WA109, all collected on 15 June 1982 [15].

The Leviathan Mine primarily mined sulfur. The native sulfur was used to make sulfuric acid to leach metals from rock excavated at the Yerington Pit in Nevada. Although it was a sulfur deposit, there was considerable cryptocrystalline pyrite [17] that produced AMD. Oxidation of the native sulfur did not contribute much to the AMD because the isotopic composition and trace element content clearly indicated that the pyrite was the main source of the AMD [18]; marcasite and chalcopyrite were also present but are thought to be less important than pyrite in the generation of AMD [18].

A white aluminous mineral that was observed to precipitate below the confluence of Leviathan and Mountaineer Creeks in 1981 and 1982 can be seen in Figure 2. Remediation on site since this time has eliminated these precipitates. No water with pH < 5 now leaves the mine site.

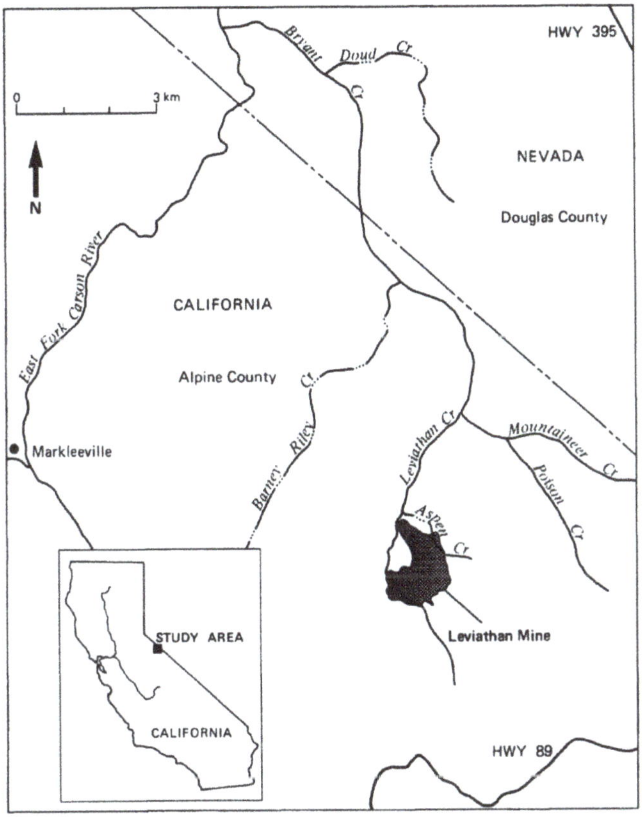

Figure 1. Location of Leviathan mine in eastern California (insert) and the main tributaries receiving acid mine drainage [15]. In the early 1980s aluminum precipitates were found immediately below the confluence of Mountaineer Creek with Leviathan Creek.

Figure 2. White aluminum-rich precipitate found below the confluence of Leviathan and Mountaineer Creeks, near Markleeville, California. Photo: D. Kirk Nordstrom, U.S. Geological Survey, June 1982.

Details of sampling and analytical procedures and site locations can be found in the respective references cited above and in [19] for Leviathan data. Bottles for cations were acid washed and rinsed in distilled water. Anion bottles were thoroughly rinsed with distilled water only. The Leviathan samples were all filtered on site through 0.1 μm pore size Millipore™ membranes. Field measurements included pH, temperature, specific conductance, electromotive force (EMF, for calculation of Eh), and discharge. Cation samples were acidified with ultrapure concentrated nitric acid (2 mL per 250 mL sample). A separate bottle for Fe(II/III) and As(III/V) was collected and acidified with ultrapure (redistilled) concentrated hydrochloric acid (4 mL per 500 mL sample). All cation samples had a pH ≤ 1.5 after acidification. Anion samples were filtered and kept chilled.

Analytical results were of high quality because three different techniques were used to confirm the metal and metalloid concentrations: direct-current plasma spectrometry, inductively coupled plasma spectrometry, and Zeeman-corrected graphite furnace atomic absorption spectrometry in a comprehensive analytical evaluation [20]. Charge balances were between −5% and +6% and mostly <±3%. Dissolved iron redox species were determined by FerroZine colorimetry and arsenic redox species by hydride generation atomic absorption spectrometry. Six U.S. Geological Survey (USGS) standard reference water samples were analyzed to check the accuracy and precision including two samples that were typical of acid mine water compositions.

Discharge data were collected using a Pygmy current meter or standard size current meter to record velocity and cross-sectional area was measured from the width and depth of the streams with tape measurements at appropriate intervals.

As a test of a titration simulation with PHREEQC, a sulfuric acid solution was titrated with sodium hydroxide. A typical strong acid–strong base titration of this type is shown in Figure 3 for 5 mL of 0.1 m H_2SO_4 with 0.1 m NaOH. Using the REACT command in PHREEQC only the proportion of the two solutions is needed and the graphical titration results should be the same with the same moles of base per mole of acid. As can be seen by the results in Figure 4, the titration curve is the same with the same inflection point at a pH = 7 and the same number of milliequivalents.

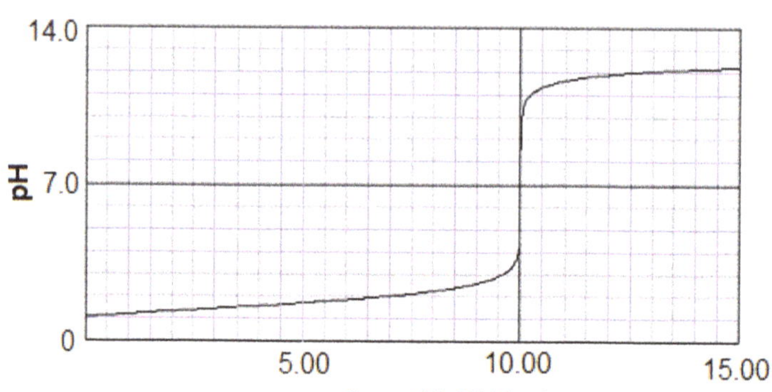

Figure 3. Typical titration graph of 5 mL 100 millimolal H_2SO_4 with 100 millimolal NaOH.

Choices must be made about what minerals should be allowed to precipitate during the simulation of AMD titrations and what solubility product constants should be used. For these simulations "amorphous" $Fe(OH)_3$ with a log K_{sp} = 3.0 [21,22], and "amorphous" basaluminite with log K_{sp} = 24.0 [23,24] were used. The effect of changing these solubility product constants to other phases such as goethite and schwertmannite for the iron precipitate and gibbsite for the aluminum precipitate was tested to determine that effect on the results for the field data. Because of the possibility of gypsum, calcite, and portlandite precipitating in the base titrations, these phases were forced to precipitate if they reached saturation.

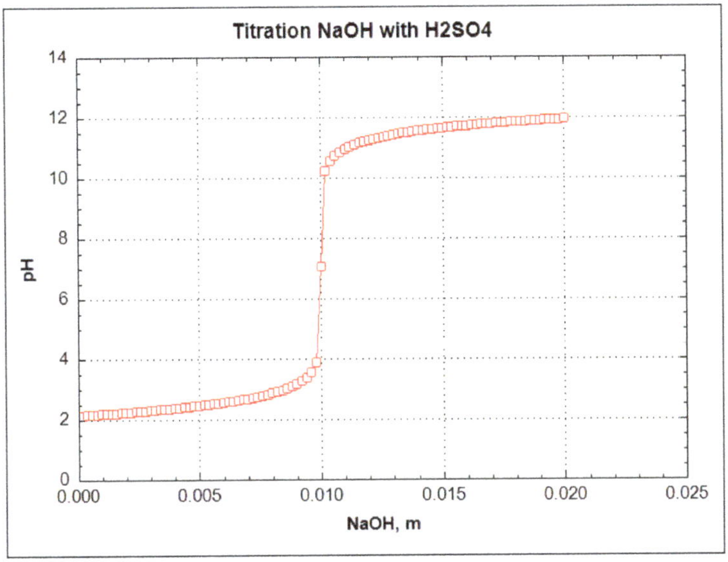

Figure 4. Simulation of pH titration of 5 millimolal H_2SO_4 with 20 millimolal NaOH added in small increments using REACT command and USER GRAPH to plot results and to check accuracy of computation. The titration is plotted in terms of total number of moles used in the titration.

The presence or absence of $CO_{2(g)}$ also makes a difference to the titration curve at pH values above 7 so the log P_{CO2} was set at −3.5 (near atmospheric value). Today's atmospheric log P_{CO_2} is actually closer to −3.38 and is increasing over time but this small difference has negligible effect on the results.

3. Results and Discussion

3.1. Titration Results of Leviathan Mine Water with $Ca(OH)_2$

The first titration simulated using actual AMD compositions is the Leviathan Mine water with initial pH of 1.80, 1570 mg Fe/L, 438 mg Al/L and 7,540 mg SO_4/L. Nearly 92% of the Fe was present as Fe(II) in the immediate portal drainage water, but during an oxygenated titration it would become Fe(III) so it was assumed to be Fe(III) for the whole simulation. To set this constraint, the solution was kept in equilibrium with atmospheric oxygen. To begin with the actual redox speciation, oxidation kinetics would have had to be built into the simulation. To simulate the kinetics would require a large amount of data and information that would vary from site to site and not easily definable beforehand. Furthermore, the rate of oxidation will depend on the experimental conditions of the titration for which several different methods exist. More importantly, dissolved Fe(II) oxidizes rather quickly from microbial activity so that the assumption of the Fe(III) = Fe_T is a reasonable approximation if the sample is not filtered. A graph of the titration is shown in Figure 5.

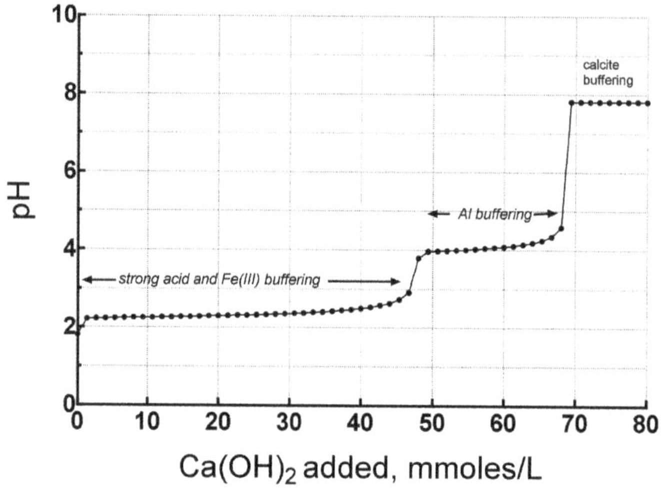

Figure 5. Titration simulation of acid mine drainage (AMD) from Leviathan Mine drainage by portlandite (slaked lime), showing the main buffering mechanisms (similar to the mine tailings model, [24–26]).

Three pH plateaus are visible at 2.2–2.6, 4, and 7.8. These plateaus indicate buffering. The plateau below pH 3 is buffering by strong acid and Fe(III) hydrolysis and precipitation. The buffering by the second hydrolysis of H_2SO_4 has a $pK_2 = 1.98$, very close to the $pK_1 = 2.19$ for Fe(III) so that their buffering reactions overlap substantially. The 45 mmoles of base required to neutralize this acidity is about 66% of the total acidity. The plateau at pH 4 reflects buffering by Al hydrolysis and precipitation which has been elucidated by [25–27] for mine tailings. Once the acidity from Al has been neutralized, the pH increases until buffering by calcite predominates at pH 7.8. Note that about 68 mmoles of portlandite were required to neutralize this AMD to a pH of 7. Next, we compare these results with those of a Berkeley Pit water titration.

3.2. Titration Results of Berkeley Pit Water with $Ca(OH)_2$

Figure 6 shows the Berkeley Pit titrated with portlandite and a few important differences with the Leviathan AMD titration are worth noting. The Berkeley Pit water had a pH of 2.56, 277 mg Al/L, 246 mg Fe/L, and 8005 mg SO_4/L. The SO_4 concentrations were nearly the same but the Fe and Al concentrations were notably lower than those for the Leviathan Mine water, and the pH was higher. The dissolved Fe concentration is only 16% of the Fe concentration in the Leviathan Mine water. The first important result is that the amount of portlandite needed to neutralize this water to a pH near 7 is about 21 mmoles, which is substantially less than the 68 mmoles of base needed to neutralize the Leviathan Mine water. The reason is the amount of strong acid in the Leviathan water was greater and the amount of hydrolyzable Fe and Al was greater for the Leviathan water. The amount of base to neutralize Al at pH 4 took only about 10 mmoles of base for the Berkeley Pit water whereas the Leviathan water took about 20 mmoles, which is consistent with the Berkeley Pit having 50% of the Al concentration of the Leviathan sample.

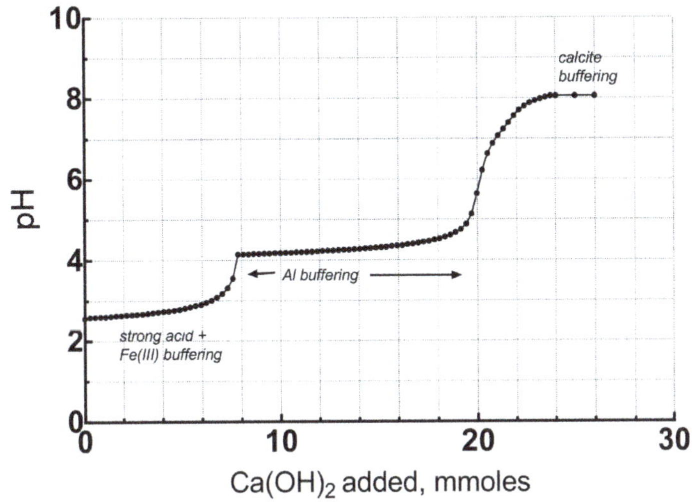

Figure 6. Titration of slaked lime or portlandite with AMD from Berkeley Pit water, showing the main buffering mechanisms with lower overall acidity than the Leviathan Mine water.

The Ca concentrations (or any other solute concentration) can be followed with the USER GRAPH routine in PHREEQC as shown in Figure 7 for the Berkeley Pit titration. The water is at saturation with respect to gypsum initially so that as portlandite is added, more gypsum precipitates and the Ca concentration decreases until all the SO_4 that can react with the portlandite is used up. After that the Ca concentration increases until it reaches calcite solubility equilibrium at about pH 8.

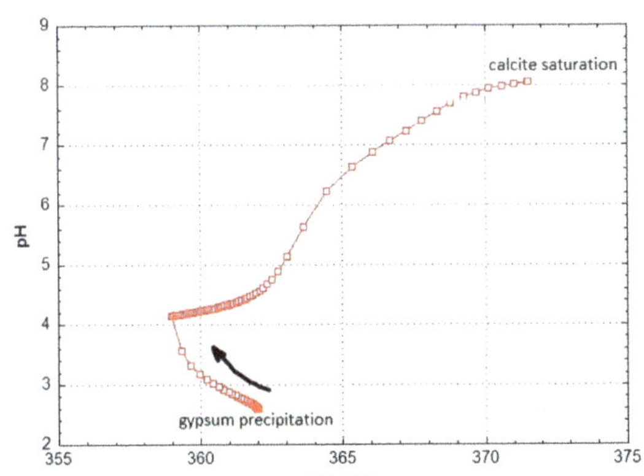

Figure 7. Change in Ca concentration during titration of Berkeley Pit water with portlandite.

3.3. *Mixing of Leviathan AMD with a Clean Tributary*

The AMD from the Leviathan mine flowed into Leviathan Creek which was diluted by some clean tributaries before it reached its confluence with Mountaineer Creek. Just before the confluence, the pH was 3.25 on 15 June 1982 and the water composition is shown in Table 1 along with the composition of Mountaineer Creek just upstream of the confluence.

Table 1. Analyses of Leviathan and Mountaineer Creeks collected on 15 June 1982.

Parameters	Leviathan Creek	Mountaineer Creek
Temperature, °C	19.5	12.5
Discharge, m^3/s	0.071 (44%)	0.091 (56%)
pH	3.25	8.85
Specific conductance, μS/cm	1100	150
Constituents (mg/L)		
Alkalinity, as HCO_3	0.0	94.3
Al	19.8	0.045
Ba	0.048	0.039
Ca	82.2	13.7
Cd	0.0079	0.0002
Cl	1.1	1
Cu	0.231	0.001
F	0.52	0.04
Fe (II)	9.01	0.0086
Fe (total dissolved)	18.4	0.0099
Mg	23.6	5.78
Mn	3.04	0.022
K	4.57	2.29
SiO_2	46.4	42.6
Na	11.8	6.83
Sr	0.708	0.237
SO_4	483	1.89

A sequence of simulations or scenarios was chosen, beginning with simple mixing with no reaction—the conservative mixing scenario. With increasing complexity and sophistication of each reaction modeling scenario, the results from each model simulation were compared to the actual field data. The conservative model is scenario A; scenario B is mixing with crystalline goethite and crystalline gibbsite as precipitates for Fe and Al; scenario C is mixing with precipitation of amorphous $Fe(OH)_3$ and amorphous $Al(OH)_3$ as precipitates; scenario D is mixing with log K_{sp} = 3.0 for amorphous $Fe(OH)_3$ [21,22] and amorphous basaluminite with log K_{sp} = 24.0 [23,28]; scenario E is the same as scenario D but substituting a schwertmannite log K_{sp} = 18.5 which agrees generally with the results of [1,24,29,30] and the formula would be $Fe_8O_8(OH)_{5.48}(SO_4)_{1.26}$ (See Supplementary Materials for an example of PHREEQCI input for this calculation). Results for these simulations are shown in Table 2, where "Measured" indicates the composition of Leviathan Creek below the confluence with Mountaineer Creek.

Any difference in concentration between the mixed water in scenario A and the "Measured" data set is an indication of the amount of reaction that must have occurred from precipitation or sorption. Several constituents are close enough in concentration (within analytical error) between scenario A and "Measured" that they can be considered "conservative," in that they are not entering into detectable reaction: Ba, Ca, Cd, F, Mg, and Sr. Those constituents that are clearly "non-conservative" are pH, Al, HCO_3 (not alkalinity), and Fe. Then there are a few constituents that are borderline: Cu, K, Mn, Na, SiO_2, and SO_4; these shall be addressed later. The other important aspect to note about scenario A is that there is much more acidity stored in reacting constituents than results from simple mixing and this acidity comes from Fe and Al hydrolysis as seen by the titration plots shown previously.

Table 2. Results of simulations resulting from mixing and reaction (see text for explanation).

Constituent (mg/L, Except pH)	A	B	C	D	E	Measured
pH	6.13	4.41	5.36	4.55	4.57 (4.64)	4.90
Al	8.75	4.80	5.90	4.16	3.67 (2.61)	5.06
Ba	0.043	0.043	0.043	0.043	0.042 (0.043)	0.042
HCO_3	37.5	0.87	0.87	0.88	0.88 (0.88)	0.0
Ca	43.8	43.8	43.8	43.8	43.8 (42.4)	44.7
Cd	0.0036	0.0036	0.0036	0.0036	0.0036 (0.0034)	0.0040
Cu	0.102	0.102	0.102	0.102	0.102 (0.098)	0.093
F	0.25	0.25	0.25	0.25	0.25 (0.24)	0.30
Fe (total)	8.13	0.00000054	0.0181	0.00167	0.0254 (0.0223)	4.72
K	3.30	3.29	3.30	3.29	3.29 (3.25)	3.21
Mg	13.6	13.6	13.6	13.6	13.6 (13.3)	13.5
Mn	1.35	1.35	1.35	1.35	1.35 (1.29)	1.26
Na	9.03	9.02	9.03	9.02	9.02 (8.89)	8.60
SiO_2	44.3	44.2	44.3	44.2	44.2 (44.1)	42.6
Sr	0.44	0.44	0.44	0.44	0.44 (0.44)	0.44
SO_4	214	214	214	209	207 (197)	206

Scenario B computes a pH that is too low because crystalline goethite and gibbsite were chosen to precipitate, requiring more base to neutralize these phases because they are less soluble than the phases normally found precipitating in AMD. With more Fe and Al hydrolysis, more acidity is produced, and the resultant pH is lower than the actual measured value. The calculated Fe concentration is also much too low compared to the measured value.

Scenario C is the other extremum where the chosen precipitating phases are the more soluble amorphous $Fe(OH)_3$ and amorphous $Al(OH)_3$ phases. Now the pH is too high because a much smaller quantity of Al precipitated and the dissolved Al is too high, although the Fe is still too low.

Scenario D uses an adjusted log K_{sp} for amorphous $Fe(OH)_3$ which is consistent with the field data reported by [21,22] as previously mentioned. It also uses amorphous basaluminite instead of a gibbsite-type phase. Now the pH is more reasonable and the SO_4 concentration does decrease slightly because of basaluminite precipitation.

In scenario E, conditions were kept the same as in scenario D except that schwertmannite was substituted for amorphous $Fe(OH)_3$. Schwertmannite has a somewhat variable stoichiometry which makes it difficult to designate a K_{sp} [31], so a commonly found value was used as mentioned above primarily to see what effect it had on the results. Indeed, the results were the same as scenario D except that the Fe concentration was a bit higher, the Al concentration was a bit lower, and the SO_4 concentration a bit lower. The change in SO_4 concentration, although small, is consistent with the precipitation of Fe and Al hydroxysulfate minerals. To show the effect of a small change in the mixing proportions, the discharge of Leviathan Creek was changed from 44% of the total to 42% of the total so that the pH might make a better match. These values are shown in parentheses in scenario E of Table 2. Changing the relative discharges does not really improve the overall mass balances. The important constituent to note is the SO_4 concentration which is worse and would argue against any changes in relative discharges.

Recalling the "borderline" constituents Cu, K, Mn, Na, SiO_2, and SO_4 the following can be suggested. Copper may be exhibiting a real decrease through sorption and has been modeled for the overall drainage by [19]. The change in K concentration is considered to be within analytical error. The small change in Mn concentration is likely real and reflecting a precipitate or co-precipitate of some oxide or hydroxide of Mn. The small decrease in Na concentration could be real or could be analytical error. The small decrease in silica concentration is considered real because there is some association of silica with secondary Fe and Al precipitates [32]. Indeed, silica is required to precipitate from mass balance considerations [33] during the oxidation of pyrite and often reaches silica solubility limits. Caraballo et al. [34] have identified a silica coating on hydrobasaluminite precipitating from the Paradise Portal in Colorado. The form of the silica is not usually known. The match for the decrease in SO_4 concentration is considered as corroboratory evidence that the proposed reaction model (E) is a reasonable interpretation of the reactive processes during mixing.

With regard to the higher concentrations of Al and Fe than simulated after precipitation, the mismatch is likely to be an artifact of field collection limitations rather than the simulation. First, fresh precipitation of Fe(III) forms colloids in the nanometer to tens of nanometer size range that can easily escape filtration (e.g., [31]). This problem is pronounced at pH values at or above 4 [33]. Second, of the measured dissolved total Fe 4.44 mg/L was Fe(II) leaving 0.28 mg/L measured Fe(III). If the Eh value of 0.692 volts is used to calculate the Fe(III) it would be about 2 orders of magnitude oversaturated with respect to amorphous $Fe(OH)_3$. If the calculation is redone (because the EMF measurement was very weak and drifting) assuming equilibrium with amorphous $Fe(OH)_3$ rather than trusting the EMF value, the dissolved Fe(III) would be about 0.006 mg/L. It is safe to say that the difference between 0.28 mg/L and 0.006 mg/L can be attributed to colloidal iron particles getting through the filter [33]. Similar nanometer size ranges have been documented for hydrobasaluminite [34].

4. Conclusions

Geochemical modeling of water-rock interactions has a substantial number of applications. However, the modeler has several assumptions to make for any given environmental setting. These assumptions include whether minerals reach equilibrium or not, which minerals should be considered, what redox conditions to choose and how to set them, and which database or thermodynamic properties should be used. Often, constraints from field studies are insufficient to avoid some of these assumptions and further work is needed to confirm or discredit the dominant reactions and processes.

Titration simulation of AMD with a commonly used base, portlandite or $Ca(OH)_2$, is a useful application when planning a neutralization plant to treat acid waters. Simulations for two different acid mine waters show clearly that sulfuric acid, Fe(III), and Al not only play the main role in buffering, but quantify the amount needed to neutralize AMD. One example from Leviathan mine water data required more than twice the amount of base than the other mine water from the Berkeley Pit because of the substantial difference in acid, Fe, and Al concentrations. Strong acid and Fe(III) buffering occurs in the pH range of 2.2–2.8 and Al buffering occurs at a pH close to 4.

In this study, a well-constrained mixing phenomenon of two streams, one with AMD at a pH of 3.25 with a clean tributary increased the pH to 4.90 by dilution and carbonate buffering. This example is a field titration in a single mix. Immediate Al precipitation and some Fe precipitation occurred with mixing. Full chemistry and discharges measured above and below the mixing point provided the opportunity to simulate the results of mixing with a geochemical model and compare them to the field measurements. The effect of no reaction (conservative mixing), reaction with the least soluble and the most soluble Fe and Al hydroxide precipitates, and with colloidal Fe and Al hydroxysulfates, schwertmannite and basaluminite, were all tested and amorphous $Fe(OH)_3$ (with $\log K_{sp} = 3$) and amorphous basaluminite ($\log K_{sp} = 24$) gave the best comparison to the water composition below the mixing point.

More testing of geochemical models like this field example will help to decrease the uncertainties and number of assumptions one has to make when applying models to environmental conditions.

Supplementary Materials: The following are available online at http://www.mdpi.com/2075-163X/10/6/547/s1. A supplement is available with an example of PHREEQC input data for the mixing calculation.

Funding: This research received no external funding.

Acknowledgments: This study would not have been possible without the support of the National Research Program of the U.S. Geological Survey and the USGS Nevada District Office (now the Nevada Water Science Center). The author is very grateful to Jim Ball (USGS, retired) for his excellent analytical expertise as well as assistance with field planning and sampling and his extraordinary patience and perseverance. The author thanks

Terry Duaime of the Montana Bureau of Mines and Geology for the spreadsheet information used in this study. Some of the original funding for the field work was also provided by the Lahontan Regional Water Quality Control Board. The author is grateful to Charles N. Alpers for his careful review and to the three anonymous reviewers who helped in improving this manuscript. Any use of trade, firm, or product names is for descriptive purposes only and does not imply endorsement by the U.S. Government.

Conflicts of Interest: The author declares no conflict of interest.

References

1. Bigham, J.M.; Nordstrom, D.K. Iron and aluminum hydroxysulfates from acid sulfate waters. In *Sulfate Minerals—Crystallography, Geochemistry, and Environmental Significance*; Reviews in Mineralogy and Geochemistry Series; Alpers, C.N., Jambor, J.L., Nordstrom, D.K., Eds.; Mineralogical Society of America and Geochemistry Society: Washington, DC, USA, 2000; Volume 40, pp. 351–403.
2. Druhan, J.; Tournassat, C. (Eds.) *Reactive Transport in Natural and Engineered Systems*; Reviews in Mineralogy and Geochemistry Series; Mineralogical Society of America and Geochemistry Society: Washington, DC, USA, 2019; Volume 85.
3. Bigham, J.M.; Schwertmann, U.; Carlson, L. Mineralogy of precipitates formed by the biogeochemical oxidation of Fe(II) in mine drainage. *Catena Suppl.* **1992**, *21*, 219–232.
4. Flynn, C.M., Jr. Hydrolysis of inorganic iron(III) salts. *Chem. Rev.* **1984**, *84*, 31–41. [CrossRef]
5. Valente, T.M.; Gomes, C. Occurrence, properties and pollution potential of environmental minerals in acid mine drainage. *Sci. Total Environ.* **2009**, *407*, 283–294. [CrossRef] [PubMed]
6. Carbone, C.; Dinelli, E.; Marescotti, P.; Gasparotto, G.; Lucchetti, G. The role of AMD secondary minerals in controlling environmental pollution: Indications from bulk leaching tests. *J. Geochem. Explor.* **2013**, *132*, 188–200. [CrossRef]
7. Acero, P.; Ayora, C.; Torrentó, C.; Nieto, J.-M. The behavior of trace elements during schwertmannite precipitation and subsequent transformation into goethite and jarosite. *Geochim. Cosmochim. Acta* **2006**, *70*, 4130–4139. [CrossRef]
8. Schwertmann, U.; Carlson, L. The pH-dependent transformation of schwertmannite to goethite. *Clay Mineral.* **2005**, *40*, 63–66. [CrossRef]
9. Sanchez-España, J. Crystallization in acidic media: From nanoparticles to macrocrystals. *Seminar. Soc. Españ. Mineral.* **2017**, *13*, 15–34.
10. Henderson, G.S.; Neuville, D.R.; Downs, R.T. (Eds.) *Spectroscopic Methods in Mineralogy and Materials Science*; Reviews in Mineralogy and Geochemistry Series; Mineralogical Society of America Geochemistry Society: Washington, DC, USA, 2014; Volume 78.
11. Beran, A.; Libowitzky, E. *Spectroscopic Methods in Mineralogy*; European Mineralogical Union Notes in Mineralogy; Eötvös University Press: Budapest, Hungary, 2004; Volume 6.
12. Adams, F.; Hajek, B.F. Effects of solution sulfate, hydroxide, and potassium concentrations on the crystallization of alunite, basaluminite, and gibbsite from dilute aluminum solutions. *Soil Sci.* **1978**, *126*, 169–173. [CrossRef]
13. Parkhurst, D.L.; Appelo, C.A.J. Description and Input and Examples for PHREEQC Version 3—A Computer Program for Speciation, Batch-Reaction, One-Dimensional Transport, and Inverse Geochemical Calculations. In *USGS Techniques and Methods 6-A43*; USGS: Denver, CO, USA, 2013.
14. Ball, J.W.; Nordstrom, D.K. User's Manual for WATEQ4F, with Revised Database and Test Cases for Calculating Speciation of Major, Trace, and Redox Elements in Natural Waters. In *U.S. Geological Survey Open-File Report 91-183*; USGS: Menlo Park, CA, USA, 1991.
15. Ball, J.W.; Nordstrom, D.K. Final Revised Analyses of Major and Trace Elements from Acid Mine Waters in the Leviathan Mine Drainage Basin, California and Nevada—October 1981 to October 1982. In *U.S. Geological Survey Water-Resources Investigation Report 89-4138*; USGS: Menlo Park, CA, USA, 1989.
16. Duaime, T.E.; McGrath, S.F. Butte, Montana: The Berkeley Pit, Changes in Water Quality and Water Sampling Methods, 1982–2017. In *Montana Bureau of Mines and Geology Bulletin 138*; Montana Bureau of Mines and Geology: Butte, MT, USA, 2019.
17. Pabst, A. Cryptocrystalline pyrite from Alpine County, California. *Am. Miner.* **1940**, *25*, 425–431.

18. Ball, J.W.; Nordstrom, D.K. Major and Trace Element Analyses of Acid Mine Waters in the Leviathan Mine Drainage Basin, California and Nevada—October 1981 to October 1982. In *U.S. Geological Survey Water-Resources Investigations Report 85-4169*; USGS: Menlo Park, CA, USA, 1985.
19. Webster, J.G.; Nordstrom, D.K.; Smith, K.S. Transport and natural attenuation of Cu, Zn, As, and Fe in the acid mine drainage of Leviathan and Bryant Creeks. In *Environmental Geochemistry of Sulfide Oxidation*; Alpers, C.N., Blowes, D.W., Eds.; American Chemical Society Symposium Series 550; American Chemical Society: Washington, DC, USA, 1994; pp. 244–260.
20. Ball, J.W.; Nordstrom, D.K. A Comparison of Simultaneous Plasma, Atomic Absorption, and Iron Colorimetric Techniques for the Determination of Major and Trace Constituents in Acid Mine Waters. In *U.S. Geological Survey Water-Resources Investigations Report 93-4122*; USGS: Menlo Park, CA, USA, 1994.
21. Nordstrom, D.K. Advances in the hydrogeochemistry and microbiology of acid mine waters. *Int. Geol. Rev.* **2000**, *42*, 499–515. [CrossRef]
22. Nordstrom, D.K.; Campbell, K.M. Modeling low-temperature geochemical processes. In *Surface and Ground Water, Weathering, and Soils*, 2nd ed.; Drever, J.I., Ed.; Treatise on Geochemistry Series; Elsevier: New York, NY, USA, 2014; Volume 7, pp. 27–68.
23. Nordstrom, D.K. The effect of sulfate on aluminum concentrations in natural waters: Some stability relations in the system $A1_2O_3$-SO_3-H_2O at 298 K. *Geochim. Cosmochim. Acta* **1982**, *46*, 681–692. [CrossRef]
24. Sanchez-España, J.; Yusta, I.; Diez-Ercilla, M. Schwertmannite and hydrobasaluminite: A re-evaluation of their solubility and control on the iron and aluminum concentration in acidic pit lakes. *Appl. Geochem.* **2011**, *26*, 1752–1774. [CrossRef]
25. Blowes, D.W.; Ptacek, C.J. Acid-neutralization mechanisms in inactive mine tailings. In *The Environmental Geochemistry of Sulfide Mine-Wastes*; Blowes, D.W., Jambor, J.L., Eds.; Mineralogical Association Canada: Toronto, ON, Canada, 1994; Volume 22, pp. 271–292.
26. Jurjovec, J.; Ptacek, C.J.; Blowes, D.W. Acid neutralization mechanisms and metal release in mine tailings: A laboratory column experiment. *Geochim. Cosmochim. Acta* **2002**, *66*, 1511–1523. [CrossRef]
27. Blowes, D.W.; Ptacek, C.J.; Jambor, J.L.; Weisener, C.G.; Paktunc, D.; Gould, W.D.; Johnson, D.B. The geochemistry of acid mine drainage. In *Environmental Geochemistry*, 2nd ed.; Sherwood Lollar, B., Ed.; Treatise on Geochemistry Series; Elsevier: New York, NY, USA, 2014; Volume 11, pp. 149–204.
28. Adams, F.; Rawajfih, Z. Basaluminite and alunite: A possible cause of sulfate retention by acid soils. *Soil Sci. Soc. Amer. J.* **1977**, *41*, 686–692. [CrossRef]
29. Bigham, J.M.; Schwertmann, U.; Traina, S.J.; Winland, R.L.; Wolf, M. Schwertmannite and the chemical modeling of iron in acid sulfate waters. *Geochim. Cosmochim. Acta* **1996**, *60*, 2111–2121. [CrossRef]
30. Regenspurg, S.; Brand, A.; Peiffer, S. Formation and stability of schwertmannite in acidic mining lakes. *Geochem. Cosmochim. Acta* **2004**, *68*, 1185–1197. [CrossRef]
31. Caraballo, M.A.; Rimstidt, J.D.; Macías, F.; Nieto, J.M.; Hochella, M.F., Jr. Metastability, nanocrystallinity and pseudo-solid solution effects on the understanding of schwertmannite solubility. *Chem. Geol.* **2013**, *360–361*, 22–31. [CrossRef]
32. Carlson, L.; Schwertmann, U. Natural ferrihydrites in surface deposits from Finland and their association with silica. *Geochim. Cosmochim. Acta* **1981**, *45*, 421–425, 427–429. [CrossRef]
33. Nordstrom, D.K. Hydrogeochemical processes governing the origin, transport and fate of major and trace elements from mine wastes and mineralized rock to surface waters. *Appl. Geochem.* **2011**, *26*, 1777–1791. [CrossRef]
34. Caraballo, M.A.; Wanty, R.B.; Verplanck, P.L.; Navarro-Valdivia, L.; Ayora, C.; Hochella, M.F., Jr. Aluminum mobility in mildly acidic mine drainage: Interactions between hydrobasaluminite, silica and trace metals from the nano to the meso-scale. *Chem. Geol.* **2019**, *519*, 1–10. [CrossRef]

minerals

MDPI

Article

Lead Mobilization and Speciation in Mining Waste: Experiments and Modeling

Clémentine Drapeau [1], Rabei Argane [2,3], Cécile Delolme [4], Denise Blanc [5], Mostafa Benzaazoua [2,6], Rachid Hakkou [3,6], Thomas Baumgartl [7], Mansour Edraki [8] and Laurent Lassabatere [9,*]

1 Direction Régionale de L'environnement, de L'aménagement et du Logement (DREAL), Auvergne Rhône-Alpes, Unité Départementale du Rhône, 63 Avenue Roger Salengro, 69100 Villeurbanne, France; clementine.drapeau@developpement-durable.gouv.fr
2 Research Institute of Mines and Environment, Université du Québec en Abitibi-Témiscamingue, 445 boul. de l'Université, Rouyn-Noranda, QC J9X 5E4, Canada; rabei.argane@uqat.ca (R.A.); Mostafa.Benzaazoua@uqat.ca (M.B.)
3 Imed-Lab, Faculty of Sciences and Technology, Cadi Ayyad University (UCA), Abdelkarim Elkhattabi Avenue, Gueliz, P.O. Box 549, Marrakech 40000, Morocco; r.hakkou@uca.ma
4 Université de Lyon, ENTPE, 3 Rue Maurice Audin, 69518 Vaulx-en-Velin, France; cecile.delolme@entpe.fr
5 Déchets Eau Environnement Pollutions (DEEP), INSA-Lyon, Université de Lyon, 7 Rue de la Physique, 69621 Villeurbanne, France; denise.blanc@insa-lyon.fr
6 Mining Environment & Cicrular Economy Program (EMEC), Mohammed VI Polytechnic University (UM6P), Lot 660.Hay Moulay Rachid, Ben Guerir 43150, Morocco
7 Geotechnical and Hydrogeological Engineering Research Group (GHERG), Federation University Australia, Churchill, VIC 3841, Australia; t.baumgartl@federation.edu.au
8 Centre for Mined Land Rehabilitation, Sustainable Minerals Institute (SMI), The University of Queensland, St. Lucia, QLD 4072, Australia; m.edraki@cmlr.uq.edu.au
9 Université de Lyon, UMR5023 Ecologie des Hydrosystèmes Naturels et Anthropisés, Université Lyon 1, ENTPE, CNRS, 3 Rue Maurice Audin, 69518 Vaulx-en-Velin, France
* Correspondence: laurent.lassabatere@entpe.fr

Citation: Drapeau, C.; Argane, R.; Delolme, C.; Blanc, D.; Benzaazoua, M.; Hakkou, R.; Baumgartl, T.; Edraki, M.; Lassabatere, L. Lead Mobilization and Speciation in Mining Waste: Experiments and Modeling. *Minerals* **2021**, *11*, 606. https://doi.org/10.3390/min11060606

Academic Editor: Juan Antelo

Received: 11 March 2021
Accepted: 31 May 2021
Published: 5 June 2021

Abstract: Mining produces significant amounts of solid mineral waste. Mine waste storage facilities are often challenging to manage and may cause environmental problems. Mining waste is often linked to contaminated mine drainage, including acidic waters with more or less elevated concentrations of trace metals such as lead. This work presents a study on the mobilization of lead from waste from two typical mining sites: Zeida and Mibladen, two now-closed former Pb–Zn mines in the Moulouya region of Morocco. Our research investigates the mobilization potential of Pb from the waste of these mines. The study involved acid–base neutralization capacity tests (ANC–BNC) combined with geochemical modeling. Experimental data allowed for the quantification of the buffering capacity of the samples and the mobilization rates of lead as a function of pH. The geochemical model was fitted to experimental results with thermodynamic considerations. The geochemical model allowed for the identification of the mineral phases involved in providing the buffering capacity of carbonated mining waste (Mibladen) and the meager buffering capacity of the silicate mining waste (Zeida). These cases are representative of contaminated neutral drainage (CND) and acid mine drainage (AMD), respectively. The results highlight the consistency between the ANC–BNC experimental data and the associated modeling in terms of geochemical behavior, validating the approach and identifying the main mechanisms involved. The modeling approach identifies the dissolution of the main solid phases, which impact the pH and the speciation of lead as a function of the pH. This innovative approach, combining ANC–BNC experiments and geochemical modeling, allowed for the accurate identification of mineral phases and surface complexation phenomena, which control the release of lead and its speciation in drainage solutions, as well as within solid phases, as a function of pH.

Keywords: mining drainage; lead; mobilization; modeling; speciation

1. Introduction

Open-pit and underground mining operations produce a significant amount of solid mineral waste. Whether it is mine waste rock in the form of waste rock piles or concentrator rejects discharged in the form of tailings, above-ground waste storage facilities are difficult to manage and expensive to rehabilitate. These voluminous and unique discharges are the origin of many environmental problems worldwide because they generate effluents containing significant quantities of trace metals, sometimes accompanied by strong acidification of the environment [1,2]. Acid mine drainage (AMD) represents the phenomenon of acidification of the environment due to the oxidation of sulfidic mining waste, while contaminated neutral drainage (CND) refers to drainage from mining waste with acid-neutralizing capacity and water pH close to neutrality. Both water types may have high concentrations of metals, metalloids, and sulfate [3,4].

In Morocco, the mining industry has always been an important contributor to the economy. This sector represents more than 27% of all exports and nearly 6% of gross domestic product [1]. Unfortunately, this industry has too often ignored the need to rehabilitate mining sites in the past, including tailing storage facilities and waste rock dumps. More than 200 abandoned sites have been identified across the country [5,6]. The intense mining activity of the Zeida (1972–1985) and Mibladen (1938–1983) mines in the upper Moulouya region generated large volumes of mining waste in the form of waste rock and mining residues from ore concentration stages. They are also used as uncontrolled construction materials: finishing and surfacing mortar for residential walls [1,7,8]. After the cessation of operations, this mining waste significantly contaminated water resources and adjacent soils [9–12]. Indeed, in addition to the substantial contamination of surface waters and soils, the weathering of this waste triggers the release of some pollutants, including toxic trace metals, e.g., [13–15].

In this context, understanding weathering processes and related contaminated neutral drainage (CND) or acid mine drainage (AMD) is fundamental for quantifying trace metal release into the environment and related environmental issues. The mobilization of trace metals is complex. It depends on geochemical processes at the solid/liquid interface: precipitation/dissolution, ion exchange, surface complexation, surface precipitation, sorption processes driven by water repellency, absorption (incorporation of the solute in the solid matrix), and diffusion [16]. All of these interactions are featured by the chemical, physicochemical, mineralogical, and physical properties of water, dissolved species, and surfaces. pH remains one of the key factors controlling the dissolution and desorption of mineral elements [16–18].

Numerous studies have provided information on the total concentrations of trace metals, and the role of physicochemical conditions on pollutant release via batch approaches or selective extractions [18–25]. The characterization of trace metal speciation is still carried out sporadically by micro-analysis or indirectly by sequential extractions. However, the environmental and human health risks depend not only on the total content but also on the trace metal speciation [26–28], and there is still a lack of information on the main solid phases and geochemical processes that control the potential mobility of major elements and trace metals in these complex matrices.

This paper presents a study on lead release from mining waste due to pH change for the mines of Zeida and Mibladen located in Morocco. We aimed at identifying the geochemical mechanisms responsible for the lead release and speciation in surface waters from those sites. Geochemical models are built as the first step towards modeling the fate of trace metals for operational objectives, including (i) better management of large quantities of mining waste, (ii) providing information on the risk of chemical contamination of the waste repository in the surrounding environment, and (iii) improving analysis of the reuse of waste for the construction sector. This study combines experimental data and geochemical modeling to fill in the gap of understanding of lead release in mining waste [17,18,29]. The experimental part is mainly based on geochemical characterization and acid–base neutralization capacity tests (ANC–BNC). The modeling approach, making use of PHREEQC geochemical software, is calibrated against experimental data from the ANC–BNC tests to simulate lead release and speciation in the liquid phase as a function of pH [30]. Lead was chosen, since it enlists heavy metals with high potential risks to human health [31], and it has been found at significant concentrations in the two studied mine tailings [1,32] with potential risks for the surrounding and downstream environments [33]. These steps allowed for the identification of the main phases bearing lead in the two studied types of mining waste and the main mechanisms responsible for their dissolution and related lead release.

2. Materials and Methods

2.1. Sampling Site, Geochemical Characterization, and ANC–BNC Tests

Mining waste characterization has already been performed by Argane et al. [1], and more details can be found in their paper. The studied samples were taken from the "tailings pond" of the Zeida and Mibladen mines in the Moulouya region of Morocco, used by locals as a supply of building materials in a completely uncontrolled manner. Large amounts of samples were extracted for this study from below the surface to avoid weathered material. Several representative locations of the tailing ponds were sampled. The climate of the region is semiarid, the annual precipitation varies from 100 to 400 mm, and the mean annual temperature ranges from 12 to 14 °C [34]. The samples (five from Zeida and two from Mibladen) correspond to composite samples over 1 m and were taken with a mechanical shovel over an area of approximately 1 m × 2 m before being dried and homogenized and before geochemical characterization [7,32].

The samples were analyzed by X-ray diffraction (XRD) in order to determine the proportions of the crystallized mineral phases. XRD analyses were performed using a Bruker A.X.S D8 Advance diffractometer equipped with a copper anticathode that scans over a diffraction angle (2θ) from 5° to 60°. Scan settings were a 0.005° 2θ step size and a 1 s counting time per step. The "Diffrac-Plus" EVA software was used to identify mineral species, and "TOPAS" software implementing Rietveld refinement allowed for the quantification of the abundance of all identified mineral species. The samples (polished sections) were also observed with a scanning electron microscope (SEM) using backscattered electrons (BSEs). The microscope was a Hitachi S-3500N variable pressure microscope equipped with an X-ray energy dispersive spectrometer from Oxford (EDS; Silicon drift spectrometer X-Max 20 mm^2). The operating conditions involved an electron beam of 20 keV, z100 mA, and a 15 mm working distance. The INCA software provided spectra and X mapping. More details can be found in [7,32].

2.2. ANC–BNC Experiments

The acid–base neutralization capacity test (ANC–BNC) was designed to quantify the mobilization of pollutants and buffering capacities of solid matrices in contact with a source of protons and alkalinity [18,35–37]. In this study, we used the experimental results from Argane et al. [1]. These authors performed batch tests according to CEN/TS 14,429 [38]. The test involved a series of extractions using acid (HNO_3) and base (NaOH) solutions of varying concentrations to cover the pH range between 2 and 12, as suggested by Chatain et al. [17]. In each batch test, solid samples (10 g) were put into contact with 100 mL solutions (liquid/solid ratio of 10 mL/g), and the reactors were agitated for 8 days using an end-over-end tumbler. Such contact time allowed for the attainment of stabilized pH values. Leachates were then filtered with 0.45 μm membranes before pH measurement. The samples then underwent acidification before ICP-AES analyses to determine trace metal concentrations. More details can be found in [7,32].

The ANC–BNC results were treated as follows. We determined the amount of protons (H^+) added for the acid part (ANC) or the amount of OH− multiplied by −1 for the basic part (BNC). Eq_H^+ has an algebraic value determining the inputs of acidity or alkalinity to the systems. The experimental curve plotted the evolution of pH as a function of the added H^+, Eq_H^+ [17]. The pH (Eq_H^+) curve was then used to compute a transformed curve by differentiating the inverse function Eq_H$^+$ (pH) with regard to pH [36,37]. The transformed curve helps to determine the buffer zones and to relate these to the dissolution of specific mineral phases, thus allowing for a better understanding of element release and mineral dissolution [37]. Such an experimental approach was also validated for the characterization of AMD and CND using synthetic pure mineral assemblages [39].

2.3. Geochemical Modelling

The geochemical modeling was performed using PHREEQC software (Interactive, version alpha 3.1.2.8538, developed by USGS, United States Geological Survey), making it possible to simulate the ANC–BNC experiments conceptually [37,39–41]. PHREEQC is based on a deterministic approach where digital tools solve all chemical reactions and associated equilibrium equations [30]. In this study, the *llnl* (Lawrence Livermore National Laboratory) database was considered. It is one of the international reference databases that is very well documented and contains all the studied mineral phases. The geochemical modeling presented in this article has been adjusted to the ANC–BNC experimental results to select the geochemical processes involved in the release of the elements and predict the element speciation as a function of pH.

ANC–BNC experiments were modeled as closed batch reactors, considering the actual conditions of experimentation, i.e., the actual mass of solid, the actual volume of solution, and the actual air volume in the closed reactors. The latter was deduced from the knowledge of the volumes of the solutions and the vessels. The system was then characterized by its gas/liquid ratio (the "Vol_Air_fix" parameter required by PHREEQC). The gas phase's initial pressure was set at atmospheric pressure at sea level, corresponding to a ratio pressure/reference pressure of unity (the "value of P_Air_0" parameter required by PHREEQC). The initial partial pressures of nitrogen, oxygen, and carbon dioxide were fixed at their typical atmospheric values. The mineral assemblages were simulated as the combination of solids with related initial amounts and solubility constants, as defined by the *llnl* database. The mixture of solid, liquid, and gas phases was then simulated for the solutions used to perform ANC–BNC experiments: neutral (water at pH 7), acidic (an addition of HNO_3), or alkaline (an addition of NaOH). PHREEQC numerically resolved the system considering mass conservation, the law of mass action, and Henry's law for mass transfer between the liquid and gas phases.

Regarding optimization, we multiplied modeling scenarios corresponding to variable mineral compositions (in terms of selected phases and initial amounts) and the variable compositions of sorption sites in terms of site density and sorption constants, with ~700 simulations for the Mibladen site and ~900 for the Zeida site. These numerous iterations allowed us to conduct sensitivity analysis, to identify critical parameters, to identify compensation effects, and to invert the observations (final fit). These also allowed for the definition of an optimum model after the following steps: (i) an initial estimate of the composition of the mineral assemblages (solid samples) and the gas phase with an indication of the minerals' propensity to dissolve or precipitate; (ii) modeling of this assemblage in contact with the acid and alkaline solutions, assuming chemical equilibrium; (iii) a comparison of predicted against observed pH and element concentrations; (iv) a trial-and-error-based forward–backward procedure, with a repetition of steps (i)–(iii) until a consistent model with accurate fits could be obtained; and finally, (v) the identification of the more likely mineral composition and a report of modeled element speciation in the liquid and solid phases as a function of pH. Note that the same initial mineral composition is considered for all acid and alkaline solutions (i.e., for all ANC–BNC experiments).

Regarding the quality of the fit, two goodness of fit indicators were considered. First, the coefficient of determination, R^2, was computed as follows:

$$R^2 = \left\{ \frac{\sum_{t=1}^{N} (y_{o,t} - \overline{y_o})(y_{s,t} - \overline{y_s})}{\left[\sum_{t=1}^{N} (y_{o,t} - \overline{y_o})^2\right]^{0.5} \left[\sum_{t=1}^{N} (y_{s,t} - \overline{y_s})^2\right]^{0.5}} \right\}^2 \quad (1)$$

where N is the number of observations; $y_{s,t}$ and $y_{0,t}$ are the simulated (modeled) and observed values at instant t, respectively; and, $\overline{y_s}$ and $\overline{y_o}$ are their respective means. The normalized root mean square error (NRMSE) was also computed, as follows [42]:

$$NRMSE = abs\left\{\frac{1}{\overline{y_o}}\left[\frac{1}{N}\sum_{t=1}^{N}(y_{s,t} - y_{o,t})^2\right]^{\frac{1}{2}}\right\} \quad (2)$$

The coefficient of determination quantifies the correlation between the modeled and observed data. The NRMSE quantifies the errors between modeled and observed data.

3. Results and Discussion

3.1. Preliminary Results

3.1.1. Characterization of the Mining Waste

The results of the XRD diffraction (Table 1) reveal that the two mine tailings have high concentrations of silicates: (in decreasing order) quartz, orthoclase, albite, chamosite, and kaolinite at Zeida and only quartz and chamosite at Mibladen. Barite and fluorite could also be found in the two mine tailings. Note that the XRD analysis does not detect the amorphous phases. However, we assumed that these results gave good insight into the relative proportions of all types of minerals, including both crystallized and amorphous minerals. The main difference between the two sites relates to their carbonate content, with significant concentrations at the Mibladen site versus zero content at the Zeida site (see Table 1).

Table 1. Mineral characterization and mass percentage of the different crystallized phases as determined by X-ray diffraction (in % of dry matter, w/w) (adapted from [32], amorphous phases undetected).

Mineral Phase		Formula	% Zeida	% Mibladen
Silicates	Quartz	SiO_2	48.8	7.16
	Orthoclase	$KAlSi_3O_8$	31.6	-
	Chamosite	$Fe_2Al_2SiO_5(OH)_4$	2.6	2.72
	Albite	$NaAlSi_3O_8$	5.5	-
	Kaolinite	$Al_2Si_2O_5(OH)_4$	1.5	-
Carbonates	Dolomite	$CaMg(CO_3)_2$	-	51.35
	Calcite	$CaCO_3$	-	22.2
Others	Barite	$BaSO_4$	7.9	16.57
	Fluorite	CaF_2	2.1	3

SEM analysis provided additional information on the phases. First, SEM analyses confirmed the presence of the minerals found with XRD, with, for instance, the occurrence of grains of quartz, barite, and fluorite, illustrated in Figure 1a. SEM analyses also revealed the presence of minerals that were not detected using XRD. For instance, small grains of pyrite could be found, often in association with silicate phases (Figure 1a). We expect that pyrite concentrations were too low to be detectable using XRD analyses (concentrations probably lower than 1%). From the presence of pyrite in the two mine tailings, as revealed by SEM analyses, we can conclude that the waste from Zeida is representative of mining waste prone to the acid mine drainage (AMD) process. Conversely, Mibladen mining waste is expected to have a substantial buffering capacity and therefore neutralize the acidity produced by pyrite oxidation and dissolution, leading to a contaminated neutral drainage (CND) process.

Regarding trace elements and trace metals, SEM also provided relevant insights by revealing phases that were not detected with XRD analyses. For lead, SEM analyses detected mineral forms of massicot (Figure 1b), galena (Figure 1c), cerussite in Zeida mine tailings and of galena, and cerussite in Mibladen mine tailings (Figure 1d). More details and images can be found in [7]. The concentrations of the crystalline forms of these minerals may be too low to be detected with X-ray diffraction (<1%).

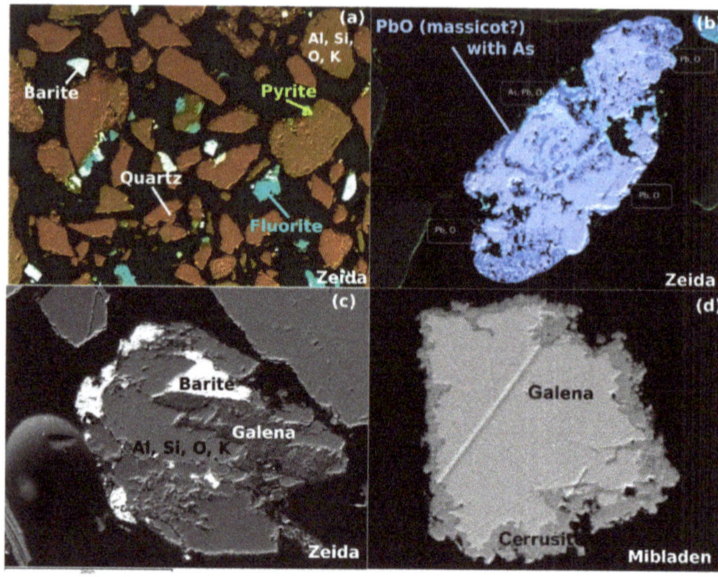

Figure 1. SEM images of the mining tailings at the Zeida and Mibladen sites; minerals are indicated if correctly identified, and the atomic composition is provided otherwise, (**a**) global view of Zeida mining waste, (**b**) grain of massicot, (**c**) grains of galena and barite in association with aluminosilicates, (**d**) grain of galena which alteration provide cerussite.

3.1.2. Modeling Approach and Validation of the Geochemical Model

The geochemical model was based on the results detailed in the previous section and, particularly, on the lists of detected minerals using XRD analyses and SEM observations. In addition to that list, the analysis of the different buffer capacities as a function of pH was analyzed to derive the main constituents of the mineral assemblages on the basis of previous works [37,39]. This method is illustrated below. In addition, the minerals expected to form, precipitate, or dissolve given thermodynamic considerations and preliminary modeling results were also added. The final selection of the mineral composition involved in the geochemical processes requires a forward–backward approach, combining modeling, mineralogical characterization, and literature. The main equations impacting element speciation in the solid, liquid, and gas phases are described in Table 2.

For lead, we considered galena ($PbS_{(s)}$) and cerussite ($PbCO_{3(s)}$), as the SEM analysis revealed them. Alamosite ($PbSiO_{3(s)}$) and lanarkite ($Pb_2(SO_4)O)_{(s)}$ had to be added to allow accurate fits of lead release, even if these were not identified using XRD or SEM (Table 2, IM). The fit of lead release allowed us to characterize the lead sorption in terms of cationic exchange and surface complexation onto iron hydroxides ($Fe(OH)_{3(s)}$). In addition to sorption/desorption processes, lead may be released when the phases bearing the sorption sites dissolve (such as iron hydroxides). Note that only the site density and the amounts of minerals were optimized, whereas constants for the Pb complexation were unchanged from the *llnl* database (considering the non-electrostatic model).

Table 2. Mineral assemblages obtained by inverse modeling (the "Phases" column), dissolution/precipitation reactions with thermodynamic constants ("Chemical Reactions" and "Equilibrium Constants"), and initial concentrations in the assemblage; surface complexation and cationic exchange sites are described in the three bottom rows. For thermodynamic constants, refer to the *llnl* database [30].

Phases	Chemical Reactions	Equilibrium Constants (pK)	Initial Concentrations (mol/L)	
			Zeida	Mibladen
	Main Mineral Phases			
	Carbonate			
Calcite	$CaCO_{3(s)} + H^+ \leftrightarrow Ca^{2+} + HCO_3^-$	1.8487	-	0.014326 (MC)
	Content of Pb in inclusion		-	0.001188 (SEM/EDS)
Dolomite	$CaMg(CO_3)_{2(s)} + 2\ H^+ \leftrightarrow Ca^{2+} + Mg^{2+} + 2\ HCO_3^-$	2.5135	-	0.0222198 (MC)
	Silicate			
Albite (DO)	$NaAlSi_3O_{8(s)} + 4H^+ \leftrightarrow Al^{3+} + Na^+ + 2H_2O + 3SiO_2$	2.7645	0.0003013 (MC)	-
Chamosite (DO)	$Fe_2Al_2SiO_5(OH)_{4(s)} + 10H^+ \leftrightarrow SiO_2 + 2Al^{3+} + 2Fe^{2+} + 7\ H_2O$	32.8416	0.0002005 (MC)	0.0003767 (MC)
Kaolinite (DO)	$Al_2Si_2O_5(OH)_{4(s)} + 6H^+ \leftrightarrow 2\ Al^{3+} + 2\ SiO_2 + 5\ H_2O$	6.8101	0.001822 (MC)	-
Quartz (DO)	$SiO_{2(s)} \leftrightarrow SiO_2$	−3.9993	0.0129952 (MC)	0.008307 (MC)
Orthoclase (DO)	$KAlSi_3O_{8(s)} + 4H^+ \leftrightarrow Al^{3+} + K^+ + 2H_2O + 3SiO_2$	−0.2753	0.00908 (MC)	-
	Others			
Pyrite	$FeS_{2(s)} + H_2O \leftrightarrow 0.25H^+ + 0.25SO_4^{2-} + Fe^{2+} + 1.75HS^-$	−24.6534	0.0001 (SEM/EDS)	0.000025 (SEM/EDS)
Barite	$BaSO_{4(s)} \leftrightarrow Ba^{2+} + SO_4^{2-}$	−9.9711	0.003385 (MC)	0.00284 (MC)
Fluorite	$CaF_{2(s)} \leftrightarrow Ca^{2+} + 2\ F^-$	−10.0370	0.000269 (MC)	0.00457 (MC)
Iron hydroxydes	$Fe(OH)_{3(s)} + 3\ H^+ \leftrightarrow Fe^{3+} + 3\ H_2O$	5.6556	0.001 (SEM/EDS)	0.00743 (SEM/EDS)
Hematite (DO)	$Fe_2O_{3(s)} + 6\ H^+ \leftrightarrow 2\ Fe^{3+} + 3\ H_2O$	0.1086	0.0001 (IM)	0.001 (IM)
Fluorapatite	$Ca_5(PO_4)_3F_{(s)} + 3\ H^+ \leftrightarrow F^- + 3\ HPO_4^{2-} + 5\ Ca^{2+}$	−24.9940	0.0001 (IM)	-
	Gas phases			
$CO_2(g)$	$CO_{2(g)} + H_2O \leftrightarrow H^+ + HCO_3^-$	Vol_Air_fix * = 2.5	In equilibrium with neutral batch solution	
$O_2(g)$	$NO_{2(g)} + 0.5H_2O + 0.25O_{2(g)} \leftrightarrow H^+ + NO_3^-$	P_Air_0 ** = 1		
	Mineral phases bearing lead			
Alamosite (DO)	$PbSiO_{3(s)} + 2\ H^+ \leftrightarrow H_2O + Pb^{2+} + SiO_2$	5.6733	0.00043 (IM)	0.00000396 (IM)
Cerussite (DO)	$PbCO_{3(s)} + H^+ \leftrightarrow HCO_3^- + Pb^{2+}$	−3.2091	0.00009 (SEM/EDS)	1.098×10^{-5} (SEM/EDS)
Galena (DO)	$PbS_{(s)} + H^+ \leftrightarrow HS^- + Pb^{2+}$	−14.8544	0.00071 (SEM/EDS)	9.794×10^{-6} (SEM/EDS)
Pb (DO)	$P + 2\ H^+ + 0.5\ O_2 \leftrightarrow H_2O + Pb^{2+}$	47.1871	-	2.9×10^{-6} (SEM/EDS)
Lanarkite	$Pb_2(SO_4)O_{(s)} + 2\ H^+ \leftrightarrow H_2O + SO_4^{2+} + 2\ Pb^{2+}$	−0.4692	0.000001 (IM)	-
	Surface complexation site for sorbing lead			
≡Hfo_sOH	≡Hfo_sOH + $Pb^{2+} \leftrightarrow$ ≡Hfo_sOPb$^+$ + H^+	4.65	0.5	0.002
≡Hfo_wOH	≡Hfo_wOH + $Pb^{2+} \leftrightarrow$ ≡Hfo_wOPb$^+$ + H^+	0.3	1.3	0.0015
	X (ion exchange) sorption site for lead			
X^-	$Pb^{2+} + 2\ X^- \leftrightarrow PbX_2$	1.05	0.009	0.0003

(DO): dissolutions only; (MC): mineral characterization; (IM): inverse modeling. * Fixed air volume in contact with 1 L of water. ** Air partial pressure initial value (set at the atmospheric pressure).

According to Biondi et al. [42], various points need to be considered to assess the models' relevance and performance objectively. First, the fits must be of good quality, guaranteeing the goodness of fit. Second, the final inverse solution must rely on a unique set of selected equations and related parameters. Indeed, the uniqueness and plausibility of the estimates guarantee the consistency and soundness of the proposed model, e.g., [43]. In our case, these two main criteria are fulfilled. The coefficients of determination R^2 are close to 1, always between 0.729 and 0.965, and the NRMSE scored values are always below 25% (Table 3). As shown in Sections 3.2 and 3.3, pH and lead release are somewhat less well resolved in the Zeida case. However, the two fits can be considered accurate. Regarding non-uniqueness, no other variants (based on alternative hypotheses regarding mineral composition, selected dissolution, and sorption processes) can improve the proposed simulations. The sensitivity analysis proved that the proposed model was among the most likely models.

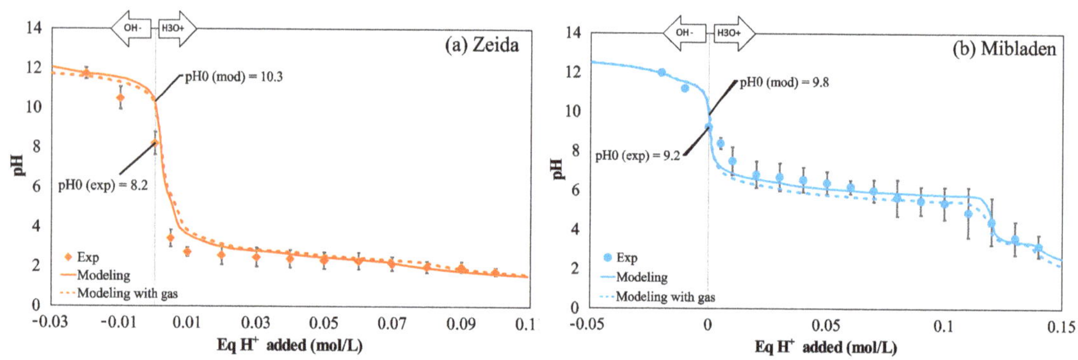

Figure 2. Evolution of the pH as a function of the H^+ added to the mining waste from Zeida (**a**) and Mibladen (**b**): experimental points (marks, adapted from [1,32])) and geochemical modeling (solid lines indicate the modeling with the gas phase, and the dashed lines indicate the modeling without the gas phase).

Table 3. Performance measurement of the final geochemical model.

Figure Number	Modeled data	NRMSE-Normalization Root Mean Square Error	R^2 Coefficient of Determination
Figure 2a	pH = f(Eq H^+ added) for Zeida tails	21.6%	0.965
Figure 2b	pH = f(Eq H^+ added) for Mibladen tails	7.50%	0.963
Figure 4a	Pb = f(pH) for Zeida tails	18.6%	0.729
Figure 4b	Pb = f(pH) for Mibladen tails	7.11%	0.825
Zeida Tails Mean		20.1%	0.847
Mibladen Tails Mean		7.30%	0.894

Since all observations were accurately fitted, including pH and element release, and since the proposed model seems unique, we assumed that the modeled results are indicative of the mineral assemblages and the geochemical processes involved in the closed batch reactors. Consequently, the modeled results were used to conclude on processes and lead speciation in the liquid and solid phases.

3.2. Characterization of Buffer Capacity with ANC–BNC Experiments

3.2.1. Observed and Modeled Buffer Capacity

Figure 2 shows the change of pH as a function of the added H^+ (Eq_H^+) corresponding to the contact with all acid and alkaline solutions for the two mine tailings of Zeida (Figure 2a) and Mibladen (Figure 2b). In each of the graphs, the markers denote the experimental results of the ANC–BNC tests, whereas the lines denote the geochemical models. The experimental value of pH at equilibrium with water has a lower value for Zeida, with a pH of 8.2 versus 9.2 for Mibladen. In the acidic zone, the pH decreases much more in the Zeida case, with values dropping to 2 (Figure 2a), whereas the decrease is much more progressive in the Mibladen case with a plateau around 6 (Figure 2b).

These differences result from the composition of the waste from the two mining sites. The mining waste from the Zeida site is more than 90% (*w*/*w*) silicates, whereas that of the Mibladen mine is less than 10% (*w*/*w*) silicates but almost 75% carbonates (Table 1). The modeling helps us to understand the role of carbonates in the bulk buffer capacity and relates the contrasting evolution of pH to the mineral compositions of the waste from the two mining sites. Indeed, each buffer zone corresponds to the dissolution of a specific mineral. The selection of the proper minerals dissolved at a given pH with appropriate amounts allows for accurate modeling of the curve pH (Eq_H^+) (see below, Figure 2) and the identification of the geochemical mechanisms involved. As proposed in previous work [37,39], the transformed pH curve d Eq_H^+/d pH as a function of pH improves the identification of minerals. This point is discussed in more detail below (Section 3.2.2).

The geochemical modeling clarified the importance of the gas phase and its components in closed batch reactors. The gas phase's contribution is usually neglected in geochemical approaches, e.g., [37], whereas it may interact with the geochemical processes. Indeed, several of its constituents may significantly impact the element speciation and distribution between the solid, liquid, and gas phases. At first, oxygen may interfere with redox processes and thus with the oxidation of pyrite and induced acidification. In addition, other species in the gas phase may interact with solutes and species in water. The dissolution of $CO_{2(g)}$ produces carbonic acid (H_2CO_3) that acidifies the water by transforming into the bicarbonate ion (HCO_3^-). This may, in turn, promote the dissolution of carbonated phases, including those involving trace metals, as detailed for lead in the following:

$$CO_{2(g)} + H_2O \Leftrightarrow H^+ + HCO_3^- \quad (3)$$

$$PbCO_{3(s)} + CO_{2(g)} + H_2O \Leftrightarrow 2\,HCO_3^- + Pb^{2+} \quad (4)$$

The geochemical modeling was then performed with and without the gas phase for the specific case of our batch reactors. The comparison of the modeled results with and without the gas phase showed no significant differences. The difference is slightly more visible for the Mibladen site, with lower pH values in the absence of gas. This result shows that neglecting the gas phase may not change the results drastically, at least for reactors in which the gas and the liquid phases have similar volumes. However, if the batch reactors are opened, the geochemical model should consider this contact with an infinite atmospheric reservoir. In the following, the results correspond to the modeling with the gas phase, its main constituents, and the related features (partial pressures and the volume of gas in the reactors).

3.2.2. Identification of the Main Mineral Phases by Modeling Buffer Capacity

Previous results have allowed for a global understanding of the impact of mining waste on pH as a function of the added Eq_H^+. The model makes it possible to refine these results by determining the effect of each of the significant phases on pH. Indeed, the models include the buffer capacity, the element release, and the solid phase simultaneously, allowing us to establish correspondence between these variables. The modeled results were obtained with the inverse method detailed in Section 3.1.2, and are depicted in Figure 3. Related mineral composition and amounts are described in Table 2. Geochemical modeling allows for the superposition of the transformed curves (derivatives d Eq_H^+/d pH) with the solid phase composition (minerals).

Prior to modeling, the buffer curves were derived from Figure 2 by transforming the curve pH (Eq_H^+), i.e., by differentiating the experimental curve Eq_H^+ (pH). The peaks more precisely delineate the different buffer zones (Figure 3). For the Zeida and Mibladen cases, the buffer zones are mainly in the acidic zone (Z1 and Z2 in Figure 3a and Z4, Z5, and Z6 in Figure 3b). These values of pH are specific to the matrices and their mineral compositions. In the alkaline area, a similar slight buffering capacity is present at pH $\simeq$ 11.5 in waste from both sites (Z3 in Figure 3a and Z7 in Figure 3b).

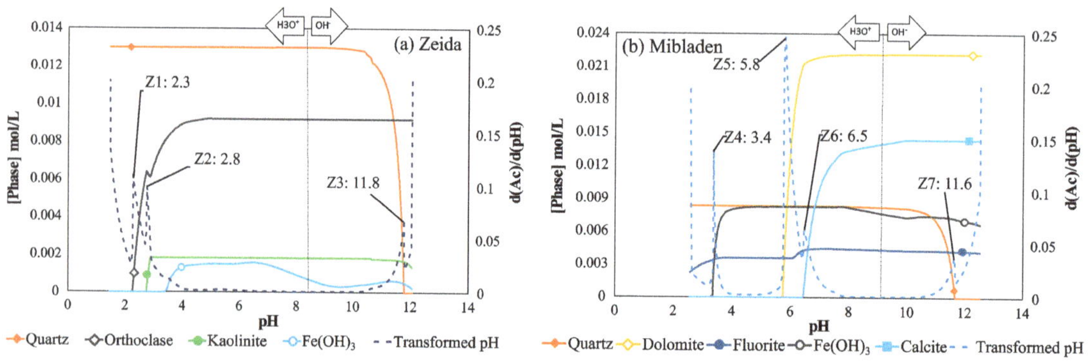

Figure 3. Transformed curve (d Eq_H$^+$/d pH = d (Ac)/d (pH), right axis) as a function of pH obtained from observations, along with modeled compositions of mineral phases (left axis), in the Zeida (**a**) and Mibladen (**b**) mining waste.

The position of the buffer zones can be related to a specific mineral, and it is possible to identify the effect of each of the mineral phases on the pH by analyzing the modeled results (see Figure 3a,b, "Quartz" versus "Transformed curve"). Thus, in both the Zeida and the Mibladen mining waste, the buffering capacity in the basic zones (Z3 and Z7) is linked to the dissolution of part of the quartz. The modeled data proves that these zones correspond to the following equation:

$$SiO_{2(s)} + H_2O \Leftrightarrow HSiO_3^- + H^+ \quad (5)$$

The main phases corresponding to the observed buffer zones in the acidic zone of the Zeida mining waste are orthoclase ($KAlSi_3O_{8(s)}$), kaolinite ($Al_2Si_2O_5(OH)_{4(s)}$), and iron hydroxides ($Fe(OH)_{3(s)}$). Indeed, the transformed curves point to buffer zones at pH values corresponding to the dissolution of these minerals (Figure 3a). The integration of the transformed curves indicates the amount of protons required to counter the buffer zones, and scores a total value of 0.0921 mol/L. This addition corresponds to a respective dissolution of orthoclase, kaolinite, and iron hydroxides of 0.0908, 0.001222, and 0.0010 mol of minerals per liter of solution. These amounts are equal to the dissolution of 8% and 21% of the initial content, respectively, of orthoclase and kaolinite. The latter is potentially a secondary mineral resulting from the weathering of Al–silicate phases. The dissolution of these phases consumes protons, as explained by the following reactions (Equations (6)–(8)):

$$KAlSi_3O_{8(s)} + 4\,H^+ \Leftrightarrow Al^{3+} + K^+ + 2\,H_2O + 3\,SiO_2 \quad (6)$$

$$Al_2Si_2O_5(OH)_{4(s)} + 6\,H^+ \Leftrightarrow 2\,Al^{3+} + 2\,SiO_2 + 5\,H_2O \quad (7)$$

$$Fe(OH)_{3(s)} + 3\,H^+ \Leftrightarrow Fe^{3+} + 3\,H_2O \quad (8)$$

It is also important to note that iron hydroxides impact the pH through their dissolution and surface complexation sites, as illustrated below in Section 3.3.2 (see Equations (14) and (15)).

The main phases that impact the pH in the acidic zone of the Mibladen mining waste are dolomite ($CaMg(CO_3)_{2(s)}$), calcite ($CaCO_{3(s)}$), fluorite ($CaF_{2(s)}$), and iron hydroxides ($Fe(OH)_{3(s)}$). Dolomite and calcite buffer the solution at a pH of approximately 6 (at respective values of 5.8 and 6.5; see Z4 and Z5 in Figure 3b). Fluorite adds a buffer zone at a pH of 3.4 (see Z4 in Figure 3b). The integration of the transformed curves indicates that it takes 0.114 mol of acid to pass the first buffer zone (Z5) and 0.024 mol of additional acid to pass the next buffer zone (Z4). This corresponds to the dissolution of the major phases of 0.017 mol of dolomite, 0.013 mol of calcite, and 0.000082 mol of fluorite per liter of solution, i.e., 8%, 7%, and 0.2% of their initial concentrations. The reaction of mineral dissolution by the consumption of protons can be expressed as follows (Equations (9)–(11)):

$$CaMg(CO_3)_{2(s)} + 2\,H^+ \Leftrightarrow Ca^{2+} + Mg^{2+} + 2\,HCO_3^- \quad (9)$$

$$Ca(CO_3)_{(s)} + H^+ \Leftrightarrow Ca^{2+} + HCO_3^- \quad (10)$$

$$CaF_{2(s)} + H_2O + H^+ \Leftrightarrow CaOH^+ + 2\,H_2F_2 \quad (11)$$

These findings show that the model appropriately captures the bulk buffer capacity. They also show how the geochemical model can be used (and was used) to derive the composition of the mineral assemblages. In more concrete terms, the transformed curve can be aligned with several models simulating the dissolution of plausible mineral phases to detect the phase that corresponds the best. Such a method may allow for the identification of each of the active mineral phases, buffer zone by buffer zone, leading to the final composition.

3.2.3. Validation of the Proposed Mineral Composition with the Tailings' Characterization

The final composition of the mineral assemblages deduced from inverse modeling (Table 2) agrees well with the characterization of the mine tailings (using XRD analysis and SEM). Indeed, most of the phases identified by the inverse modeling (Table 2) were detected either by XRD (Table 1) or SEM (see Section 3.1). In particular, pyrite, which is crucial regarding geochemical processes in mine tailings and identified by inverse modeling, was also recognized by SEM for the two mine tailings. This consistency between mineral characterization and inverse modeling consolidates the proposed final composition of the mineral assemblage. However, a few minerals had to be considered, even though they were not detected either by XRD or by SEM (i.e., alamosite, hematite, and fluorapatite). Numerical sensitivity analysis proved that these phases were needed to fit the observed evolution of pH with Eq_H$^+$ accurately. Conversely, other minerals were detected with SEM and are not part of the final mineral assemblage identified by inverse modeling. For instance, massicot ($PbO_{(s)}$) was observed with the MEB but is not part of the final assemblage (see Table 2).

Several reasons may explain the discrepancy between mineral characterization and ANC–BNC experiments. First, the differentiating step that allows for the detection of buffer zones may be impacted by the lack of precision of the experimental pH curves (Eq_H$^+$). Uncertainty in the measure of pH and the amounts of added protons may affect the quality of the derivative. Second, ANC–BNC experiments detect the main contributors to the bulk buffer capacity. Minerals with meager concentrations may contribute to a tiny proportion of the bulk buffer capacity, making them undetectable. Third, ANC–BNC experiments do not involve strong enough conditions to dissolve all the minerals, with most of them requiring very acidic conditions along with high temperature and pressure conditions to dissolve and thus exhibit any buffering capacity [44,45]. Despite potentially significant concentrations, these minerals cannot be detected using ANC–BNC experiments. Fourth, ANC–BNC experiments are conducted over reasonable durations (48 h), allowing only the quickest processes to occur. Kinetically limited dissolution processes do not have enough time to complete. Consequently, ANC–BNC must be combined with additional information to complete the characterization of the solid phase. Our study combined our data with the knowledge derived from previous work to obtain the most probable composition of the mining waste (see Table 2).

3.2.4. Comparison of Waste from the Two Mining Sites: Two Types of Mine Drainage

The geochemical model reveals the common and contrasting features between the two mine tailings. While only a tiny amount of alkaline solution is needed to reach the asymptote at pH = 12, more acid is required to reach the final asymptote in the acid zone at pH = 2.35 in the waste from the two mining sites. For Zeida mine tailings, it takes less than 0.07 mol, while it takes more than 0.15 mol for Mibladen tailings (Figure 2). Such a difference indicates a significant difference in terms of mineral composition. The waste from both mining sites is composed of the same types of mineral phases (including sulfides), except that the Mibladen tailings are strongly carbonated. The presence of carbonates (calcite and dolomite) explains the significant difference observed in the ANC–BNC experiments, particularly the tremendous buffering capacity zone observed between pH 5 and 7 (Figure 3). Calcite and dolomite are expected to make a significant difference during weathering processes and related mining drainage. Their buffer capacities are expected to counteract the acidification produced by the weathering of sulfide phases such as pyrite (FeS_2). In both mine tailings, thanks to SEM analysis, we also detected traces of pyrite (see Table 2), the impact of which is so significant that even small quantities may significantly acidify the environment. However, calcite and dolomite in the Mibladen tailings' assemblage are expected to buffer such acidification processes and to lead to contaminated neutral drainage. Conversely, acid mining drainage is expected for the Zeida assemblage.

The investigation of waste from the Zeida and Mibladen mining sites showed the two main drainage scenarios: acid mine drainage (AMD) and contaminated neutral drainage (CND) [1,39]. AMD, the best-known phenomenon, is due to autocatalysis by ferrous iron of the dissolution of pyrite that triggers very strong acidification of the solution [46]. On the other hand, CND occurs under basic conditions when some carbonate phases are in sufficient quantity to buffer the solution while dissolving the sulfide phases [39]. The type of mine drainage is expected to dictate the pollutant release [1,47,48].

3.3. *Mobilization and Speciation of Lead*

3.3.1. Mobilization of Lead as a Function of pH

The samples from the studied sites contain lead in significant quantities. The Zeida mine samples are the least loaded, with total concentrations of lead between 3610 and 4970 mg/kg. The Mibladen mine samples are more concentrated, containing between 4640 and 5140 mg/kg of lead [1]. We remind the reader that the samples were homogenized for each mine tailing before performing the ANB-BNC experiments.

First, in the neutral zone, when the pH is between 6 and 8, lead is very poorly mobilized (Figure 4). Conversely, lead solubilization is greatest in the acid zone for the two samples (Figure 4). In addition, lead solubilization is significantly greater in Zeida samples in comparison to Mibladen samples. In terms of proportion, the mobilized quantities approach 77% on average for Zeida, compared to 1% for Mibladen of the total lead content. This difference in lead mobilization depends mainly on the carrier phases but also the pH. Due to a lower buffering capacity, lower pH values were attained for Zeida (Figure 2a vs. Figure 2b). Further acidification (by providing larger amounts of protons) may have increased the quantity of lead mobilized for Mibladen. However, we decided to apply the same standardized protocol to the two different types of mining waste. By doing so, we simulated experimentally similar conditions of exposure to air and water to be more representative of actual environmental conditions (leaching by acid rain or even the use of waste in the construction industry). By revealing this significant difference between the two types of mining waste, we could demonstrate the effect of dolomite and calcite on acidification and lead mobilization. However, the increased buffering capacity of Mibladen is not sufficient to completely prohibit lead mobilization.

As stated above, given the agreement between observed and modeled data (see Figure 4 and Section 3.1.2), the model is considered as representative of lead mobilization. Thus, it is possible to conclude on geochemical processes and lead speciation in both the liquid and the solid phases based on the modeled results, at least for the studied closed reactors. The extension of our findings to the prediction of lead release in the field is addressed in Section 3.4.

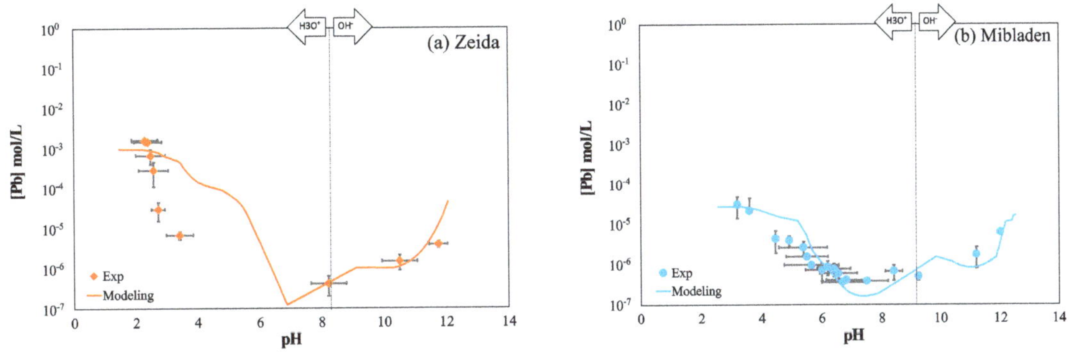

Figure 4. Solubilization of lead (total dissolved concentration), as a function of pH, from Zeida (**a**) and Mibladen (**b**) mine tailings; experimental curves (cross, adapted from [1,32]) and geochemical modeling (solid line).

3.3.2. Speciation of Lead as a Function of pH

The benefit of the modeling approach is fully illustrated by the provision of the characterization of lead speciation as a function of pH (Figure 5). The study here focuses on the lead carrier phases that are impacted by the ANC–BNC tests. The other carrier phases that are slightly or not impacted by the pH are not represented here since they play no role regarding lead mobilization. For the sake of precision, the whole range of pH meshes into 200 increments in the acidic zone, 200 increments in the basic zone, plus 1 point for the neutral zone (contact with distilled water). For all of these modeled points, we provide the mineral composition of the solid and liquid phases and lead speciation, including lead in solution, lead sorbed onto surface complexation sites (iron hydroxides) or cationic exchange sites, and lead precipitated or associated with lead-bearing minerals (Figure 5).

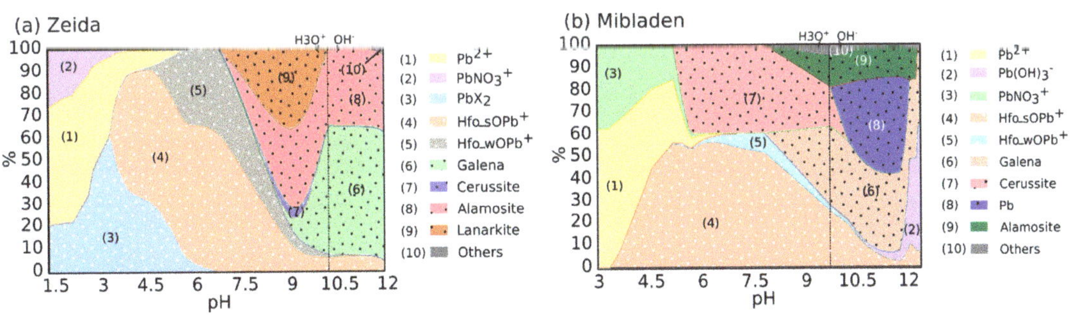

Figure 5. Speciation of lead, as a function of pH, for Zeida (**a**) and Mibladen tail (**b**) mining waste, obtained using the geochemical model (black dots: crystalline phases, white dots: surfaces, plain: in solution). Note that the range of the x-axis corresponds to the range of pH obtained for the same ANC–BNC tests with the same amounts of acidity and alkalinity. Consequently, the pH ranges differ between the two cases.

In Figure 5, the crystalline phases carrying lead are represented by black dots, while the iron hydroxide phases and ion exchange phases are represented by white dots. Lead in solution is represented by a solid color. The mineral phases that impact the mobilization of lead are galena, cerussite, alamosite, and lanarkite. These phases are relatively stable in neutral and acidic zones and control the mobilization of lead in these zones. In Mibladen tailings, calcite and dolomite promote the precipitation of calcareous phases such as cerussite, which plays a more important role here than it does in Zeida tailings (Figure 5b vs. Figure 5a).

Lead sorption to hydroxides and cationic exchange sites also contributes significantly to its speciation. This is particularly clear for Zeida (Figure 5a). Iron hydroxides sorb lead onto strong (s) and weak (w) sorption sites, referred to as ≡Hfo_sOH and ≡Hfo_wOH, respectively:

$$\equiv Hfo\_sOH + Pb^{2+} \Leftrightarrow \equiv Hfo\_sOPb^{+} + H^{+} \quad (12)$$

$$\equiv Hfo\_wOH + Pb^{2+} \Leftrightarrow \equiv Hfo\_wOPb^{+} + H^{+} \quad (13)$$

$$\equiv Hfo\_sOH \Leftrightarrow \equiv Hfo\_sO^{-} + H^{+} \quad (14)$$

$$\equiv Hfo\_wOH \Leftrightarrow \equiv Hfo\_wO^{-} + H^{+} \quad (15)$$

When present in the solid phase, iron hydroxides can sorb lead. Conversely, they release the sorbed lead when they dissolve. These results are in agreement with the literature, and sequential extractions were carried out on these types of matrices. Previous works attest that lead is mainly linked to the crystalline fraction and to iron hydroxides [49,50]. The distribution and the impact of mineral assemblages on mobilization depend very strongly on the total content, the type of soil, and its properties [49,51].

In solution, lead is not very well complexed. It is mainly found in the form of Pb^{2+} in acidic conditions. Since nitrate is added when acidifying with HNO_3, lead binds to nitrate to form $PbNO_3^{+}$. In the basic zone, it is mainly associated with the OH^{-} anion, in excess in solution, and forms $Pb(OH)_3^{-}$. These results of lead speciation as a function of pH are consistent with the literature [52]. Figure 5 may indicate lead speciation in the field, for any given pH, provided that ANC–BNC results are representative of element release in the field (this point is addressed in Section 3.4). This work's significant contribution is that it provides more insight on the lead speciation that can be combined with experimental methods of sequential extraction. These are known to be insufficient for a comprehensive understanding of all the mechanisms involved in lead and trace metal mobilization [53].

These results are representative and consistent with the phenomena of mine drainage, both for the levels of mobilization of lead and speciation in the liquid and solid phases as a function of the pH. We have significant amounts of lead in both Zeida and Mibladen mine tailings, meaning significant pollution (Figure 4). These two examples represent two distinct cases of mine drainage, i.e., AMD and CND. AMD is the most well-known phenomenon that impacts the mobilization of elements and produces very high acidity. This acidity logically promotes the release of trace metals. However, CND also impacts the chemical quality of the leachate even though the buffering capacity minimizes acidity. Indeed, even if calcareous phases buffer the solution, they do not prevent the dissolution of sulfides and other lead-carrying phases, resulting in lead release [1,47,48]. In these two conditions (AMD and CND), iron hydroxides appear to be a crucial contributor to lead release.

3.4. Representativeness of ANC–BNC Experiments

ANC–BNC runs were performed to characterize pH evolution and element release (lead) as a function of the amounts of added acidity or alkalinity, as suggested by [39] for mine tailings. Geochemical modeling was deployed to determine the amounts of minerals that were involved in the observed pH trends and element release. The mineral assemblages were characterized in terms of selected minerals and related amounts and validated against mineral characterization (XRD analysis and SEM). The number of sorption sites onto iron oxides and clay (cationic exchange) was also deduced from fitting experimental data, whereas the equilibrium constants were fixed at their regular values (the *llnl* database). From these data, speciation panels were proposed as a function of pH (Figure 5), assuming that the knowledge of pH was enough to derive the fate of lead in those mine tailings under all circumstances.

It should be noted that these modeled data were inferred from modeling specific ANC–BNC experiments that quantify buffer capacity and element release under specific experimental conditions. These may significantly differ from field conditions with a collection of contrasting conditions [54], including the following features:

- In the field, the contact with the atmosphere strongly differs between field and batch conditions, with closed conditions for batch experiments versus utterly open conditions (contact with a quasi-infinite reservoir of gas). In the case of batch reactors, only the gaseous species present in the volume of gas may interact with solutes. The consumption of oxidants (such as oxygen) may limit oxidation processes, whereas oxidants are instantaneously renewed in the case of contact with the open atmosphere. We then expect more pyrite oxidation [55] and induced acidification, mineral dissolution, and lead release under field conditions (no limitation of $O_{2(g)}$), e.g., [2,53]. In other words, if the three phases remain the same between batch and field conditions, the gas phase does not have the same composition.
- In the field, very slow processes that are kinetically limited may govern element release for a very long period of time, whereas the contact time in batch reactors is restricted to a few days, thus selecting mostly the instantaneous mechanisms. The addition of acid or alkalinity that is performed to boost dissolution and element release may not compensate for the very long time needed for some minerals to dissolve. For instance, we expect pyrite dissolution to be kinetically limited [56,57] and thus much less important under batch conditions than under field conditions. The same expectations may be stated for calcite and dolomite dissolutions that are prone to significant kinetic limitations [58].
- The liquid/solid ratio is all but the same between batch and field conditions. Indeed, in batch reactors, the solid phase is dispersed into the liquid solution with a very high liquid/solid (L/S) ratio, in the order of 10 L/kg in our cases and in most studies [54]. In the field, assuming usual values of soil bulk and mineral densities, i.e., 1.75 g/cm^3 and 2.65 g/cm^3, nominal values of saturated water content and the corresponding L/S ratio can be estimated, revealing a value of 0.2 L/kg in the field versus 10 L/kg in the batch reactors. We expect that higher L/S ratios would promote solubilization and element release.
- In addition to the difference in the liquid/solid ratio, field conditions are characterized by dynamic conditions (flow-induced solute transport), whereas batch reactors simulate static conditions with optimal contact between the solid grains and solutes. Numerous previous works have already demonstrated that the elements' fate significantly differs between static and dynamic conditions [59–61]. In columns, the local hydrodynamic conditions make the access of solutes to sorption sites more difficult and thus may derive from optimal sorption [62]. In addition, the soil structure may be prone to preferential flows, in particular for soils with aggregates and macropores, e.g., [63,64], which may, in turn, reduce the access of solutes to the reactive particles and decrease solute sorption [65–67].

In conclusion, ANC–BNC experiments combined with modeling can be seen as a promising tool for characterizing mineral compositions and sorption sites. However, it should not be considered the only representative solution of element release in the field. It provides generic trends, and proved in our case that Mibladen mine tailings have a high carbonate content, which should counteract any source of acidification (including that provided by pyrite oxidation) and lower, without preventing it completely, the release of lead (and other trace metals). The prediction of metal trace release in the field should not be purely inferred from ANC–BNC experiments, but rather predicted by modeling the field's actual conditions and considering knowledge support using ANC–BNC experiments on mineral composition and sorption sites. For our specific study case, we may also advise conducting additional experiments or monitoring in the field to consolidate our findings.

4. Conclusions

This study's objective was to characterize the mechanisms of lead mobilization and its speciation as a function of pH in waste from two mining sites, representative of AMD and CND, to better manage mining waste [68]. The main challenge was to set up and apply a methodology combining experimentation and modeling. This new and promising methodology provides crucial information for understanding the dominant mechanisms responsible for major and trace elements, including trace metal (e.g., lead) release and speciation in water. Such an approach has already been validated with pure phases mimicking mineral assemblages prone to generating AMD and CND, but it required further validation with real matrices [39].

The acid–base neutralization capacity tests revealed buffering powers specific to the two types of mining waste and strongly depended on carbonates. The geochemical modeling associated with the experimental results proved very satisfactory and confirmed our hypotheses based on our interpretation of the experimental results. The association between experimental results and geochemical modeling proved very relevant for studying the impact of solid phases on the buffering capacity, mobilization, and speciation of lead as a function of pH.

The carbonate phases (dolomite and calcite), which have a strong buffering capacity, are buffer acidity sources that do not totally prevent contaminated neutral drainage. The dissolution of the crystallized mineral phases containing lead and the desorption of lead from surface complexation sites (iron hydroxides) are the predominant mechanisms for lead release in solution. Despite different buffering powers, the mobilization of lead follows the same trends as a function of pH, but with much higher concentrations in the case of AMD. The results of lead speciation as a function of pH are consistent with the literature. Most notably, they allow for an accurate description of the evolution of the various lead-bearing phases as a function of pH. They also make it possible to determine the lead's speciation in the liquid phase and thus shed light on the potential toxicity of the leachate produced by mining waste.

This study provides a detailed understanding of the state of the system, assuming that ANC–BNC experiments simulate real conditions accurately. To better understand these mechanisms under natural conditions, it would be attractive to combine our findings with investigations adapted to dynamic conditions accounting for the kinetics of mineral dissolution and slow processes. Indeed, in the field, the contact between mine tailings, water, and atmospheric gases is much more complex with dynamic conditions, variable solid/liquid ratios, much longer contact times, etc. We may advise combining laboratory results with field monitoring to strengthen the statements on lead release and the related toxicity. In addition, while this study focuses on the mobilization of lead, other trace metals should be considered [33]. ANC–BNC and geochemical modeling appear to be very promising tools for the characterization of mining waste and related mining drainage. However, as with any tool, other types of information, in particular field investigations, and other information sources should be considered as well.

Author Contributions: Conceptualization, C.D. (Clémentine Drapeau), L.L., and C.D. (Cécile Delolme); methodology, C.D. (Clémentine Drapeau), L.L., R.A., M.B., R.H., and C.D. (Cécile Delolme); software, C.D. (Clémentine Drapeau) with the help of L.L., C.D. (Clémentine Drapeau), D.B., M.E., M.B., and T.B.; validation, C.D. (Clémentine Drapeau), L.L., and C.D. (Cécile Delolme); resources, C.D. (Cécile Delolme) and L.L.; data curation, R.A., R.H., M.B., and C.D. (Clémentine Drapeau); writing—original draft preparation, C.D. (Clémentine Drapeau) with the help of L.L. and C.D. (Cécile Delolme); writing—review and editing, all authors. All authors have read and agreed to the published version of the manuscript.

Funding: This work received the support of the INFILTRON project funded by the French National Research Agency (ANR-17-CE04-010).

Institutional Review Board Statement: Not applicable.

Informed Consent Statement: Not applicable.

Data Availability Statement: Not applicable.

Acknowledgments: The authors would like to thank the reviewers and editors for improving the manuscript.

Conflicts of Interest: The authors declare no conflict of interest.

References

1. Argane, R.; El Adnani, M.; Benzaazoua, M.; Bouzahzah, H.; Khalil, A.; Hakkou, R.; Taha, Y. Geochemical Behavior and Environmental Risks Related to the Use of Abandoned Base-Metal Tailings as Construction Material in the Upper-Moulouya District, Morocco. *Environ. Sci. Pollut. Res.* **2016**, *23*, 598–611. [CrossRef]
2. Dold, B. Evolution of Acid Mine Drainage Formation in Sulphidic Mine Tailings. *Minerals* **2014**, *4*, 621–641. [CrossRef]
3. Bussiere, B. Colloquium 2004: Hydrogeotechnical Properties of Hard Rock Tailings from Metal Mines and Emerging Geoenvironmental Disposal Approaches. *Can. Geotech. J.* **2007**, *44*, 1019–1052. [CrossRef]
4. Coussy, S. *Stabilisation de Rejets Miniers Pollués à l'arsenic à l'aide de Sous-Produits Cimentaires: Étude de l'influence de La Cristallochimie Sur Le Risque de Mobilisation des Polluants*; INSA de Lyon: Lyon, France, 2011.
5. Benzaazoua, M.; Bussière, B.; Dagenais, A.-M.; Archambault, M. Kinetic Tests Comparison and Interpretation for Prediction of the Joutel Tailings Acid Generation Potential. *Environ. Geol.* **2004**, *46*, 1086–1101. [CrossRef]
6. Hakkou, R.; Benzaazoua, M.; Bussière, B. Acid Mine Drainage at the Abandoned Kettara Mine (Morocco): 2. Mine Waste Geochemical Behavior. *Mine Water Environ.* **2008**, *27*, 160–170. [CrossRef]
7. Argane, R.; Benzaazoua, M.; Hakkou, R.; Bouamrane, A. A Comparative Study on the Practical Use of Low Sulfide Base-Metal Tailings as Aggregates for Rendering and Masonry Mortars. *J. Clean. Prod.* **2016**, *112*, 914–925. [CrossRef]
8. Argane, R.; Benzaazoua, M.; Hakkou, R.; Bouamrane, A. Reuse of Base-Metal Tailings as Aggregates for Rendering Mortars: Assessment of Immobilization Performances and Environmental Behavior. *Constr. Build. Mater.* **2015**, *96*, 296–306. [CrossRef]
9. Baghdad, B.; Naimi, M.; Bouabdli, A.; Sonnet, P.; Lutts, S. Heavy Metals in Tailings, Soils and Vegetation. *Ecol. Mediterr.* **2006**, *32*, 86.
10. Bouabdli, A.; Saidi, N.; M'rabet, S.; Escarre, J.; Leblanc, J.M. Heavy Metal Transport by the Moulouya River (Morocco). *Rev. Sci. Eau* **2005**, *18*, 199–213.
11. El Hachimi, M.; El Founti, L.; Bouabdli, A.; Saidi, N.; Fekhoui, M.; Tassé, N. Pb et As Dans Des Eaux Alcalines Minières: Contamination, Comportement et Risques (Mine Abandonnée de Zeïda, Maroc). *Rev. Sci. Eau J. Water Sci.* **2007**, *20*, 1–13. [CrossRef]
12. El Kheir, S.B.; Oubbih, J.; Saidi, N.; Bouabdli, A. Uptake and Fixation of Zn, Pb, and Cd by Thlaspi Caerulescens: Application in the Cases of Old Mines of Mibladen and Zaida (West of Morocco). *Arab. J. Geosci.* **2008**, *1*, 87–95. [CrossRef]
13. Li, Z.; Ma, Z.; van der Kuijp, T.J.; Yuan, Z.; Huang, L. A Review of Soil Heavy Metal Pollution from Mines in China: Pollution and Health Risk Assessment. *Sci. Total Environ.* **2014**, *468–469*, 843–853. [CrossRef]
14. Sheoran, A.S.; Sheoran, V. Heavy Metal Removal Mechanism of Acid Mine Drainage in Wetlands: A Critical Review. *Miner. Eng.* **2006**, *19*, 105–116. [CrossRef]
15. Fashola, M.O.; Ngole-Jeme, V.M.; Babalola, O.O. Heavy Metal Pollution from Gold Mines: Environmental Effects and Bacterial Strategies for Resistance. *Int. J. Environ. Res. Public. Health* **2016**, *13*, 1047. [CrossRef]
16. Sigg, L.; Stumm, W.; Behra, P. *Chimie des Milieux Aquatiques*; Dunot: Paris, France, 2000.
17. Chatain, V.; Benzaazoua, M.; Cazalet, M.L.; Bouzahzah, H.; Delolme, C.; Gautier, M.; Blanc, D.; de Brauer, C. Mineralogical Study and Leaching Behavior of a Stabilized Harbor Sediment with Hydraulic Binder. *Environ. Sci. Pollut. Res.* **2013**, *20*, 51–59. [CrossRef] [PubMed]
18. Drapeau, C.; Delolme, C.; Chatain, V.; Gautier, M.; Blanc, D.; Benzaazoua, M.; Lassabatere, L. Spatial and Temporal Stability of Major and Trace Element Leaching in Urban Stormwater Sediments. *Open J. Soil Sci.* **2017**, *7*, 19. [CrossRef]
19. Benvenuti, M.; Mascaro, I.; Corsini, F.; Lattanzi, P.; Parrini, P.; Tanelli, G. Mine Waste Dumps and Heavy Metal Pollution in Abandoned Mining District of Boccheggiano (Southern Tuscany, Italy). *Environ. Geol.* **1997**, *30*, 238–243. [CrossRef]
20. Gao, L.; Chen, J.; Tang, C.; Ke, Z.; Wang, J.; Shimizu, Y.; Zhu, A. Distribution, Migration and Potential Risk of Heavy Metals in the Shima River Catchment Area, South China. *Environ. Sci. Process. Impacts* **2015**, *17*, 1769–1782. [CrossRef] [PubMed]
21. Izquierdo, M.; Tye, A.M.; Chenery, S.R. Lability, Solubility and Speciation of Cd, Pb and Zn in Alluvial Soils of the River Trent Catchment UK. *Environ. Sci. Process. Impacts* **2013**, *15*, 1844–1858. [CrossRef]
22. Marsalek, J.; Watt, W.E.; Anderson, B.C. Trace Metal Levels in Sediments Deposited in Urban Stormwater Management Facilities. *Water Sci. Technol.* **2006**, *53*, 175–183. [CrossRef]
23. Roessler, J.G.; Townsend, T.G.; Ferraro, C.C. Use of Leaching Tests to Quantify Trace Element Release from Waste to Energy Bottom Ash Amended Pavements. *J. Hazard. Mater.* **2015**, *300*, 830–837. [CrossRef] [PubMed]
24. Shu, W.S.; Ye, Z.H.; Lan, C.Y.; Zhang, Z.Q.; Wong, M.H. Acidification of Lead/Zinc Mine Tailings and Its Effect on Heavy Metal Mobility. *Environ. Int.* **2001**, *26*, 389–394. [CrossRef]
25. Zuo, X.; Fu, D.; Li, H. Speciation Distribution and Mass Balance of Copper and Zinc in Urban Rain, Sediments, and Road Runoff. *Environ. Sci. Pollut. Res.* **2012**, *19*, 4042–4048. [CrossRef] [PubMed]
26. Alloway, B.J. *Heavy Metals in Soils. Blackie Academic and Professional*; Chapman and Hall: London, UK, 1995.
27. Barceloux, D.G.; Barceloux, D. Chromium. *J. Toxicol. Clin. Toxicol.* **1999**, *37*, 173–194. [CrossRef]

28. World Health Organization. *Guidelines for Drinking-Water Quality*; World Health Organization: Geneva, Switzerland, 2004; Volume 1.
29. Van der Sloot, H.A.; van Zomeren, A.; Dijkstra, J.J.; Meeussen, J.C.L.; Comans, R.N.J.; Scharff, H. Prediction of the Leaching Behaviour of Waste Mixtures by Chemical Speciation Modelling Based on a Limited Set of Key Parameters. In Proceedings of the Tenth International Waste Management and Landfill Symposium, Sardinia, Italy, 3–7 October 2005.
30. Parkhurst, D.L.; Appelo, C. User's Guide to PHREEQC (Version 2): A Computer Program for Speciation, Batch-Reaction, One-Dimensional Transport, and Inverse Geochemical Calculations. 1999. *Water-Resour. Investig. Rep.* **1999**, *99*, 312.
31. Gidlow, D.A. Lead Toxicity. *Occup. Med.* **2004**, *54*, 76–81. [CrossRef]
32. Argane, R. Valorisation Des Rejets Miniers à Faible Teneur En Sulfures Comme Granulats Pour Mortiers. Ph.D. Thesis, INSA, Lyon, France, 2015.
33. El Azhari, A.; Rhoujjati, A.; El Hachimi, M.L.; Ambrosi, J. Pollution and Ecological Risk Assessment of Heavy Metals in the Soil-Plant System and the Sediment-Water Column around a Former Pb/Zn-Mining Area in NE Morocco. *Ecotoxicol. Environ. Saf.* **2017**, *144*, 464–474. [CrossRef]
34. Combe, M.; Simonot, M. La Haute Moulouya, Le Sillon d'Itzer-Enjil et Le Massif de Bou-Mia-Aouli. *Notes Mém. Serv. Géol. Domaines Rif. Maroc. Orient. Ed. Serv. Géol. Maroc.* **1971**, *231*, 193–201.
35. Coussy, S.; Benzaazoua, M.; Blanc, D.; Moszkowicz, P.; Bussière, B. Arsenic Stability in Arsenopyrite-Rich Cemented Paste Backfills: A Leaching Test-Based Assessment. *J. Hazard. Mater.* **2011**, *185*, 1467–1476. [CrossRef]
36. Glass, G.; Buenfeld, N. Differential Acid Neutralisation Analysis. *Cem. Concr. Res.* **1999**, *29*, 1681–1684. [CrossRef]
37. Peyronnard, O.; Blanc, D.; Benzaazoua, M.; Moszkowicz, P. Study of Mineralogy and Leaching Behavior of Stabilized/Solidified Sludge Using Differential Acid Neutralization Analysis: Part II: Use of Numerical Simulation as an Aid Tool for Cementitious Hydrates Identification. *Cem. Concr. Res.* **2009**, *39*, 501–509. [CrossRef]
38. AFNOR CEN/TS 14429 Characterization of Waste. Leaching Behavior Tests. Influence of PH on Leaching with Initial Acid/Base Addition. 2015. Available online: http://www.coal-ash.co.il/research/Yoetz_Deutsch_2008_full.pdf (accessed on 1 May 2021).
39. Drapeau, C.; Delolme, C.; Vézin, C.; Blanc, D.; Baumgartl, T.; Edraki, M.; Lassabatère, L. Anc–Bnc Titrations and Geochemical Modeling for Characterizing Calcareous and Siliceous Mining Waste. *Minerals* **2021**, *11*, 257. [CrossRef]
40. Engelsen, C.J.; van der Sloot, H.A.; Wibetoe, G.; Petkovic, G.; Stoltenberg-Hansson, E.; Lund, W. Release of Major Elements from Recycled Concrete Aggregates and Geochemical Modelling. *Cem. Concr. Res.* **2009**, *39*, 446–459. [CrossRef]
41. Halim, C.E.; Short, S.A.; Scott, J.A.; Amal, R.; Low, G. Modelling the Leaching of Pb, Cd, As, and Cr from Cementitious Waste Using PHREEQC. *J. Hazard. Mater.* **2005**, *125*, 45–61. [CrossRef]
42. Biondi, D.; Freni, G.; Iacobellis, V.; Mascaro, G.; Montanari, A. Validation of Hydrological Models: Conceptual Basis, Methodological Approaches and a Proposal for a Code of Practice. *Phys. Chem. Earth Parts ABC* **2012**, *42*, 70–76. [CrossRef]
43. Pollacco, J.A.P.; Nasta, P.; Soria-Ugalde, J.M.; Angulo-Jaramillo, R.; Lassabatere, L.; Mohanty, B.P.; Romano, N. Reduction of Feasible Parameter Space of the Inverted Soil Hydraulic Parameter Sets for Kosugi Model. *Soil Sci.* **2013**, *178*, 267–280. [CrossRef]
44. Sposito, G. *The Chemistry of Soils*; Oxford University Press: Oxford, UK, 2008.
45. Cornell, R.M.; Schwertmann, U. *The Iron Oxides: Structure, Properties, Reactions, Occurrences and Uses*; John Wiley & Sons: Hoboken, NJ, USA, 2003.
46. Blowes, D.; Ptacek, C.; Jambor, J.; Weisener, C. The Geochemistry of Acid Mine. *Environ. Geochem.* **2005**, *9*, 149.
47. Lapakko, K.A.; Antonson, D.A.; Wagner, J.R. Mixing of Limestone with Finely-Crushed Acid Producing Rock. In Proceedings of the the Fourth International Conference on Acid Rock Drainage, Vancouver, BC, Canada, 31 May–6 June 1997; pp. 1345–1360.
48. Singer, P.C.; Stumm, W. Acidic Mine Drainage: The Rate-Determining Step. *Science* **1970**, *167*, 1121–1123. [CrossRef]
49. Ramos, L.; Hernandez, L.M.; Gonzalez, M.J. Sequential Fractionation of Copper, Lead, Cadmium and Zinc in Soils from or near Doñana National Park. *J. Environ. Qual.* **1994**, *23*, 50–57. [CrossRef]
50. Marguı, E.; Salvadó, V.; Queralt, I.; Hidalgo, M. Comparison of Three-Stage Sequential Extraction and Toxicity Characteristic Leaching Tests to Evaluate Metal Mobility in Mining Wastes. *Anal. Chim. Acta* **2004**, *524*, 151–159. [CrossRef]
51. Othmani, M.A.; Souissi, F.; Bouzahzah, H.; Bussière, B.; Da Silva, E.F.; Benzaazoua, M. The Flotation Tailings of the Former Pb-Zn Mine of Touiref (NW Tunisia): Mineralogy, Mine Drainage Prediction, Base-Metal Speciation Assessment and Geochemical Modeling. *Environ. Sci. Pollut. Res.* **2015**, *22*, 2877–2890. [CrossRef] [PubMed]
52. Asavapisit, S.; Fowler, G.; Cheeseman, C. Solution Chemistry during Cement Hydration in the Presence of Metal Hydroxide Wastes. *Cem. Concr. Res.* **1997**, *27*, 1249–1260. [CrossRef]
53. Fanfani, L.; Zuddas, P.; Chessa, A. Heavy Metals Speciation Analysis as a Tool for Studying Mine Tailings Weathering. *J. Geochem. Explor.* **1997**, *58*, 241–248. [CrossRef]
54. Limousin, G.; Gaudet, J.-P.; Charlet, L.; Szenknect, S.; Barthes, V.; Krimissa, M. Sorption Isotherms: A Review on Physical Bases, Modeling and Measurement. *Appl. Geochem.* **2007**, *22*, 249–275. [CrossRef]
55. Long, H.; Dixon, D.G. Pressure Oxidation of Pyrite in Sulfuric Acid Media: A Kinetic Study. *Hydrometallurgy* **2004**, *73*, 335–349. [CrossRef]
56. Williamson, M.A.; Rimstidt, J.D. The Kinetics and Electrochemical Rate-Determining Step of Aqueous Pyrite Oxidation. *Geochim. Cosmochim. Acta* **1994**, *58*, 5443–5454. [CrossRef]
57. Bierens de Haan, S. A Review of the Rate of Pyrite Oxidation in Aqueous Systems at Low Temperature. *Earth-Sci. Rev.* **1991**, *31*, 1–10. [CrossRef]

58. Sjöberg, E.L.; Rickard, D.T. Temperature Dependence of Calcite Dissolution Kinetics between 1 and 62 °C at PH 2.7 to 8.4 in Aqueous Solutions. *Geochim. Cosmochim. Acta* **1984**, *48*, 485–493. [CrossRef]
59. Lassabatere, L.; Winiarski, T.; Galvez Cloutier, R. Retention of Three Heavy Metals (Zn, Pb, and Cd) in a Calcareous Soil Controlled by the Modification of Flow with Geotextiles. *Environ. Sci. Technol.* **2004**, *38*, 4215–4221. [CrossRef]
60. Kuechler, R.; Noack, K. Comparison of the Solution Behaviour of a Pyrite–Calcite Mixture in Batch and Unsaturated Sand Column. *J. Contam. Hydrol.* **2007**, *90*, 203–220. [CrossRef]
61. Zhou, L.; Martin, S.; Cheng, W.; Lassabatere, L.; Boily, J.-F.; Hanna, K. Water Flow Variability Affects Adsorption and Oxidation of Ciprofloxacin onto Hematite. *Environ. Sci. Technol.* **2019**, *53*, 10102–10109. [CrossRef]
62. Bai, H.J.; Lassabatere, L.; Lamy, E. Colloid Transport in Aggregated Porous Media with Intra- and Interaggregate Porosities. *Ind. Eng. Chem. Res.* **2018**, *57*, 6553–6567. [CrossRef]
63. Lassabatere, L.; Yilmaz, D.; Peyrard, X.; Peyneau, P.E.; Lenoir, T.; Šimůnek, J.; Angulo-Jaramillo, R. New Analytical Model for Cumulative Infiltration into Dual-Permeability Soils. *Vadose Zone J.* **2014**, *13*, 1–15. [CrossRef]
64. Lassabatere, L.; Di Prima, S.; Bouarafa, S.; Iovino, M.; Bagarello, V.; Angulo-Jaramillo, R. BEST-2K Method for Characterizing Dual-Permeability Unsaturated Soils with Ponded and Tension Infiltrometers. *Vadose Zone J.* **2019**, *18*. [CrossRef]
65. Lassabatere, L.; Spadini, L.; Delolme, C.; Février, L.; Galvez Cloutier, R.; Winiarski, T. Concomitant Zn-Cd and Pb Retention in a Carbonated Fluvio-Glacial Deposit under Both Static and Dynamic Conditions. *Chemosphere* **2007**, *69*, 1499–1508. [CrossRef]
66. Batany, S.; Peyneau, P.-E.; Lassabatère, L.; Béchet, B.; Faure, P.; Dangla, P. Interplay between Molecular Diffusion and Advection during Solute Transport in Macroporous Media. *Vadose Zone J.* **2019**, *18*. [CrossRef]
67. Raimbault, J.; Peyneau, P.-E.; Courtier-Murias, D.; Bigot, T.; Gil Roca, J.; Béchet, B.; Lassabatère, L. Investigating the Impact of Exit Effects on Solute Transport in Macropored Porous Media. *Hydrol. Earth Syst. Sci. Discuss.* **2020**, *2020*, 1–20. [CrossRef]
68. Tayebi-Khorami, M.; Edraki, M.; Corder, G.; Golev, A. Re-Thinking Mining Waste through an Integrative Approach Led by Circular Economy Aspirations. *Minerals* **2019**, *9*, 286. [CrossRef]

minerals

MDPI

Article

Geochemical Controls on Uranium Release from Neutral-pH Rock Drainage Produced by Weathering of Granite, Gneiss, and Schist

Elliott K. Skierszkan [1,*], John W. Dockrey [2], K. Ulrich Mayer [1], Viorica F. Bondici [3], Joyce M. McBeth [4] and Roger D. Beckie [1]

1 Department of Earth, Ocean and Atmospheric Sciences, University of British Columbia, 2020–2207 Main Mall, Vancouver, BC V6T 1Z4, Canada; umayer@eoas.ubc.ca (K.U.M.); rbeckie@eoas.ubc.ca (R.D.B.)

2 Lorax Environmental Services Ltd., 2289 Burrard St, Vancouver, BC V6J 3H9, Canada; john.dockrey@lorax.ca

3 Canadian Light Source Inc., 44 Innovation Blvd, Saskatoon, SK S7N 2V3, Canada; ibi.bondici@lightsource.ca

4 Department of Geology, University of Regina, 3737 Wascana Parkway, Regina, SK S4S 0A2, Canada; Joyce.McBeth@uregina.ca

* Correspondence: eskiersz@eoas.ubc.ca

Received: 14 August 2020; Accepted: 4 December 2020; Published: 9 December 2020

Abstract: We investigated geochemical processes controlling uranium release in neutral-pH ($pH \geq 6$) rock drainage (NRD) at a prospective gold deposit hosted in granite, schist, and gneiss. Although uranium is not an economic target at this deposit, it is present in the host rock at a median abundance of 3.7 μg/g, i.e., above the average uranium content of the Earth's crust. Field bin and column waste-rock weathering experiments using gneiss and schist mine waste rock produced circumneutral-pH (7.6 to 8.4) and high-alkalinity (41 to 499 mg/L as $CaCO_3$) drainage, while granite produced drainage with lower pH (pH 4.7 to >8) and lower alkalinity (<10 to 210 mg/L as $CaCO_3$). In all instances, U release was associated with calcium release and formation of weakly sorbing calcium-carbonato-uranyl aqueous complexes. This process accounted for the higher release of uranium from carbonate-bearing gneiss and schist than from granite despite the latter's higher solid-phase uranium content. In addition, unweathered carbonate-bearing rocks having a higher sulfide-mineral content released more uranium than their oxidized counterparts because sulfuric acid produced during sulfide-mineral oxidation promoted dissolution of carbonate minerals, release of calcium, and formation of calcium-carbonato-uranyl aqueous complexes. Substantial uranium attenuation occurred during a sequencing experiment involving application of uranium-rich gneiss drainage into columns containing Fe-oxide rich schist. Geochemical modeling indicated that uranium attenuation in the sequencing experiment could be explained through surface complexation and that this process is highly sensitive to dissolved calcium concentrations and pCO_2 under NRD conditions.

Keywords: uranium; sorption; neutral-rock drainage; mine waste rock; geochemical modeling; weathering; aqueous complexation

1. Introduction

Uranium contamination in water is a global concern due to this element's chemical toxicity towards humans and other living organisms [1]. Water-quality guidelines for U in Canada are 20 μg/L for drinking water and 15 μg/L for the protection of aquatic life (long-term exposure) [2,3]. Contamination of water by U frequently involves U mining and milling [4–6] but can also arise through natural weathering of rocks enriched in U through magmatic differentiation, such as granites and rhyolites [7–11].

Uranium(aq) mobility is controlled by pH, redox conditions, aqueous complexation, interactions with mineral surfaces, and dissolution/precipitation of minerals [12–14]. The two dominant oxidation states of U are U(VI) and U(IV), with U(IV) being generally insoluble pH > 3 [14]. Under reducing conditions, U(VI) is converted to U(IV) via microbial and abiotic reduction [14,15]. Uranium(VI) forms the uranyl aquo cation (UO_2^{2+}), which is soluble at a wider pH range but can be attenuated via sorption onto clays, organic matter, and hydrous ferric oxides (HFO) [13,16–22]. Hydrous ferric oxides are a particularly important uranyl sorbent in mine wastes [23]. Recent work suggests that HFO sequesters uranyl via a bidentate inner-sphere uranyl complex at pH 3 to pH 9 [20]. From pH 7 to pH 9 and in waters containing dissolved inorganic carbon (DIC), uranyl dicarbonate $UO_2(CO_3)_2^{2-}$] also sorbs onto HFO via outer-sphere complexation [20]. However, uranyl sorption is considerably weakened by formation of the calcium-carbonato-uranyl aqueous complexes $Ca_2UO_2(CO_3)_3$ and $CaUO_2(CO_3)_3^{2-}$ (henceforth referred to as CCU) under neutral to alkaline-pH conditions, where U mobility can present a concern [18,24,25]. Analogous Mg-uranyl-carbonate species also exist [25], although their influence on U mobility is less-studied.

Under oxidizing conditions in mine wastes, U mobility depends upon a series of geochemical reactions involving dissolution of sulfide and carbonate minerals after the exposure of waste rock or tailings to oxygen and water. These reactions ultimately govern solution pH, redox, aqueous U complexation, and sorption site availability. First, sulfide minerals in the presence of atmospheric O_2 and water undergo oxidation, which produces HFO and sulfuric acid (Reaction (1)). Second, in the presence of carbonate minerals, acidity is buffered via carbonate-mineral dissolution (Reaction (2)) [23]:

(1) pyrite oxidation

$$4FeS_{2\,pyrite} + 15O_{2(aq)} + 14H_2O \rightarrow 4Fe(OH)_{3\,ferrihydrite} + 8SO_4^{2-}{}_{(aq)} + 16H^+{}_{(aq)} \tag{1}$$

(2) carbonate-mineral dissolution by acid attack

$$CaCO_{3\,calcite} + 2H^+{}_{(aq)} \rightleftharpoons Ca^{2+}{}_{(aq)} + H_2CO_{3(aq)} \tag{2}$$

Dissolved inorganic carbon (DIC) and Ca released in Reaction (2) can become available to complex uranyl and increase its mobility through Reaction (3), below [18]. Calcium release could also theoretically increase U mobility by competing for HFO sorption sites, although experimental evidence suggests this effect is negligible [26].

$$U(VI)O_2^{2+}{}_{(aq)} + x\,Ca^{2+}{}_{(aq)} + 3\,HCO_3^-{}_{(aq)} \rightleftharpoons Ca_xU(VI)O_2(CO_3)_3^{2x-4} + 3\,H^+{}_{(aq)};\ (x = 1 \text{ or } 2) \tag{3}$$

Conversely, generation of HFO from sulfide-mineral oxidation might decrease U mobility by increasing sorption-site availability (Reaction (1)) [16,17,27].

The interplay of Reactions (1) to (3) has important ramifications for U behavior in sulfide- and carbonate-bearing mine wastes, yet their cumulative outcome on net U mobility is uncertain. Although there are many studies of U geochemistry in mine wastes, these have predominantly focused on acidic drainage (pH < 6) where CCU complexation is negligible [12,18,19,27–37]. Neutral-rock drainage (NRD) (e.g., pH ≥ 6 [38]) is typically characterized by high alkalinity (hundreds of mg/L as $CaCO_3$) and dissolved Ca reaching hundreds of mg/L, i.e., orders-of-magnitude above typical fresh groundwaters [39–44]. Under these conditions, CCU complexation (Reaction (3)) might exacerbate U release. The role of Ca in mobilizing U has only received attention since the relatively recent discoveries of the importance of CCU in controlling U sorption [15,18,25] but this process has not yet been considered in NRD despite its Ca-rich characteristics.

Additionally, past studies on U behavior in a mining context have largely focused on wastes derived from U mining [12,28,30,36,45–47], yet magmatically differentiated rocks hosting a wide variety of ores (e.g., REE, Li) can also contain elevated U that could be mobilized through mine-waste weathering [48]. Thus, NRD-producing magmatically differentiated rocks may be particularly susceptible to U leaching due to the combination of a solid-phase source of U and leachate chemistry that enhances CCU complexation.

The objective of this work is, therefore, to evaluate the geochemical processes that control U mobility in Ca-rich NRD resulting from weathering of granitic and metamorphic rocks enriched in U. A combination of geochemical data gathered from exploration drill-core and kinetic leaching tests on a proposed mine's waste rock are analyzed to assess the outcomes of Reactions (1) through (3) above on U release in this setting.

2. Study Site and Geological Setting

The study site is located at the Coffee gold deposit in subarctic Yukon, northwest Canada (Figure 1). This deposit formed ca. 97 million years ago through interaction of Au-As-Sb-S bearing fluid with metamorphic and granitic country rock along structurally controlled faults and fractures [49,50] The deposit is located in the Dawson Range, wherein plutonic and metamorphic rocks are regionally enriched in U relative to typical crustal rocks [51]. Weathering of these rocks produces baseline U concentrations that can exceed 500 μg/L in groundwater and 300 μg/L in surface water, i.e., an order of magnitude above water-quality guidelines, making U an element of regional environmental concern [51]. The dominant local geological units are the Permian-aged Sulphur Creek orthogneiss, Klondike schist, and Snowcap schist, which were intruded by the Cretaceous-aged Coffee Creek biotite granite [49]. Minor expressions of marble and younger dykes are also present. Extensively weathered bedrock and the absence of Pleistocene glaciation has produced a deep oxidation front in the deposit, reaching depths of up to 300 m.

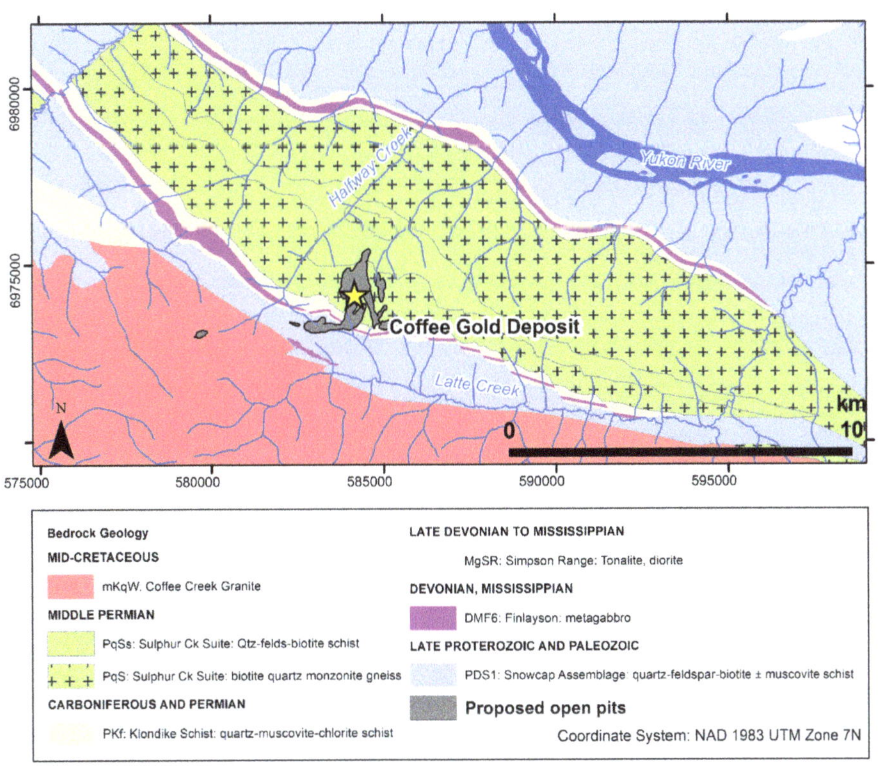

Figure 1. Geological map showing the bedrock geology underlying the Coffee deposit and the proposed mine pit locations. Bedrock geology from ref. [50].

The extent of oxidation in drill-core samples at the Coffee deposit is used to categorize rocks into three weathering facies: the "oxide" facies corresponds to rocks with >95% of surfaces showing visible oxidation and the presence characteristic limonite/hematite stains or cubic vugs that remain after pyrite oxidation; the "transition" facies corresponds to 5% to 50% of surface with visible oxidation on pyrite grains, and the "fresh facies" is defined by samples where <5% of surfaces are oxidized and there is only minor joint-controlled oxidation that does not extend into adjacent rock.

3. Methods

3.1. Drill-Core Characterization

Geochemical characteristics of the rock types and weathering facies were established using a dataset of 479 drill-core samples collected at the Coffee deposit. All samples were classified by field exploration geologists based on rock type (gneiss, schist, or granite) and weathering facies (oxide, transition, or fresh). In the laboratory, samples were pulverized for geochemical analyses of elemental abundances using an aqua regia digestion followed by analysis via inductively coupled plasma mass spectrometry (ICP-MS), total inorganic carbon (TIC), and total sulfur. Analyses were either conducted at a commercial lab (ALS, North Vancouver, BC, Canada), or at the Pacific Centre for Isotopic and Geochemical Research (University of British Columbia, Vancouver, BC, Canada).

The acid-neutralization potential ratio (NPR), which is a standard parameter used to assess the likelihood that mine wastes produce acidic-rock drainage [38], was calculated from TIC and total sulfur results. In this calculation, the acid-generation potential is calculated from total sulfur assuming that all S can be released as H_2SO_4, which consumes 1 mol of $CaCO_3$ per mol H_2SO_4. Acid-neutralization potential is calculated from TIC assuming that all TIC is $CaCO_3$. The NPR reflects the ratio of acid-neutralization potential to acid-generation potential, both expressed as kg $CaCO_3$ per ton rock as per the NPR equation [38]:

$$\text{NPR} = \frac{\text{TIC}\left[\text{wt.\% C}\right] \times 83.3}{\text{total S}\left[\text{wt.\% S}\right] \times 31.25} \tag{4}$$

Rocks with a NPR < 1 are classified as potentially acid generating (PAG), rocks with a NPR >> 1 are classified as non-acid generating (NAG), and rocks with a NPR near or slightly above 1 (e.g., 1 < NPR < 2) are classified as uncertain [38].

The oxidation state of U in select rock samples representing each major lithology (gneiss, schist, and granite) and weathering facies (oxide, transition, fresh) was determined by X-ray absorption near-edge spectroscopy (XANES). Data were collected on the Hard X-ray Micro-Analysis beamline (HXMA, 061D-1) at the Canadian Light Source (CLS, Saskatoon, SK, Canada). The X-ray beam was calibrated to a Y foil at the first inflection point of the Y K-edge (17,038 eV) and U spectra were collected at the L_3 edge (17,166 eV) over 5 energy regions: −200 eV to −160 eV, 10 eV steps; −160 eV to −100 eV, 0.5 eV steps (covering the yttrium foil calibration standard); −100 eV to −30 eV, 10 eV steps; −30 eV to +40 eV (U L_3 edge), 0.5 eV steps; and +40 eV to 14.2 K, 0.05 K steps. The dwell time was 1 s per step. Data were collected in fluorescence mode with a 32-element Ge detector. Multiple spectra were collected for each sample and to correct for potential beam drift. Each analysis featured a Y-foil calibration measurement. We also collected XANES spectra of uranyl nitrate and uraninite standards using the same beamline settings as samples. To estimate the relative proportion of U redox species in each sample, linear combination fitting was conducted using the ATHENA software suite (Demeter v. 0.9.22) [52] and the uranyl nitrate and uraninite as standards.

3.2. Rock-Weathering Experiments

3.2.1. Experimental Design

Relationships between lithological characteristics, aqueous geochemistry, and U release were assessed with rock weathering experiments that included "field bins" and columns wherein rock was exposed to water and oxygen and the resulting leachate was characterized. These experiments were conducted under water-unsaturated conditions to mimic field waste-rock weathering conditions where fluid exchange with atmospheric oxygen leads to the oxidation of sulfide minerals as per Reaction (1). Field-bin experiments have a longer water residence time and lower water/rock ratio than columns, which leads to better representation of secondary sorption and mineral precipitation reactions that occur in full-scale waste-rock weathering environments in comparison with columns. Composite rock samples representing the various lithologies (granite, gneiss, schist) and weathering facies (oxide, transition, fresh) were prepared for these experiments (Supplementary Table S1).

The field bins consisted of 119 L HDPE barrels that were filled with approximately 200 kg of dry crushed rock, sieved to a grain size < 6 mm, and exposed from 2013 to 2018 to outdoor temperature and precipitation conditions in Whitehorse, Yukon, Canada (Supplementary Figure S1) [53,54]. Field bins were initially irrigated with 28 L of de-ionized water, after which atmospheric precipitation was the only source of water. Leachates were collected into closed plastic containers that were sampled approximately monthly for geochemical analyses.

The column experiments involved application of 400 mL of Nanopure™ water purified through distillation and de-ionization (DDI water) onto to Plexiglas® reactors containing 10 kg of crushed and sieved (<6 mm) waste rock every 14 days. Columns were 21 cm in inner diameter and 20.5 cm tall and were operated at a lab temperature of 4 °C to reflect the subarctic climate of the field site. A nylon mesh overlying a perforated Plexiglas® disk at the base of the columns was used to retain the rock in the columns.

Three "Phases" of column experiments were performed. Phase 1 involved application of DDI water to the columns every 2 weeks between 2014 and 2018, with the leachate collected for chemical analysis. During a 6-month period in Phase 1 and beginning in May 2015, the effluent from the base of the columns was collected and then re-applied to the top columns once. Upon the second pass of water through the column, samples were withdrawn for chemical analyses. This recirculation procedure was included to increase the water contact time with the rock such that it is closer to field conditions in waste-rock storage facilities (WRSF). After January 2016, chemical concentrations stabilized and thus the recirculation procedure was removed: thereafter, DDI water was applied as influent once and the resulting leachate collected for analysis without recirculation.

Phases 2 and 3 were designed to investigate U mobility controls during interaction of leachate produced by one rock type with another rock's solids. In Phase 2, leachate from high-alkalinity and U-poor schist-bearing columns was used as influent for U-rich gneiss-bearing columns, which was hypothesized to drive U desorption via CCU complexation. In Phase 3, this order was reversed, to assess whether the Fe-rich schist could attenuate U from gneiss leachate via sorption. Phases 2 and 3 were conducted using gneiss and schist pairings, with one experimental set devoted to oxide-facies rocks and another to transition-facies rocks (Supplementary Table S1).

3.2.2. Geochemical Analyses

Column and field bin leachates were sampled for pH, alkalinity, anions, and metals analyses at SGS Canada Inc. (Burnaby, BC, Canada). Alkalinity and anions were determined on filtered (0.45 µm) sample aliquots using titration with H_2SO_4 and ion chromatography, respectively. Metals were analyzed on filtered (0.45 µm) and acidified (HNO_3 to pH < 2) sample aliquots using inductively coupled plasma mass spectrometry. Each water-sample batch was accompanied by a method blank, a duplicate sample analysis to monitor precision, and a matrix spike to monitor accuracy. Lab QA/QC criteria at SGS include matrix spike recoveries better than 70% for metals and 75% for anions, and reproducibility between duplicate samples better than 20%.

The grain-size distribution and geochemical composition of rocks used in weathering experiments was also determined by SGS Canada. Metals were analyzed by ICP-MS after aqua-regia digestion of pulverized samples. Analyses of duplicate and reference materials (OREAS 260 and OREAS502B, Melbourne Australia) yielded ICP-MS reproducibility and accuracy were ≤5%. Total inorganic carbon (TIC) was measured by treating rock sample powders with $HClO_4$ to convert C to CO_2, which was measured by coulometry. Rock total sulfur content was determined by Leco combustion analysis. Sulfide was measured by first leaching sulfate with $NaCO_3$ and analyzing the residual sample with the Leco analyzer. Shake-flask extractions were conducted by SGS on rock samples using distilled de-ionized water at a 3:1 water:rock ratio shaken for 24 h, with the leachate analyzed for U content by ICP-MS.

Mineralogical abundances were quantified by powder X-ray diffraction (XRD) at the CLS (analytical details in Section S.1. of the Supplementary Materials) or by SGS Canada Inc (Burnaby, BC, Canada). Quantitative evaluation of materials by scanning electron microscopy (QEMSCAN) analysis was used at SGS to quantify the proportions of Fe-minerals found as oxides, and of S-minerals as sulfides, respectively. QEMSCAN analyses were conducted on samples pulverized to 80% passing <106 μm and then graphite-impregnated into polished thin sections.

At the end of column experiments, residues were also characterized for elemental abundances (ICP-MS) and shake-flask extractable U as described above. Sequential chemical extractions (SCEs) were conducted on the residues from the two schist columns (C3-ScO and C4-ScT) using a 5-step protocol based on the method of Tessier et al. [55] (procedural details in Section S.2. of the Supplementary Materials).

3.3. Geochemical Modeling

The geochemical code PHREEQC [56] was used to calculate aqueous speciation and mineral saturation indices (SI) on field-bin and column effluents. We also conducted a series of simulations in PHREEQC to investigate U sorption during Phase 3 of the column experiments. A detailed explanation of the modeling approach is given in the Supplementary Materials (Section S.4). Briefly, U sorption was assumed to be dominated by surface complexation and it was modeled using a 2-step approach. In the first step, HFO availability was calibrated for the schist columns using Phase 1 major-ion chemistry data, the rock/water ratio, QEMSCAN quantification of Fe-oxides, and U solid-phase abundance data from aqua regia and sequential chemical extractions. These calibrated HFO compositions for the schist were then used in simulations of Phase 3, which involved batch reactions with gneiss influent solution and residual porewater in the schist column in the presence of HFO, calcite, and H_2SO_4 (assumed to be sourced from sulfide-mineral oxidation). Changes in major-ion chemistry (pH, alkalinity, Ca, sulfate) were considered by allowing dissolution/precipitation of calcite, dissolution/exsolution of CO_2, and addition of H_2SO_4 along with surface complexation with HFO (full details in Supplementary Materials Section S.4.2).

PHREEQC simulations were conducted using the wateq4f.dat database, which was amended with association constants for the Ca-carbonato-uranyl and Mg-carbonato-uranyl aqueous complexes from Dong and Brooks [24] and the surface complexation constants for HFO-uranyl and HFO-carbonate species as per Mahoney et al. [18,57] (Supplementary Table S2). Surface complexation was modeled using the Dzombak–Morel diffuse-double layer with a 50:1 ratio of strong and weak sorption sites [57,58].

4. Results and Discussion

4.1. Uranium, Sulfur, and Carbonate Content of the Coffee Deposit

Drill-core geochemical analyses indicate that rock at the Coffee deposit is generally NAG owing to excess of TIC over sulfur, which average 0.5 wt.% and <0.07 wt.%, respectively. 97.5% of samples are classified as NAG with NPR > 1 (Supplementary Figure S2). Rock types can generally be characterized as follows: schist is typically the most enriched in sulfur and TIC, granite has low sulfur and low TIC, and gneiss has intermediate sulfur and TIC content (Figure 2). Sulfide contents are minimal in oxide-facies samples of all rock types: median values are 0.005 wt.% in granite-oxide and gneiss-oxide, and 0.01 wt.% for schist-oxide. Gneiss-oxide and especially granite-oxide rocks are also depleted in TIC relative to their less-weathered counterparts, with median TIC values of 0.2 wt.% and 0.1 wt.%, respectively, although they generally are NAG (Supplementary Figure S2). Weathering of the deposit over geological timescales has therefore created an oxidized zone that is depleted in sulfide minerals yet has maintained excess TIC to create NAG rock.

Rocks at the Coffee deposit are modestly enriched in U relative to typical crustal rocks, with a median abundance of 3.7 μg/g and a maximum of 144 μg/g (Table 1), both above the 2.7 μg/g average of Earth's upper crust [59]. The deposit is also enriched in U in comparison with regional country rocks: Sulphur Creek suite orthogneiss and Whitehorse suite granite samples collected outside the Coffee deposit have median abundances of 3.3 and 3.2 μg/g, respectively [51]. The lack of clear relationships between U and hydrothermally sourced elements As, S, and Sb in fresh-facies rocks in the deposit suggests that its U was not sourced from the hydrothermal system (Supplementary Figure S3).

Table 1. Uranium abundance at the Coffee deposit by rock type and weathering facies (all units μg/g).

	All Rocks	Granite				Gneiss				Schist			
Facies		**All**	**Fresh**	**Transition**	**Oxide**	**All**	**Fresh**	**Transition**	**Oxide**	**All**	**Fresh**	**Transition**	**Oxide**
n	479	97	31	52	14	232	59	127	46	111	36	50	25
median	3.7	6.5	6.5	6.5	8.9	3.8	2.5	3.9	7.6	2.0	1.3	1.9	2.2
mean	6.1	10	7.5	10	16	6.5	3.8	5.5	12	2.2	1.9	2.1	2.9
2SD	18	20	10	21	28	21	8.0	9.0	42	3.0	3.0	1.7	4.2
max	140	53	29	53	52	140	20	22	140	10	6.2	4.3	10
min	0.20	1.8	1.9	2.4	1.8	0.27	0.27	0.29	1.3	0.34	0.34	0.81	0.63

The strongest enrichment in U is found the oxide facies of each major lithology (granite, gneiss and schist; Table 1), where median values are 8.9, 7.6, and 2.2 μg/g, respectively (Supplementary Table S1). This enrichment of oxide-facies rocks in U may indicate of retention of U transported by the groundwater flow system. This hypothesis is supported by significantly higher U/Ti in the oxide facies ($p < 0.05$, Mann–Wilcox test) in comparison with the fresh facies across all lithologies (Figure 2) because Ti is relatively immobile in groundwater but not U.

Uranium redox speciation of select gneiss and schist samples determined by linear combination fitting (LCF) of the XANES spectra show that samples from all weathering facies have an appreciable U(VI) content, with proportionally more U(VI) in oxide-facies rocks than in transition- and fresh-facies rocks (Figure 3 and Supplementary Table S3). These results suggest the presence of U(IV) phases in unoxidized portions of the deposit, while rock oxidation has liberated U that was transported and re-deposited as U(VI) in oxide-facies rocks.

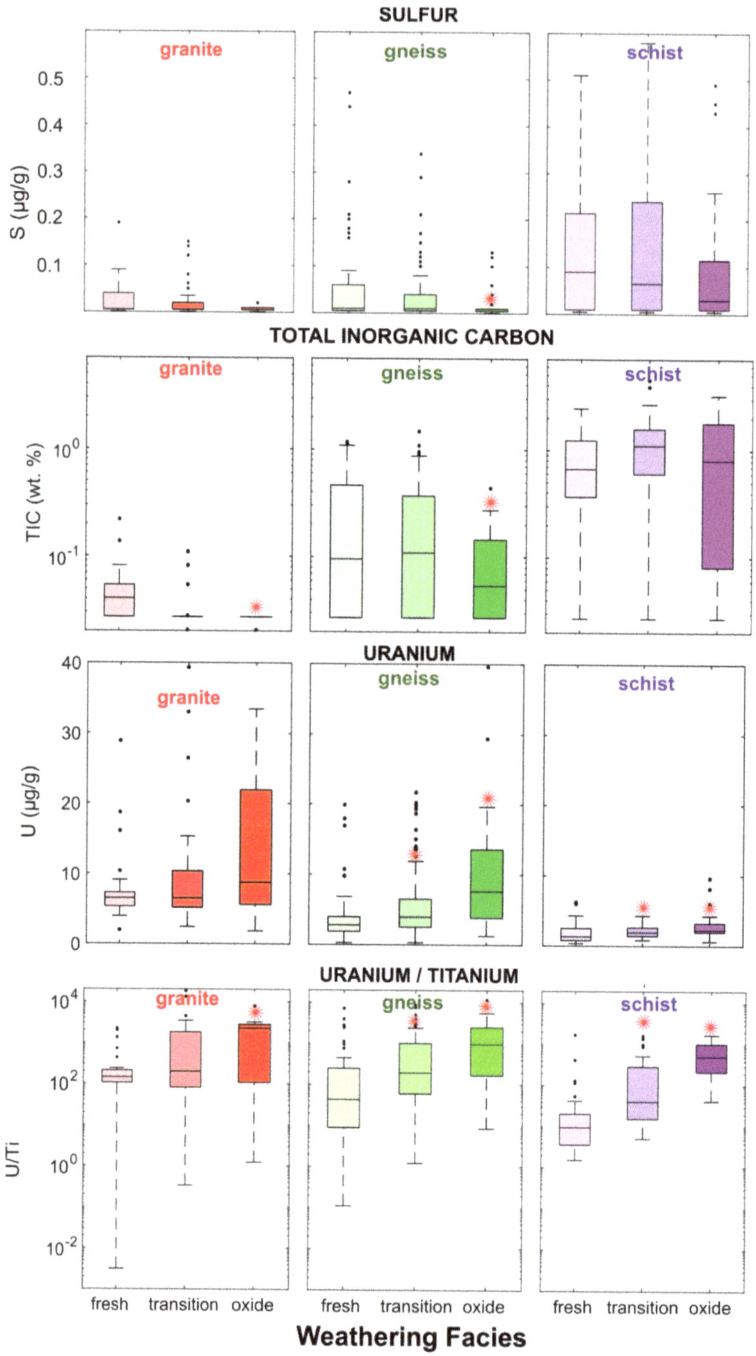

Figure 2. Sulfur, total inorganic carbon (TIC), U, and U/Ti by rock type and by weathering facies in Coffee drill-core samples. Red stars indicate median U or U/Ti is significantly higher ($p < 0.05$); or TIC and sulfur content is significantly lower ($p < 0.05$) in transition or oxide facies than in fresh facies for a given rock type (Mann–Wilcox Test).

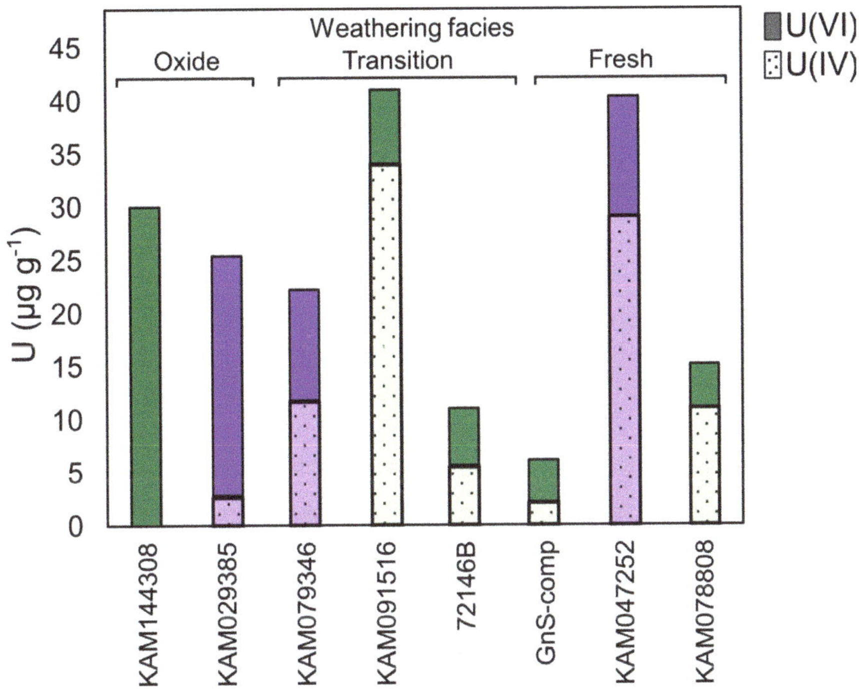

Figure 3. Uranium redox speciation in select Coffee drill-core samples determined by XANES-LCF. Solid fill = U(VI); hatched fill = U(IV); green = gneiss; purple = schist. Full LCF model results and quality of LCF fits (R-factor) are presented in Supplementary Table S3. The LCF error is estimated at ±10%.

4.2. Rock Weathering Experiments

Fresh-facies rocks for a given rock type used in kinetic tests contained the highest sulfide and carbonate mineral content, while oxide-facies rocks were depleted in sulfide and carbonate minerals, reflecting observations from the larger drill-core dataset (Figure 4). All rocks used in kinetic tests were NAG, with NPRs ranging from 6.5 to 81, except the granite-transition which had an NPR of 0.7 (Supplementary Table S1). Carbonate minerals included calcite, dolomite, and ankerite, the sum of which varied in the order schist > gneiss > granite (Supplementary Table S1). Carbonates comprised 9.7 to 24 wt.% in schist and ranged from below XRD detection limits to 4.3 wt.% in gneiss, while no carbonates were detected in granite (Supplementary Tables S1 and S4). Pyrite and arsenopyrite were the dominant sulfide minerals observed by XRD and QEMSCAN analyses with abundances below 1 wt.%. The Fe-oxide content determined by QEMSCAN was notably higher in the schist-oxide at 2.0 wt.% in comparison with other rocks which had 0.20 to 0.56 wt.% Fe-oxide (Supplementary Table S1). Fe-oxide abundance was generally greater in oxide-facies rocks for a given lithology than in their transition-facies and fresh-facies counterparts. No U minerals were identified by XRD. The initial solid-phase U content was higher in the gneiss and granite (5.5 to 7.6 μg/g) than in the schist (2.6 to 3.2 μg/g) (Figure 4).

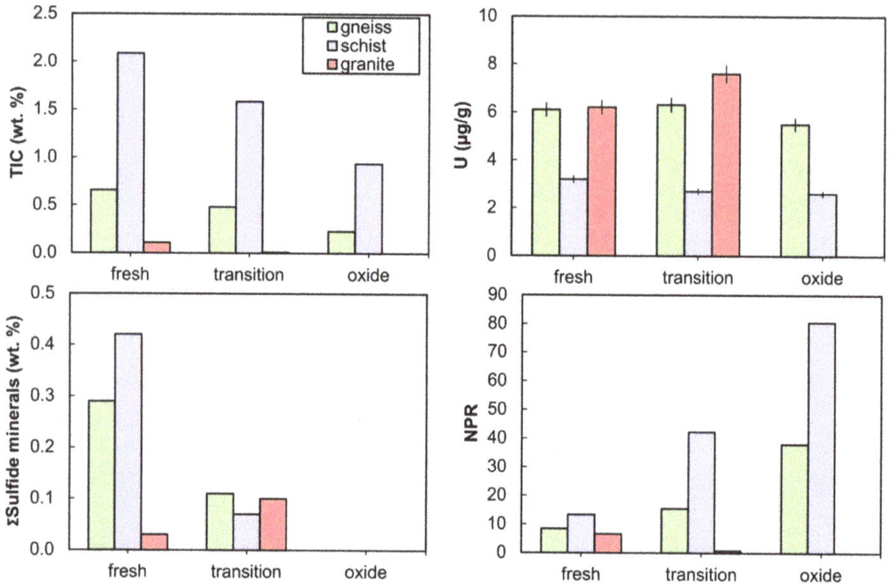

Figure 4. Solid-phase composition of rocks used in kinetic tests: total inorganic carbon (TIC), uranium, sum of sulfide minerals (from quantitative evaluation of materials by scanning electron microscopy, QEMSCAN), and neutralization potential ratio (NPR) as a function of rock type and weathering facies. Error bars on U abundance are ±5%.

Leachate pH and alkalinity were distinctly lower in granite weathering experiments in comparison with gneiss and schist experiments (Figure 5), owing to the minimal carbonate content in the granite. Granite-transition rock produced leachate pH ca. 7.5 in the column, and pH < 6 in the field bin, with alkalinity concentrations being close to detection limits. In contrast, the carbonate-rich gneiss and schist produced drainage in the pH 7.6 to 8.4 range, and several hundred mg/L of alkalinity (as $CaCO_3$) (Figure 5). Geochemical modeling indicated that leachates were near calcite saturation in all column experiments, with mineral saturation indices (SI) typically ranging between −0.25 and +0.50, except the granite-transition column where calcite was undersaturated (SI −1 to −3). All columns were undersaturated with respect to gypsum (SI < −0.8). Thus, gneiss and schist drainage can generally be described as high-alkalinity NRD, while granite drainage is circum-neutral to moderately acidic with low alkalinity.

Within a given weathering facies, U concentrations followed the order gneiss > schist > granite (Figure 5). Leachates in the gneiss and schist field bins contained tens to hundreds of µg/L U, while all granite U concentrations remained <10 µg/L. This lithological control was also observed in column experiments: in the gneiss-transition and schist-transition columns, U concentrations stabilized between ca. 180 µg/L and ca. 30 µg/L respectively, while they rapidly declined below 1 µg/L in the granite columns (Figure 5). Comparable lithological controls and U concentration ranges occur in groundwater around the Coffee deposit, where monitoring wells screened in gneiss can produce water containing several hundred µg/L U while those in schist and granite typically have lower concentrations that range from ca. 30 to 80 µg/L, and ca. 4 to 75 µg/L, respectively [60]. There was an additional effect of weathering facies, with oxide-facies rocks releasing the least U. In gneiss-oxide and schist-oxide field bins, U concentrations were <10 µg/L, while they ranged from 23 to 750 µg/L in the transition- and fresh-facies experiments (Figure 5).

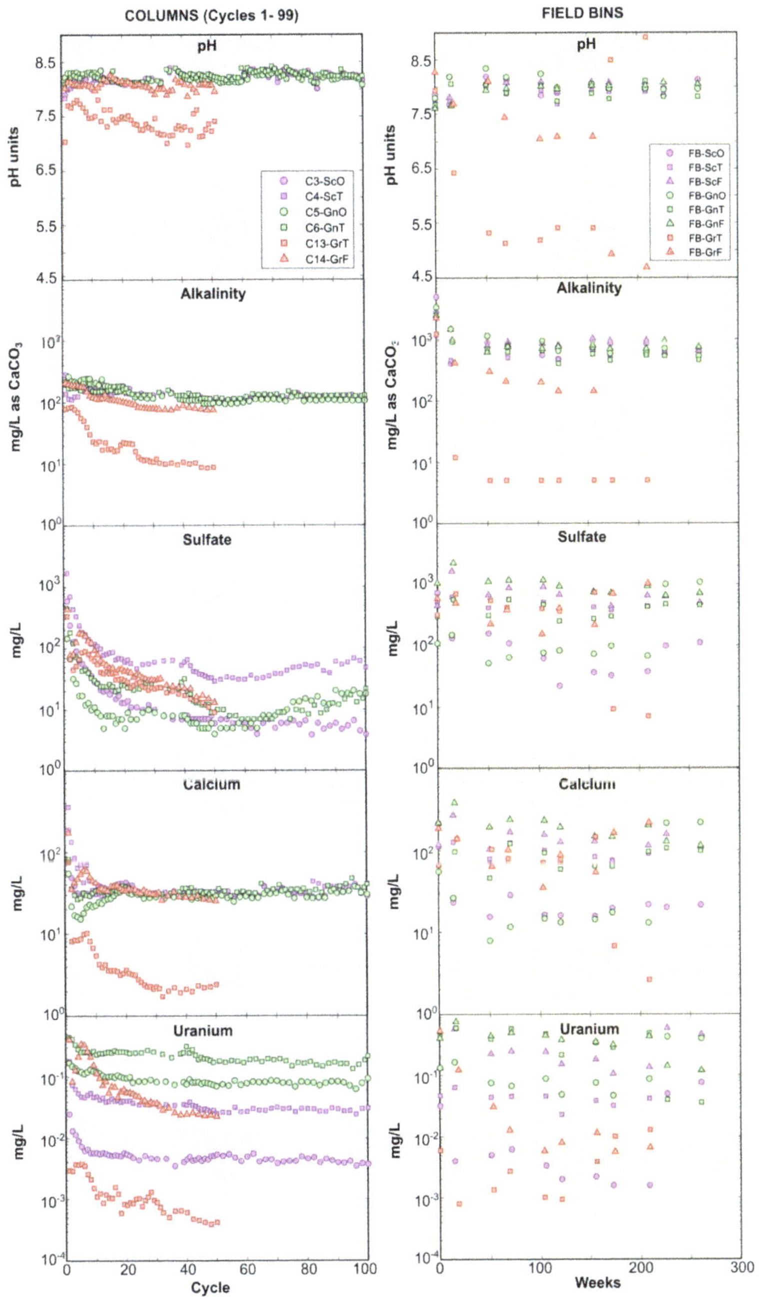

Figure 5. Column (**left panel**) and field bin (**right panel**) leachate pH, alkalinity, sulfate, Ca, and U. For column tests, only cycles representing Phase 1, i.e., prior to the sequencing experiments, are shown (i.e., cycles 1 to 99). Note differences in y-axis scale between left and right panels. Symbol shape indicates weathering facies: triangles = fresh; squares = transition; circles = oxide. Color indicates lithology: green = gneiss; red = granite; purple = schist.

Uranium, Ca, and sulfate release rates in field bins also correlated with rock weathering facies, with the highest rates produced in the least oxidized rocks (Figure 6). This was not the case for alkalinity, which was released at the lowest rate in the fresh facies for each rock type (Figure 6). Lower alkalinity release rates in fresh-facies rocks can be explained by their higher sulfide-mineral content, which led to sulfide-mineral oxidation, sulfuric acid (and sulfate) generation, and alkalinity consumption. (Figure 6). Calcium release rates followed sulfate release rates because Ca was released from carbonate-mineral dissolution reactions that buffered H_2SO_4 (Figure 6). While there is longstanding knowledge of the relationship between alkalinity and U release due to aqueous complexation of U by bicarbonate [13], the lack of correlation between alkalinity and U release rates in our experiments indicate that alkalinity was not a limiting factor in forming mobilizing U. Rather, greater U release from fresh-facies rocks may be explained by their higher Ca release that promoted formation of weakly sorbing CCU complexes (Reaction (3)) [18]. This hypothesis is supported by the higher proportion of U associated with CCU complexes in fresh- and transition-facies field bins in comparison with their oxide-facies counterparts (Supplementary Figure S4).

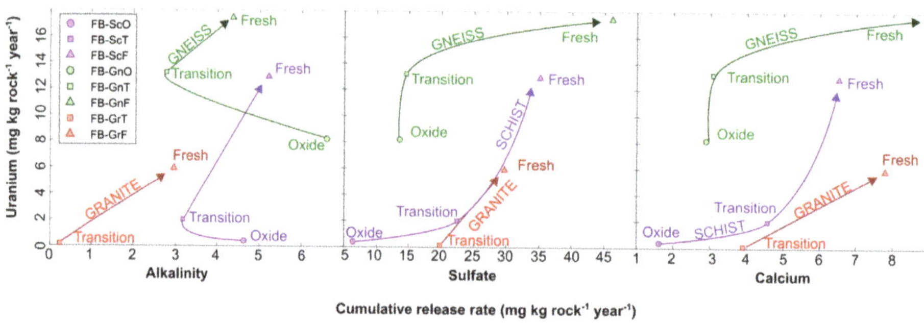

Figure 6. Cumulative U release rate in field-bin experiments against release rates of alkalinity (**left**), sulfate (**middle**) and calcium (**right**). Arrows indicate trend from oxide to transition to fresh weathering facies in each rock type.

Uranium release rates were also decoupled from solid-phase U content: despite granite-transition and granite-fresh rock having the highest U abundances, they released substantially less U than their gneiss and schist counterparts (Figure 6). For the granite-transition field bin, the disproportionally low U release rate relative to the U abundance can be explained by its pH (ca. 5) and low alkalinity (<10 mg/L as $CaCO_3$). Under these conditions, speciation calculations in PHREEQC suggest that a substantial proportion of U was present as UO_2^{2+} and UO_2OH^+, which can effectively sorb onto HFO through inner-sphere complexation [20] (Supplementary Figure S4). In the granite-fresh experiment, the concentration of these species was negligible but between 1% and 27% of $U_{(aq)}$ was in the form of uranyl carbonates, predominantly as $UO_2(CO_3)_2^{2-}$, which also sorbs onto HFO through outer-sphere complexation [20]. In contrast, in the higher-alkalinity schist and gneiss field bins, 91% to 99% of $U_{(aq)}$ was speciated as CCU, which are essentially unavailable for U sorption [18,20], while only a minor proportion of more sorption-reactive UO_2^{2+}, UO_2OH^+, and uranyl carbonate species were present, promoting higher U release rates (Supplementary Figure S4).

4.2.1. Controls on U Surface Complexation during Column Sequencing Experiments

Upon application of leachate from the *schist*-transition column to the *gneiss*-transition column in Phase 2, a pronounced increase in U concentrations in gneiss effluent was observed, reaching up to 272 µg/L (Figure 7). In the transition-facies columns and prior to Phase 2 and 3 sequencing experiments, drainage from the schist-transition column had produced considerably higher major-ion concentrations, including dissolved Ca and sulfate, than the gneiss-transition column (Figure 7).

Alkalinity concentrations were slightly lower for the schist than for the gneiss. Uranium concentrations were notably higher for the gneiss—ca. 150 µg/L—in comparison with values of approximately 30 µg/L for the schist. Given the lower alkalinity in the schist-transition leachate, alkalinity was not responsible for driving the U released during the schist-to-gneiss sequencing experiment. The increase in U concentrations during Phase 2 is instead hypothesized to be attributed to elevated Ca in the schist effluent that promoted CCU formation and U desorption after this effluent was circulated through the gneiss column.

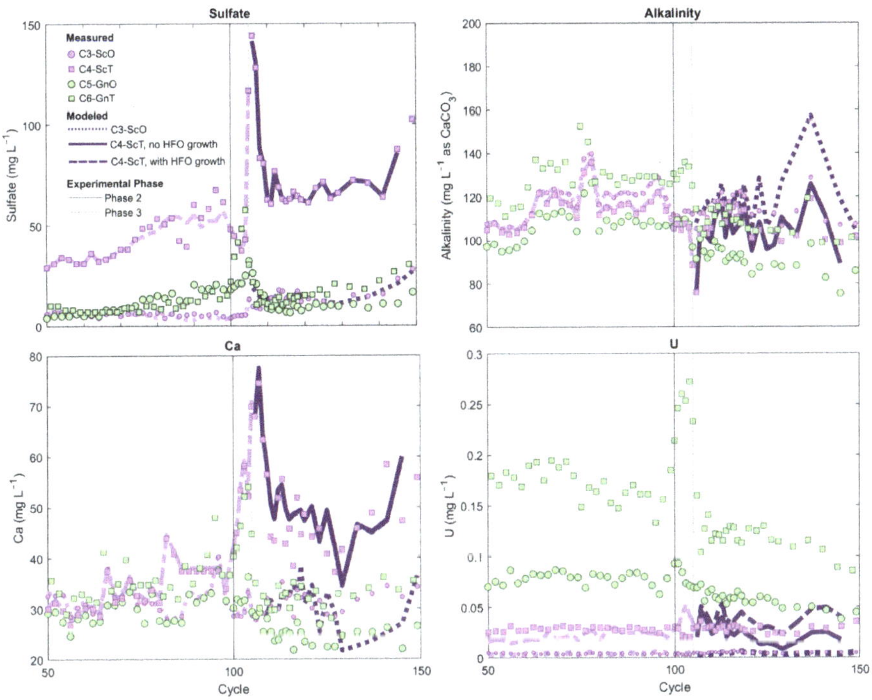

Figure 7. Column leachate composition during the sequencing experiments for transition (square symbols) and oxide (circular symbols) gneiss and schist columns. Sequencing experiments are described in Supplementary Table S1. Lines represent modeled concentrations: dotted lines = schist-oxide model; solid lines = schist-transition model with hydrous ferric oxides (HFO) growth; dashed line = schist-transition model with a fixed HFO composition. Light and dark line colors differentiate Phase 1 and Phase 3 model outputs.

In the analogous oxide-facies columns, circulation of leachate from the *schist*-oxide column through the U-rich *gneiss*-oxide column also produced an increase in U concentrations in the gneiss-oxide effluent, albeit more modest than in the transition-facies column, from ca. 80 µg/L prior to the sequencing experiment up to 93 µg/L in the first two cycles of Phase 2. A greater Fe-oxide content in the oxide-facies gneiss might explain the more muted release of U observed in comparison with the gneiss-transition column during Phase 2 of the sequencing experiment (Supplementary Table S1).

Upon reversal of the order of flow in Phase 3, marked U attenuation from the U-rich gneiss column was observed in both transition-facies and oxide-facies sequencing experiments (Figure 7). In the *transition-facies* experiment, despite application of *gneiss* leachate with U concentrations of >100 µg/L into the *schist* column, schist U effluent steadily remained in the 30 to 40 µg/L range (i.e., similar to concentrations prior to sequencing experiments) with no sign of elevated U breakthrough. Although it is impossible to explicitly discriminate between U mass release within the schist and U sourced from the gneiss feed solution, mass-balance calculations show that at least 83% of the U originating from the gneiss solution was retained if U inputs from the schist are assumed to be zero. If the per-cycle U release rate from within the schist is assumed to have remained unchanged during Phase 3 relative to the rate observed at the end of Phase 1 (cycles 90 to 99, when U concentrations were stable), this estimate increases to >100%. As U retention cannot in fact exceed 100%, these calculations suggest that the 30 to 40 μg/L U in schist effluent reflected an equilibrium control threshold concentration that was maintained even after application of the U-rich gneiss leachate to the schist column.

In the *oxide-facies* Phase 3 sequencing experiment, U attenuation was even greater: U concentrations in the U-rich gneiss-oxide effluent declined from ca. 80 μg/L to ca. 5 μg/L after circulation through the schist-oxide column (Figure 7). Schist-oxide effluent U concentrations were also unchanged during Phase 2 in comparison with the stable concentrations observed toward the end of Phase 1 in that column, similar to the *transition-series* Phase 3 result (Figure 7). Mass-balance calculations give a lower limit of 94% U retention within the schist-oxide column during Phase 3; this estimate increases to 99% if per-cycle U release from schist solids is assumed to have remained unchanged relative to the rate observed at the end of Phase 1 (cycles 90–99). Additional net U release within the schist during Phase 3 of the oxide sequencing experiment was, therefore, likely negligible. The lack of U breakthrough from the schist-oxide is consistent with its exceptionally high Fe-oxide content relative to other rocks, making it capable of retaining U through sorption despite elevated porewater Ca and alkalinity (Supplementary Table S1).

Shake-flask and SCE of schist-transition and schist-oxide column residues at the end of the sequencing experiments provided an indication of U attenuation via sorption processes. Shake-flask extractable U increased from < 0.014 μg/g prior to column experiments to values ranging from 0.29 to 0.49 μg/g (Supplementary Figure S5). Residue SCE analyses indicated that between 16% and 24% of the U at the end of the experiments was soluble in the chemical reagents $MgCl_2$, Na-acetate, and hydroxylamine-hydrochloride which nominally target weakly sorbed, carbonate-bound and exchangeable, and crystalline phases [55]. In both columns, U contents recovered in the $MgCl_2$ and Na-acetate extractions were highest in the upper 10 cm of column residues and decreased in deeper intervals of the columns, suggesting limited downward transport of U in the column after it received gneiss-oxide U-rich leachate (Supplementary Figures S5 and S6).

4.2.2. Modeling Sorption during the Application of U-Rich Gneiss Effluent to Schist Columns

The role of sorption on U attenuation during Phase 3 of column-sequencing experiments was further investigated with geochemical models. In these models, the HFO composition of schist-oxide and schist-transition rocks were calibrated from QEMSCAN Fe-oxide results, scaled to reproduce Phase 1 U concentration data (calibration details in Supplementary Materials Section S.4 and presented visually in Supplementary Figures S7 and S8), and yielded 21 g/L HFO and 4.7 g/L HFO, respectively. These HFO abundances correspond to the Fe-oxide content determined by QEMSCAN in these rocks multiplied by a calibration factor of 0.08 and 0.15, respectively (Supplementary Materials Section S.4.1) and they are consistent with the expectation of more HFO in more oxidized rock facies (Supplementary Table S1).

To simulate aqueous geochemistry of the schist-oxide sequencing experiment during Phase 3, the calibrated HFO concentration in the schist was allowed to interact with a mixture of influent gneiss-oxide leachate and residual porewater in the schist, the proportions of which were estimated from water mass balance. This solution-HFO mixture was forced to match the pCO_2 and calcite SI observed in schist effluent at each cycle in batch reactions by allowing calcite dissolution/precipitation and CO_2 dissolution/exsolution. These reactions ensured that the measured pH, Ca, and alkalinity concentrations were reproduced in the model (Supplementary Materials Section S.4.2). These simulations indicated that the marked U attenuation observed in Phase 3 of the *schist-oxide* experiment could be explained through a surface-complexation model that provided a reasonable fit of observed major ion, pH and U concentrations [root-mean square error (RMSE) 0.0011 for U measured vs. U modeled] (Figure 7).

A similar modeling approach was used to examine U attenuation during Phase 3 of *transition-facies* experiment, except that this model also required consideration of processes related to sulfide-mineral oxidation. Sulfate concentrations were approximately one order of magnitude higher in the schist-transition column experiment in comparison with its schist-oxide counterpart (Figure 1). Sulfate concentrations also showed a steady increase after cycle 60 (Figure 1 and Supplementary Figure S9). These trends in sulfate concentration, along with the higher initial sulfide-mineral content of the schist-transition rock as seen in QEMSCAN analyses (Supplementary Table S1), suggested that sulfide-mineral oxidation was occurring in the schist-transition column as the experiment progressed. Calcium concentrations from the schist-transition effluent were also consistently above those of the gneiss-transition feed solution during Phase 3, suggesting that sulfide-mineral oxidation may have driven calcite dissolution within the schist after the application of gneiss leachate, because of H_2SO_4 production associated with sulfide oxidation (Figure 7). Therefore, we accounted for calcite dissolution and sulfide-mineral oxidation in a model wherein gneiss influent solution and residual schist porewater were mixed in the presence of HFO, while any excess sulfate in schist-transition effluent above that expected from conservative mixing was attributed to H_2SO_4 gained from pyrite oxidation (full details in Supplementary Materials Section S.3.2). The HFO-solution-H_2SO_4 system was forced to match schist effluent calcite saturation indices and pCO_2 by allowing calcite dissolution-precipitation and CO_2 dissolution-exsolution.

Similar to the schist-oxide modeling results, these simulations closely reproduced effluent pH, alkalinity, Ca, and sulfate concentrations in the schist-transition experiment (Figure 7). The decrease in U concentrations observed between column influent (gneiss-transition leachate) and effluent was also reproduced in this model, substantiating the hypothesis that sorption was driving U attenuation during the experiment. However, U concentrations in the model were on average 35% above measured effluent concentrations, and the model fit (RMSE = 0.012) was poorer than in the schist-oxide experiment.

We considered an additional series of simulations to assess whether the oxidation of sulfide minerals in the schist-transition column could have led to growth in HFO over time (as per Reaction (1)), and thus increased U sorption capacity during the experiment. Quantifying HFO growth as pyrite oxidation progresses is highly complex given the evolution in surface area/volume and the possible Fe-oxide phase transformation and recrystallization reactions that occur with ageing [61]. Growth in HFO was modeled by assuming that the rise in sulfate concentrations in the schist-transition column between cycles 60 and 99, which was highly linear (Supplementary Figure S9), translated to proportional and linear growth in HFO availability over time. By applying an HFO growth rate that was 1/100th the rate of per-cycle sulfate concentration increase between cycles 60 and 99, and assuming that this growth rate remained constant during Phase 3, the PHREEQC model provided a better match to experimental U data (RMSE = 0.012) (Figure 7). These simulations suggest that addition of Fe(III)-(oxyhydr)oxides during oxidation of this rock might have led to increasing U sorption capacity, while this effect was unlikely in the schist-oxide experiment where the lack of sulfide-mineral weathering precluded HFO genesis.

4.3. Simulating the Effect of Pyrite-Oxidation/Calcite-Dissolution Reactions on U Concentrations in Neutral-Rock Drainage (NRD)

Column experiments produced leachate with lower dissolved Ca and sulfate concentrations and pCO_2 than those which commonly occur in full-scale waste-rock storage facilities (WRSF) that contain carbonate and sulfide minerals. Higher rock/water ratios and longer water residence times in NRD-producing WRSFs can yield Ca and sulfate concentrations in the hundreds and thousands of mg/L, respectively [23,39–43]. Under these conditions, gypsum ($CaSO_4{\cdot}2H_2O$) precipitation can limit the concentrations of Ca. Additionally, pCO_2 can reach up to two orders of magnitude above atmospheric pCO_2 in WRSFs because of carbonate-mineral dissolution and limitations on gas exchange with the atmosphere [44,62–66]. Given the dependence of U sorption on Ca concentrations and pCO_2 [16,19,20,67], we assessed how a U-HFO system would respond in hypothetical scenarios wherein pyrite oxidation and carbonate-mineral dissolution reactions progressed to a higher degree, i.e., to the point at which Ca concentrations became controled by gypsum mineral saturation.

These scenarios correspond to a theoretical sulfide- and carbonate-rich oxidative weathering setting without any effects of transport (e.g., mixing, dilution), reaction kinetics, or other site-specific considerations (temperature, hydrology, geochemical heterogeneity, particle size or facility construction and design). Simulations were run in PHREEQC in a series of batch reactions in which a U-HFO system was titrated by incremental steps of FeS_2 oxidation with infinite access to atmospheric O_2 and carbonate-mineral dissolution until $Ca_{(aq)}$ concentrations became limited by gypsum saturation. The HFO used in these simulations and initial U(VI) abundance was that of the the gneiss-transition column because this rock released the most U in column experiments (Figure 5). The HFO of this rock was calibrated in the same way as described above for the schist sorption model (HFO calibration results in Supplementary Materials Section S.4.1). The HFO-U system was equilibrated with leachate chemistry representative of steady-state concentrations in the gneiss-transiton column (i.e., at cycle 89 of the experiment) prior to pyrite and calcite titration simulations.

Four simulations are presented (Table 2). The base case (Case 1) involved pyrite oxidation in the presence of U(VI) and HFO under atmospheric pCO_2 conditions (10^{-3} atm) and with calcite as the carbonate mineral. In Cases 2A and 2B, we replaced calcite with a hypothetical Mg-bearing calcite of the same solubility constant as calcite (log K = −8.48) to investigate whether Mg-uranyl-carbonate complexes could sustain elevated aqueous U after Ca concentrations become controlled by gypsum saturation. In Case 2A the Ca/Mg molar ratio of the Mg-bearing calcite was fixed at 9:1, and in Case 2B this ratio was decreased to 1.5:1 to match the Ca/Mg ratios observed in the gneiss-transition leachates. In Case 4C, the effect of CO_2 buildup was investigated by fixing the pCO_2 to $10^{-2.7}$ atm (ca. 2000 ppm).

Table 2. Hypothetical PHREEQC models examining U sorption behavior as a function of pyrite oxidation in a pyrite-carbonate-HFO system initially containing 0.077 mmol of U(VI) and 3.29 g/L HFO (surface area 600 m^2/g, 40:1 ratio of weak:strong sorption sites).

Case	Carbonate Phase	pCO_2 (atm)
1. Base Case	$CaCO_3 \rightleftharpoons Ca^{2+} + CO_3^{2-}$, log K = −8.48	$10^{-3.5}$
2A. CaMg system, high Ca/Mg	$Ca_{0.9}Mg_{0.1}CO_3$ * $\rightleftharpoons 0.9Ca^{2+}\ 0.1Mg^{2+} + CO_3^{2-}$, log K = −8.48	$10^{-3.5}$
2B. CaMg system, low Ca/Mg	$Ca_{0.6}Mg_{0.4}CO_3$ * $\rightleftharpoons 0.6Ca^{2+}\ 0.4Mg^{2+} + CO_3^{2-}$, log K = −8.48	$10^{-3.5}$
4C. High pCO_2	$CaCO_3 \rightleftharpoons Ca^{2+} + CO_3^{2-}$, log K = −8.48	$10^{-2.7}$

* Hypothetical phase used to examine effect of dissolution of Mg-bearing mineral on U mobility.

Progressive pyrite oxidation and carbonate-mineral disolution in these simulations produced NRD conditions (pH 8.15 to pH 7.75) with $Ca_{(aq)}$ reaching >580 mg/L before gypsum saturation was attained (Figure 8). Alkalinity concentrations rose in all simulations to several hundred mg/L (as $CaCO_3$), yet this rise was decoupled from U(VI) desorption. Instead, U desorption directly followed $Ca_{(aq)}$ concentrations, which increased until gypsum saturation was reached (Figure 8). The relationship between Ca and U release corresponds with observations and sorption modeling of the column and field-bin experiments. Reactive-transport processes in actual WRSFs tend to yield effluent with higher sulfate/Ca ratios and lower alkalinity than is presented in these models owing to the more conservative behavior of sulfate over alkalinity and $Ca_{(aq)}$. Nevertheless, under oxidizing conditions and in the concentration ranges of 100 to 580 mg/L Ca and 100 to 400 mg/L alkalinity (as $CaCO_3$), these simulation highlight the role of $Ca_{(aq)}$ in driving U(VI) release through desorption, consistent with kinetic experiments as well as previous laboratory studies of U mobility aqueous environments [18,19,68].

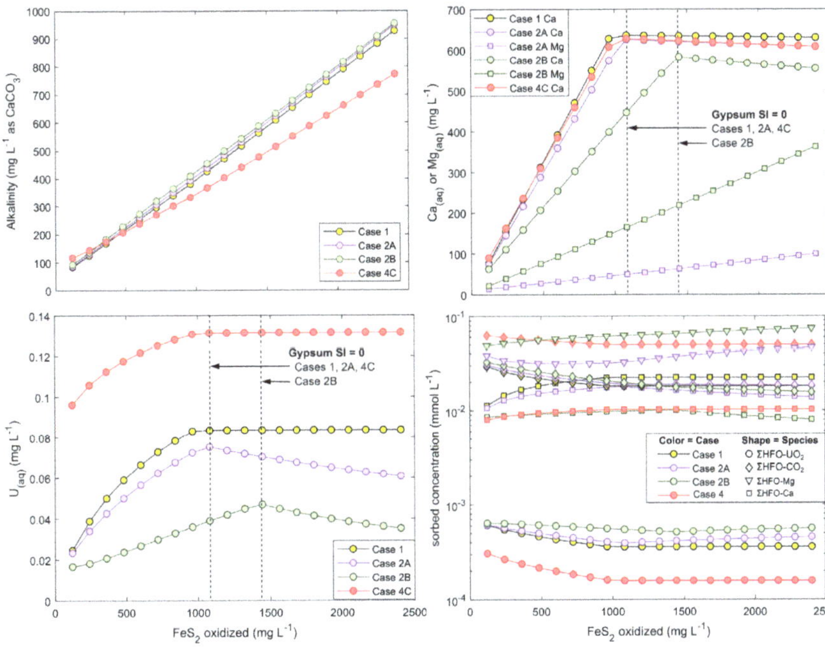

Figure 8. Modeled alkalinity (**top left**), Ca and Mg (**top right**), and U (**bottom left**) concentrations, and sorbed species (**bottom right**) onto HFO in hypothetical batch reactions involving FeS_2 oxidation in the presence of excess carbonate. Arrows indicate approximately when gypsum mineral saturation index reaches 0. The different modeling scenarios are presented in Table 2. The HFO composition that are shown at bottom right are the respective millimolar sums of all U, DIC, Mg, and Ca species that are sorbed in each simulation.

Modeled aqueous U speciation in all simulations was overwhelmingly dominated by CCU complexes (>97%). Cases 2A and 2B (with Mg-carbonate) further demonstrated that when $Ca_{(aq)}$ was controlled by gypsum saturation, U desorption was not exacerbated by Mg^{2+} release and formation of Mg-uranyl-carbonate complexes. These complexes constituted <1% of aqueous U while CCU complexes overwhelmingly dominated U speciation even after $Ca_{(aq)}$ concentrations were controlled by gypsum precipitation. Rather than increasing U mobility through Mg-uranyl-carbonate complexation, replacing calcite with a Mg-calcite had the opposite effect of decreasing U mobility because of lower $Ca_{(aq)}$ concentrations than in Case 1 in which pure (Mg-free) calcite was used (Figure 8).

Case 4 shows that at higher pCO_2, U mobilization is also dependent on $Ca_{(aq)}$ until the point of gypsum saturation and that U sorption is subsantially weaker at higher pCO_2. This result is consistent with laboratory studies that show a narrowing of the uranyl-HFO sorption envelope under elevated pCO_2 and neutral pH conditions (pH ca. 7 to 8.5) [16,19,20,67].

Overall, these models emphasized the role of $Ca_{(aq)}$ and pCO_2 on U mobility under well-buffered and oxidizing conditions.

5. Conclusions and Implications for U mobility in NRD

Weathering experiments on granite, schist, and gneiss mine wastes indicated that waste rock containing μg/g levels of U can present a source for U release, in particular under NRD conditions. Weathering experiments and geochemical simulations suggested that U release in NRD was more sensitive to dissolved Ca concentrations than to alkalinity. The combined effects of sulfide-mineral oxidation, H_2SO_4 generation, and carbonate-mineral dissolution generated circumneutral to alkaline-pH drainage and released Ca that promoted formation of weakly sorbing calcium-carbonato-uranyl complexes. So long as the carbonate mineral dissolution rate matches the rate of acid generation, CCU complexation and U mobilization can be sustained. Thus, fresh (unweathered) rocks sustained higher U release rates when exposed to oxidizing conditions because they released more Ca and thus enhanced CCU complexation while also having lower sorption-site availability. Conversely, oxidized rocks effectively attenuated U because of a combination of higher sorption-site availability and lower carbonate and sulfide-mineral abundances. Sequencing experiments indicated that U concentrations could, therefore, be decreased by circulating U-rich NRD into HFO-rich oxidized rocks. Overall, these results indicate that the availability of carbonate and sulfide minerals and HFO can better predict U release than solid-phase U abundances alone. Future studies of U in NRD should also consider the potential for U desorption under higher pCO_2 conditions, such as those often found in full-scale WRSFs.

Supplementary Materials: The following are available online at http://www.mdpi.com/2075-163X/10/12/1104/s1, Supplementary Sections: Section S.1. Detailed methods of powder X-ray diffraction analyses [69,70]. Section S.2. Sequential chemical extraction protocol and results. Section S.3. Additional supplementary figures and tables. Section S.4. PHREEQC model description for simulation of column experiments. Supplementary Tables: Supplementary Table S1. Solid-phase composition of rocks used in column and field bin experiments. Supplementary Table S2. Geochemical reactions added to the wateq4f.dat database for PHREEQC simulations [71]. Supplementary Table S3. Uranium redox speciation determination by XANES-LCF using uranyl nitrate and uraninite as U(VI) and U(IV) standards, respectively. Supplementary Table S4. Mineralogy of rocks used in kinetic tests determined by pXRD. Supplementary Figures: Supplementary Figure S1. Schematic diagram of field-bin experiment apparatus. Supplementary Figure S2. Acid-generating potential (calculated from total-sulfur) against carbonate neutralizing potential (calculated from TIC) in drill-core samples. Supplementary Figure S3. Bivariate plots of U against As, Sb, and S by rock type for Coffee drill-core samples from the fresh weathering facies. Supplementary Figure S4. Aqueous U speciation in field-bin leachates. Supplementary Figure S5. Uranium recovered by shake-flask extractions (SFE) in schist columns. Supplementary Figure S6. Sequential chemical extraction results for residues at the end of column experiments. Supplementary Figure S7. Schematic description of equilibrium sorption modeling of column experiments in PHREEQC. Supplementary Figure S8. Measured and modeled U concentrations in schist and gneiss effluent during Phase 1. Supplementary Figure S9. Sulfate concentrations in column experiments.

Author Contributions: Conceptualization, J.W.D. and E.K.S.; Methodology, J.W.D., E.K.S., V.F.B., J.M.M.; Data Analysis, E.K.S., J.W.D., J.M.M., V.F.B., K.U.M., R.D.B.; Writing—Original Draft Preparation, E.K.S.; Writing—Review & Editing, all co-authors. All authors have read and agreed to the published version of the manuscript.

Funding: This research was conducted with contributions from Mitacs (project IT5928 and IT12366) and Lorax Environmental Services Ltd. (Vancouver, BC, Canada), the Government of Yukon, and Newmont, and an NSERC Discovery Grant awarded to J. McBeth (RGPIN-2014-03719).

Acknowledgments: The authors thank Newmont for access to samples and data used in this research. X-ray absorption spectroscopy analyses were conducted at the Canadian Light Source (CLS), which is supported by the Canada Foundation for Innovation, NSERC, the University of Saskatchewan, the Government of Saskatchewan, Western Economic Diversification Canada, the National Research Council Canada, and the Canadian Institutes of Health Research. Ning Chen and Weifeng Chen are thanked for their assistance in the operation of the HXMA beamline at the CLS. Three anonymous reviewers and David Flather provided comments to improve the manuscript.

Conflicts of Interest: E.K.S. and J.W.D. have been involved as consultants for the Government of Yukon and Newmont. The funders had no role in the design of the study; in the collection, analyses, or interpretation of data; in the writing of the manuscript; or in the decision to publish the results.

References

1. Goulet, R.R.; Spry, D.J.; Fortin, C. Uranium. *Fish Physiol.* **2011**, *31*, 391–428. [CrossRef]
2. Health Canada Uranium in Drinking Water—Document for Public Consultation. 2017, p. 81. Available online: https://www.canada.ca/en/health-canada/programs/consultation-uranium-drinking-water/document.html (accessed on 15 October 2018).
3. Canadian Council of Ministers of the Environment Canadian Water Quality Guidelines for the Protection of Aquatic Life—Uranium. Canadian Environment Quality Guidelines. 1999. Available online: http://st-ts.ccme.ca/en/index.html?lang=en&factsheet=225 (accessed on 15 October 2018).

4. Wang, Y.; Bagnoud, A.; Suvorova, E.; Mcgivney, E.; Chesaux, L.; Phrommavanh, V.; Descostes, M.; Bernier-latmani, R. Geochemical Control on Uranium(IV) Mobility in a Mining-Impacted Wetland. *Environ. Sci. Technol.* **2014**. [CrossRef] [PubMed]
5. Smedley, P.L.; Smith, B.; Abesser, C.; Lapworth, D. Uranium occurrence and behaviour in British groundwater. In *British Geological Survey Groundwater Systems & Water Quality Programme Commissioned Report CR/06/050*; British Geological Survey: Keyworth, Nottigham, UK, 2006; p. 48.
6. Waseem, A.; Ullah, H.; Rauf, M.K.; Ahmad, I. Distribution of natural uranium in surface and groundwater resources: A review. *Crit. Rev. Environ. Sci. Technol.* **2015**, *45*, 2391–2423. [CrossRef]
7. Riedel, T.; Kübeck, C. Uranium in groundwater—A synopsis based on a large hydrogeochemical data set. *Water Res.* **2018**, *129*, 29–38. [CrossRef] [PubMed]
8. Kim, Y.S.; Park, H.S.; Kim, J.Y.; Park, S.K.; Cho, B.W.; Sung, I.H.; Shin, D.C. Health risk assessment for uranium in Korean groundwater. *J. Environ. Radioact.* **2004**, *77*, 77–85. [CrossRef] [PubMed]
9. Banks, D.; Røyset, O.; Strand, T.; Skarphagen, H. Radioelement (U, Th, Rn) concentrations in Norwegian bedrock groundwaters. *Environ. Geol.* **1995**, *25*, 165–180. [CrossRef]
10. Gascoyne, M. Hydrogeochemistry, groundwater ages and sources of salts in a granitic batholith on the Canadian Shield, southeastern Manitoba. *Appl. Geochem.* **2004**, *19*, 519–560. [CrossRef]
11. Frengstad, B.; Midtgård Skrede, A.K.; Banks, D.; Reidar Krog, J.; Siewers, U. The chemistry of Norwegian groundwaters: III. The distribution of trace elements in 476 crystalline bedrock groundwaters, as analysed by ICP-MS techniques. *Sci. Total Environ.* **2000**, *246*, 21–40. [CrossRef]
12. Abdelouas, A. Uranium Mill Tailings: And Environmental Impact. *Elements* **2006**, *2*, 335–342. [CrossRef]
13. Langmuir, D. Uranium solution-mineral equilibria at low temperatures with applications to sedimentary ore deposits. *Geochim. Cosmochim. Acta* **1978**, *42*, 547–569. [CrossRef]
14. Cumberland, S.A.; Douglas, G.; Grice, K.; Moreau, J.W. Uranium mobility in organic matter-rich sediments: A review of geological and geochemical processes. *Earth Sci. Rev.* **2016**, *159*, 160–185. [CrossRef]
15. Brown, S.T.; Basu, A.; Ding, X.; Christensen, J.N.; Depaolo, D.J. Uranium isotope fractionation by abiotic reductive precipitation. *Proc. Natl. Acad. Sci. USA* **2018**, *115*, 8688–8693. [CrossRef] [PubMed]
16. Waite, T.D.; Davis, J.A.; Payne, T.E.; Waychunas, G.A.; Xu, N. Uranium(VI) adsorption to ferrihydrite: Application of a surface complexation model. *Geochim. Cosmochim. Acta* **1994**, *58*, 5465–5478. [CrossRef]
17. Dublet, G.; Lezama Pacheco, J.; Bargar, J.R.; Fendorf, S.; Kumar, N.; Lowry, G.V.; Brown, G.E. Partitioning of uranyl between ferrihydrite and humic substances at acidic and circum-neutral pH. *Geochim. Cosmochim. Acta* **2017**, *215*, 122–140. [CrossRef]
18. Stewart, B.D.; Mayes, M.A.; Fendorf, S.S. Impact of Complexes on Uranium(VI) Adsorption to Synthetic and Natural Sediments. *Environ. Sci. Technol.* **2010**, *44*, 928–934. [CrossRef] [PubMed]
19. Fox, P.M.; Davis, J.A.; Zachara, J.M. The effect of calcium on aqueous uranium(VI) speciation and adsorption to ferrihydrite and quartz. *Geochim. Cosmochim. Acta* **2006**, *70*, 1379–1387. [CrossRef]
20. Kobayashi, Y.; Fukushi, K.; Kosugi, S. A Robust Model for Prediction of U(VI) Adsorption onto Ferrihydrite Consistent with Spectroscopic Observations. *Environ. Sci. Technol.* **2020**, *54*, 2304–2313. [CrossRef]
21. Tournassat, C.; Tinnacher, R.M.; Grangeon, S.; Davis, J.A. Modeling uranium(VI) adsorption onto montmorillonite under varying carbonate concentrations: A surface complexation model accounting for the spillover effect on surface potential. *Geochim. Cosmochim. Acta* **2018**, *220*, 291–308. [CrossRef]
22. Bachmaf, S.; Planer-Friedrich, B.; Merkel, B.J. Effect of sulfate, carbonate, and phosphate on the uranium(VI) sorption behavior onto bentonite. *Radiochim. Acta* **2008**, *96*, 359–366. [CrossRef]
23. Blowes, D.W.; Ptacek, C.J.; Jambor, J.L.; Weisener, C.G.; Paktunc, D.; Gould, W.D.; Johnson, D.B. The Geochemistry of Acid Mine Drainage. In *Treatise on Geochemistry*, 2nd ed.; Elsevier Ltd.: Amsterdam, The Netherlands; Oxford, UK; Waltham, MA, USA, 2014; Volume 11, ISBN 9780080983004.
24. Dong, W.; Brooks, S. Determination of the Formation Constants of Ternary Complexes of Uranyl and Carbonate with Alkaline Ba Earth Metals (Mg^{2+}, Ca^{2+}, Sr^{2+}, and Ba^{2+}) Using Anion Exchange Method. *Environ. Sci. Technol.* **2006**, *40*, 4689–4695. [CrossRef]
25. Bernhard, G.; Geipel, G.; Brendler, V.; Nitsche, H. Speciation of uranium in seepage waters of a mine tailing pile studied by tine-resolved laser-induced fluorescence spectroscopy (TRLFS). *Radiochim. Acta* **1996**, *74*, 87–91. [CrossRef]
26. Langmuir, D. Adsorption of uranyl onto ferric oxyhydroxides: Application of the surface complexation site-binding model. *Geochim. Cosmochim. Acta* **1985**, *49*, 1931–1941. [CrossRef]

27. Noubactep, C.; Chen-Braucher, D.; Schlothauser, T. Uranium release from a natural rock under near-natural oxidizing conditions. *Eng. Life Sci.* **2006**, *267*, 591–602. [CrossRef]
28. Blake, J.M.; Avasarala, S.; Ali, A.M.S.; Spilde, M.; Lezama-Pacheco, J.S.; Latta, D.; Artyushkova, K.; Ilgen, A.G.; Shuey, C.; Nez, C.; et al. Reactivity of As and U co-occurring in Mine Wastes in northeastern Arizona. *Chem. Geol.* **2019**, *522*, 26–37. [CrossRef] [PubMed]
29. Bain, J.G.; Mayer, K.U.; Blowes, D.W.; Frind, E.O. Modelling the closure-related geochemical evolution of groundwater at a former uranium mine. *J. Contam. Hydrol.* **2001**, *52*, 109–135. [CrossRef]
30. Boekhout, F.; Gérard, M.; Kanzari, A.; Michel, A.; Déjeant, A.; Galoisy, L.; Calas, G.; Descostes, M. Uranium migration and retention during weathering of a granitic waste rock pile. *Appl. Geochem.* **2015**, *58*, 123–135. [CrossRef]
31. Gómez, P.; Garralón, A.; Buil, B.; Turrero, M.J.; Sánchez, L.; de la Cruz, B. Modeling of geochemical processes related to uranium mobilization in the groundwater of a uranium mine. *Sci. Total Environ.* **2006**, *366*, 295–309. [CrossRef] [PubMed]
32. Larson, L.N.; Kipp, G.G.; Mott, H.V.; Stone, J.J. Applied Geochemistry Sediment pore-water interactions associated with arsenic and uranium transport from the North Cave Hills mining region, South Dakota, USA. *Appl. Geochem.* **2012**, *27*, 879–891. [CrossRef]
33. Singh, S.P.; Hendry, M.J. Solid-phase distribution and leaching behaviour of nickel and uranium in a uranium waste-rock piles. *Water Air Soil Pollut.* **2013**, *224*. [CrossRef]
34. Tayal, A.; Conradson, S.D.; Kanzari, A.; Lahrouch, F.; Descostes, M.; Gerard, M. Uranium speciation in weathered granitic waste rock piles: An XAFS investigation. *RSC Adv.* **2019**, *9*, 11762–11773. [CrossRef]
35. Troyer, L.D.; Stone, J.J.; Borch, T. Effect of biogeochemical redox processes on the fate and transport of As and U at an abandoned uranium mine site: An X-ray absorption spectroscopy study. *Environ. Chem.* **2014**, *11*, 18–27. [CrossRef]
36. Yin, M.; Dong, X.; Li, H.; Chen, Y.; Xiao, T.; Sun, J.; Song, G.; Liu, J.; Belshaw, N.; Xiao, E.; et al. Mechanism of uranium release from uranium mill tailings under long-term exposure to simulated acid rain: Geochemical evidence and environmental implication. *Environ. Pollut.* **2018**, *244*, 174–181. [CrossRef]
37. Campbell, K.M.; Gallegos, T.J.; Landa, E.R. Biogeochemical aspects of uranium mineralization, mining, milling, and remediation. *Appl. Geochemistry* **2015**, *57*, 206–235. [CrossRef]
38. International Network for Acid Prevention (INAP) Global Acid Rock Drainage Guide (GARD Guide). Available online: http://www.gardguide.com (accessed on 2 November 2020).
39. Blowes, D.W.; Jambor, J.L.; Hanton-Fong, C.J.; Lortie, L.; Gould, W.D. Geochemical, mineralogical and microbiological characterization of a sulphide-bearing carbonate-rich gold-mine tailings impoundment, Joutel, Quebec. *Appl. Geochem.* **1998**, *13*, 687–705. [CrossRef]
40. Vriens, B.; Peterson, H.; Laurenzi, L.; Smith, L.; Aranda, C.; Mayer, K.U.; Beckie, R.D. Chemosphere Long-term monitoring of waste-rock weathering at the Antamina mine, Peru. *Chemosphere* **2020**, *215*, 858–869. [CrossRef] [PubMed]
41. Majzlan, J.; Števko, M.; Chovan, M.; Luptáková, J.; Milovská, S.; Milovský, R.; Jeleň, S.; Sýkorová, M.; Pollok, K.; Göttlicher, J.; et al. Mineralogy and geochemistry of the copper-dominated neutral mine drainage at the Cu deposit L'ubietová-Podlipa (Slovakia). *Appl. Geochem.* **2018**, *92*, 59–70. [CrossRef]
42. Shahhosseini, M.; Doulati Ardejani, F.; Baafi, E. Geochemistry of rare earth elements in a neutral mine drainage environment, Anjir Tangeh, northern Iran. *Int. J. Coal Geol.* **2017**, *183*, 120–135. [CrossRef]
43. Abongwa, P.T.; Atekwana, E.A.; Puckette, J.O. Hydrogeochemical Investigation of Metal Evolution in Circum-Neutral Mine Discharge. *Water Air Soil Pollut.* **2020**, *231*. [CrossRef]
44. Kříbek, B.; Sracek, O.; Mihaljevič, M.; Knésl, I.; Majer, V. Geochemistry and environmental impact of neutral drainage from an uraniferous coal waste heap. *J. Geochem. Explor.* **2018**, *191*, 1–21. [CrossRef]
45. Kanzari, A.; Gérard, M.; Boekhout, F.; Galoisy, L.; Calas, G.; Descostes, M. Impact of incipient weathering on uranium migration in granitic waste rock piles from former U-mines (Limousin, France). *J. Geochem. Explor.* **2017**, *183*, 114–126. [CrossRef]
46. Yin, M.; Tsang, D.C.W.; Sun, J.; Wang, J.; Shang, J.; Fang, F.; Wu, Y.; Liu, J.; Song, G.; Xiao, T.; et al. Critical insight and indication on particle size effects towards uranium release from uranium mill tailings: Geochemical and mineralogical aspects. *Chemosphere* **2020**, *250*, 126315. [CrossRef] [PubMed]
47. Liu, B.; Peng, T.; Sun, H.; Yue, H. Release behavior of uranium in uranium mill tailings under environmental conditions. *J. Environ. Radioact.* **2017**, *171*, 160–168. [CrossRef] [PubMed]

48. Li, Z.; Hadioui, M.; Wilkinson, K.J. Conditions affecting the release of thorium and uranium from the tailings of a niobium mine. *Environ. Pollut.* **2019**, *247*, 206–215. [CrossRef] [PubMed]
49. MacWilliam, K. The Geology and Genesis of the Coffee Gold Deposit in West-Central Yukon, Canada: Implications for the Structural, Magmatic, and Metallogenic Evolution of the Dawson Range and Gold Exploration Models. Ph.D. Thesis, University of British Columbia, Vancouver, BC, Canada, 2018; 497p.
50. Wainwright, A.J.; Simmons, A.T.; Finnigan, C.S.; Smith, T.R.; Carpenter, R.L. Geology of new gold discoveries in the Coffee Creek area, White Gold District, west-central Yukon. In *Yukon Exploration and Geology*; Yukon Geological Servey: Whitehorse, YT, Canada, 2010; pp. 233–248. Available online: http://data.geology.gov.yk.ca/Reference/50358#InfoTab (accessed on 8 December 2020).
51. Skierszkan, E.K.; Dockrey, J.W.; Mayer, K.U.; Beckie, R.D. Release of geogenic uranium and arsenic results in water-quality impacts in a subarctic permafrost region of granitic and metamorphic geology. *J. Geochem. Explor.* **2020**, *17*. [CrossRef]
52. Ravel, B.; Newville, M. ATHENA, ARTEMIS, HEPHAESTUS: Data analysis for X-ray absorption spectroscopy using IFEFFIT. *J. Synchrotron Radiat.* **2005**, *12*, 537–541. [CrossRef]
53. Vriens, B.; Skierszkan, E.K.; St-Arnault, M.; Salzsauler, K.; Aranda, C.; Mayer, K.U.; Beckie, R.D. Mobilization of Metal(oid) Oxyanions through Circumneutral Mine Waste-Rock Drainage. *ACS Omega* **2019**, *4*, 10205–10215. [CrossRef]
54. Hirsche, D.T.; Blaskovich, R.; Mayer, K.U.; Beckie, R.D. A study of Zn and Mo attenuation by waste-rock mixing in neutral mine drainage using mixed-material field barrels and humidity cells. *Appl. Geochem.* **2017**, *84*, 114–125. [CrossRef]
55. Tessier, A.; Campbell, P.G.C.; Bisson, M. Sequential Extraction Procedure for the Speciation of Particulate Trace Metals. *Anal. Chem.* **1979**, *51*, 844–851. [CrossRef]
56. Parkhurst, D.L.; Appelo, C.A.J. PHREEQC Version 3—A Computer Program for Speciation, Batch-Reaction, One-Dimensional Transport, and Inverse Geochemical Calculations. Available online: https://wwwbrr.cr.usgs.gov/projects/GWC_coupled/phreeqci/ (accessed on 15 November 2017).
57. Mahoney, J.J.; Cadle, S.A.; Jakubowski, A.T. Uranyl adsorption onto hydrous ferric oxides—A re-evaluation for the diffuse layer model database. *Environ. Sci. Technol.* **2009**, *43*, 9260–9266. [CrossRef]
58. Dzombak, D.A.; Morel, F. *Surface Complexation Modeling: Hydrous Ferric Oxide*; John Wiley & Sons: New York, NY, USA, 1990; ISBN 978-0-471-63731-8.
59. Rudnick, R.L.; Gao, S. Composition of the Continental Crust. In *Treatise on Geochemistry*, 2nd ed.; Elsevier Ltd.: Amsterdam, The Netherlands; Oxford, UK; Waltham, MA, USA, 2014; Volume 4, ISBN 9780080983004.
60. Dockrey, J.; Flather, D.; Findlater, L.; Jackson, S.; Helsen, J.; Scott, J.; Gjertsen, J. Elevated Uranium and Arsenic Concentrations in Baseline Water Quality at the Coffee Gold Project: Implications for Geochemical Predictions. In Proceedings of the 24th Annual BC MEND ML/ARD Workshop, Vancouver, BC, Canada, 29–30 November 2017.
61. Cornell, R.M.; Schwertmann, H.C.U. *The Iron Oxides: Structure, Properties, Reactions, Occurences and Uses*, 1st ed.; Wiley-VCH Verlag GmbH & Co. KGaA: Weinheim, Germany, 2003.
62. Lorca, M.E.; Mayer, K.U.; Pedretti, D.; Smith, L.; Beckie, R.D. Spatial and Temporal Fluctuations of Pore-Gas Composition in Sulfidic Mine Waste Rock. *Vadose Zone J.* **2016**, *15*. [CrossRef]
63. Appels, W.M.; Wall, S.N.; Barbour, S.L.; Hendry, M.J.; Nichol, C.F.; Chowdhury, S.R. Pyrit Verwitterung im rekultivierten Abraum aus Schiefergestein eines Ölsandförderbetriebes in der Nähe von Fort McMurray, Kanada. *Mine Water Environ.* **2017**, *36*, 479–494. [CrossRef]
64. Birkham, T.K.; Hendry, M.J.; Wassenaar, L.I.; Mendoza, C.A.; Seok Lee, E. Characterizing geochemical reactions in unsaturated mine waste-rock piles using gaseous O_2, CO_2, $12CO_2$, and $13CO_2$. *Environ. Sci. Technol.* **2003**, *37*, 496–501. [CrossRef] [PubMed]
65. Vriens, B.; Arnault, M.S.; Laurenzi, L.; Smith, L.; Mayer, K.U.; Beckie, R.D. Localized Sulfide Oxidation Limited by Oxygen Supply in a Full-Scale Waste-Rock Pile. *Vadose Zone J.* **2018**, *17*, 180119. [CrossRef]
66. Heikkinen, P.M.; Räisänen, M.L.; Johnson, R.H. Geochemical characterisation of seepage and drainage water quality from two sulphide mine tailings impoundments: Acid mine drainage versus neutral mine drainage. *Mine Water Environ.* **2009**, *28*, 30–49. [CrossRef]
67. Morrison, S.J.; Spangler, R.R.; Tripathi, V.S. Adsorption of uranium(VI) on amorphous ferric oxyhydroxide at high concentrations of dissolved carbon(IV) and sulfur(VI). *J. Contam. Hydrol.* **1995**, *17*, 333–346. [CrossRef]

68. Stewart, B.D.; Amos, R.T.; Nico, P.S.; Fendorf, S. Influence of uranyl speciation and iron oxides on uranium biogeochemical redox reactions. *Geomicrobiol. J.* **2011**, *28*, 444–456. [CrossRef]
69. Toby, R.H.; Von Dreele, R.B. GSAS-II: The genesis of a modern open-source all purpose crystallography software package. *J. Appl. Crystallogr.* **2013**, *46*, 544–549. [CrossRef]
70. Rietveld, H.M. A profile refinement method for nuclear and magnetic structures. *J. Appl. Crystallogr.* **1969**, *2*, 65–71. [CrossRef]
71. Appelo, C.A.J.; Van der Weiden, M.J.J.; Tournassat, C.; Charlet, L. Surface complexation of ferrous iron and carbonate on ferrihydrite and the mobilization of arsenic. *Environ. Sci. Technol.* **2002**, *36*, 3096–3103. [CrossRef]

Publisher's Note: MDPI stays neutral with regard to jurisdictional claims in published maps and institutional affiliations.

minerals

Article

Mineralogical and Geochemical Characterization of Gold Mining Tailings and Their Potential to Generate Acid Mine Drainage (Minas Gerais, Brazil)

Mariana Lemos [1,2], Teresa Valente [1,*], Paula Marinho Reis [1,3], Rita Fonseca [4], Itamar Delbem [5], Juliana Ventura [2] and Marcus Magalhães [2]

1 Institute of Earth Sciences, Pole of University of Minho, Campus de Gualtar, Universidade do Minho, 4710-057 Braga, Portugal; id8548@alunos.uminho.pt or mglemos@anglogoldashanti.com.br (M.L.); pmarinho@dct.uminho.pt (P.M.R.)
2 Anglogold Ashanti, Mining & Technical, COO International, Nova Lima 34000-000, Brazil; jdventura@AngloGoldAshanti.com.br (J.V.); mfmagalhaes@AngloGoldAshanti.com.br (M.M.)
3 GEOBIOTEC, Departmento de Geociências, Campus Universitário de Santiago, Universidade de Aveiro, 3810-193 Aveiro, Portugal
4 Institute of Earth Sciences, Pole of University of Évora, University of Évora, 7000-345 Évora, Portugal; rfonseca@uevora.pt
5 Microscopy Center, Universidade Federal de Minas Gerais, Belo Horizonte 31270-013, Brazil; ssgerais@microscopia.ufmg.br
* Correspondence: teresav@dct.uminho.pt

Abstract: For more than 30 years, sulfide gold ores were treated in metallurgic plants located in Nova Lima, Minas Gerais, Brazil, and accumulated in the Cocoruto tailings dam. Both flotation and leaching tailings from a deactivated circuit, as well as roasted and leaching tailings from an ongoing plant, were studied for their acid mine drainage potential and elements' mobility. Detailed characterization of both tailings types indicates the presence of fine-grain size material hosting substantial amounts of sulfides that exhibit distinct geochemical and mineralogical characteristics. The samples from the ongoing plant show high grades of Fe in the form of oxides, cyanide, and sulfates. Differently, samples from the old circuit shave higher average concentrations of Al (0.88%), Ca (2.4%), Mg (0.96%), and Mn (0.17%), present as silicates and carbonates. These samples also show relics of preserved sulfides, such as pyrite and pyrrhotite. Concentrations of Zn, Cu, Au, and As are higher in the tailings of the ongoing circuit, while Cr and Hg stand out in the tailings of the deactivated circuit. Although the obtained results show that the sulfide wastes do not tend to generate acid mine drainage, leaching tests indicate the possibility of mobilization of toxic elements, namely As and Mn in the old circuit, and Sb, As, Fe, Ni, and Se in the tailings of the plant that still works. This work highlights the need for proper management and control of tailing dams even in alkaline drainage environments such as the one of the Cocoruto dam. Furthermore, strong knowledge of the tailings' dynamics in terms of geochemistry and mineralogy would be pivotal to support long-term decisions on wastes management and disposal.

Keywords: geochemistry and environmental mineralogy; tailings; mobility of toxic elements; acid mine drainage; Minas Gerais—Brazil

Citation: Lemos, M.; Valente, T.; Reis, P.M.; Fonseca, R.; Delbem, I.; Ventura, J.; Magalhães, M. Mineralogical and Geochemical Characterization of Gold Mining Tailings and Their Potential to Generate Acid Mine Drainage (Minas Gerais, Brazil). *Minerals* **2021**, *11*, 39. https://doi.org/10.3390/min11010039

Received: 5 December 2020
Accepted: 24 December 2020
Published: 31 December 2020

1. Introduction

Mining of metals throughout the world produces high volumes of wastes represented by different types of materials [1]. Although highly variable as a function of the characteristics of the ore deposit and the beneficiation processes, as an example [2] estimated around 50,000 Mt of wastes, with 33% of them in tailing dams. In plants for Au metallurgy, the ratio tailings/concentrate can achieve 200:1 [3]. In South Africa, one of the largest Au producing countries, [2] reported a production of 7.4×10^5 t of tailings from 1997 to 2006.

These waste materials are discharged in tailing dams and/or ditches in conditions that could be dangerous to the environment. Even though the tailings treatment and control processes use alkaline products [4], the main sources of Au are often associated with sulfides, such as pyrite (FeS_2), arsenopyrite (FeAsS), and pyrrhotite ($Fe_{1-x}S$). Deposition of these waste materials in subaerial storage facilities allows the oxidation of sulfides, which may promote the generation of acid mine drainage (AMD) [5–8] with the consequent mobilization of potentially toxic elements (PTE). Unregulated mining also deserves a special mention for its contribution to AMD pollution in regions with great influence of artisanal and small-scale mining operations [9,10].

In order to assure safe closure conditions, it is important to evaluate the AMD potential in the long term and understand the factors that may control the mobility of PTE. Several parameters, in addition to the presence of sulfides, can contribute to generation of AMD and contamination of the surrounding environment [11–15]. Tailings can have a wide variety of chemical and mineralogical compositions and can suffer geochemical evolution controlled not only by the original ore paragenesis, but also depending on the mineral processing techniques and weathering conditions [15]. Supergenic evolution of the tailings results in a secondary paragenesis with high diversity of mineral phases, including soluble sulfates, oxides and iron oxyhydroxides, and arsenates, among others. These secondary minerals, occurring as efflorescences, crusts and other iron-rich precipitates, play important roles in the dynamics of the sulfide oxidation and in mobility of PTE in surface and groundwater [16–18]. Another important factor is the dimension of the tailing particles, as the relationship between size and exposure to weathering can directly influence the oxidation rate of sulfides [7,19].

In general, AMD develops when the neutralization capacity of carbonate minerals or alkaline reagents used in the mineral processing plants (as lime) is depleted due to ongoing sulfide-mineral oxidation. In addition to carbonates, or when they are absent, silicate minerals can provide some level of acid buffering [19–22]. Nevertheless, major and trace elements, such as Se and As can be leached from the tailing dams, contributing to the contamination of water bodies, even if the global geochemical balance is neutral or even alkaline [8]. Furthermore, [23] presents a review of the geochemistry of these elements, emphasizing their sources and mobility in naturally contaminated rocks, namely due to desorption from iron-rich minerals.

Once the production of AMD starts, it is very complex to stop the process. In addition, the remediation techniques based on passive or active processes [24,25] are often inefficient or expensive as they require long-term application of the treatment strategies. Moreover, as mentioned by [26], these classical approaches demand an adequate control and disposal of the resulted PTE-enriched sludge.

Thus, for the management and control of the tailings, it is crucial to assess the AMD potential as well as the mobility of toxic elements such as As, Ni or Se. In general, a good understanding of the environmental behavior of the tailings involves a set of procedures, including acid base accounting and leaching tests (e.g., batch, column and/or in situ) [3,14,15,27–35].

In this work, results of chemical, mineralogical and environmental evaluation of two different set of tailings from Au beneficiation are presented and discussed. The tailings represent different scenarios regarding mineralogy and geochemistry as well as mineral processing techniques and temporal evolution. Therefore, there are tailings from an old circuit (already deactivated) that are accumulated in an abandoned dam (Cocoruto dam) and tailings from an active plant, both located in Minas Gerais, Brazil. These two types represent distinctive beneficiation processes. The old circuit was a sulfide flotation plant and direct leaching. The ongoing circuit includes stages of calcination of sulfide concentrates and leaching.

The general objective is to compare the two types of tailings, using detailed geochemical and mineralogical characterization, namely for determining the AMD potential and the ability to mobilize PTE. This comparison intends to (i) understand the supergenic evolution of different tailings and (ii) support monitoring and identify possible opportunities for environmental improvements and management of both types of tailings.

2. Location and Characterization of the Study Area

The study area is in the Iron Quadrangle (QF), one of the most gold-producing regions in Brazil [36,37]. The gold world-class deposits are part of the Rio das Velhas Greenstone Belt, largely located in State of Minas Gerais (Figure 1a), with an estimated 4.5% of the world's ore reserves, almost half, about 936 tonnes [38–40]. The orebodies, hosted in Archaean rocks, are structurally associated and controlled by hydrothermal alteration. The mineralization is related to different typologies such as iron-banded formation, clastic and turbiditic metasedimentary rocks, and quartz veins enriched with sulfides, mainly pyrite, arsenopyrite, and pyrrhotite [38,40–42].

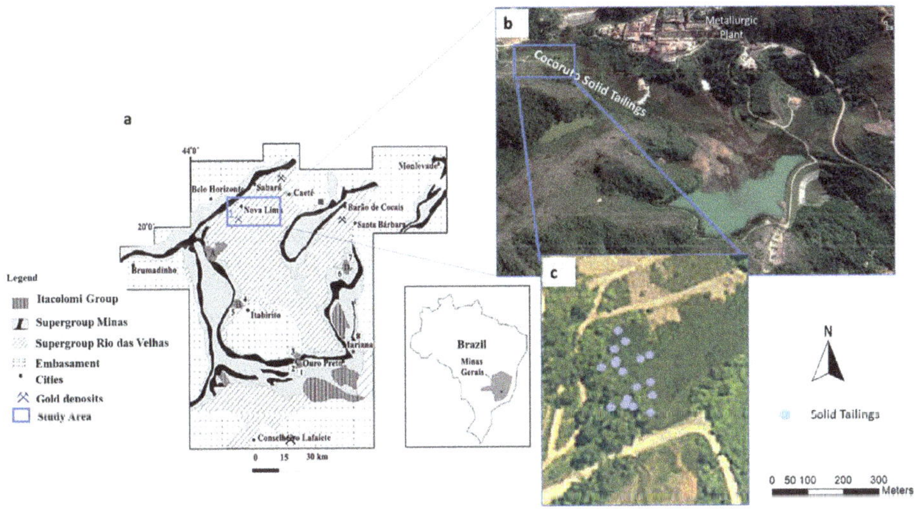

Figure 1. Location of the study area and Iron Quadrangle map (**a**) [43,44]; (**b**,**c**) Images of the Cocoruto Dam sampled area (SIRGAS2000—10-09-2019).

These sulfide Au ores have been treated for over 30 years in metallurgic plants located in Nova Lima, in the northern part of the QF, Minas Gerais, 25 km away from its capital, Belo Horizonte (Figure 1a). The region has a warm and temperate climate, according to the Cfa classification (humid subtropical climate). The average rainfall is around 1390 mm per year, with the month of August being the driest and the month of December having the highest rainfall, with an average of 302 mm. The average temperature is 23.3 °C, with January the hottest month of the year while the lowest annual temperature is in June (average temperature of 17.6 °C) [45].

The circuits of the Nova Lima plant are and were fed by ores from the Raposos (Raposos circuits (Figure 2a)) and Cuiaba mines (Cuiaba circuit (Figure 2b) and ongoing roasted plant (Figure 2c)). The Raposos circuit (Figure 2a) is a plant that treated non-refractory sulfide (pyrite, pyrrhotite and arsenopyrite) ore. The circuit reached 90% of Au recovery and was divided into grinding, gravity concentration, conventional leaching and CIP (carbon in leach), elution and electrorecovery. This part of the plant was deactivated in 1998 with the deactivation of the Raposos underground mining works, [46].

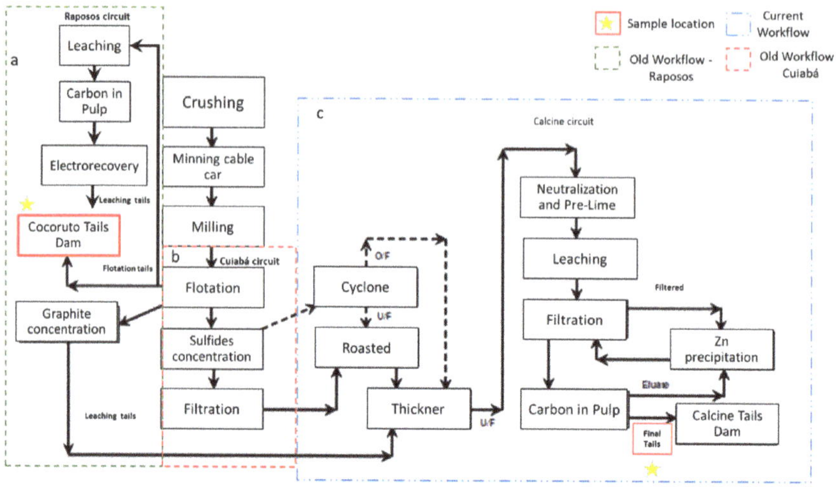

Figure 2. Old and ongoing workflow of Nova Lima´s plant (**a**,**b**). Old circuit sources of Cocoruto Dam. (**c**) Ongoing workflow [46]. The yellow stars represent sampling points.

The Cuiaba circuit (Figure 2b) treats refractory ore (mainly Au enclosed in pyrite) with 92% of Au recovery and was divided into grinding, gravity concentration, sulfide flotation, roasting, neutralization, calcined leaching, CIP, elution and electrodeposition [46]. Currently at the Nova Lima plant, the material is treated from the roasting stage (Figure 2c), as the previous steps are carried out directly at the plant located in the Cuiaba mine in Sabará, Minas Gerais. The tailings from the Raposos and the Cuiaba flotation circuits were deposited in a dam, already deactivated, named by Cocoruto (Figure 2a,b). The tailings from the ongoing circuit (roasting and leaching) are available in a dam known as calcined (Figure 2c).

Both structures are downstream type, monitored and declared safe according to the mining national agency [47]. In this study, the tailings samples representing the old circuits were collected directly at the Cocoruto dam, while the tailings from the ongoing circuit were collected directly at the calcined dam discharge. Therefore, the Nova Lima facility provides the opportunity to examine the potential environmental impact promoted by leaching and mobilization of PTE [48].

3. Materials and Methods

The sampling campaign was performed in the spring, a season (September 2019) in which weather conditions vary greatly. Typically, during this month it is common to alternate between dry and humid periods, with temperatures between 19 °C and 25 °C [45]. A total of 15 sites were selected in the Cocoruto dam (CoT samples) to represent the old circuit. Sampling, using percussion and auger methods, was performed at a maximum of 10 m. Another set of fresh tailings samples (CaT samples) was collected over 28 days in September 2019 during the production stage, representing a total of 40 samples from the active plant (Figure 2c). All samples were immediately sealed and refrigerated until analysis. Additional material was transferred to polypropylene bags and frozen until analysis. Refrigerated and frozen samples were packaged and shipped to the chemical laboratory for analysis.

The chemical composition was analyzed by inductively coupled plasma mass spectrometry (ICP-MS, PerkinElmer SCIEX, Waltham, MA, USA—detection limit 0.001 mg/kg) after acid digestion (nitric acid, hydrogen peroxide and hydrochloric acid) in accordance with Method 3050B published by the U.S. Environmental Protection Agency (USEPA) [49]. In addition, Au analyses were performed by atomic absorption spectroscopy (AAS, Varian,

Palo Alto, CA, USA—detection limit 0.05 mg/kg), using the fire assay method. All metal solutions were prepared from concentrated stock solutions (Sigma Aldrich). High-purity water produced with a Millipore Milli-Q system was used throughout the analytical process. Quality assurance and quality control (QC/QA) procedures included duplicates, blanks and standard reference materials for ICP-MS. Certified Reference Materials Si81 (Rocklabs) for solid tails were selected to represent a wide range of total elemental concentrations. Results of blanks were always below detection limits. Values for precision (expressed as RSD %) were less than 5% for all elements.

The particle size distribution (PSD) was obtained by vibratory sieving between 212-38 μm. In addition to the geochemical and PSD data, polished sections were prepared in composed samples for mineralogical characterization. The mineralogical study was carried out through optical microscopy and scanning electron microscopy (SEM—Field Electron and Ion Company–FEI; Hillsboro, OR, USA) at the UFMG, Belo Horizonte. The samples were analyzed in an FEI electronic microscope, Quanta 600 FEG, high vacuum mode, coupled to the automated analyzer software (MLA—mode GXMAP and SPL-DZ) and the EDS Espirit Bruker (20 Kve) microanalysis system.

The evaluation of acid mine drainage potential consisted of the comparison of the Modified Acid-Base Accounting (MABA) and Net Acid Generation Test (NAG-test) [50,51]. The acid base accounting (ABA) test is conducted to assess the static potential of generating acidic drainage from a material. It was conducted in accordance with the MEND guide [50] and aims to determine the balance between acid production and consumption from the mineral components of a given sample. Acid production and consumption were measured, respectively, using the parameters acidity potential (AP) and neutralization potential (NP). The NAG-test was carried out in accordance with the MEND guide [51], in which hydrogen peroxide was used to oxidize an aliquot (2.5 g) of comminuted sample (<75 mm). Final pH values (NAG-pH), and electrical conductivity were noted, and the liquor, filtered, was titrated with NaOH solution to defined pH values (4.5 and 7.0) in order to obtain indications on the global amount of non-neutralized acidity, per sample weight. The results of the NAG and MABA tests are compared with each other, allowing a refinement of the interpretation of the latter, and a more improved classification (according to both criteria) regarding the sample's AMD potential [50,51]. In addition, leaching test derived (SPLP) was carried out in accordance with the USEPA 1312 method [52].

4. Results

4.1. Chemical Composition

Table 1 and Figure 2 show the composition of the tailings. The elements were selected primarily based on the environmental relevance and inclusion in the Brazilian legal framework as well as potential economic interest. The samples from the ongoing process (CaT) comprise the following major elements, according to the decreasing order of their average concentration: Fe > Ca > Al > Mg > K > Mn, Na. The samples from the old Cocoruto dam (CoT) contain higher levels of Mn than CaT tailings: Fe > Ca > Mg > Al > Mn > K, Na (Table 1 and Figure 2). A relative depletion of the iron concentrations could be verified for the tailings deposited historically in the Cocoruto Dam. The elements Na and, especially K, on the other hand, had an enrichment of the order of 3× and 10×, respectively (Table 1 and Figure 3).

The average concentrations of sulfur and cyanide in the ongoing process samples were also an order of magnitude higher than that of the CoT samples (Figure 3). Based only on the sulfur contents, this result could suggest that the CaT samples may have greater acid generation potential, compared to the samples from the Cocoruto Dam.

The trace elements with environmental interest [40] presented the following decreasing order of their average concentrations for CaT: (As > Zn) 10 mg/kg > (Cu > Ni > Pb > Co) 102 mg/kg > (Sb, Cr, Ba, Cd) 101 mg/kg and CoT (As) 103 mg/kg > (Zn) 102 mg/kg > (Cr > Cu > Ni > Pb > Hg) 101 mg/kg. The As followed by Zn remained as the main trace elements of environmental interest in the samples of tailings collected both in mineral processing and in the Cocoruto Dam (Table 1 and Figure 4). The element Hg was not detected in the CaT samples, but it had an average content of 21 mg/kg in the samples from the Cocoruto Dam.

Table 1. Statistical summary of the major and trace elements. CoT—old circuit; CaT—ongoing circuit; N—number of samples.

Element	Source	N	Average	Standard Deviation	Minimum	Maximum
Ag(mg/kg)	CoT	30	0.5000	0.0000	0.5000	0.5000
	CaT	40	3.260	0.5478	2.970	4.960
Al (%)	CoT	30	0.8790	0.4150	0.5670	1.4820
	CaT	40	0.622	0.0569	0.516	0.799
Sb (mg/kg)	CoT	30	0.5000	0.0000	0.5000	0.5000
	CaT	40	49.060	10.4800	29.600	70.300
As (mg/kg)	CoT	30	1799	171.2	1610	2018
	CaT	40	7905	878.0	3553	9275
Ba (mg/kg)	CoT	30	17.1500	7.3800	12.0000	28.1000
	CaT	40	23.427	2.4770	18.600	30.300
Cd (mg/kg)	CoT	30	2.9325	0.0670	2.8700	3.0000
	CaT	40	14.083	1.2010	11.000	17.000
Ca (%)	CoT	30	2.3850	0.4850	1.9190	2.8720
	CaT	40	1.672	0.1177	1.404	1.903
Pb (mg/kg)	CoT	30	21.1000	4.4800	15.9000	26.7000
	CaT	40	258.480	32.0600	215.000	343.000
Cyanide (mg/kg)	CoT	30	2.9850	1.2920	1.6200	4.2600
	CaT	40	51.890	21.1700	13.900	87.600
Co (mg/kg)	CoT	30	4.0000	0.0000	4.0000	4.0000
	CaT	40	116.330	11.7400	95.300	144.000
Cu (mg/kg)	CoT	30	78.0500	15.4100	56.6000	93.0000
	CaT	40	670.520	56.4700	583.000	824.000
Cr (mg/kg)	CoT	30	85.2000	20.5000	59.0000	109.0000
	CaT	40	35.020	8.4700	26.900	81.400
Fe (%)	CoT	30	6.7600	0.5520	6.3970	7.5660
	CaT	40	24.859	2.0270	21.557	28.974
P (%)	CoT	30	0.0002	0.0000	0.0002	0.0002
	CaT	40	0.047	0.0075	0.027	0.061
Mg (%)	CoT	30	0.9570	0.2720	0.6340	1.2040
	CaT	40	0.450	0.0424	0.317	0.520
Mn (%)	CoT	30	0.1728	0.0307	0.1396	0.2119
	CaT	40	0.079	0.0072	0.059	0.088
Hg (mg/kg)	CoT	30	0.0688	0.0338	0.0250	0.1000
	CaT	40	0.036	0.0274	0.025	0.160
Ni (mg/kg)	CoT	30	72.2000	15.2200	52.9000	90.1000
	CaT	40	392.630	51.7200	269.000	506.000
K (%)	CoT	30	0.0411	0.0178	0.0226	0.0644
	CaT	40	0.377	0.0310	0.326	0.471
Na (%)	CoT	30	0.0091	0.0010	0.0086	0.0106
	CaT	40	0.049	0.0050	0.041	0.062
Sulfates (%)	CoT	30	0.2175	0.1763	0.0400	0.4400
	CaT	40	1.505	0.1333	1.230	1.810
Zn (mg/kg)	CoT	30	170.5	38.3	136	0.5000
	CaT	40	2724	318.8	1941	1.430
Au (mg/kg)	CoT	30	1.055	0.6620	0.4800	2.010
	CaT	40	1.510	0.3731	0.3100	2.430

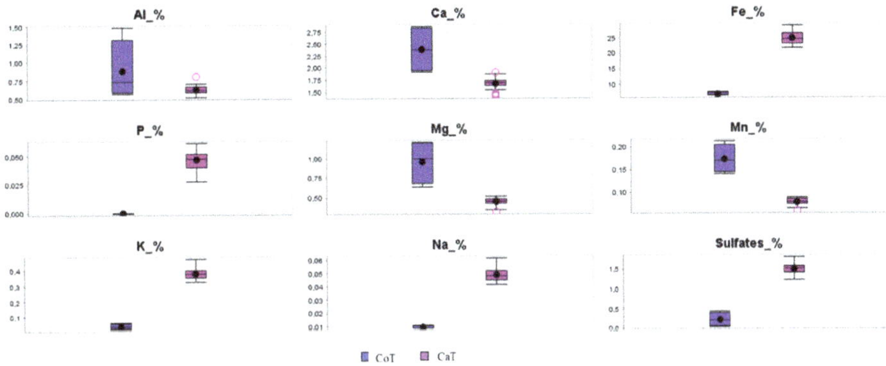

Figure 3. Basic statistics and comparison of major element (%log) for the two sets of tailings samples.

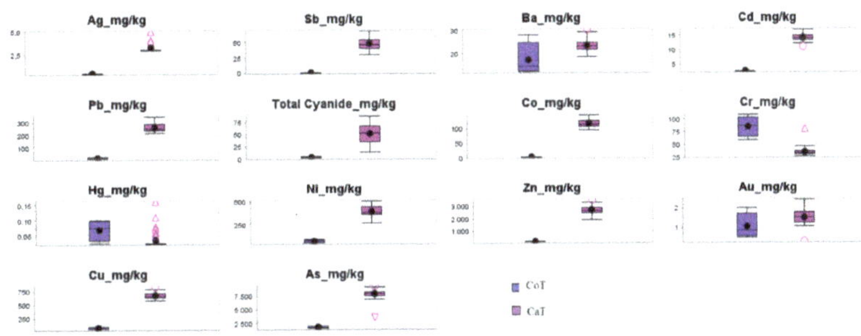

Figure 4. Basic statistics and comparison of trace element (mg/kg log) for the two sets of tailings samples.

4.2. Size Particle Distribution

In general, the grain size distribution (Figure 5) is similar to other dams that store tailings from Au exploitation [7,8,27,53]. The tailings presented 80% of their particles classified as silt-sized (Figure 5). The CoT samples are shown to have a thicker distribution than the ongoing process tailings (CaT). The difference can be explained by changes in the grinding process and different features of the ore sources and their variability [30]. In addition, the samples that represent CoT originate from older circuits, which have low grinding efficiency when compared to the ongoing one (CaT).

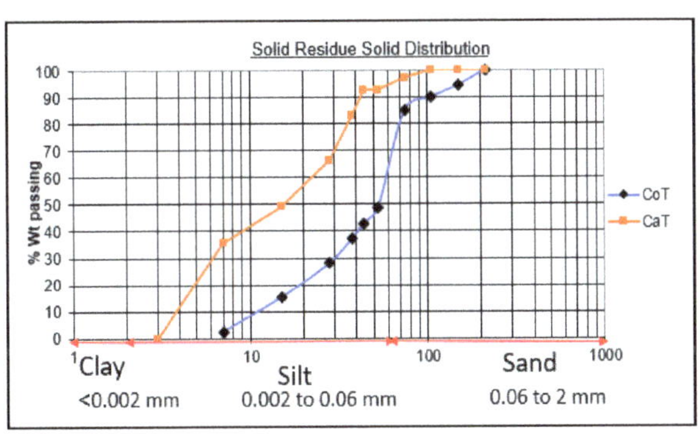

Figure 5. Size distribution (Krumbein phi scale) of solid tailings.

4.3. Mineralogy

Table 2 shows the mineralogical composition of both tailings. The samples from the Cocoruto dam (old circuit) consist mainly of quartz, carbonates, iron oxides and phyllosilicates such as muscovite and chlorite (Figure 6). This is in accordance with higher Al, Ca and Mg concentrations. Sulfides are present, but they look well-preserved, mainly in the form of iron sulfides with sphalerite, covellite and arsenopyrite subordinated.

Table 2. Mineralogical Composition of CaT and CoT.

Minerals	Chemical Formula	CaT (Wt%)	CoT (Wt%)
Quartz	SiO_2	15.6	55.8
Feldspar Group			
Albite	$NaAlSi_3O_8$	1.50	0.370
Anorthite	$CaAl_2Si_2O_8$	0.053	0.01
K feldspar	$KAlSi_3O_8$		0.390
Phyllosilicates			
Biotite	$KMg_{2.5}Fe_{0.5}AlSi_3O_{10}(OH)_{1.75}F_{0.25}$	1.00	0.16
Smectite	$(Si,Al)(Mg,Fe)O(OH)NaH_2O.$	1.80	0.13
Muscovite Group	$KAl_3Si_3O_{10}(OH)_{1.9}F_{0.1}$	11.0	5.56
Chlorite	$(Mg,Fe)_3(Si,Al)_4O_{10}(OH)_2.(Mg,Fe)_3(OH)_6$	3.30	6.12
Oxides			
Iron Oxides/Hydroxides	$Fe_2O_3/FeOOH$	56.8	8.86
Rutile/Anathase	TiO_2	0.599	0.49
Carbonates			
Ankerite	$Ca(Fe,Mg,Mn)(CO_3)$	1.00	11.2
Siderite	$FeCO_3$	-	7.25
Calcite	$CaCO_3$	0.200	2.25
Sulfates			
Gypsum	$CaSO_4$ $2H_2O$	7.00	0.030
Sulfides			
Pyrite	FeS_2	0.002	0.500
Pyrrhotite	$Fe_{(1-x)}S$	0.004	0.790
Arsenopyrite	FeAsS	0.056	0.240
Gesdorffite	NiAsS	0.010	-
Covellite	CuS	0.100	0.070
Sphalerite	ZnS	-	0.010
Gold Minerals			
Native Gold	Au > 80%, Ag, Cu, Hg	(526)	(364)
Electrum	Au = 80%, Ag = 20%	(42)	(10)

The samples from the ongoing process (CaT) are characterized by high concentrations of iron oxides, in addition to sulfates and silicates (mainly quartz). The high concentration of oxides justifies higher Fe contents than the samples of Cocoruto. Sulfides are reliquaries, associated with oxidized phases, and are rare. In Figure 6b, is identified, in false images, the presence of As, Cu, Ni, and Zn with oxidized phases, in addition to Fe. In both types of samples there are still fine Au grains. These particles are associated with iron oxides in the CaT and with sulfides in CoT samples. (Figure 6a–d). The optical microscopy images (Figure 6e–f) support these observations, showing the dominant hematite in CaT samples and the presence of Au in association with pyrite in CoT samples, respectively.

4.4. AMD Potential

The results of the MABA tests are presented graphically in Figure 7. Calcined samples collected in mineral processing showed relatively small variability in acidity (AP) and neutralization potentials (NP): In the first case, it ranged between 1.5 and 2.5 kg/$CaCO_3$ t and, in the second, it varied between 8.3 and 15.5 kg/$CaCO_3$ t. All these samples were classified as potentially non-acid-forming (PNA) by the MABA assay criterion, as they presented NP/AP ratio (NPR) > 2. The average NPR value for CaT corresponded to 6.1 kg/$CaCO_3$ t.

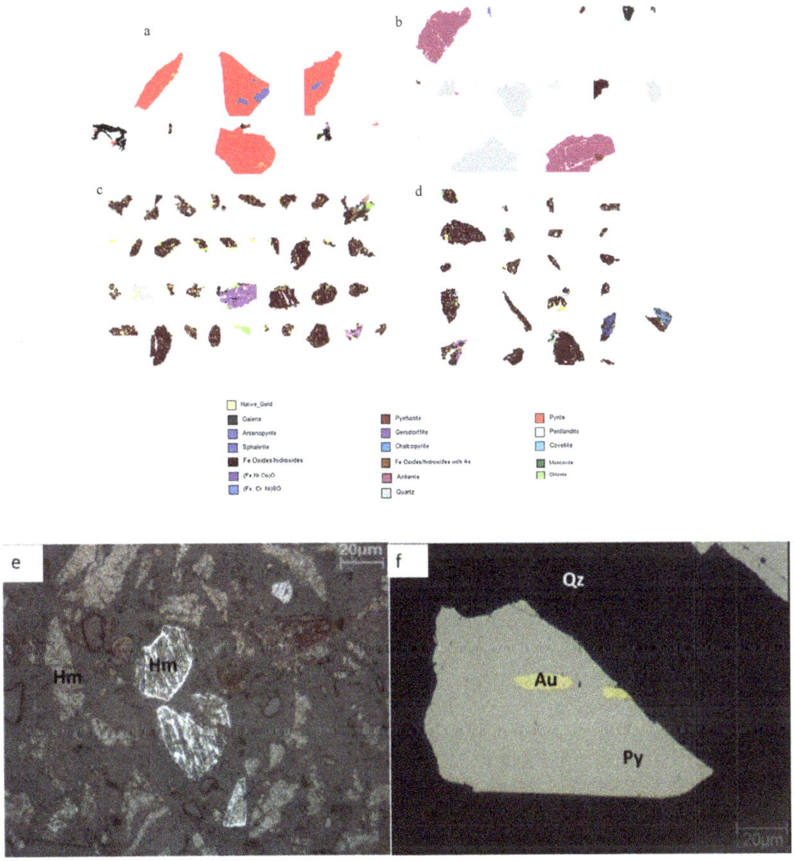

Figure 6. False Images (**a**) sulfides associated with iron oxide and Au particles—CoT, (**b**) quartz, carbonates and Fe oxides particles—CoT, (**c**,**d**) Fe oxides associated with Au, silicates and Fe oxides with As, Ni, Co, Al, Zn, Pb—CaT samples. Photomicrographs taken under reflected light and uncrossed nico (**e**) hematite (Ht) from CaT samples and (**f**) Au associated with pyrite (Py)—CoT samples.

The tailings samples collected at the Cocoruto Dam showed relevant dispersion of acidity and neutralization potentials, in accordance with the mineralogical features. The AP values oscillated mainly in the range of 5–150 kg/$CaCO_3$ t. NP values between 44 to 123 kg/$CaCO_3$ t. This dispersion resulted in the following relative potentials for the generation of acid drainage, according to the MABA test criteria: Nonacid-forming samples (NPR > 2) represent 45%, samples in the uncertainty zone (NPR 1 to 2) are 50%, and samples acid-forming (NPR < 1) are 5%. Vertically, the samples in the uncertainty zone corresponded mainly to depths of 4.0 to 5.5 m and 8.0 to 11.5 m. The non-acid-forming samples occurred mainly between 0.0 and 1.5 m and from 6.0 to 7.5 m in depth (Figure 7b).

The NAG test (net acid generation) was another method used in order to compare with the results obtained by MABA, mainly because of the S grades. Figure 8 presents the graph NAGpH versus NPR, allowing to correlate the results of the MABA and NAG-tests, and to better refine the classification of acid generation potential of the samples under analysis. The results of the NAG test showed a pH value well above the limit of 4.5, indicating all samples as non-acid-forming, regardless of the result of the NPR. The pH fluctuated from 7.4 to 8.5 for samples from the Cocoruto Dam, and from 8.1 to 9.5 for process samples. These results indicate less potential for liquid acidity than previously indicated, only by the NPR (MABA) criterion. The summary of the assessment of the potential for acidity generation, based on the MABA and NAG, therefore, demonstrates that all process samples resulted in potentially non-acid forming (NAF). Only 45% of the samples collected at the Cocoruto Dam resulted in potentially non-acid forming, using the MABA criterion. However, this number was 100% using the NAG criterion.

a

b

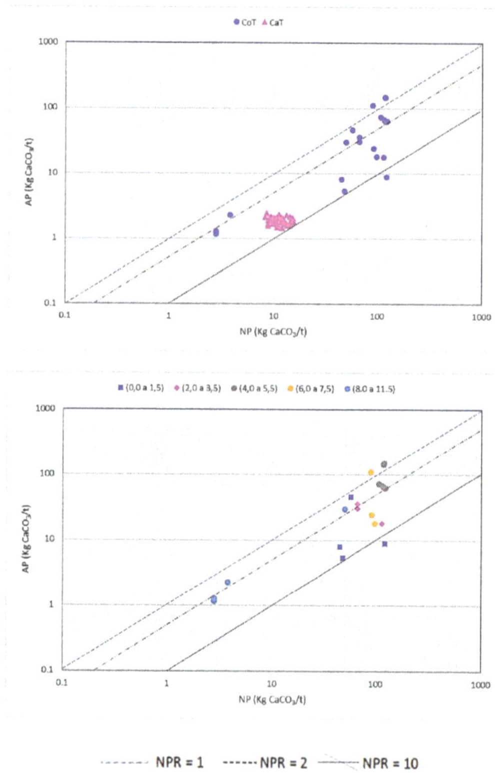

Figure 7. Results of Modified Acid-Base Accounting (MABA) tests (**a**) global and (**b**) for CoT, by sampling depth.

4.5. SPLP Leaching

In addition to the tests already presented, the SPLP leaching test (USEPA 1312 method) was carried out in order to assess the potential for leaching toxic elements. Data from leaching tests can be used as indicators of potentially contaminating substances, even though the test conditions are very limited in view of the environmental geochemical conditions. A basic criterion used to assess the magnitude of the potential impact is the comparison with benchmarks in the context of the legal framework. The Brazilian law, through CONAMA Resolution 396/2008, was used as reference for the permitted limits of the following elements: Al, Sb, As, Ba, Be, Cd, Pb, Co, Cu, Cr, Fe, Mn, Hg, Ni, Ag and Zn. Since there is no definition for the predominant industrial use, recreation was defined as the predominant use.

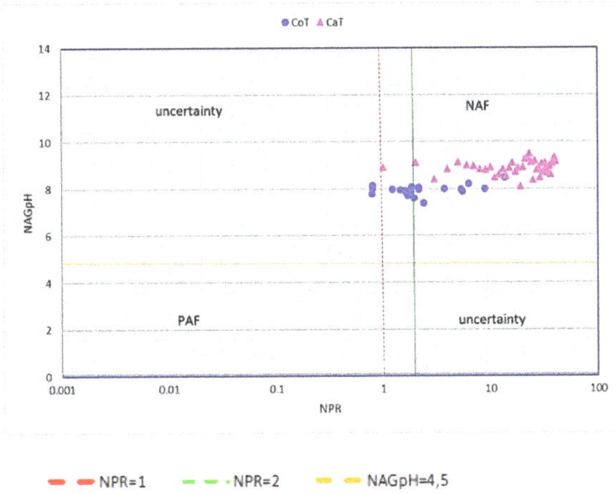

Figure 8. Relationship between the results of the MABA and Net Acid Generation (NAG) pH tests (NAF: region of non-acid-forming samples; PAF: region of potentially acid-forming samples).

The elements Cd, Pb, Hg, and Zn were not detected in the leaching extracts (Table 3 and Figure 9), with two exceptions: Pb equal to 0.02 mg/L, and Zn equal to 0.1 mg/L in a sample from the Cocoruto Dam. In the extracts of this dam, only the elements As and Mn showed concentrations above the respective legal reference (VMP) (i.e., 0.05 mg As/L, and 0.1 mg Mn/L). The average concentrations were around an order of magnitude higher than the VMP, in these cases (Figure 9).

Table 3. VMP—maximum allowed values From Brazilian Law—CONAMA [54].

Samples	Not Detected Elements	Parameters with Values AboveVMP Conama 2005 Reference
CoT	Cd, Hg, Pb, Zn	As (VMP 0.05 mg/L) Mn (VMP 0.1 mg/L)
CaT	Cd, Hg, Pb, Zn	As (VMP 0.05 mg/L) Fe (VMP 0.3 mg/L) Ni (VMP 0.1 mg/L) Se (VMP 0.01 mg/L)

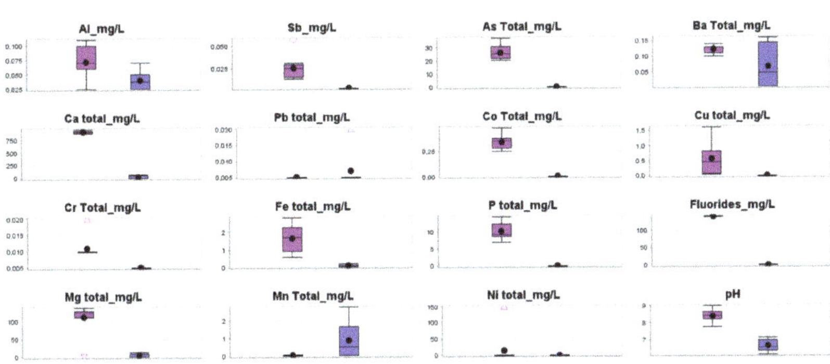

Figure 9. Comparison of the average levels of metals, metalloids, fluorides and pH in SPLP leaching extracts. VMP Al = 0.2 mg/L; VMP Sb = NA; VMP As = 0.05 mg/L; VMP Ba = 1.0 mg/L; VMP Ca = NA; VMP Pb = 0.05 mg/L; VMP Co = NA; VMP Cu = 1.0 mg/L; VMP Cr = 0.05 mg/L; VMP Fe = 0.3 mg/L; VMP P = NA; VMP Fluorides = NA; VMP Mg = NA; VMP Mn = 0.1 mg/L; VMP Ni = 0. 1 mg/L. NA—Not applied for recreation use.

On the other hand, in the calcined ongoing process the elements As, Fe, Ni, and Se showed values above the CONAMA references, in at least one of the leaching extracts.

5. Discussion

The differences in particle distribution could contribute to differentiate the two types of tailings regarding the speed and magnitude of AMD. It is known that the rate of oxidation of sulfides is directly related to the size of the particles and the availability of the surface for reaction [20,55]. The mineralogical characteristics confirmed the distinction between tailings that were subjected to different processes, in accordance with the geochemical signatures (Figure 1). Such geochemical and mineralogical diversity could result in different potentials for mobilizing pollutants in AMD. For example, the higher amounts of sulfides in the Cocoruto tailings (old circuit) could suggest higher potential for AMD generation. However, Figure 6 suggests that inhibition of the oxidative dissolution processes occurs, since the sulfide grains are preserved. Moreover, AMD generation could be controlled by the presence of carbonates due to the ore paragenesis and metallurgical ore treatment [7]. Neutralization conditions also occur in CaT samples due to the addition of lime.

The chemical and mineralogical study highlights the occurrence of high grades of Au in these two sets of tailings. In CoT samples, the concentration reaches 1.0 mg/kg with Au crystals associated with sulfides and rock minerals such as carbonates and silicates (Figure 6a,b). In the ongoing process, in CaT samples, the concentration reaches 1.70 mg/kg, with grains enclosed in iron oxides (Figure 6c). Both Au features portray limitations and problems in the beneficiation old and ongoing processes. However, this inefficiency of the mineral separation processes may represent an opportunity for creating profit through valorization of the mine wastes. The reuse of tailings is now an important research line, as discussed in [53,56–60]. Especially in view of the scenario experienced in Brazil after the mediatized disasters of the Mariana (2015) and Brumadinho (2019) dams [61–64], it is crucial to find sustainable solutions for the closure stage. Therefore, in the present cases, the obtained results suggest the possibility of transforming an environmental liability into an economic outcome.

The tests for prediction of AMD indicate that minimum pH is around 8, reaching a maximum of 9.6, reflecting the alkaline products used for extraction of Au with cyanide solutions. However, the levels of S present in the forms of sulfate and sulfide, thin PSD, and the presence of toxic leachable elements above permitted limits do not rule out the generation of AMD in practice if the tailings are stored improperly. These results, in addition to geochemical and mineralogical features, imply that the samples from the Cocoruto Dam are not potentially acid-forming under the conditions in which they are disposed. The net potential for acidity generation is low. However, in the case of removal of these materials, they must be disposed properly, without long-term exposure (months/years).

Under pH conditions close to neutrality, a series of trace elements can remain solubilized [8], as verified for As, Fe, Ni, Mn and Se. This behavior was also observed by [23], specifically for As and Se leached out in underground rock excavation works. The Al for the sample of the CoT tailings was an exception. Thus, the generation of effluents with metals or metalloids does not require long periods of exposure of sulfide minerals [7]. The As stood out in SPLP leaching test, with an average concentration (26 mg As/L), 3×of magnitude above the permitted limits by the Brazilian law (VMP). This could be explained by the presence of arsenopyrite in CoT samples and iron oxide in CaT. For the other parameters mentioned, the averages values cofound were up to an order of magnitude higher than the corresponding VMP.

6. Conclusions

An integrated characterization of tailings from different mineral beneficiation processes was performed to evaluate the environmental risk of mobilizing pollutants from the mine drainage.

The tailings presented 80% of their particles classified as silt-sized. Sulfides are present in the samples from the historic Cocoruto dam, but they appear well-preserved. The tailings consist mainly of quartz (55.8%), carbonates (20.7%), iron oxides (8.86%), muscovite (5.56%) and chlorite (6.12%). Therefore, the potential acidity associated with sulfides could be neutralized by the dissolution of carbonates, such as calcite and silicates such as chlorite, besides the reagents used during the process. This fact contributes to make it a less dangerous waste when compared to the current one (CaT).

The tailings from the ongoing process (CaT) are characterized by high contents of iron oxides (56.8%), gypsum (7%) and silicates (mainly quartz—15.6%). In both types of tailings, the chemical composition allowed to identify elements with environmental relevance by their toxicity, namely Mn in the old tailings and As, Fe, and Ni in the more recent tailings. In addition to these elements of environmental concern, it should be noted the interesting values of Au (up to 2.4 mg/kg in the recent tailings), suggesting a potential for reuse during closures of dam facilities or tailings mass-rearrangement needs.

The geochemical tests, MABA and NAG, indicate that these tailings are not potentially acid-forming, under the conditions in which they are disposed. The net potential for acidity generation is low. However, the results of the SPLP leaching tests indicated a leaching potential for As and Mn in the Cocoruto Dam. Samples that are subject to calcination indicate leaching potential for other elements, namely Sb, As, Fe, Ni, and Se. It is of note that As concentrations in these samples were two orders of magnitude higher than the ones found in samples from Cocoruto Dam. Moreover, the concentrations of S present in the forms of sulfate and sulfide, together with thin PSD and the presence of leachable PTE above permitted limits by the Brazilian law are important issues of concern. In the case of tailings mobilization or inappropriate storage, the generation of AMD and contamination by PTE is likely to occur.

Author Contributions: Conceptualization, M.L., T.V., P.M.R. and R.F.; Data curation, M.L., T.V., I.D., J.V. and M.M.; Funding acquisition, M.L., T.V. and J.V.; Investigation, M.L., T.V., P.M.R., R.F., I.D., J.V. and M.M.; Resources, T.V. and J.V., Supervision, T.V., Visualization, M.L. and T.V., Writing—original draft, M.L., T.V., P.M.R. and R.F. All authors have read and agreed to the published version of the manuscript.

Funding: This work was funded by FCT—Fundação para a Ciência e a Tecnologia through projects UIDB/04683/2020 e UIDP/04683/2020 and Nano-MINENV 029259 (PTDC/CTA-AMB/29259/2017, and by AngloGold Ashanti Brazil.

Institutional Review Board Statement: Not applicable.

Informed Consent Statement: Not applicable.

Data Availability Statement: Not applicable.

Acknowledgments: We thank our colleagues from ICT, microscopy center from Universidade Federal de Minas Gerais (CM-UFMG), and AngloGold Ashanti who provided insight and expertise that greatly assisted the research. The authors are grateful to the two anonymous reviewers and to the editor Dr Carlito Tabelin for their valuable contributions to improving the manuscript.

Conflicts of Interest: Authors declare no conflict of interest.

References

1. Mohapatra, D.; Kirpalani, D. Process effluents and mine tailings: Sources, effects and management and role of nanotechnology. *Nanotechnol. Environ. Eng.* **2017**, 2, 1–12. [CrossRef]
2. Blight, G. Mine waste: A brief overview of origins, quantities, and methods of storage. In *Waste: A Handbook for Management*, 2nd ed.; Letcher, T., Vallero, D., Eds.; Academic Press: Cambridge, MA, USA, 2011; pp. 77–88.
3. Lottermoser, B. *Mine Wastes: Characterization, Treatment and Environmental Impacts*; Springer: New York, NY, USA, 2012; pp. 1–400. ISBN 978-3-662-05133-7.
4. Park, I.; Tabelin, C.B.; Jeon, S.; Li, X.; Seno, K.; Ito, M.; Hiroyoshi, N. A review of recent strategies for acid mine drainage prevention and mine tailings recycling. *Chemosphere* **2019**, *219*, 588–606. [CrossRef] [PubMed]

5. Blowes, D.W.; Jambor, J.L.; Hanton-Fong, C.J.; Lortie, L.; Gould, W.D. Geochemical, mineralogical and microbiological characterization of a sulphide-bearing carbonate-rich gold mine tailings impoundment, Joutel, Quebec. *Appl. Geochem.* **1998**, *13*, 687–705. [CrossRef]
6. Johnson, D.B. Chemical and microbiological characteristics of mineral spoils and drainage waters at abandoned coal and metal mines. *Water Air Soil Pollut.* **2003**, *3*, 47–66. [CrossRef]
7. Lindsay, M.B.; Condon, P.D.; Jambor, J.L.; Lear, K.G.; Blowes, D.W.; Ptacek, C.J. Mineralogical, geochemical, and microbial investigation of a sulfide-rich tailings deposit characterized by neutral drainage. *J. App. Geochem.* **2009**, *24*, 2212–2221. [CrossRef]
8. Bissacot, L.; Ciminelli, V.; Logsdon, M. Arsenic Mobility under a Neutral Mine Drainage Environment in a Gold-Mine Tailings Dam—In Agreeing on solutions for more sustainable mine water management. In Proceedings of the 10th International Conference on Acid Rock Drainage & IMWA Annual Conference (ICARD & IMWA-2005), Santiago, Chile, 21–24 April 2005; pp. 170–174.
9. Opiso, E.; Aseneiro, J.; Banda, M.; Tabelin, C.B. Solid-phase partitioning of mercury in artisanal gold mine tailings from selected key areas in Mindanao, Philippines, and its implications for mercury detoxification. *Waste Manag. Res.* **2018**, *36*, 269–276. [CrossRef] [PubMed]
10. Tabelin, C.B.; Silwamba, M.; Paglinawan, F.C.; Mondejar, A.J.S.; Duc, H.G.; Resabal, V.J.; Opiso, E.M.; Igarashi, T.; Tomiyama, S.; Ito, M.; et al. Solid-phase partitioning and release-retention mechanisms of copper, lead, zinc and arsenic in soils impacted by artisanal and small-scale gold mining (ASGM) activities. *Chemosphere* **2020**, *260*, 127574. [CrossRef]
11. Nordstrom, K. Mine Waters: Acidic to Circumneutral. *Elements* **2011**, *7*, 393–398. [CrossRef]
12. Nordstrom, D.K. Hydrogeochemical processes governing the origin, transport and fate of major and trace elements from mine wastes and mineralized rock to surface waters. *Appl. Geochem.* **2011**, *26*, 1777–1791. [CrossRef]
13. Smuda, J.; Dold, B.; Spangenberg, J.E.; Friese, K.; Kobek, M.R.; Bustos, C.A.; Pfeifer, H.R. Element cycling during the transition from alkaline to acidic environment in an active porphyry copper tailings impoundment, Chuquicamata, Chile. *J. Geochem. Explor.* **2014**, *140*, 23–40. [CrossRef]
14. Guseva, O.; Opitz, A.K.B.; Broadhurst, J.L.; Harrison, S.T.L.; Bradshaw, D.J.; Becker, M. Fe-Sulfide Liberation and Association as a Proxy for the Interpretation of Acid Rock Drainage (ARD) Test Results. In *Mine Water—Risk to Opportunity*; Wolkersdorfer, C.H., Sartz, L., Weber, A., Burgess, J., Tremblay, G., Eds.; Tshwane University of Technology: Pretoria, South Africa, 2018; pp. 345–351.
15. Wolkersdorfer, C.; Nordstrom, D.K.; Beckie, R.D.; Cicerone, D.S.; Elliot, T.; Edraki, M.; Valente, T.; França, S.C.A.; Kumar, P.; Oyarzún, R. Guidance for the Integrated Use of Hydrological, Geochemical, and Isotopic Tools in Mining Operations. *Mine Water Environ.* **2020**, *39*, 204–228. [CrossRef]
16. Tabelin, B.C.; Corpuz, R.D.; Igarashi, T.; Villacorte-Tabelin, M.; Alorro, R.D.; Yoo, K.; Raval, S.; Ito, M.; Hiroyoshi, N. Acid mine drainage formation and arsenic mobility under strongly acidic conditions: Importance of soluble phases, iron oxyhydroxides/oxides and nature of oxidation layer on pyrite. *J. Hazard. Mater.* **2020**, *399*, 122844. [CrossRef] [PubMed]
17. Tomiyama, S.; Igarashi, T.; Tabelin, C.B.; Tangviroon, P.; Li, H. Acid mine drainage sources and hydrogeochemistry at the Yatani mine, Yamagata, Japan: A geochemical and isotopic study. *J. Contam. Hydrol.* **2019**, *225*, 103502. [CrossRef] [PubMed]
18. Tomiyama, S.; Igarashi, T.; Tabelin, C.B.; Tangviroon, P.; Ii, H. Modeling of the groundwater flow system in excavated areas of an abandoned mine. *J. Contam. Hydrol.* **2020**, *230*, 103617. [CrossRef]
19. Lapakko, K.A.; Engstrom, J.N.; Antonson, D.A. Effects of particle size on drainage quality from three lithologies. In Proceedings of the 7th International Conference on Acid Rock Drainage (ICARD), St. Louis, MO, USA, 26–30 March 2006; pp. 1026–1050.
20. Eary, L.; Williamson, M. Simulations of the neutralizing capacity of silicate rocks in acid mine drainage environments. *J. Am. Soc. Min. Reclam.* **2006**, *2*, 564–577. [CrossRef]
21. Blowes, D.W.; Ptacek, C.J.; Jurjovec, J. Mill tailings: Hydrogeology and geochemistry. In *Environmental Aspects of Mine Wastes*; Jambor, J.L., Ritchie, A.I.M., Eds.; Mineral Association of Canada, Short Course: Québec, QC, Canada, 2003; Volume 31, pp. 95–116.
22. Ritcey, G.M. Tailings Management in Gold Plants. *Hydrometallurgy* **2005**, *78*, 3–20. [CrossRef]
23. Tabelin, C.B.; Igarashi, T.; Villacorte-Tabelin, M.; Park, I.; Opiso, E.M.; Ito, M.; Hiroyoshi, N. Arsenic, selenium, boron, lead, cadmium, copper, and zinc in naturally contaminated rocks: A review of their sources, modes of enrichment, mechanisms of release, and mitigation strategies. *Sci. Total Environ.* **2018**, *645*, 1522–1553. [CrossRef]
24. Moodley, I.; Sheridan, C.M.; Kappelmeyer, U.; Akcil, A. Environmentally sustainable acid mine drainage remediation: Research developments with a focus on waste/by-products. *Miner. Eng.* **2018**, *126*, 207–220. [CrossRef]
25. Dutta, M.; Islam, N.; Rabha, S.; Narzary, B.; Bordoloi, M.; Saikia, D.; Silva, L.F.O.; Saikia, B.K. Acid mine drainage in an Indian high-sulfur coal mining area: Cytotoxicity assay and remediation study. *J. Hazard. Mater.* **2020**, *389*, 304–389. [CrossRef]
26. Igarashi, T.; Herrera, P.S.; Uchiyama, H.; Miyamae, H.; Iyatomi, N.; Hashimoto, K.; Tabelin, C.B. The two-step neutralization ferrite-formation process for sustainable acid mine drainage treatment: Removal of copper, zinc and arsenic, and the influence of coexisting ions on ferritization. *Sci. Total Environ.* **2020**, *715*, 136877. [CrossRef]
27. Nengovhela, A.C.; Yibas, B.; Ogola, J.S. Characterisation of gold tailings dams of the Witwatersrand Basin with reference to their acid mine drainage potential Johannesburg, South Africa. *Water* **2006**, *32*, 4–8. [CrossRef]
28. Valente, T.; Antunes, M.; Sequeira, B.M.A.; Prudêncio, M.I.; Marques, R.; Pamplona, J. Mineralogical attenuation for metallic remediation in a passive system for mine water treatment. *Environ. Earth Sci.* **2012**, *66*, 39–54. [CrossRef]

29. Valente, T.; Grande, J.A.; De La Torre, M.L. Extracting value resources from acid mine drainages and mine wastes in the Iberian Pyrite Belt. In Proceedings of the International Mine Water Association Symposium 2016—Mining Meets Water—Conflicts and Solutions, Leipzig, Germany, 11–15 July 2016; pp. 1339–1340.
30. Foli, G.; Gawu, S.K.Y. Modifed acid–base accounting model validation and pH buffer trend characterization in mine drainage at the AngloGold-Ashanti Obuasi mine in Ghana, West Africa. *Environ. Earth Sci.* **2017**, *76*, 663–682. [CrossRef]
31. Becker, M.; Charikinya, E.; Ntlhabane, S.; Voigt, M.; Broadhurst, J.; Harrison, S.T.L.; Bradshaw, D. An Integrated Mineralogy-based Modelling Framework for the Simultaneous Assessment of Plant Operational Parameters with Acid Rock Drainage Potential of Tailings. In *Mine Water—Risk to Opportunity*; Wolkersdorfer, C.H., Sartz, L., Weber, A., Burgess, J., Tremblay, G., Eds.; Tshwane University of Technology: Pretoria, South Africa, 2018; Volume 1, pp. 309–315.
32. Borba, R.P.; Figueiredo, B.R.; Rawlins, B. Geochemical distribution of arsenic in waters, sediments and weathered gold mineralized rocks from Iron Quadrangle, Brazil. *Environ. Geol.* **2003**, *44*, 39–52. [CrossRef]
33. Tabelin, C.B.; Hashimoto, A.; Igarashi, T.; Yoneda, T. Leaching of boron, arsenic and selenium from sedimentary rocks: I. Effects of contact time, mixing speed and liquid-to-solid ratio. *Sci. Total Environ.* **2014**, *472*, 620–629. [CrossRef] [PubMed]
34. Tabelin, C.B.; Sasaki, R.; Igarashi, T.; Park, I.; Tamoto, S.; Arima, T.; Ito, M.; Hiroyoshi, N. Simultaneous leaching of arsenite, arsenate, selenite and selenate, and their migration in tunnel-excavated sedimentary rocks: I. Column experiments under intermittent and unsaturated flow. *Chemosphere* **2017**, *186*, 558–569. [CrossRef] [PubMed]
35. Tamoto, S.; Tabelin, C.B.; Igarashi, T.; Ito, M.; Hiroyoshi, N. Short and long term release mechanisms of arsenic, selenium and boron from a tunnel-excavated sedimentary rock under in situ conditions. *J. Contam. Hydrol.* **2015**, *175–176*, 60–71. [CrossRef]
36. Porto, C.G. A Mineralização Aurífera do Depósito Córrego do Sítio e Sua Relação Com o Enxame de Diques Metamáficos No Corpo CACHORRO BRAVO—Quadrilátero Ferrífero—Minas Gerais. Master's Thesis, Universidade Federal de Minas Gerais, Belo Horizonte, Spain, 2008; pp. 1–117. (In English).
37. Goldfarb, R.J.; Groves, D.; Gardoll, S. Orogenic gold and geologic time: A global synthesis. *Ore Geol. Rev.* **2001**, *18*, 1–75. [CrossRef]
38. Lobato, L.M.; Ribeiro-Rodrigues, L.C.; Vieira, F.W.R. Brazil's premier gold province. Part II: Geology and genesis of gold deposits in the Archean Rio das Velhas greenstone belt, Quadrilátero Ferrífero. *Miner. Depos.* **2001**, *36*, 249–277. [CrossRef]
39. IBRAM—Instituto Brasileiro de Mineração. Available online: http://www.ibram.org.br/ (accessed on 20 September 2020).
40. Deschamps, E.; Ciminelli, V.S.T.; Lange, F.T. Soil and sediment geochemistry of the iron quadrangle, Brazil the case of arsenic. *J. Soils Sediments* **2002**, *2*, 216–222. [CrossRef]
41. Vieira, F.W.R.; Biasi, E.E.; Lisboa, L.H. Geology of and excursion to the Morro Velho and Cuiabá mines. In *Field and Mine Trip to Quadrilátero Ferrífero. Brazil Gold'91 Internat Symp on the Geology of Gold*; Fleischer, R., Grossi, S.J.H., Fuzuikawa, K., Ladeira, E.A., Eds.; Departamento Nacional de Produção Mineral (DNPM): Minas Gerais, Brazil, 1991; pp. 87–99.
42. Vitorino, A.L.A.; e Silva, R.C.F.; Lobato, L.M. Shear-zone-related gold mineralization in quartz-carbonate veins from metamafic rocks of the BIF-hosted world-class Cuiabá deposit, Rio das Velhas greenstone belt, Quadrilátero Ferrífero, Brazil: Vein classification and structural control. *Ore Geol. Rev.* **2020**, *127*, 103–789. [CrossRef]
43. Almeida, F.F.M. *Origem e Evolução da Plataforma Brasileira*; Departamento Nacional de Produção Mineral (DNPM): Rio de Janeiro, Brazil, 1967; pp. 1–36.
44. Alkmin, F.F.; Marshak, S. Transamazonian Orogeny in the Souther São Francisco Craton Region, Minas Gerais, Brazil: Evidence for Paleoproterozoic Collision and Collapse in the Quadrilátero Ferrífero. *Precambrian Res.* **1998**, *90*, 29–58. [CrossRef]
45. IBGE—Instituto Brasileiro de Geografia e Estatística. Available online: https://www.ibge.gov.br/ (accessed on 30 September 2019).
46. Moura, W. Especiação de Cianeto para Redução do Consumo no Circuito de Lixiviação de Calcinado da Usina do Queiróz. Dissertação de Mestrado, Departamento de Engenharia Metalúrgica c de Minas—UFMG, Belo Horizonte, Brazil, 2005; pp. 1–138. (In English).
47. ANM—Agencia Nacional de Mineração. Available online: https://www.gov.br/anm/pt-br (accessed on 30 September 2019).
48. Ciminelli, V.S.T. Arsenic in mining: Sources and stability. In *One Century of the Discovery of Arsenicosis in Latin America (1914–2014), Proceedings of the 5th International Congress on Arsenic in the Environment, Buenos Aires, Argentina, 2014*; Taylor & Francis Group: London, UK, 2014; pp. 3–7.
49. USEPA—United States Environmental Protection Agency. *Test Methods for Evaluating Solid Waste—Physical/Chemical Methods (SW-846)*; USEPA: Washington, DC, USA, 1996.
50. MEND—The Mine Environment Neutral Drainage Program. *Coastech Research, Acid Rock Drainage Prediction Manual, MEND Project Report*; MEND: Ottawa, ON, Canada, 1991.
51. MEND—The Mine Environment Neutral Drainage Program. *Prediction Manual for Drainage Chemistry from Sulphidic Geologic Materials. MEND Report. 1.20.1*; MEND: Ottawa, ON, Canada, 2009.
52. USEPA—United States Environmental Protection Agency. *Acid Mine Drainage Prediction: Technical Document, US EPA, Office of Solid Waste Special Waste Branch, EPA530-R94-036, NTIS PB94-201829*; USEPA: Washington, DC, USA, 2018.
53. Oliveira, W.S.; Neves, M.F.; Brito, M.M.B.; Mesquita, R.M.M.; Costa, D.S. Uso da flotação para o aproveitamento de um rejeito fino de minério de ouro. In Proceedings of the XXVI Encontro Nacional de Tratamento de Minérios e Metalurgia Extrativa 2015, Poços de Caldas-MG, Brasil, 18–22 October 2015.

54. CONAMA (CONSELHO NACIONAL DE MEIO AMBIENTE). *Resolução nº 396, de 03 de abril de 2008. Dispõe Sobre a Classificação e Diretrizes Ambientais Para o Enquadramento das Águas Subterrâneas e dá Outras Providências*; CONAMA: Brasília, Brazil, 2008.
55. Erguler, G.K.; Erguler, Z.A.; Akcakoca, H.; Ucar, A. The effect of column dimensions and particle size on the results of kinetic column test used for acid mine drainage (AMD) prediction. *Miner. Eng.* **2014**, *55*, 18–29. [CrossRef]
56. Falagán, C.; Grail, B.M.; Johnson, D.B. New approaches for extracting and recovering metals from mine tailings. *Miner. Eng.* **2016**, *106*, 71–78. [CrossRef]
57. Li, H.; Ma, A.; Srinivasakannan, C.; Zhang, L.; Li, S.; Yin, S. Investigation on the recovery of gold and silver from cyanide tailings using chlorination roasting process. *J. Alloy. Comp.* **2018**, *763*, 241–249. [CrossRef]
58. Lu, H.; Qi, C.; Li, C.; Gan, D.; Du, Y.; Li, S. A light barricade for tailings recycling as cemented paste backfill. *J. Clean. Prod.* **2020**, *247*, 119388. [CrossRef]
59. Araya, N.; Kraslawski, A.; Cisternas, L.A. Towards mine tailings valorization: Recovery of critical materials from Chilean mine tailings. *J. Clean. Prod.* **2020**, *263*, 121555. [CrossRef]
60. Alcalde, J.; Kelm, U.; Vergara, D. Historical assessment of metal recovery potential from old mine tailings: A study case for porphyry copper tailings, Chile. *Miner Eng.* **2018**, *127*, 334–338. [CrossRef]
61. Girotto, L.; Espíndola, E.L.G.; Gebara, R.C.; Freitas, J.S. Acute and Chronic Effects on Tadpoles (*Lithobates catesbeianus*) Exposed to Mining Tailings from the Dam Rupture in Mariana, MG (Brazil). *Water Air Soil Pollut.* **2020**, *231*, 1–15. [CrossRef]
62. Aires, U.R.V.; Santos, B.S.M.; Coelho, C.D.; Silva, D.D.; Calijuri, M.L. Changes in land use and land cover as a result of the failure of a mining tailings dam in Mariana, MG, Brazil. *Land Use Policy* **2018**, *70*, 63–70. [CrossRef]
63. Thompson, F.; Oliveira, B.C.; Cordeiro, M.C.; Masi, B.P.; Rangel, T.P.; Paz, P.; Freitas, T.; Lopes, G.; Silva, B.S.; Cabral, A.S.; et al. Severe impacts of the Brumadinho dam failure (Minas Gerais, Brazil) on the water quality of the Paraopeba River. *Sci. Total Environ.* **2020**, *705*, 135914. [CrossRef] [PubMed]
64. Rotta, L.H.S.; Alcântara, E.; Park, E.; Negri, R.G.; Lin, Y.N.; Bernardo, N.; Mendes, T.S.G.; Filho, C.R.S. The 2019 Brumadinho tailings dam collapse: Possible cause and impacts of the worst human and environmental disaster in Brazil. *Int. J. Appl. Earth Obs. Geoinf.* **2020**, *90*, 102119. [CrossRef]

minerals

MDPI

Article

ANC–BNC Titrations and Geochemical Modeling for Characterizing Calcareous and Siliceous Mining Waste

Clémentine Drapeau [1], Cécile Delolme [2], Clément Vézin [3], Denise Blanc [4], Thomas Baumgartl [5], Mansour Edraki [6] and Laurent Lassabatere [3,*]

1 Direction Régionale de L'environnement, de L'aménagement et du Logement (DREAL), Auvergne Rhône-Alpes, Unité Départementale du Rhône, 63 Avenue Roger Salengro, 69100 Villeurbanne, France; clementine.drapeau@developpement-durable.gouv.fr
2 ENTPE, 3 rue Maurice Audin, F-69518 Vaulx-en-Velin, France; cecile.delolme@entpe.fr
3 Université de Lyon, UMR5023 Ecologie des Hydrosystèmes Naturels et Anthropisés, Université Lyon 1, ENTPE, CNRS, 3 rue Maurice Audin, F-69518 Vaulx-en-Velin, France; clement.vezin@gmail.com
4 Université de Lyon, INSA-Lyon, Déchets Eau Environnement Pollutions (DEEP), 7 rue de la Physique, F-69621 Villeurbanne, France; denise.blanc@insa-lyon.fr
5 Geotechnical and Hydrogeological Engineering Research Group (GHERG), Federation University Australia, Churchill, VIC 3841, Australia; t.baumgartl@federation.edu.au
6 Centre for Mined Land Rehabilitation, Sustainable Minerals Institute (SMI), The University of Queensland, St Lucia, QLD 4072, Australia; m.edraki@cmlr.uq.edu.au
* Correspondence: laurent.lassabatere@entpe.fr

Abstract: Pyrite and calcite are mineral phases that play a major role in acid and neutral mine drainage processes. However, the prediction of acid mine drainage (AMD) or contaminated neutral drainage (CND) requires knowledge of the mineral composition of mining waste and the related potential for element release. This paper studies the combination of acid–base neutralizing capacity (ANC–BNC) with geochemical modeling for the characterization of mining waste and prediction of AMD and CND. The proposed approach is validated with three synthetic mineral assemblages: (1) siliceous sand with pyrite only, representing mining waste responsible for AMD, (2) siliceous sand with calcite and pyrite, representing calcareous waste responsible for CND, and (3) siliceous sand with calcite only, simulating calcareous matrices without any pyrite. The geochemical modeling approach using PHREEQC software was used to model pH evolution and main element release as a function of the added amount of acid or base over the entire pH range: 1 < pH < 13. For calcareous matrices (sand with calcite), the results are typical of a carbonated environment, the geochemistry of which is well known. For matrices containing pyrite, the results identify different pH values favoring the dissolution of pyrite: pH = 2 in a pyrite-only environment and pH = 6 where pyrite coexists with calcite. The neutral conditions can be explained by the buffering capacity of calcite, which allows iron oxyhydroxide precipitation. Major element release is then related to the dissolution and precipitation of the mineral assemblages. The geochemical modeling allows the prediction of element speciation in the solid and liquid phases. Our findings clearly prove the potential of combined ANC–BNC experiments along with geochemical modeling for the characterization of mining waste and the assessment of risk of AMD and CND.

Keywords: pyrite; calcite; pH; speciation; geochemical modeling

Citation: Drapeau, C.; Delolme, C.; Vézin, C.; Blanc, D.; Baumgartl, T.; Edraki, M.; Lassabatere, L. ANC–BNC Titrations and Geochemical Modeling for Characterizing Calcareous and Siliceous Mining Waste. *Minerals* **2021**, *11*, 257. https://doi.org/10.3390/min11030257

Academic Editor: Juan Antelo

Received: 24 January 2021
Accepted: 23 February 2021
Published: 28 February 2021

1. Introduction

Mining and mineral processing activities produce large volumes of byproducts known as mining waste. The impacts of tailing ponds and waste rock piles can potentially lead to serious environmental, health, socioeconomic, and geotechnical issues [1–6]. Responsible mining activities demand sustainable management of mineral mine waste. The primary objective is to reduce waste production, followed by recycling and reusing waste as much as possible [7]. Remediation is considered as the last resort for managing the impacts of mining waste. In any case, it is necessary to understand and predict the geochemical processes underlying elemental release (major and trace) from mining waste. Pyrite and calcite are mineral phases that play a major role in two mine drainage processes: acid mine drainage (AMD) [8,9] and contaminated neutral drainage (CND) [10–12]. According to the literature, pyrite dissolution is catalyzed by ferric oxidation acidifying the solution, also described as the AMD process. Contaminated neutral drainage can occur when sulfide oxidation is weak or when neutralization is strong enough [13,14]. In most investigations of closed mines, high concentrations of trace metals and sulfates were mobilized by leaching from tailings [15,16]. Under these conditions, effluents are generated and some toxic metals and metalloids, such as Ni, Zn, Co, As, and Sb, may be solubilized [1,10]. In both drainage situations (AMD or CND), pyrite and other mineral phase dissolutions pose a crucial environmental problem [17]. Pyrite is the most common sulfur-containing mineral phase. It can contain many trace elements (Ag, Au, Te (tellurium), Tl (thallium), Zn, etc.), at times in significant concentrations (e.g., up to 10% for arsenic) [18].

There are many studies that addressed major and trace element release during AMD processes, but most relied on bulk measures of those processes and very few investigated element release over the whole pH range [19–21]. Many papers are based on an experimental characterization of mining waste without including a detailed identification of mineral and pollutant-bearing phases. Some presented the role of the carbonate phase on leachate pH near 6 [1,10]. Some highlighted the impact of the carbonate phase on geochemical mechanisms of pyrite dissolution, particularly the precipitation of iron oxyhydroxides [22–24]. However, speciation of soluble major and trace elements is rarely characterized. Most studies focused on the kinetics and total release of trace elements [1,9,17,25]. It is argued that the carbonate phase is not sufficient to limit the release of elements, even if it buffers the acidity produced by the dissolution of pyrite [1,10]. Thus, there is some uncertainty regarding its classification, management, and reuse in contrast to AMD [8]. It is important to identify this aspect even more precisely, to characterize the mineral assemblages, and to predict element speciation in order to fully understand the potential elements released and involved in geochemical mechanisms during AMD and CND processes.

In addition to the approaches mentioned above, acid–base neutralizing capacity (ANC–BNC) titration tests have proven to be accurate and interesting tools for the characterization of buffer capacity and element release of solid matrices [26]. ANC–BNC involves placing the studied matrix in contact with an acidic or alkaline solution in a closed reactor during a 24–48 h period. Then, pH and element release are monitored and reported as a function of the amount of acidity or alkalinity. Such experiments were used to assess element release from several types of cement and concrete matrices [20,27] into contaminated soils [28], including even natural soils [29]. Several protocols were developed for these titration experiments [30], but the main principles remain the same. Recently, the European Council proposed a standardize protocol that is considered as a benchmark (European standard NF EN 14429) [31].

Coupled with geochemical modeling, ANC–BNC titrations provide a quite efficient tool for characterizing solid matrices in terms of mineral assemblages and element speciation [20,27]. Geochemical modeling has several benefits in the identification of minerals compared with other techniques. It allows the identification of geochemical processes and mechanisms, particularly through access to all geochemical information for each model step iteration (e.g., element speciation, physicochemical parameters, and solid phase quantity). These numerical data supplement the intensive experimental work and avoid labor-intensive and cumbersome experimental procedures such as sequential extraction (SSE). SSE, which is known to be expensive and time-consuming, partitions elements into exchangeable, reducible, oxidizable, residual, carbonate organic matter fractions that are not necessarily representative of the real element speciation in the solid phase [32,33].

To the authors' knowledge, the use of ANC–BNC titrations along with geochemical modeling to characterize mining waste and element release has not been investigated in depth, particularly the capability of such titration experiments to identify the right mineral composition of mining waste and element speciation in the solid and liquid phases. Our study used the ANC–BNC approach (titrations and geochemical modeling) to characterize buffer capacity and major element release from contrasting mining wastes, including waste with high or very low carbonate content and pyrite. We relate buffer capacity and element release to the mineral assemblage composing the studied matrices and the dissolution–precipitation processes. In order to avoid side mechanisms and uncontrolled composition of mineral phases, we used mineral assemblages made of pure phases (standardized sand, pure calcite, and pyrite) instead of real matrices. The use of mineral assemblages appears to be an efficient way to remove uncertainty related to the complexity of real matrices and allows perfect knowledge of tested solid matrices. To carry out ANC–BNC experiments, we used the European standard NF EN 14429 [31]. The geochemical model was implemented in PHREEQC to precisely model the experimental protocol (added amounts and volumes, vessel used, etc.) to predict solid and liquid phase compositions as a function of the amount of acid or alkalinity added. Several compositions in terms of minerals and amounts were tested until a good fit of the observed pH and element release was obtained. The optimum composition of the predicted mineral assemblage was compared to the synthetic matrices, and then the modeled data were discussed with regard to the element speciation in the liquid and solid phases as a function of pH. Our findings produce an adequate model that correctly represents the experimental element release and buffer capacity and provides relevant insight into the interplay of pyrite and calcite in element release.

2. Materials and Methods

2.1. Acid and Base Neutralization Capacity (ANC–BNC)

Acid and base neutralizing capacity tests characterize the physicochemical stability and buffer capacity of a solid phase when it is placed in contact with a source of acidity or alkalinity. ANC–BNC titrations were conducted according to the European standard NF EN 14429 [31], developed by the European Committee for Standardization (*Comité Européen de Normalisation*, CEN). This protocol proved to be efficient for different types of matrices [20,23,26]. The experimental protocol was divided into four steps: (1) pre-study, with simple titration to estimate the amount of acid and base to be introduced to cover the whole range of pH, (2) batch preparation using a solid/liquid ratio of around 1/10, (3) mechanical stirring for 48 h, and (4) measurement and determination of element concentration after filtration at 0.45 μm.

Calcite (calcium carbonate; Sigma-Aldrich, Ultrapure, Merck) and pyrite (iron disulfide; Aldrich, Ultrapure, Merck) solid phases were chosen for the ANC–BNC titrations. Extra-pure sea sand (siliceous sand), SiO_2 (Fisher Chemical), was chosen as the background solid phase. Three types of mineral assemblages were tested as follows: (1) calcite in siliceous sand, (2) pyrite in siliceous sand, and (3) calcite–pyrite assemblage in siliceous sand. The compositions of the three synthetic mixtures were fixed as follows: (1) 3 g of calcite and 7 g of SiO_2, (2) 0.1 g of pyrite and 9.9 g of SiO_2, and (3) 3 g of calcite, 0.1 g of pyrite, and 6.9 g SiO_2. The solids were sieved at 1 mm. The three mineral assemblages were referred to as calcite/SiO_2, pyrite/SiO_2, and calcite/pyrite/SiO_2. Mixtures 2 and 3 represent common types of mining waste responsible for AMD and CND, respectively [1,23]. Mixture 1 was added to complete the experimental plan and determine the role of calcite.

These solid matrices (10 g) were put in contact with 100 mL of solution in vessels with a volume of 250 mL. Acid and base were added by means of HNO_3 and NaOH solutions at 5 mol/L. The volumes of distilled water and HNO_3 or NaOH solution were determined to maintain the same liquid volume, i.e., 100 mL. The quantity of acid or base added was adapted to the shape of the curves, with more points in the vicinity of the buffering zone. To this point, the pre-study phase provided a clear idea of the zones to refine. The closed reactors were agitated with a specific device at 7 rpm. These had to be opened from time to time to allow leakage of CO_2 and reduce the risk of over-pressurization. At the beginning and the end, the closed reactors were opened for leachate measurement at t = 0,

44, and 48 h. Leachate measurements included the following physicochemical parameters: pH (Fisher Scientific), conductivity (sensION+, Hach), and oxidation–reduction potential (Intellical MTC101, Hach). At the end, solutions were filtered through a 0.45 μm filter and split into three subsamples: one to determine total concentrations (calcium, iron, and sulfur), one to measure cations (Ca^{2+}, Fe^{2+}), and one to measure anions ($SO_4{}^{2-}$). The first subsamples were analyzed for total calcium and iron using flame atomic adsorption (AA) spectrometry (PinAAcle 900T, Perkin Elmer) and for total sulfur using inductively coupled plasma–optical emission spectrometry (ICP-OES; ULTIMA 2). The second subsamples were acidified using HNO_3 (5 mol/L) when too alkaline (i.e., pH < 7) before being analyzed by ion chromatography (Dionex ICS 1100). The third subsamples were directly analyzed for $SO_4{}^{2-}$ using ion chromatography (Dionex ICS 900).

To characterize the buffer capacity and element release, the pH variation was plotted as a function of the amount of acid and base added as follows: pH = f(Eq H^+), where Eq H^+ is the amount of H^+ ions added and is a negative value when OH^- is added. To ease the comparison between pH and the redox potential (pe), pH was plotted as pe. Lastly, element concentration was plotted as a function of pH.

2.2. Geochemical Modeling

The geochemical modeling was implemented using PHREEQC software (interactive version alpha 3.1.2.8538, United States Geological Survey). This makes it possible to conceptually represent the experiments. PHREEQC, on the basis of a deterministic approach, proposes a thermodynamic model. All chemical reactions and related equations are resolved by the software's numerical tools [34]. The llnl reference database (Lawrence Livermore National Laboratory) was used, as it enrolls many mineral phases. The geochemical modeling was used to fit the ANC–BNC titrations and, thus, test several hypotheses on the geochemical processes and element release and speciation [35]. Several scenarios were tested regarding the amounts of mineral phases submitted to ANC–BNC titrations and the phases expected to form following the dissolution of the original mineral assemblages.

The modeling methodology was inspired from previous studies on urban stormwater sediments [19–21]. There were three steps: (1) estimate the mineral assemblage (minerals and amounts) initially present that can precipitate or dissolve (literature review study and inverse modeling) [36], (2) directly model the complete ANC–BNC titration experiments (adding all acid/base amounts in the modeled closed reactors), and (3) compare observed and modeled pH and element concentrations in the liquid phase and analyze the predicted mineral assemblage. These three steps were repeated for several initial mineral assemblages until the modeled data got close to the observations. Once the optimum scenario was identified, the predicted element speciation in the liquid and solid phases was plotted as a function of pH and considered as indicative of the real speciation. In this study, only the final optimal model is presented, although it resulted from a large number of runs (~100 models for calcite/SiO_2 and pyrite/SiO_2 alone and 200 models for calcite/pyrite/SiO_2). The thermodynamic constants implemented in the llnl database were never changed; only the initial mineral assemblage and the list of phases allowed to form or dissolve were optimized.

The final model was defined by the solid and gaseous phase and their initial contents (mineral assemblage for the solid phase), the amount of base or acid added to the system, and the list of phases that were not initially in the mixture but were expected to form and potentially dissolve, defining transitional states (from a thermodynamic point of view). For the gaseous phase, as the batches were opened during the experiment to avoid overpressurization and the risk of explosion, the batch reactors were considered as partially opened. The gaseous phase was modeled with a constant volume and very low partial CO_2 pressure to mimic CO_2 leakage during the opening of the batch reactors. The phases that were allowed to precipitate included gypsum, foshagite [37], and iron hydroxides ($Fe(OH)_2$, $Fe(OH)_3$) [38].

3. Results and Discussion

3.1. Physicochemical Parameters

Figure 1a illustrates the difference in the buffer capacity between the three types of matrices. Indeed, the pH curve for the pyrite/SiO_2 matrix shows very low buffering capacity (approximately pH = 2). Buffering capacity is revealed by the stabilization of pH, despite the addition of acid or base. In contrast, pH curves for calcite/SiO_2 and calcite/pyrite/SiO_2 reveal an important buffer zone between 0.05 mol H^+ and 0.55 mol H^+ added, corresponding to pH $\approx$ 6. This can be directly explained by the buffer capacity of calcite (see also the discussion on modeling below).

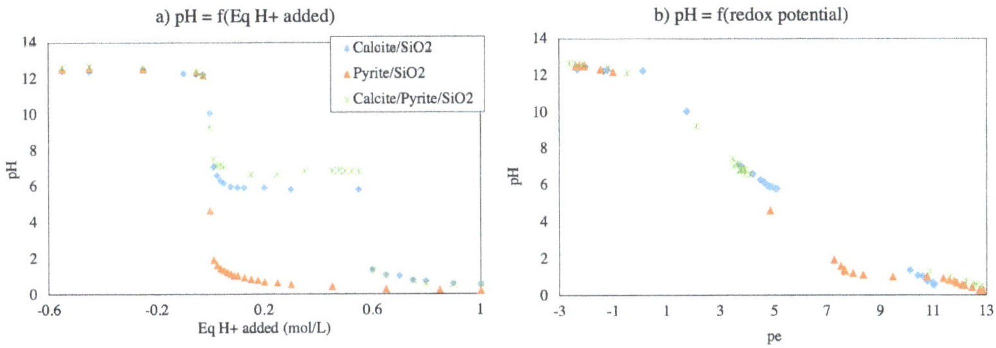

Figure 1. pH (**a**) and redox potential (**b**) of three synthetic matrices: calcite/SiO_2, pyrite/SiO_2, and calcite/pyrite/SiO_2. The vertical line indicates the point of equilibrium with pure water (in Figure 1a).

The variation in redox potential as a function of pH was similar among the three matrices in terms of both extreme values and slopes (Figure 1b). However, the decrease in redox potential with pH was more important for pyrite/SiO_2 (pH $\approx$ 2 and pe $\approx$ 7–9; Figure 1b), which is characteristic of pyrite oxidation. In addition, the decrease in redox potential was less marked in the presence of calcite, showing its effect in reducing the effect of pyrite dissolution on the redox potential. Indeed, according to the literature, the dissolution of pyrite impacts the redox potential, but to a lesser extent in the presence of calcite [39,40].

The pH variations between t = 44 and 48 h were low with regard to the precision of the pH meter for all mixtures (data not shown here): 0.17 pH units for calcite/SiO_2 only, 0.08 pH units for pyrite/SiO_2, and 0.24 pH units for calcite/pyrite/SiO_2. Consequently, we considered that short-term thermodynamic equilibrium was reached, which justified the use of the PHREEQC geochemical model based on chemical equilibrium. However, Table 1 presents the percent of maximum element release, highlighting that pyrite dissolution was not complete. This finding is consistent with the literature; pyrite dissolution kinetics are very slow (suspension equilibrium time is 300 h) [41], which means that only a fraction of pyrite dissolved over the course of the experiment. Such a result indicates the difference between short- and long-term chemical stabilization (or equilibrium). To account for this, the initial amounts of pyrite allowed to react in the batch reactors were reduced in the geochemical model. Limiting the initial amount was the chosen way to introduce the kinetic limitation of pyrite dissolution and distinguish between short- and long-term equilibrium.

Table 1. Contents of release elements in solution as a function of mixture and pH area (mol/mol).

	Calcium Release		**Iron Release**		**Sulfur Release**	
Matrix	Calcite/SiO_2	Calcite/pyrite/SiO_2	Pyrite/SiO	Calcite/pyrite/SiO_2	Pyrite/SiO	Calcite/pyrite/SiO_2
Acid area	97%	99%	45%	25%	52%	37%
Basic area	<1%	<1%	<1%	<1%	6%	11%

3.2. Modeling of Mineral Phase Dissolution

First, the proposed inversion (based on iterating steps 1–3 of the geochemical modeling approach, as described in Section 2) proved quite efficient to retrieve the initial composition of the synthetic matrices. The method pointed out the right minerals quite quickly, and the amounts of involved minerals were close to the targeted amounts. We should note that we had to reduce the quantity of mineral assemblages in contact with water, probably in relation to the kinetic limitations, as explained above. Figure 2 shows the experimental data along with the modeled data of pH and redox potential for the three synthetic matrices. The ANC/BNC titrations were well represented with the initial conditions and chemical reactions selected in the final geochemical model.

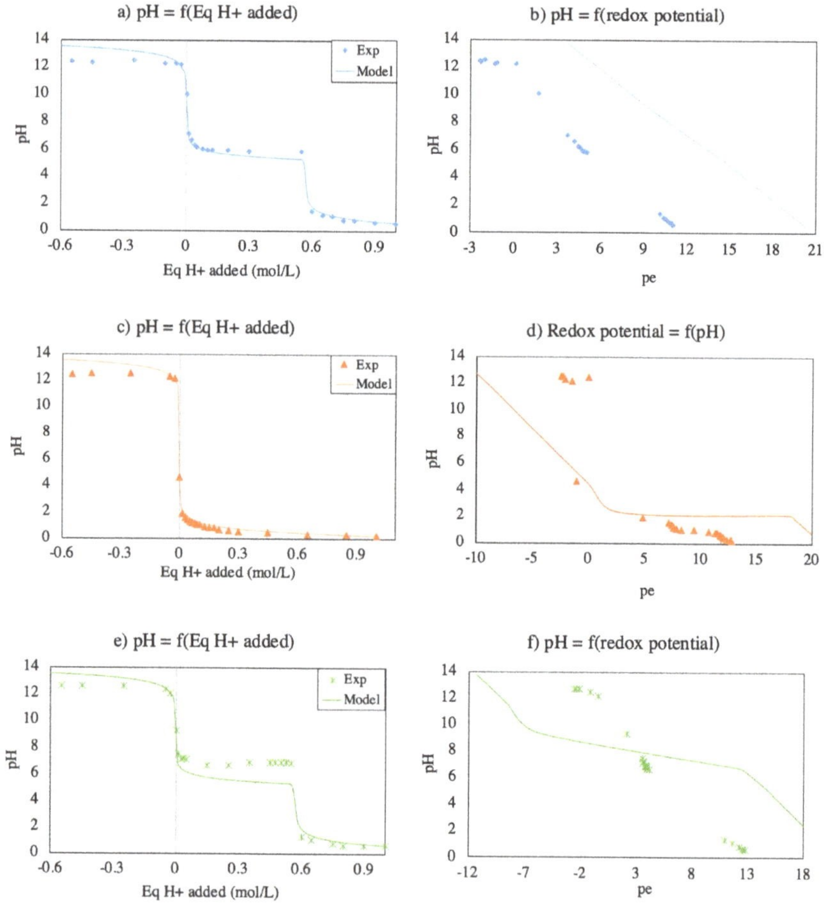

Figure 2. pH as a function of H^+ equivalents added for (**a**) calcite/SiO_2, (**c**) pyrite/SiO_2, and (**e**) calcite/pyrite/SiO_2 assemblages, and relationship between redox potential (pe) and pH for (**b**) calcite/SiO_2, (**d**) pyrite/SiO_2, and (**f**) calcite/pyrite/SiO_2. The vertical line in (**b**) indicates the point of equilibrium with pure water.

The pH curve of calcite only has three distinct areas (Figure 2a): (i) basic, (ii) buffer, and (iii) acidic. In the basic area, the curve reaches an asymptote at approximately pH = 12, which indicates the presence of quartz and shows the contribution of the SiO_2 in the calcite/SiO_2 assemblage. The total duration of the experiment (48 h) was sufficient to initiate quartz dissolution [42]. The same trend was observed for the three synthetic matrices. The pH buffer zone was situated at approximately pH = 6 (Figure 2a), which is the buffer capacity of calcite for open systems [43,44]. The dissolution of calcite consumes protons. This pH range is important because it represents the range of natural waters [45]. In the acidic area, the calcite was completely dissolved. After 0.55 mol/L of acid was added, the pH curve decreased sharply until it reached the asymptote at approximately pH = 1.

The redox potential of the calcite/SiO_2 matrix decreased linearly when the pH increased (Figure 2b). According to the literature and theory, these parameters are interdependent: an increase in pH is accompanied by a decrease in redox potential (pe). This linearity is related to the low redox reactivity of calcite [46]. The modeling was satisfactory with an accurate simulation of trends, despite the observed difference between observations and simulation (Figure 2b, line versus points). The offset between the model and observed data is expected to be related to the pyrite dissolution process (see below). In addition, we guess that the perturbation of the closed reactors (open during titrations to avoid over-pressurization) may have added an additional source of errors.

For pyrite/SiO_2, the pH curve shows no buffer capacity (Figure 2c). Indeed, pyrite strongly acidified the solution; when [Eq H^+] = 0, the pH value was 4.65. According to the literature, pyrite dissolution in an oxidizing environment occurs in two steps: (1) oxidation of sulfides (Equation (1)), and (2) oxidation of ferrous ion to ferric iron (Equation (2)), which is kinetically limiting [45,47]. These reactions are self-catalyzed by the direct oxidation of pyrite by ferric iron (Equation (3)) for pH < 3.5. This is the predominant reaction in the AMD process. It is this reaction that strongly acidifies the solution through the large production of protons [48].

$$FeS_2 + 7/2\,O_2 + H_2O \rightarrow Fe^{2+} + 2\,SO_4{}^{2-} + 2\,H^+. \tag{1}$$

$$Fe^{2+} + 1/4\,O_2 + H^+ \rightarrow Fe^{3+} + 1/2\,H_2O. \tag{2}$$

$$FeS_2 + 14\,Fe^{3+} + 8\,H_2O \rightarrow 15\,Fe^{2+} + 2\,SO_4{}^{2-} + 16\,H^+. \tag{3}$$

The difference in value between modeled and measured pe demonstrates the oxidative impact of pyrite dissolution, mainly through Equation (1), which was represented by the geochemical model [46].

The curve for calcite/pyrite/SiO_2 shows a pH evolution similar to that for calcite/SiO_2 (Figure 2a,c). The pH level, stable at approximately pH = 6, can be associated with the buffer capacity of calcite. The same content of acid (0.55 mol/L) was required to complete the calcite dissolution and decrease the pH. The potential redox curve is similar to that of pyrite/SiO_2 in terms of shape (Figure 2d,f). However, there is a shift of the step change from pH $\approx$ 2 for pyrite/SiO_2 only to pH $\approx$ 8 for calcite/pyrite/SiO_2 (Figure 2f). Such a shift supports the hypothesis on the impact of calcite on pyrite dissolution and the related effects on pH and pe. The model displays this impact in more detail (Figure 3). Indeed, the pyrite phase is expected to dissolve completely at pH = 7 in the calcite/pyrite/SiO_2 mineral assemblage, whereas the pyrite phase is expected to dissolve completely only at pH = 2 in the pyrite/SiO_2 assemblage. Furthermore, even in contact with pure water, pyrite dissolves less when it is the sole mineral than when it is combined with calcite (Figure 3). The parameters related to geochemical modeling are listed in Table 2 and were optimized by fitting experimental data, as explained in Section 2.

3.3. Solubility and Speciation of Elements

Below, we investigate the elemental release of calcium (provided by the dissolution of calcite), as well as iron and sulfur (provided by the dissolution of pyrite). The characterization and modeling of their release help in identifying the phases that dissolve and potentially form, as well as the geochemical processes involved in the evolution of the calcite/SiO_2, pyrite/SiO_2, and calcite/pyrite/SiO_2 assemblages.

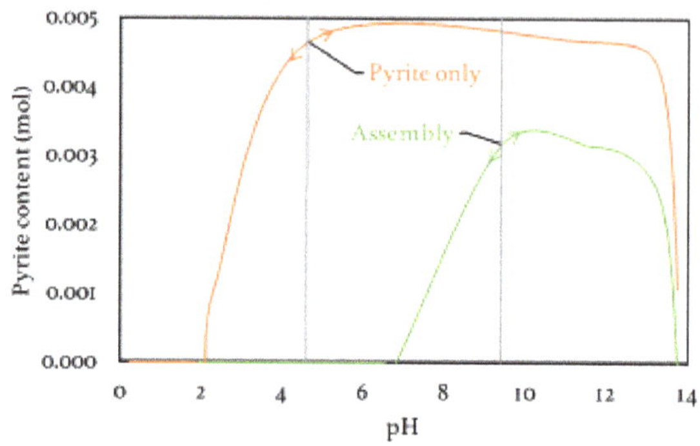

Figure 3. Impact of calcite on pyrite dissolution; arrows indicate the direction of addition or depletion of protons; the vertical line indicates the point of equilibrium with pure water (natural pH).

Table 2. Mineral and gas contents of three synthetic matrices: (1) calcite/SiO_2, (2) pyrite/SiO_2, and (3) calcite/pyrite/SiO_2; other tabulated minerals are expected to form and take part in transitional states.

Phases	Dissolution Equation	Solubility Constant logK	Initial Content (mol/L)/ Partial Pressure (atm)		
			(1)	(2)	(3)
	Mineral Phases				
Calcite (DO)	$CaCO_3 + H^+ = Ca^{2+} + HCO_3^-$	1.84	0.285	-	0.285
Foshagite	$Ca_4Si_3O_9(OH)_2{:}0.5\ H_2O + H^+ = 3\ SiO_2 + 4\ Ca^{2+} + 5.5\ H_2O$	65.9	0	-	0
Pyrite (DO)	$FeS_2 + H_2O = 0.25\ H^+ + 0.25\ SO_4^{2-} + Fe^{2+} + 1.75\ HS^-$	−24.7	-	0.0047	0.0032
Quartz (DO)	$SiO_2 = SiO_2$	−4.00	1.165	0.0165	0.0115
Gypsum	$CaSO_4{:}2H_2O = Ca^{2+} + SO_4^{2-} + 2\ H_2O$	−4.48	-	-	0
$Fe(OH)_2$	$Fe(OH)_2 + 2\ H^+ = Fe^{2+} + 2\ H_2O$	13.9	-	0	0
$Fe(OH)_3$	$Fe(OH)_3 + 3\ H^+ = Fe^{3+} + 3\ H_2O$	5.66	-	0	0
	Gas Phases				
Ar(g)	Ar = Ar	−2.86		0.0934	
CO_2(g)	$CO_2 + H_2O = H^+ + HCO_3^-$	−7.81		0.00037	
N_2(g)	$N_2 = N_2$	−3.19		0.781	
O_2(g)	$NO_2 + 0.5\ H_2O + 0.25\ O_2 = H^+ + NO_3^-$	8.37		0.00207	

DO, Dissolve only: do not allow mineral to precipitate in model.

3.3.1. Calcium

The curve of soluble calcium (Figure 4a) indicates maximum content at an acidic pH. Approximately 0.29 mol/L of calcium was released at pH < 6, representing 97% of the initial calcium introduced with the calcite (Table 1). This finding confirms the hypothesis that the buffer capacity of the solid phase of calcite results from calcite dissolution. The fast kinetics of calcite dissolution explains the significant content of calcium release [37,49]. Figure 4b shows the modeled calcium speciation as a function of pH. Soluble calcium was mainly in the form of Ca^{2+} (approximately 60%) and $CaNO3^+$ (approximately 40%), due to the addition of HNO3 for acidification of the solution. In the basic area, calcium was scarcely released, to an extent of less than 1% (Table 1).

There are few differences between calcite/SiO_2 and calcite/pyrite/SiO_2 regarding calcium release and speciation (Figure 4c,d). At an acidic pH, 99% of the calcium introduced initially was solubilized (see Table 2 for the assemblage). However, the gypsum phase ($CaSO_4$) precipitated at 4 < pH < 8 without any significant effect on calcium release, given the small amount of precipitated $CaSO_4$.

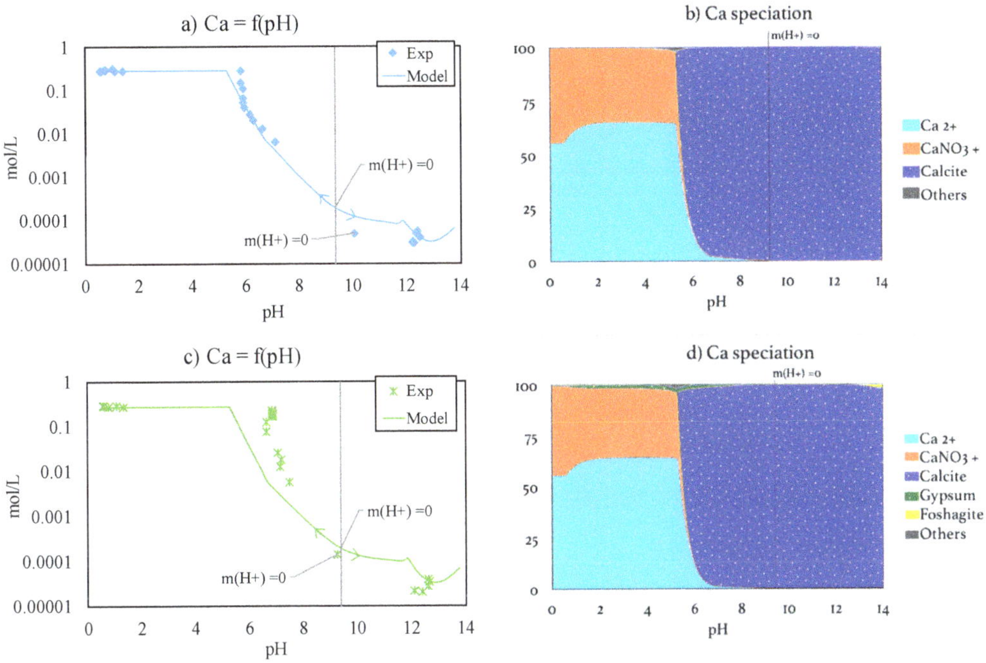

Figure 4. Total experimental soluble content and modeled speciation of calcium as a function of pH: (**a**,**b**) calcite/SiO_2 and (**c**,**d**) calcite/pyrite/SiO_2. Arrows indicate the direction of addition or depletion of protons; the vertical line indicates the point of equilibrium with pure water (natural pH).

3.3.2. Iron

Figure 5 shows the soluble content and modeled speciation curves of iron with pH for pyrite/SiO_2 and calcite/pyrite/SiO_2. Between the two, the variations in solubility with pH have similar trends. However, the modeled speciation is quite different. For pyrite/SiO_2, iron speciation represents the reaction sequence described previously (Equations (1)–(3)). Indeed, iron in solution was mainly in the form of ferrous iron. This mechanism demonstrates the importance of O_2 (g) and the catalytic oxidant impact of ferric iron [9,50]. Once the pyrite phase completely dissolved, iron in the solution was mainly in the form of ferric iron. On the other hand, speciation of iron for the calcite/pyrite/SiO_2 assemblage was different. Indeed, ferrous iron no longer appeared. The pyrite phase completely dissolved at pH < 7 with the precipitation of iron hydroxide mineral (Fe(OH)3). According to the literature, when pH > 3.5, oxidation of ferric and ferrous iron is important and iron hydroxide will precipitate, as follows [9,41,47]:

$$Fe^{3+} + 3\,H_2O \leftrightharpoons Fe(OH)_3 + 3\,H^+. \quad (4)$$

The model confirmed the impact of the carbonate phase on sulfur dissolution. Thus, when a carbonate buffers the solution up to pH 3.5, the ferric iron does not act as an oxidant. Meanwhile, the precipitation equation of $Fe(OH)_3$ (Equation (4)) acidified the solution, but the calcite dissolution compensated for it, leading to stabilization of pH over 3.5 [39]. This also explains the impact on redox potential; the step change shifted and was softer because the ferric iron no longer played its oxidation role.

Iron was released very little under basic conditions, less than 1% in both studies (Table 1). Conversely, in the acidic area, iron was released in solution at a rate of 45% for pyrite/SiO_2 and 25% for calcite/pyrite/SiO_2. These results support the chemical mechanisms previously outlined. Indeed, in a carbonaceous environment, $Fe(OH)_3$ precipitation (Equation (4)) decreased the iron content in solution.

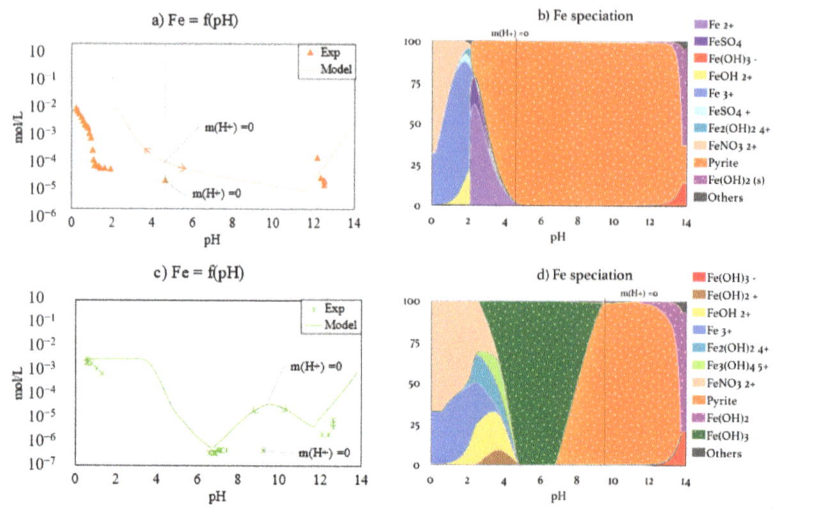

Figure 5. Total experimental soluble content and modeled speciation of iron as a function of pH: (**a**,**b**) pyrite/SiO_2 and (**c**,**d**) calcite/pyrite/SiO_2. Arrows indicate the direction of addition or depletion of protons; the vertical line indicates the point of equilibrium with pure water (natural pH).

3.3.3. Sulfur

The solubility and speciation curves of sulfur are different between pyrite/SiO_2 and calcite/pyrite/SiO_2 (Figure 6). In both cases, sulfur was mainly in the form of SO_4^{2-}, then HSO_4^{3-} when pH < 3. According to the literature, this results from the addition of protons and the acid–base properties of sulfur [41]. Moreover, the model supports the results from experiments with varying anion concentrations (SO_4^{2-}) [41]. For the assemblage, the precipitation of gypsum considerably reduced the release of sulfur in the acidic area, despite the complete dissolution of pyrite. Similar to iron, these results highlight pyrite dissolution at alkaline pH values more than calcite dissolution, which buffers the solution.

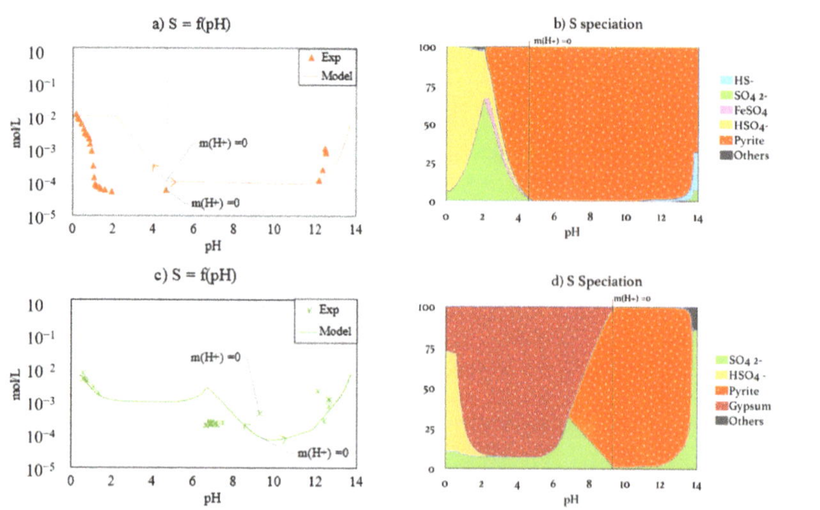

Figure 6. Total experimental soluble content and modeled speciation of sulfur as a function of pH: (**a**,**b**) pyrite/SiO_2 and (**c**,**d**) calcite/pyrite/SiO_2. Arrows indicate the direction of addition or depletion of protons; the vertical line indicates the point of equilibrium with pure water (natural pH).

In the basic area, the leaching of sulfur was significant (Table 1) with 6% for pyrite/SiO_2 versus 11% for calcite/pyrite/SiO_2. No sulfur-containing mineral precipitated under these pH conditions (Figure 6b,d). In the acidic area, sulfur was also released as follows: 52% for pyrite/SiO_2 and 37% for calcite/pyrite/SiO_2. Again, these results support the mechanisms previously presented and agree with the literature [41].

4. Discussion

Our results clearly show that carbonate phases buffer the solution and modify pyrite dissolution, without preventing it. The mechanism of neutralization of sulfuric acid by calcite may be modeled by considering the following equation, as suggested by Lapakko et al. [10]:

$$2\,CaCO_3 + H_2SO_4 \rightarrow 2\,Ca^{2+} + 2\,HCO_3^{-} + SO_4^{2-}. \tag{5}$$

This mechanism was particularly well represented by the model and the modeled results of sulfur speciation (Figure 6d). Indeed, unlike the pyrite/SiO_2 assemblage, in the calcite/pyrite/SiO_2 assemblage, HSO_4^- did not appear in solution before pH < 2, at the beginning of gypsum dissolution. At pH > 2, HSO_4^- was transformed by the buffering capacity of the carbonate phase into SO_4^{2-}. Modeling and speciation provided an important understanding of the mechanisms involved in the geochemical evolution of the assemblage of pyrite and calcite.

Therefore, the proposed geochemical model in this paper satisfactorily reports the complexity and interdependence between the dissolution of mineral phases during ANC–BNC titration. Figure 7 presents these complex and essential aspects of pyrite dissolution dependent on acid–base conditions and, thus, the presence of calcite. Whatever the acid–base conditions, oxygen allows pyrite dissolution to begin. This results in the production of ferrous iron and acid (Equation (1)). The ferrous iron is then oxidized to ferric iron through oxygen and a proton (Equation (2)). Depending on the environmental conditions, the reactions that follow are different. In very acidic conditions (pH < 3.5), ferric iron and water allow for pyrite dissolution. Ferrous iron and many protons are produced (Equation (3)). It is this quantity of protons that strongly acidifies the leachate and causes acid mine drainage. Pyrite dissolution can be very important, because, with the initial oxidation, ferric iron is auto-catalyzed (Equation (3)). In more basic conditions (pH > 3.5), iron hydroxide precipitation ($Fe(OH)_3$) dissolves the pyrite phase, without Fe^{2+}/Fe^{3+} saturation. However, because Equations (1) and (3) produce protons and SO_4^{2-}, a mineral phase with high buffer capacity is required to maintain pH > 3.5, causing the contaminated neutral drainage. The carbonate dissolution provides CO_3^{2-}, which can stabilize the pH above 3.5.

The conclusions of this study are based on experimental data that were obtained for specific proportions of calcite and pyrite. However, depending on the mining waste, the contents may vary. The geochemical model employed to fit our experimental data was used to predict pyrite dissolution for several scenarios. Figure 8 shows pyrite dissolution as a function of pH for different values of the ratio of calcite and pyrite contents (mol/mol): 1/100, 1/10, 1, 10, 100, and 1000. Note that our experimental results were obtained for a ratio of 88 (mol/mol). Very low calcite content had little influence on pyrite dissolution and pH in pure water (see crosses in Figure 8 corresponding to $m(H^+) = 0$). As soon as 1/10 calcite was introduced, the pH in pure water reached a value of 9.3. The calcite phase took the lead. When the calcite content increased, the dissolution of pyrite was complete at increasingly alkaline pH: pH = 2 for 1/100 of the nominal calcite content versus pH = 3 for the nominal content (calcite/pyrite = 1). For ratios greater than 10, similar impacts of calcite on pyrite dissolution were observed, with complete dissolution at pH $\approx$ 6.5. Sufficient calcite content is required to buffer the solution, as pyrite dissolution produces considerable acidity (Figure 7).

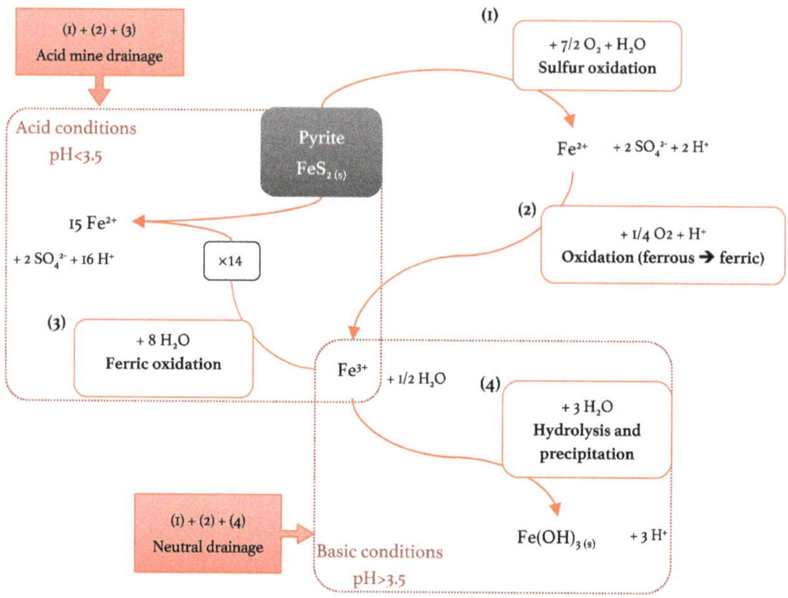

Figure 7. Processes of pyrite dissolution as a function of pH [1,9,22,39,40,51,52].

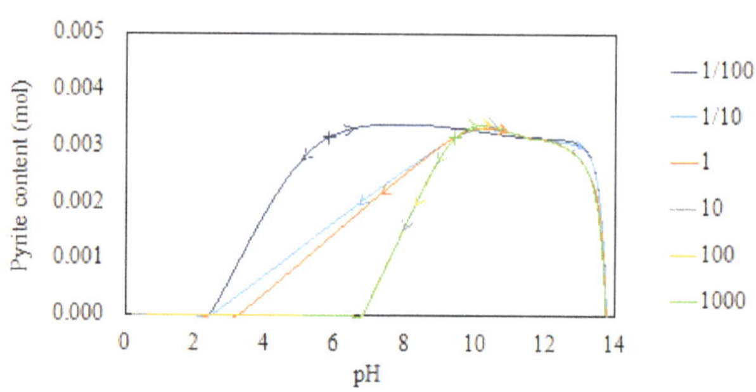

Figure 8. Models of pyrite dissolution as a function of calcite content (calcite/pyrite ratio in mol/mol) in calcite/pyrite/SiO_2 assemblage. Arrows indicate the direction of addition or depletion of protons; crosses indicate the point of equilibrium with pure water (natural pH).

5. Conclusions

This study investigated the potential of ANC–BNC titration combined with geochemical modeling for the characterization of mining waste in terms of buffer capacity and element release and identification of the main parent mechanisms. We used the European standard NF EN 14429 [31] to perform ANC–BNC titrations on three synthetic mineral assemblages representing common mining wastes (for the two last mineral assemblages): calcite/SiO_2, pyrite/SiO_2, and calcite/pyrite/SiO_2. ANC–BNC titrations were combined with geochemical modeling to interpret pH, redox potential, and element release evolution as a function of the added amounts of acid or base. Geochemical modeling was used to relate the main buffer zones and releases to mineral precipitation or dissolution.

The application of the proposed approach (ANC–BNC combined with geochemical modeling) proved efficient to (i) characterize the three synthetic matrices by approaching their real composition and (ii) identify the main processes involved in element release and the production of AMD and CND. The acid–base neutralization capacity test introduced differences among the three matrices (calcite/SiO_2, pyrite/SiO_2, and calcite/pyrite/SiO_2), which were supported and developed by the proposed geochemical model. The modeling approach very satisfactorily simulated our experimental results (pH and element release data) and provided results in line with previous studies, whereas the modeling of pe trends seemed trickier. The geochemical model explained the differences in the dissolution of pyrite depending on the presence of calcite [1,38]. Separately, pyrite and calcite had an equilibrium pH in pure water of ~4.5 and ~9.0, respectively. The addition of base did not affect the pyrite and calcite phases. The addition of acid to solution resulted in phase dissolution. However, the synthetic matrices behaved differently with acid; the addition of acid resulted in the catalysis of pyrite dissolution by the neutralizing capacity of the calcite phase. The calcite dissolution provided CO_3^- ions, which allowed for stabilization of the pH value in the basic area. At these high pH values, Fe^{3+} precipitated as iron hydroxide, which permitted pyrite dissolution without acidification of the solution. These chemical mechanisms, involved in both AMD and CND, emphasize the crucial role of the carbonate phase in pyrite and sulfide dissolution, especially the speciation of released elements. Despite the carbonate buffer capacity, the dissolution of pyrite and other potential trace element-bearing phases is not prohibited [22]. Thus, even if the leachate is not acidic, pyrite may dissolve, thereby releasing sorbed heavy metals. Consequently, the addition of calcite may not be sufficient to entirely mitigate the environmental hazard [17]. This analytical approach of associating experimentation and modeling allowed for a quite accurate understanding of the various geochemical mechanisms. The approach assumes some prospects and adaptations for other matrices.

This study investigated the geochemical processes of pure phases. However, real matrices may behave differently. First of all, the phases may be much more complex from a thermodynamic point of view, with additional complex mechanisms (precipitation and dissolution kinetics, passivation of calcite, coating of mineral surfaces due to precipitation, etc.) [53]. In addition, biological processes were not considered, and these are known to impact AMD, with Fe-oxidizing bacteria involved in the generation of ferric iron and the acceleration of pyrite dissolution [54]. Additional tests should be conducted with different types of strains to investigate the influence of bacteria on AMD and CND processes. The question of toxicity and the fate of bacterial communities in these peculiar environments should be addressed. The proposed study offers a general frame for understanding the main geochemical processes responsible for AMD and highlights the need to account for the specifics of each real matrix, its compositions (pyrite–calcite ratio), and its related environment. Ongoing research will focus on applying the proposed method to complex matrices such as real mining waste or urban sediments.

Author Contributions: Conceptualization, C.D. (Clémentine Drapeau), L.L., and C.D. (Cécile Delolme); methodology, C.D. (Clémentine Drapeau), L.L., and C.D. (Cécile Delolme); software, C.D. (Clémentine Drapeau) and C.V. with the help of L.L., C.D. (Cécile Delolme), D.B., M.E., and T.B.; validation, C.D. (Clémentine Drapeau), L.L., and C.D. (Cécile Delolme); resources, C.D. (Cécile Delolme) and L.L.; data curation, C.V. and C.D. (Clémentine Drapeau); writing—original draft preparation, C.D. (Clémentine Drapeau) with the help of L.L. and C.D. (Cécile Delolme); writing—review and editing, all authors. All authors read and agreed to the published version of the manuscript.

Funding: This research received no external funding.

Data Availability Statement: Not applicable.

Acknowledgments: The authors would like to thank Nathalie Dumont and David Lebouil (Laboratoire Déchets Eau Environnement Pollutions, DEEP) and Myriam Hammada (Laboratories d'Ecologie des Hydrosystèmes Naturels et Anthropisés, LEHNA) for technical assistance and analytical analyses. The authors are grateful to the field observatory for urban water management (OTHU) and CNRS EC2CO for their logistical and financial support.

Conflicts of Interest: The authors declare no conflict of interest.

References

1. Argane, R.; El Adnani, M.; Benzaazoua, M.; Bouzahzah, H.; Khalil, A.; Hakkou, R.; Taha, Y. Geochemical Behavior and Environmental Risks Related to the Use of Abandoned Base-Metal Tailings as Construction Material in the Upper-Moulouya District, Morocco. *Environ. Sci. Pollut. Res.* **2016**, *23*, 598–611. [CrossRef] [PubMed]
2. Hakkou, R.; Benzaazoua, M.; Bussière, B. Acid Mine Drainage at the Abandoned Kettara Mine (Morocco): 2. Mine Waste Geochemical Behavior. *Mine Water Environ.* **2008**, *27*, 160–170. [CrossRef]
3. Wolkersdorfer, C.; Bowell, R. Contemporary Reviews of Mine Water Studies in Europe. *Mine Water Environ.* **2004**, *23*, 161. [CrossRef]
4. Azam, S.; Li, Q. Tailings Dam Failures: A Review of the Last One Hundred Years. *Geotech. News* **2010**, *28*, 50–54.
5. Khalil, A.; Hanich, L.; Bannari, A.; Zouhri, L.; Pourret, O.; Hakkou, R. Assessment of Soil Contamination around an Abandoned Mine in a Semi-Arid Environment Using Geochemistry and Geostatistics: Pre-Work of Geochemical Process Modeling with Numerical Models. *J. Geochem. Explor.* **2013**, *125*, 117–129. [CrossRef]
6. Othmani, M.A.; Souissi, F.; Bouzahzah, H.; Bussière, B.; Da Silva, E.F.; Benzaazoua, M. The Flotation Tailings of the Former Pb-Zn Mine of Touiref (NW Tunisia): Mineralogy, Mine Drainage Prediction, Base-Metal Speciation Assessment and Geochemical Modeling. *Environ. Sci. Pollut. Res.* **2015**, *22*, 2877–2890. [CrossRef]
7. Yilmaz, E. Advances in Reducing Large Volumes of Environmentally Harmful Mine Waste Rocks and Tailings. *Gospod. Surowcami Miner.* **2011**, *27*, 89–112.
8. Morin, K.A.; Hutt, N.M. *Environmental Geochemistry of Minesite Drainage: Practical Theory and Case Studies, Digital Edition*; MDAG Publishing: Surrey, BC, Canada, 2001; ISBN 0-9682039-1-4.
9. Singer, P.C.; Stumm, W. Acidic Mine Drainage: The Rate-Determining Step. *Science* **1970**, *167*, 1121–1123. [CrossRef] [PubMed]
10. Lapakko, K.A.; Antonson, D.A.; Wagner, J.R. Mixing of Limestone with Finely-Crushed Acid Producing Rock. In Proceedings of the Fourth International Conference on Acid Rock Drainage, Vancouver, BC, Canada, 31 May–6 June 1997; pp. 1345–1360.
11. Plante, B.; Benzaazoua, M.; Bussière, B. Kinetic Testing and Sorption Studies by Modified Weathering Cells to Characterize the Potential to Generate Contaminated Neutral Drainage. *Mine Water Environ.* **2011**, *30*, 22–37. [CrossRef]
12. Bussiere, B. Colloquium 2004: Hydrogeotechnical Properties of Hard Rock Tailings from Metal Mines and Emerging Geoenvironmental Disposal Approaches. *Can. Geotech. J.* **2007**, *44*, 1019–1052. [CrossRef]
13. Heikkinen, P.M.; Räisänen, M.L.; Johnson, R.H. Geochemical Characterisation of Seepage and Drainage Water Quality from Two Sulphide Mine Tailings Impoundments: Acid Mine Drainage versus Neutral Mine Drainage. *Mine Water Environ.* **2009**, *28*, 30–49. [CrossRef]
14. Benzaazoua, M.; Bussière, B.; Dagenais, A.-M.; Archambault, M. Kinetic Tests Comparison and Interpretation for Prediction of the Joutel Tailings Acid Generation Potential. *Environ. Geol.* **2004**, *46*, 1086–1101. [CrossRef]
15. Shu, W.S.; Ye, Z.H.; Lan, C.Y.; Zhang, Z.Q.; Wong, M.H. Acidification of Lead/Zinc Mine Tailings and Its Effect on Heavy Metal Mobility. *Environ. Int.* **2001**, *26*, 389–394. [CrossRef]
16. Benvenuti, M.; Mascaro, I.; Corsini, F.; Lattanzi, P.; Parrini, P.; Tanelli, G. Mine Waste Dumps and Heavy Metal Pollution in Abandoned Mining District of Boccheggiano (Southern Tuscany, Italy). *Environ. Geol.* **1997**, *30*, 238–243. [CrossRef]
17. Dold, B. Evolution of Acid Mine Drainage Formation in Sulphidic Mine Tailings. *Minerals* **2014**, *4*, 621–641. [CrossRef]
18. Abraitis, P.K.; Pattrick, R.A.D.; Vaughan, D.J. Variations in the Compositional, Textural and Electrical Properties of Natural Pyrite: A Review. *Int. J. Miner. Process.* **2004**, *74*, 41–59. [CrossRef]
19. Engelsen, C.J.; van der Sloot, H.A.; Wibetoe, G.; Petkovic, G.; Stoltenberg-Hansson, E.; Lund, W. Release of Major Elements from Recycled Concrete Aggregates and Geochemical Modelling. *Cem. Concr. Res.* **2009**, *39*, 446–459. [CrossRef]
20. Peyronnard, O.; Blanc, D.; Benzaazoua, M.; Moszkowicz, P. Study of Mineralogy and Leaching Behavior of Stabilized/Solidified Sludge Using Differential Acid Neutralization Analysis: Part II: Use of Numerical Simulation as an Aid Tool for Cementitious Hydrates Identification. *Cem. Concr. Res.* **2009**, *39*, 501–509. [CrossRef]
21. Halim, C.E.; Short, S.A.; Scott, J.A.; Amal, R.; Low, G. Modelling the Leaching of Pb, Cd, As, and Cr from Cementitious Waste Using PHREEQC. *J. Hazard. Mater.* **2005**, *125*, 45–61. [CrossRef]
22. Pérez-López, R.; Castillo, J.; Quispe, D.; Nieto, J.M. Neutralization of Acid Mine Drainage Using the Final Product from CO_2 Emissions Capture with Alkaline Paper Mill Waste. *J. Hazard. Mater.* **2010**, *177*, 762–772. [CrossRef]
23. Coussy, S.; Benzaazoua, M.; Blanc, D.; Moszkowicz, P.; Bussière, B. Arsenic Stability in Arsenopyrite-Rich Cemented Paste Backfills: A Leaching Test-Based Assessment. *J. Hazard. Mater.* **2011**, *185*, 1467–1476. [CrossRef] [PubMed]
24. Tiruta-Barna, L. Using PHREEQC for Modelling and Simulation of Dynamic Leaching Tests and Scenarios. *J. Hazard. Mater.* **2008**, *157*, 525–533. [CrossRef]
25. Simate, G.S.; Ndlovu, S. Acid Mine Drainage: Challenges and Opportunities. *J. Environ. Chem. Eng.* **2014**, *2*, 1785–1803. [CrossRef]
26. Glass, G.K.; Buenfeld, N.R. Differential Acid Neutralisation Analysis. *Cem. Concr. Res.* **1999**, *29*, 1681–1684. [CrossRef]
27. Peyronnard, O.; Benzaazoua, M.; Blanc, D.; Moszkowicz, P. Study of Mineralogy and Leaching Behavior of Stabilized/Solidified Sludge Using Differential Acid Neutralization Analysis. Part I: Experimental Study. *Cem. Concr. Res.* **2009**, *39*, 600–609. [CrossRef]
28. Kania, M.; Gautier, M.; Blanc, D.; Lupsea-Toader, M.; Merlot, L.; Quaresima, M.-C.; Gourdon, R. Leaching Behavior of Major and Trace Elements from Sludge Deposits of a French Vertical Flow Constructed Wetland. *Sci. Total Environ.* **2019**, *649*, 544–553. [CrossRef] [PubMed]

29. Brahy, V.; Deckers, J.; Delvaux, B. Estimation of Soil Weathering Stage and Acid Neutralizing Capacity in a Toposequence Luvisol–Cambisol on Loess under Deciduous Forest in Belgium. *Eur. J. Soil Sci.* **2000**, *51*, 1–13. [CrossRef]
30. Wahlström, M.; Laine-Ylijoki, J.; Kaartinen, T. *Acid Neutralization Capacity of Waste-Specification of Requirement Stated in Landfill Regulations*; Nordic Council of Ministers: Copenhagen, Denmark, 2009.
31. AFNOR. *CEN/TS 14429 Characterization of Waste. Leaching Behavior Tests. Influence of PH on Leaching with Initial Acid/Base Addition*; AFNOR: Paris, France, 2015.
32. Jamali, M.K.; Kazi, T.G.; Arain, M.B.; Afridi, H.I.; Jalbani, N.; Kandhro, G.A.; Shah, A.Q.; Baig, J.A. Speciation of Heavy Metals in Untreated Sewage Sludge by Using Microwave Assisted Sequential Extraction Procedure. *J. Hazard. Mater.* **2009**, *163*, 1157–1164. [CrossRef]
33. Zuo, X.; Fu, D.; Li, H. Speciation Distribution and Mass Balance of Copper and Zinc in Urban Rain, Sediments, and Road Runoff. *Environ. Sci. Pollut. Res.* **2012**, *19*, 4042–4048. [CrossRef]
34. Parkhurst, D.L.; Appelo, C. *User's Guide to PHREEQC (Version 2): A Computer Program for Speciation, Batch-Reaction, One-Dimensional Transport, and Inverse Geochemical Calculations*; 1999.
35. Chatain, V.; Benzaazoua, M.; Cazalet, M.L.; Bouzahzah, H.; Delolme, C.; Gautier, M.; Blanc, D.; De Brauer, C. Mineralogical Study and Leaching Behavior of a Stabilized Harbor Sediment with Hydraulic Binder. *Environ. Sci. Pollut. Res.* **2013**, *20*, 51–59. [CrossRef] [PubMed]
36. Van der Sloot, H.A.; Van Zomeren, A.; Dijkstra, J.J.; Meeussen, J.C.L.; Comans, R.N.J.; Scharff, H. Prediction of the Leaching Behaviour of Waste Mixtures by Chemical Speciation Modelling Based on a Limited Set of Key Parameters. In *Proceedings of the Tenth International Waste Management and Landfill Symposium, Sardinia, Italy*; Report ECN-RX-05-164; Environmental Sanitary Engineering Centre: Cagliari, Italy; Energy Research Centre of the Netherlands: Petten, The Netherlands, 2005; Available online: www.ecn.nl (accessed on 28 February 2021).
37. Bouchelaghem, F. A Numerical and Analytical Study on Calcite Dissolution and Gypsum Precipitation. *Appl. Math. Model.* **2010**, *34*, 467–480. [CrossRef]
38. Chandra, A.P.; Gerson, A.R. The Mechanisms of Pyrite Oxidation and Leaching: A Fundamental Perspective. *Surf. Sci. Rep.* **2010**, *65*, 293–315. [CrossRef]
39. Evangelou, V.P.; Seta, A.K.; Holt, A. Potential Role of Bicarbonate during Pyrite Oxidation. *Environ. Sci. Technol.* **1998**, *32*, 2084–2091. [CrossRef]
40. Descostes, M.; Beaucaire, C.; Mercier, F.; Savoye, S.; Sow, J.; Zuddas, P. Effect of Carbonate Ions on Pyrite (FeS_2) Dissolution. *Bull. Société Géologique Fr.* **2002**, *173*, 265–270. [CrossRef]
41. Bonnissel-Gissinger, P.; Alnot, M.; Ehrhardt, J.-J.; Behra, P. Surface Oxidation of Pyrite as a Function of PH. *Environ. Sci. Technol.* **1998**, *32*, 2839–2845. [CrossRef]
42. Crundwell, F.K. The Mechanism of Dissolution of Minerals in Acidic and Alkaline Solutions: Part II Application of a New Theory to Silicates, Aluminosilicates and Quartz. *Hydrometallurgy* **2014**, *149*, 265–275. [CrossRef]
43. Morse, J.W.; Arvidson, R.S. The Dissolution Kinetics of Major Sedimentary Carbonate Minerals. *Earth-Sci. Rev.* **2002**, *58*, 51–84. [CrossRef]
44. Pokrovsky, O.S.; Golubev, S.V.; Schott, J.; Castillo, A. Calcite, Dolomite and Magnesite Dissolution Kinetics in Aqueous Solutions at Acid to Circumneutral PH, 25 to 150 °C and 1 to 55 Atm PCO2: New Constraints on CO2 Sequestration in Sedimentary Basins. *Chem. Geol.* **2009**, *265*, 20–32. [CrossRef]
45. Stumm, W.; Morgan, J.J. *Aquatic Chemistry: Chemical Equilibria and Rates in Natural Waters*; John Wiley & Sons: Hoboken, NJ, USA, 2012; Volume 126.
46. Sigg, L.; Stumm, W.; Behra, P. Chimie Des Milieux Aquatiques. *Chim. Milieux Nat.* **2000**.
47. Garrels, R.M.; Thompson, M.E. Oxidation of Pyrite by Iron Sulfate Solutions. *Am. J. Sci.* **1960**, *258*, 57–67.
48. Williamson, M.A.; Rimstidt, J.D. The Kinetics and Electrochemical Rate-Determining Step of Aqueous Pyrite Oxidation. *Geochim. Cosmochim. Acta* **1994**, *58*, 5443–5454. [CrossRef]
49. Sjöberg, E.L.; Rickard, D.T. Temperature Dependence of Calcite Dissolution Kinetics between 1 and 62 °C at PH 2.7 to 8.4 in Aqueous Solutions. *Geochim. Cosmochim. Acta* **1984**, *48*, 485–493. [CrossRef]
50. Moses, C.O.; Herman, J.S. Pyrite Oxidation at Circumneutral PH. *Geochim. Cosmochim. Acta* **1991**, *55*, 471–482. [CrossRef]
51. Deutsch, W.J.; Siegel, R. *Groundwater Geochemistry: Fundamentals and Applications to Contamination*; CRC Press: Boca Raton, FL, USA, 1997.
52. Plante, B.; Benzaazoua, M.; Bussière, B. Predicting Geochemical Behaviour of Waste Rock with Low Acid Generating Potential Using Laboratory Kinetic Tests. *Mine Water Environ.* **2011**, *30*, 2–21. [CrossRef]
53. Lasaga, A.C.; Kirkpatrick, J. *Kinetics of Geochemical Processes*; Walter de Gruyter GmbH & Co KG: Berlin, Germany, 2018; Volume 8.
54. Santelli, C.M.; Welch, S.A.; Westrich, H.R.; Banfield, J.F. The Effect of Fe-Oxidizing Bacteria on Fe-Silicate Mineral Dissolution. *Exp. Theor. Geomicrobiol.* **2001**, *180*, 99–115. [CrossRef]

minerals

Article

Numerical Modeling of a Laboratory-Scale Waste Rock Pile Featuring an Engineered Cover System

Katherine E. Raymond [1,*], Nicolas Seigneur [1,†], Danyang Su [1], Bissé Poaty [2], Benoît Plante [2], Bruno Bussière [2] and K. Ulrich Mayer [1]

1. Department of Earth, Ocean, and Atmospheric Sciences, 2207 Main Mall., Vancouver, BC V6T 1Z4, Canada; nicolas.seigneur@mines-paristech.fr (N.S.); dsu@eoas.ubc.ca (D.S.); umayer@eoas.ubc.ca (K.U.M.)
2. Institut de Recherche en Mines et en Environnement, Université du Québec en Abitibi-Témiscamingue, 445 Boulevard de l'Université, Rouyn-Noranda, QC J9X 5E4, Canada; PoatyBisse.PoatyBisse@UQAT.ca (B.P.); Benoit.Plante@uqat.ca (B.P.); Bruno.Bussiere@uqat.ca (B.B.)

* Correspondence: kraymond@eoas.ubc.ca

† Now at: MINES ParisTech, PSL University, Centre de Geosciences, 35 rue St Honore, 77330 Fontainebleau, France.

Received: 21 May 2020; Accepted: 15 July 2020; Published: 23 July 2020

Abstract: Improved design to reduce contaminant mass loadings from waste rock piles is an increasingly important consideration. In certain cases, an engineered cover system containing a flow control layer (FCL) may be used to mitigate the release of metals out of a pile using capillary barrier effects (CBEs), diverting water away from reactive materials below. In this study, a reactive transport model was calibrated to observational data from a laboratory experiment designed to evaluate a cover system. The results show that the numerical model is capable of capturing flow rates out of multiple drainage ports and matching key effluent concentrations by including the spatial distribution of hydraulic parameters and mineral weathering rates. Simulations were also useful for characterizing the internal flow pathways within the laboratory experiment, showing the effectiveness of the cover in diverting the flow away from the reactive waste rock and identifying secondary CBEs between different rock types. The numerical model proved beneficial in building an improved understanding of the processes controlling the metal release and conceptualizing the system. The model was expanded to investigate the robustness of the cover system as a function of the applied infiltration rate, supporting that water diversion will occur with infiltration rates representative of field conditions.

Keywords: reactive transport modeling; waste rock; engineered cover system; capillary barrier effects; contaminated neutral drainage; heterogeneity

1. Introduction

The long-term storage of waste rock—material considered not economically viable, which must be removed from the subsurface to access valuable resources during the mining process—is an increasingly important consideration for mine operations globally. When exposed to atmospheric conditions, sulfide minerals within waste rock oxidize and may cause the release of acid rock drainage (ARD), or if sufficient neutralizing minerals are present, contaminated neutral drainage (CND) [1–6]. To mitigate the release of metals and contaminants from ARD and CND, engineered design structures such as cover systems, subaqueous disposal, blending and layering, or co-disposal techniques may be applied as a means of prevention [7–12]. In addition to the laboratory- and field-scale tests required to assess these techniques, reactive transport modeling (RTM) may be used as a tool to investigate the processes involved in the release and attenuation of metals from waste rock piles through scenario analyses and/or to quantitatively analyze and interpret data from laboratory experiments and field observations. For example, RTM has been used in the context of mine waste management to compare outcomes under multiple scenarios, investigate the effect of material heterogeneity, contrast different storage techniques, and bracket expected behaviour for forecasting in a number of previous studies e.g., [13–22].

In this study, the reactive transport code MIN3P-HPC [23] was used to create a numerical model to supplement the ongoing research for an engineered cover system applied at the Lac Tio Mine site (Rio Tinto Iron and Titanium), located near Havre-Saint-Pierre, QC, Canada. At this site, near-neutral pH drainage may at times contain Ni concentrations above local regulatory limits (0.50 mg/L) (i.e., Quebec Province, Canada) [24,25], requiring water treatment. To control Ni and the release of other metals, a flow control layer (FCL) composed of sand and unreactive crushed anorthosite available at the site was designed to divert water around the metal-containing waste material, which is mainly composed of below-grade hematite–ilmenite (HI) ore. The characterization of Lac Tio waste rock and the effectiveness of the FCL at diverting water while minimizing erosion has been well studied in the past [25–32]. However, a quantitative two-dimensional analysis of the flow and reactive transport processes has not been conducted to date. To this end, a numerical model was constructed and calibrated for the quantitative analysis of a laboratory experiment conducted by Poaty [31], which was designed to evaluate the implementation of a FCL at the full scale waste rock pile at the Lac Tio site. The specific objectives of this work were to use RTM (1) to quantitatively evaluate the functioning of the FCL constrained by the experimental data and the measured material properties, (2) to characterize the internal unsaturated flow regime within the laboratory model, and (3) to improve the understanding of the influences of physical and chemical heterogeneities (such soil hydraulic parameters and mineral precipitation–dissolution kinetics) between different material zones, controlling the discrepancies of effluent quality between the drainage ports throughout the simulation.

Site Background

The Lac Tio Mine is an open-pit mine featuring a massive HI deposit hosted in anorthosite gangue material. The mine site has been in operation since 1950 and is located 43 km north of Havre-Saint-Pierre, Quebec, Canada. Waste rock type at this site varies between anorthosite and highly mineralized material with an HI content >70 wt. % [25]. Nickel, a primary focus of previous studies, comes primarily from Ni-bearing pyrite ($(Ni_xFe_{1-x})S_2$), where x is spatially distributed and correlated to Ni abundance, varying between 0 and 10 wt. % throughout the HI deposit. Additionally, Ni is present in the form of millerite (NiS) embedded within pyrite grains [24]. The release of Ni from waste rock is retarded through sorption mechanisms in both plagioclase and ilmenite rock [33]. The drainage from Lac Tio waste rock is maintained at near neutral conditions due to the weathering of silicate minerals, specifically: calcic plagioclase which is mainly close to labradorite in composition ($Na_{0.4}Ca_{0.6}Al_{1.6}Si_{2.4}O_8$), enstatite ($MgSiO_3$), and biotite ($K(Mg,Fe^{2+})_3(Al,Fe^{3+})Si_3O_{10}(OH,F)_2$) [34]. Silicate dissolution to maintain near-neutral pH in the absence of carbonate minerals has been demonstrated previously [35,36], and is typically incongruent, preferentially releasing cations such as Ca and Na from plagioclase minerals, Mg from pyroxenes [37,38] and Mg, Fe, and K from sheet silicates [39]. Further detailed analyses of the weathering mechanisms and specific dissolution rates have been previously determined for Lac Tio waste rock [34].

In order to improve the drainage quality, an option under investigation at the mine consists of designing an FCL using crushed anorthosite and sand. The objectives of the FCL are to divert water from reactive materials using contrast in grain size to create capillary barrier effects (CBEs) and to reduce the mass loadings of contaminants out of the pile to minimize the water treatment requirements. Previous studies [14,40,41] found that a slope of 5° was most effective for the FCL to divert precipitation while minimizing erosion. To investigate this design under controlled conditions, a heterogeneous laboratory-scale model was constructed [31] utilizing the relevant waste rock lithologies and compositions. Weekly flush experiments were performed on this laboratory model inclined at various angles between 0° and 5° over a period of 24 weeks between October 2017 and April 2018.

Information regarding the results of these transient tests can be found in previous works [31]. In this study, the system covered with the FCL inclined at a 5° angle was subjected to a constant precipitation rate with leachate collected at different sampling ports along the base of the tank. This experimental study aimed to emulate the flow dynamics for the covered Lac Tio waste rock under high infiltration rates. The objective of the experiment was to quantitatively investigate the performance of the FCL and to help in quantifying hydrogeochemical reaction rates, which can potentially be used for future upscaling purposes.

2. Materials and Methods

2.1. Laboratory Methods

2.1.1. Design of Experiment and Construction

The laboratory experiment [31] was constructed by depositing crushed waste rock material from the Lac Tio mine site using a hand-held scoop in 4 sections: 3 zones of HI material (labeled S1–S3) and 1 anorthosite zone at the end, all of which were deposited mimicking the angle of repose expected to develop during the deposition of waste rock at the field scale. Each zone was 80 cm in height and was emplaced in the 60 cm wide tank (Figure 1). Overlying these layers, a 10 cm thick sand layer and 8 cm thick crushed anorthosite layer were placed to create the FCL. Anorthosite material was crushed to <10 mm for the FCL using a crusher at the Lac Tio mine site. The emplaced cover material was compacted using a Proctor hammer with a constant mass to approach target porosities and to create a CBE, with the goal to divert water from the HI material below. Grain size distribution curves for the materials in each of the sections can be found in Appendix A. Along the base, 8 drainage ports (1 cm diameter, referred to as reservoirs 1–8) were installed and equipped with tubing to guide the drainage water into collection buckets below (Figure 1). Effluent volumes were determined for each reservoir; subsequently, the samples were collected for geochemical analysis. With the tank exposed to atmospheric conditions along the top, initial conditions throughout the experiment were assumed to be well aerated with respect to O_2 (21%) and CO_2 (0.0412%) [42]. Additional details on the laboratory model design and construction can be found in Poaty [31].

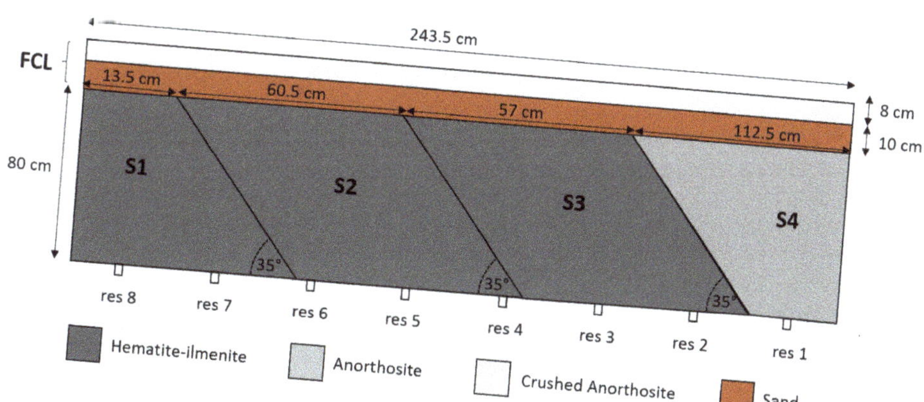

Figure 1. Schematic of the laboratory experiment with a flow control layer (FCL) rotated at 5°, adapted from Poaty [31].

2.1.2. Flow Measurements

The experiment was performed to create steady-state flow conditions, applying precipitation with a sprinkler system uniformly across the surface at a constant rate over a 17-day period in February 2019 (photos of the model setup are included in Appendix A). A rate of 180 L day^{-1} (corresponding to an application rate of 108 mm day^{-1}) was selected. This application rate represents the equivalent of one month of precipitation entering the pile under field conditions. The high-water application rate was selected in order to meet the experimental time constraints and to achieve steady state conditions quickly, while providing the benefit of testing the engineered cover system under extreme conditions. Drainage was collected in the buckets below each reservoir and weighed to determine the daily flow rates from each port. In addition, the frequency domain (FD) sensors placed throughout the experiment were used to measure the volumetric water content using dielectric constants [29]. While the FD probes were not calibrated specifically for this 17-day experiment, variations in the measured apparent dielectric constant values were used to support the bucket weight measurements and to determine when the steady-state flow conditions in the laboratory experiment were reached. While the flow rates obtained from the bucket weight measurements provided information on how much water was leaving at each reservoir, they could not be used alone to characterize the internal flow pathways within the experiment.

2.1.3. Geochemical Monitoring

Leachates were collected in buckets below reservoirs 1–8, in which pH and electrical conductivity were monitored daily. Once these parameters stabilized, water samples were collected from the buckets and filtered with 0.45 µm mesh filters into pre-acidified (with 2% HNO_3) 125 mL polyvinyl chloride (PVC) bottles for full-suite geochemical analyses using ICP–MS. Samples were collected and analyzed over a period of 10 days after the steady-state flow conditions were achieved, which was considered sufficient to represent the stabilized effluent water quality within the scope of this study. After the completion of transient flush experiments [31] and prior to the experiment completed for this study, the laboratory set-up was left inactive. During this period of inactivity, the materials were left to oxidize freely; therefore, the samples were not collected between days 1 through 7, since these samples were not considered representative for geochemically stable drainage conditions. Early time drainage was expected to represent the release of weathering products that had built up during several months of inactivity between transient flush experiments and the experiment conducted for this study. Results on the transient behaviour of the laboratory experiment have been published elsewhere [31].

2.2. *Numerical Methods*

2.2.1. Reactive Transport Code MIN3P-HPC

MIN3P-HPC [23] is built on the parallel version of MIN3P-THCm [43,44], with implementation of unstructured grid capabilities. The key features of the code include coupled multicomponent transport and biogeochemical reactions in variably saturated porous media [44–47]. The code uses Richards' equation to solve the flow in the unsaturated/saturated subsurface. A global implicit method and direct substitution approach was implemented for the solution of the multicomponent advection–dispersion equation and biogeochemical reactions [44,48]. Additional details about the code may be found in [44,45,48].

In this study, a 2D numerical model solving for the steady-state flow and transient geochemical conditions was set up to quantitatively interpret the data collected from the laboratory experiment presented above. The conceptual model of the experiment including the boundary conditions is depicted in Figure 2, while the mesh configuration used is provided in Appendix B. The versatility of the unstructured grid available in MIN3P-HPC allowed for the direct inclusion of the angled surfaces (such as the 5° rotation of the FCL and sloped edges following the angle of repose in the HI zones, as depicted in Figure 2), which could not have been captured adequately using a standard rectangular-based mesh. Details regarding the calibration process for flow and geochemistry are provided in the following subsections.

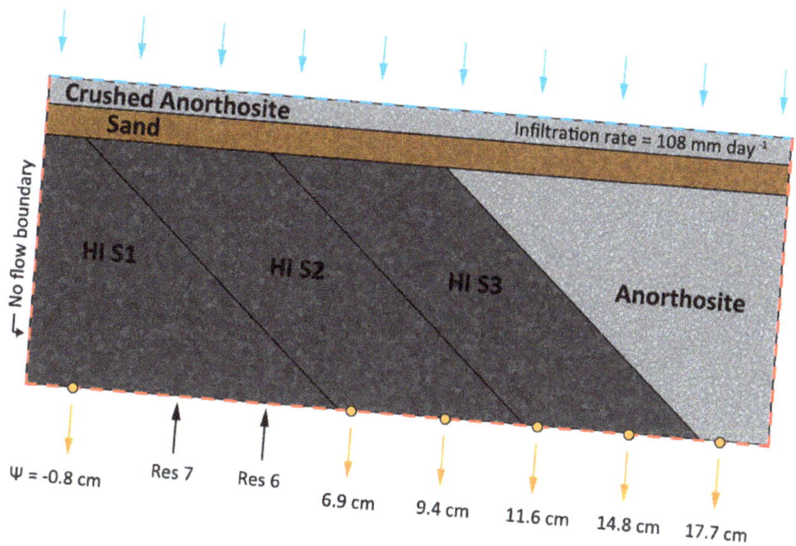

Figure 2. Conceptual model with the labeled boundary conditions and material property zones (where HI S1–HI S3 refer to 3 zones of hematite–ilmenite). Along the top surface (blue), an infiltration rate of 108 mm day^{-1} is applied. The red dashed lines surrounding the other boundaries represent no flow boundaries. Yellow symbols with arrows represent the pressure head (Ψ) boundaries where the effluent was collected from each reservoir. Pressure head boundaries were calibrated to achieve the drainage rates observed from the laboratory experiment under steady-state flow conditions. Reservoirs 6 and 7 were treated as no flow boundaries in the numerical model, since no outflow was observed at either port; however, their locations are labeled here. Corresponding hydraulic heads from left to right: −2.2, −2.3, −2.2, −2.5, −2.1, and −2.0 cm.

2.2.2. Flow Model and Calibration

Previous works [27,29] utilized suction and moisture content measurements (collected using Watermark® Granular Matrix sensors and ECH$_2$O EC-5 soil moisture sensors, respectively) from transient flow experiments to better understand the internal flow regimes. In this study, the measured flow rates from each reservoir were used to calibrate the numerical model. Figure 2 illustrates the material property zones and boundary conditions used. Although a water application of 180 L day^{-1} evenly distributed across the surface was intended from the sprinkler system, the total outflow collected from the reservoirs only averaged 158 L day^{-1} under steady-state flow conditions. This difference in mass-in versus mass-out is attributed to a combination of factors including evaporation and difficulties in maintaining pump pressure and water delivery. A steady and uniform water infiltration (158 L day^{-1}, corresponding to an infiltration rate of 108 mm day^{-1}), was applied in the model to match the total measured outflow under the steady-state conditions.

Constant pressure heads were specified to calibrate the steady-state drainage rates at each reservoir. The simulated drainage rates proved to be sensitive to specified pressure heads at each of the sampling ports. Since no flow was measured out of reservoirs 6 and 7 during the experiment, these locations were also treated as no flow boundaries in the numerical model (Figure 2). The variably saturated flow model was further constrained by measured physical parameters as summarized in Table 1. Porosity measurements were taken directly from the materials used in the laboratory experiment, while saturated hydraulic conductivity and unsaturated hydraulic soil parameters were obtained from previous column experiments and characterization of the waste rock and FCL [28,31,41,49]. Residual volumetric moisture contents reported in previous studies [41,50] were used to determine residual saturations for sand and waste rock materials. The van Genuchten-model was used to describe functional relationships between the pressure head, saturation and relative permeability [51], based on the parameters derived in previous studies [30,41,49,50]. For the HI material, zone-specific soil hydraulic function parameters were not available, and material-specific data from previous studies [30,50] were used uniformly for all zones. These parameters were used directly and were not varied during the model calibration.

Table 1. Physical parameters for a variably saturated flow separated by zone.

Zone	Saturated Hydraulic Conductivity (ms^{-1})	Porosity (-)	Residual Saturation (-)	van Genuchten α (m^{-1})	van Genuchten n (-)	Air Entry Pressure (m)
HI S1	5.5×10^{-4} [50]	0.39 [31]	0.13 [31,50]	5.96 [51]	3.61 [51]	0.04 [50]
HI S2	5.5×10^{-4} [50]	0.55 [31]	0.09 [31,50]	5.96 [51]	3.61 [51]	0.04 [50]
HI S3	5.5×10^{-4} [50]	0.58 [31]	0.09 [31,50]	5.96 [51]	3.61 [51]	0.04 [50]
Anorthosite	5.0×10^{-4} [30]	0.35 [31]	0.14 [31,50]	3.60 [50]	1.56 [50]	0.04 [31,50]
Sand	7.8×10^{-4} [50]	0.37 [31]	0.08 [31,41]	1.64 [50]	4.99 [41]	0.30 [31,41]
Crushed Anorthosite	5.0×10^{-5} [30]	0.29 [31]	0.10 [31,50]	20.0 [30]	1.56 [50]	0.04 [30]

The values from the literature (rather than direct measurements such as porosity) were considered homogeneous within the individual material zones. In reality, the hydraulic conductivity and additional soil hydraulic parameters within these zones are likely heterogeneous; influenced by the natural heterogeneity of the material and potentially also affected by compaction during the construction of the laboratory model. However, since direct measurements of these parameters in the laboratory model were not available, the outflow was calibrated using pressure heads—rather than inferring the hydraulic conductivity distribution—was considered justified. In particular, this approach was chosen because the boundary conditions at the drainage ports are not well defined, since tubing to guide drainage was attached to the ports and since it was unclear whether all drainage ports performed equally. While this approach does not allow to infer the exact hydraulic conductivity distribution throughout the experimental vessel, the hydraulic parameters were adequately constrained by the related literature data and measurements, providing a suitable approach for investigating the diversion of water as a result of CBEs, which was the main purpose of this study.

2.2.3. Reactive Transport Model and Calibration for Leachate Concentrations

The reactive transport model considered a free-phase diffusion coefficient of $D_0 = 2.4 \times 10^{-9}$ m^2 s^{-1} for all dissolved species, while longitudinal and transverse vertical dispersivities of 10^{-3} m and 5×10^{-4} m were used. In the gas phase, a free phase diffusion coefficient of 2.1×10^{-5} m^2 s^{-1} was used for all gases. Effective diffusion coefficients were calculated based on the Millington–Quirk relationship [48]. Only the most relevant chemical components were considered in the model including $SO_4{}^{2-}$, Ni^+, Fe^{2+}, Fe^{3+}, Ca^{2+}, K^+, Mg^{2+}, Na^+, Al^{3+}, H_4SiO_4, $CO_3{}^{2-}$, H^+ and O_2(aq). The measured composition of the tap water applied to the surface of the physical experiment was specified along the infiltration boundary in the numerical model as a specified flux boundary (Table 2). Concentrations in Table 2 are provided as the total component concentrations, except for the components H^+, $CO_3{}^{2-}$ and O_2(aq), which are specified in terms of pH, alkalinity and pO_2, respectively. Aqueous speciation amongst these components is computed based on the hydrolysis and complexation reactions defined in the model database, which is derived from the USGS wateq4f database [52]. Chloride is present in the tap water as a result of chlorination, although it was not measured since Cl^- has very limited reactivity and its omission has no significant effect on the simulation results. Although rainwater is more dilute than tap water, both tap water and rain water are substantially more dilute compared to pore water within the waste rock and drainage water; therefore, no significant impacts on the weathering reactions are anticipated by using tap water in the experiment. Boundary conditions in the reservoirs with active drainage were specified as free exit boundary conditions.

While the mineralogy of the various materials is complex, the mineralogy considered for the numerical model has been simplified to include only the minerals which are considered to effectively control drainage chemistry. Table 3 summarizes the volume fractions of the primary mineral phases included. The majority of these volume fractions were taken from the previous characterization of waste rock by [53]; however, pyrite volume fractions were calculated based on the results from S analyses [31]. In addition to these primary phases, the secondary phases, ferrihydrite, $Al(OH)_3$ (am), SiO_2 (am) and kaolinite were allowed to precipitate. Table 4 summarizes the reaction pathways, rate expressions and relevant equilibrium constants of the primary and secondary minerals considered in the numerical model. A simple first order rate expression was used to describe the oxidation of the Ni-bearing pyrite. More complex models were not deemed necessary, in particular considering that the published rate expressions were determined under highly controlled laboratory conditions and are not necessarily directly applicable to field situations [54].

Table 2. Composition of the infiltrating water applied to the model.

Species	Concentration (mg L^{-1})
SO_4^{2-}	65
Ni^+	0.007
Fe^{2+}	-
Fe^{3+}	0.105
Ca^{2+}	24
K^+	0.859
Mg^{2+}	3.41
Na^+	5.95
Al^{3+}	0.02
$H_4SiO_4^0$	7.84
pH	6.87
Alkalinity (as $CaCO_3$)	22
O_2(aq)	0.21 (as pO_2 in atm)

Table 3. Initial mineral volume fractions considered in the numerical model.

Mineral	Volume Fraction					
	HI S1	HI S2	HI S3	Anorthosite	Sand	Crushed Anorthosite
Ni-bearing pyrite	0.1649 [1]	0.1524 [1]	0.1060 [1]	0.0004 [1]	-	0.0096 [1]
Labradorite	0.2054 [53]	0.2054 [53]	0.2054 [53]	0.2054 [53]	-	0.2054 [53]
Enstatite	0.1158 [53]	0.1158 [53]	0.1158 [53]	0.1158 [53]	-	0.1158 [53]
Biotite	0.0086 [53]	0.0086 [53]	0.0086 [53]	-	-	-

[1] Calculated from measured S wt. % [31].

Table 4. Minerals and their dissolution–precipitation reactions considered in the numerical model. Effective rate constants ($k_{eff,m}$) were calibrated depending on material zones and are presented in the results section in Table 6.

Mineral	Reaction	Rate Expression	$\log K^m$
Ni-bearing pyrite	$Ni_{0.14}Fe_{0.86}S_2 + H_2O + 3.5O_2 \rightarrow 0.86Fe^{2+} + 0.14Ni^+ + 2SO_4^{2-} + 2H^+$	$-k_{eff,m}[O_2]$	-
Labradorite	$Na_{0.4}Ca_{0.6}Al_{1.6}Si_{2.4}O_8 + 1.6H_2O + 6.4H^+ \rightarrow 0.4Na^+ + 0.6Ca^{2+} + 1.6Al^{3+} + 2.4H_4SiO_4$	$-k_{eff,m}[H^+]^{0.55}$ [55]	-
Enstatite	$MgSiO_3 + 2H^+ + H_2O \rightarrow Mg^{2+} + H_4SiO_4$	$-k_{eff,m}[H^+]^{0.25}$ [56]	-
Biotite	$K(Mg_2Fe)AlSi_3O_{10}(OH)_2 + 10H^+ \rightarrow 2Mg^{2+} + K^+ + 2Fe^{2+} + 3H_4SiO_4$	$-k_{eff,m}[H^+]^{0.25}$ [57]	-
Ferrihydrite	$Fe(OH)_3 + 3H^+ \rightarrow Fe^{3+} + 3H_2O$	$-k_{eff,m}\left[\frac{1-IAP}{K}\right]$	-4.891 [58]
$Al(OH)_3$ (am)	$Al^{3+} + 3H_2O \rightarrow Al(OH)_3 + 3H^+$	$-k_{eff,m}\left[\frac{1-IAP}{K}\right]$	-10.80 [58]
SiO_2 (am)	$H_4SiO_4 \rightarrow SiO_2 + 2H_2O$	$-k_{eff,m}\left[\frac{1-IAP}{K}\right]$	3.018 [58]
Kaolinite	$2Al^{3+} + 2H_4SiO_4 + H_2O \rightarrow Al_2Si_2O_5(OH)_4 + 6H^+$	$-k_{eff,m}\left[\frac{1-IAP}{K}\right]$	-7.435 [58]

Nickel in the HI material is sourced from both pyrite and millerite grains (found in inclusions within the pyrite). However, the simulations considered pyrite as the only Ni source, with the dissolution rates calibrated using measured SO_4 release rates. Although previous work to characterize Lac Tio waste rock identified sorption as an important Ni-attenuation mechanism [25,33,34,53], this process was not explicitly included in the current numerical model. In this study, the mineral weathering rates in the numerical model were calibrated to match the release rates observed in the drainage of the physical model and constitute effective release rates relevant to the duration of the experiment, implicitly accounting for the effect of sorption. Accounting for sorption explicitly was beyond the scope of the present study.

Previous work has shown that both S and Ca release rates were stoichiometrically associated with sulfide (primarily in the form of pyrite, although trace amounts of chalcopyrite and millerite have also been detected) and plagioclase weathering, respectively, due to limited secondary mineral precipitation [34]. To reproduce the observed effluent chemistry of the system, rate coefficients for mineral dissolution–precipitation reactions (Table 4) were calibrated using a stepwise approach. Initially, the oxidative dissolution rate of Ni-bearing pyrite was calibrated constrained by the measured concentrations of SO_4 and Ni from each reservoir. During this step, the Ni abundance in pyrite was also adjusted within the 0–10 wt. % range observed at the Lac Tio site. The presence of Ni impurities in pyrite has been noted to increase pyrite oxidation rates in previous studies [59,60]. Then, the most reactive silicates with pH-buffering capacity were considered (labradorite, enstatite, and biotite) to reproduce pH and major ion concentrations. Labradorite weathering rates were calibrated to reproduce Ca release rates and to achieve an initial calibration for pH. Enstatite and biotite were subsequently introduced to represent additional pH-buffering reactions by silicate minerals (using observed Mg and K release rates to calibrate, respectively). The secondary minerals ferrihydrite, SiO_2 (am), $Al(OH)_3$ (am) and kaolinite have been identified [34] and were included to control the Fe, Al, and Si concentrations in the drainage waters.

2.3. *Reduced Infiltration Scenario Analysis*

Expanding upon the model calibrated for the experiment under the steady-state flow conditions, the influence of reduced infiltration on the internal flow patterns and effectiveness of the FCL was investigated through additional simulations with a 10× lower infiltration rate (10.8 mm day^{-1}). Since the outflow rates are unknown for these conditions, three different scenarios for a hypothetical water table were explored: (1) the same water table height as observed in the model for high infiltration rates; (2) a reduced water table height, which is expected for the reduced infiltration conditions; and (3) a scenario with no water table forming within the experimental vessel. This analysis was only performed to investigate the flow processes with a specific focus on the performance of the FCL; an analysis on the effect of drainage chemistry was not performed due to a lack of observational data.

2.4. *Assumptions and Limitations*

Building a conceptual model for RTM requires making assumptions and simplifications depending on what data are available and the overall purpose of the model. For example, Table 1 shows that porosity was determined for the various materials emplaced in the experimental vessel, while other input parameters were taken from other Lac Tio column and field experiments. In reality, the values for each material property zone may vary slightly compared to the parameters obtained from the companion experiments. Parameters may also vary spatially and show heterogeneities within each material property zone. In addition, the geochemical reaction network represents a simplification of the actual reactions occurring in the physical experiment and mineral reaction rates, which were assumed to be constant in time within each material property zone. This assumption is considered justified given the timescale of this experiment, and is also supported through previous work by [24] who determined through humidity cell tests that silicate weathering rates in fresh Lac Tio waste rock are very similar to weathering rates in the material that has undergone 25 years of natural weathering.

In addition, the numerical model was built in 2D, while the laboratory experiment contains features that deviate from a true 2D system. Specifically, the drainage ports in the laboratory experiment were located in the center of the base of the experiment in the y-direction. In the 2D numerical model, each reservoir had to be taken as a line of drainage rather than a 1 cm diameter central port.

Despite these simplifications, it must be emphasized that most model input data were available through measurements from the well constrained experiment and were kept constant during the model calibration process. The only parameters that varied during the modeling were the pressure heads at the drainage ports and the mineral weathering rates with the goal to reproduce drainage rates and concentrations from the various reservoirs under quasi-steady-state flow conditions.

3. Results

3.1. Drainage Rates and Internal Fluid Flow Pathways

FD probe sensor measurements showed that stable volumetric water contents were reached within the laboratory experiment after 2 days, while the measured flow rates reached a quasi-steady state after approximately 8 days. Simulated results shown in Figure 3 and Table 5 capture the drainage rates with less than 1% difference at all points. The drainage rate out of reservoir 3 (approximately 70 L day^{-1}) constitutes 44% of the total water volume leaving the model. However, based on examining the drainage rates alone, it is unclear whether a higher volume of water leaving reservoir 3 is caused by water leaking through the FCL into the HI material, or whether water is being diverted by the FCL, as intended, and then rerouted to reservoir 3.

Table 5. Results for the measured versus simulated drainage rates.

Reservoir	8	7	6	5	4	3	2	1	Total
				Flow Rate (L day^{-1})					
Measured	7.98	0.00	0.00	21.20	3.88	69.15	29.33	26.50	158.03
Simulated	7.98	0.00	0.00	21.13	3.86	69.38	29.05	26.54	157.94
Percent Difference	−0.08%	0.00%	0.00%	0.33%	0.51%	−0.33%	0.94%	−0.16%	0.06%
				Percentage of Total Flow					
Measured	5.05%	0.00%	0.00%	13.42%	2.46%	43.75%	18.56%	16.77%	100.0%
Simulated	5.05%	0.00%	0.00%	13.38%	2.45%	43.93%	18.39%	16.80%	100.0%

In Figure 4, the modeled results of Darcy flux vectors are presented by scaled arrows, which are used to conceptualize flow through the laboratory model. Simulations suggest that in the experiment, a near flat water table forms at $z \approx -2$ cm, approximately corresponding to the elevation of reservoir 8. The pressure head boundaries used to calibrate the model (Figure 2) were positive for reservoirs 1–5, facilitating the outflow. The pressure head at reservoir 8 was slightly negative; however, outflow can still occur under these conditions since a drainage tube was attached to the outflow port at the base of the tank. In the HI zones in Figure 4, the smaller arrows show that there is less water flowing downwards compared to the anorthosite zone. These simulated flow pathways suggest that while there is some water leaking through the FCL into the HI material below, the FCL angled at 5° effectively diverts a significant portion of the water under quasi-steady-state flow conditions. Furthermore, the model results suggest the development of secondary CBEs between the HI material and the uncrushed anorthosite layer (Figure 4). Based on these conditions, a nearly flat water table forms along the base of the model and causes some of the drainage water to move updip (to the left) from the uncrushed anorthosite layer back into the HI material, contributing to the larger flow rates observed at reservoirs 2 and 3. These results suggest that leakage through the FCL is limited and that the large outflows at reservoirs 3 are due to the internal rerouting of flow paths due to CBEs.

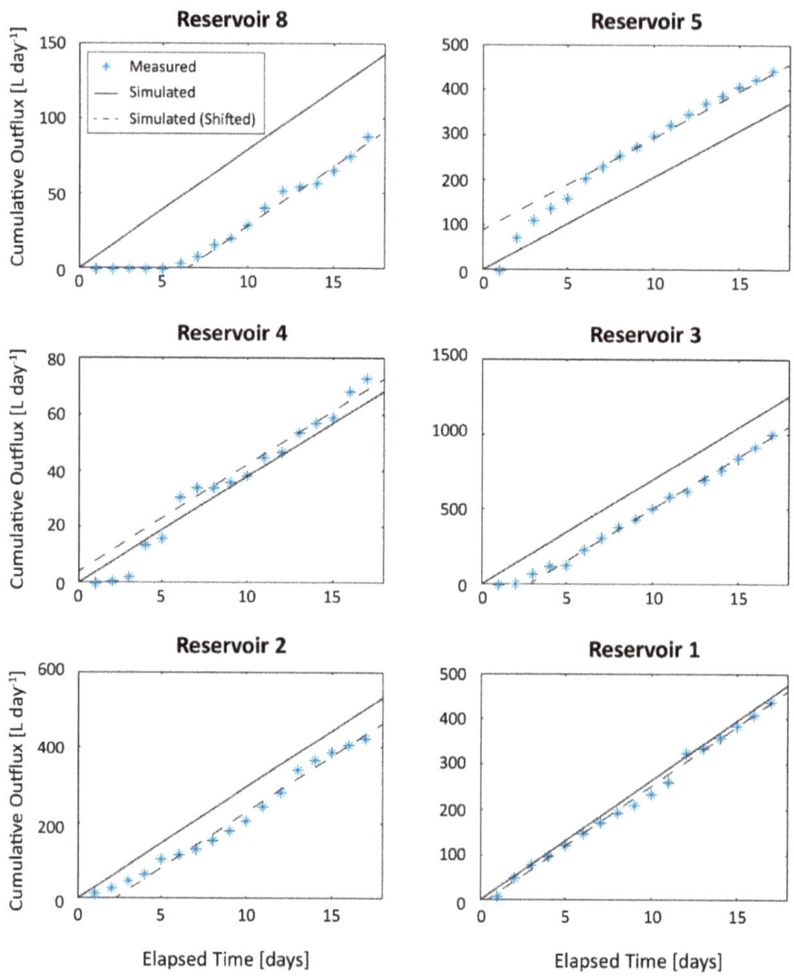

Figure 3. Measured versus simulated cumulative drainage. Blue symbols represent the measured cumulative drainage from the laboratory experiment, while the black lines represent the simulated cumulative drainage obtained from the steady-state flow simulations, calibrated using pressure heads at each reservoir. The dashed lines represent the simulated cumulative drainage shifted to account for the delay of the development of steady-state flow conditions in the physical experiment. The constant slopes at each reservoir indicate that steady-state flow conditions were reached.

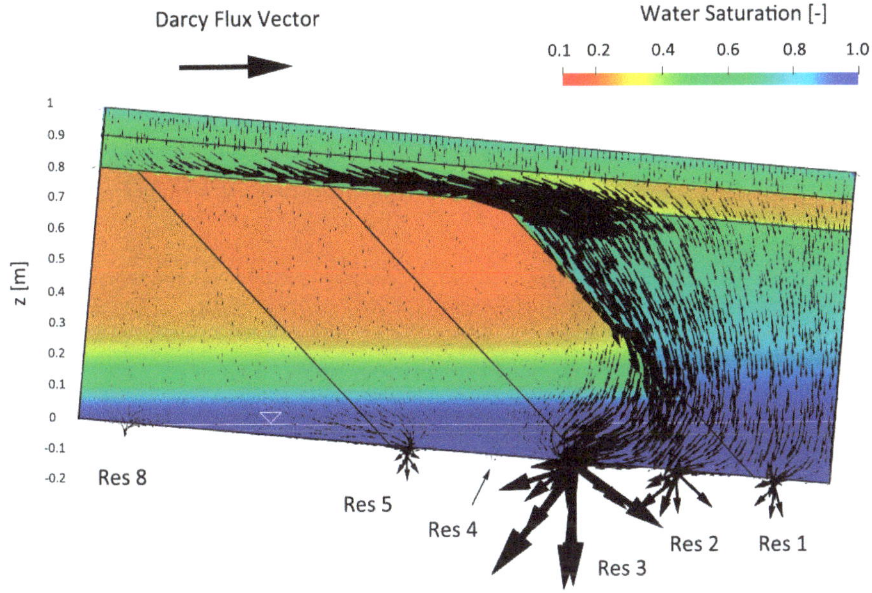

Figure 4. Simulated results of the calibrated flow model. The black arrows represent the Darcy flux vectors simulated by the numerical model indicating the direction of the fluid flow pathways and their relative magnitude, overlain on the simulated water saturations throughout the cross section of the experiment. The left bottom corner was set to z = 0 cm; the simulated water table is shown in white.

3.2. Solute Concentrations in Drainage

Results of the measured and simulated water quality at each reservoir are shown in Figure 5. The observed solute concentrations in drainage suggest that geochemically stable conditions were achieved after 12 days. Therefore, average solute concentrations between day 12 and day 17 were used to calibrate the mineral weathering rates within the numerical model (Table 6). Table A1 in Appendix C provides statistical information regarding the achievement of stable geochemical conditions for the major elements of concern (S, Ca, K, Mg, Na and Si), justifying this approach. Observations between day 8 and day 12 show the elevated solute concentrations and decreased pH, interpreted as capturing the tail end of the flushing of weathering products originating from oxidation reactions that took place while the experimental setup was idle. Early time behavior in the model also shows variations in concentrations, due to the specified initial conditions. Simulated concentrations in drainage between day 0 and day 10 are affected by the displacement and mixing of resident pore water with infiltrating water and weathering reactions. These early time results are not relevant for the model calibration and interpretation, performed for geochemically stable conditions, and are therefore represented with dashed lines (Figure 5).

The simulations included the diffusion of O_2 through the gas phase. Results suggest that O_2 was not depleted within the model domain over the duration of the experiment, and subsequently the oxidation of Ni-bearing pyrite was not inhibited. Pyrite oxidation rate coefficients were calibrated to reproduce the SO_4-release observed in the experiment. Pyrite grains have been observed to typically contain between 0–10 wt. % Ni at the Lac Tio site [25], and a pyrite composition of $Ni_{0.14}Fe_{0.86}S_2$ (corresponding to 7 wt. % Ni) was found to represent the observed Ni release rates most accurately, consistent with the observed Ni contents in drainage. The current approach utilizing effective Ni-release rates, implicitly accounting for sorption, was able to adequately reproduce the observed Ni-release, although previous work has shown that sorption processes are significant in fresh rock at the Lac Tio site and important for controlling the release of Ni [24,25]. It is possible that the effect of sorption in this experiment was less pronounced than in previous studies, since multiple pore volumes had passed through the material during previous experiments [31], which may have contributed to saturating available sorption sites. In addition, high volumetric flow rates were applied to the laboratory model for this study, which may have reduced the sorption processes typically occurring in the Lac Tio waste rock due to rapid flushing. Another possibility is that some sorption-processes may have been accounted for implicitly by the calibrated Ni-content in pyrite used in the numerical model. As previously mentioned, Ni may range from 0–10 wt. % content in pyrite in Lac Tio waste rock [25], therefore it is possible that the HI material in the laboratory model is greater than the 7 wt. % used here. Additional laboratory testing may be able to deduce the actual Ni-release rates and the contribution of sorption towards effective Ni-release rates in this experiment; however, this is beyond the scope of the present study.

As mentioned in the methods section, labradorite, enstatite, and biotite were the primary minerals considered for pH buffering. The volume fractions of minerals presented in Table 3 were not depleted during the 17 days simulated by the numerical model and therefore did not affect the progress towards geochemically stable conditions. The rate coefficients for these phases were calibrated to match the release of major ions for each mineral (specifically: the pyrite was calibrated using SO_4 release; labradorite to Ca, biotite to K, and the remaining Mg to enstatite, while considering these influences on effluent pH). These rate coefficients for pyrite and silicates are summarized for each zone in Table 6.

While simulated concentrations in drainage waters reproduce the measured values for SO_4, Ca, K, Mg, Na and pH fairly well for most reservoirs (Figure 5), the Si concentrations were above the measured values. The simulations suggested that the pore water was undersaturated with respect to SiO_2 (am) throughout the domain. Although the precipitation of kaolinite was considered in the simulations, its precipitation was not able to control Si to the observed concentrations. Previous studies [33,34] provide evidence for the presence of additional silicate minerals in the HI material such as pigeonite, muscovite, chlorite, orthoclase, and spinel. These minerals, with different Si contents and incongruent weathering processes, may also contribute to the silicate weathering and may at least in part explain the discrepancies between the simulated and measured results for Si. However, considering these additional mineral phases would result in increased model complexity and non-uniqueness; these phases were omitted from the analysis. In addition, simulated concentrations for Al and Fe were both below the measured values of approximately 3×10^{-6} mol L^{-1} and 5×10^{-6} mol L^{-1}, respectively. In the simulations, the precipitation of $Al(OH)_3$ (am) and ferrihydrite were considered as quasi-equilibrium reactions, while the speciation calculations for observed water chemistry indicate the supersaturated conditions with respect to these phases, implying that their precipitation is kinetically inhibited. Despite these simplifications, the model was able to reproduce the release of solutes of key interest (SO_4 and major cations) from the different reservoirs, meeting the study objectives (Figure 5).

Table 6. Calibrated effective rate coefficients for the mineral weathering reactions (units in (mol dm^{-3} bulk s^{-1}), except for pyrite, which is provided in units of (L H_2O dm^{-3} bulk s^{-1}); these units account for the scaling from pore volume to bulk volume for the first-order rate oxidation rate).

Mineral	$k_{eff,m}$					
	HI S1	**HI S2**	**HI S3**	**Anorthosite**	**Sand**	**Crushed Anorthosite**
Ni-bearing pyrite	9.6×10^{-7}	3.0×10^{-7}	3.5×10^{-7}	1.0×10^{-7}	-	8.0×10^{-7}
Labradorite	2.3×10^{-6}	8.0×10^{-7}	8.0×10^{-7}	3.5×10^{-8}	-	9.0×10^{-8}
Enstatite	7.0×10^{-11}	2.0×10^{-11}	1.0×10^{-11}	1.0×10^{-11}	-	1.0×10^{-11}
Biotite	3.0×10^{-11}	3.0×10^{-11}	3.0×10^{-11}	3.0×10^{-11}	-	4.5×10^{-11}
Ferrihydrite	7.0×10^{-9}	7.0×10^{-9}	7.0×10^{-9}	7.0×10^{-9}	7.0×10^{-9}	7.0×10^{-9}
$Al(OH)_3$ (am)	1.0×10^{-8}	1.0×10^{-8}	1.0×10^{-8}	1.0×10^{-8}	1.0×10^{-8}	1.0×10^{-8}
SiO_2 (am)	1.0×10^{-8}	1.0×10^{-8}	1.0×10^{-8}	1.0×10^{-8}	1.0×10^{-8}	1.0×10^{-8}
Kaolinite	1.0×10^{-8}	1.0×10^{-8}	1.0×10^{-8}	1.0×10^{-8}	1.0×10^{-8}	1.0×10^{-8}

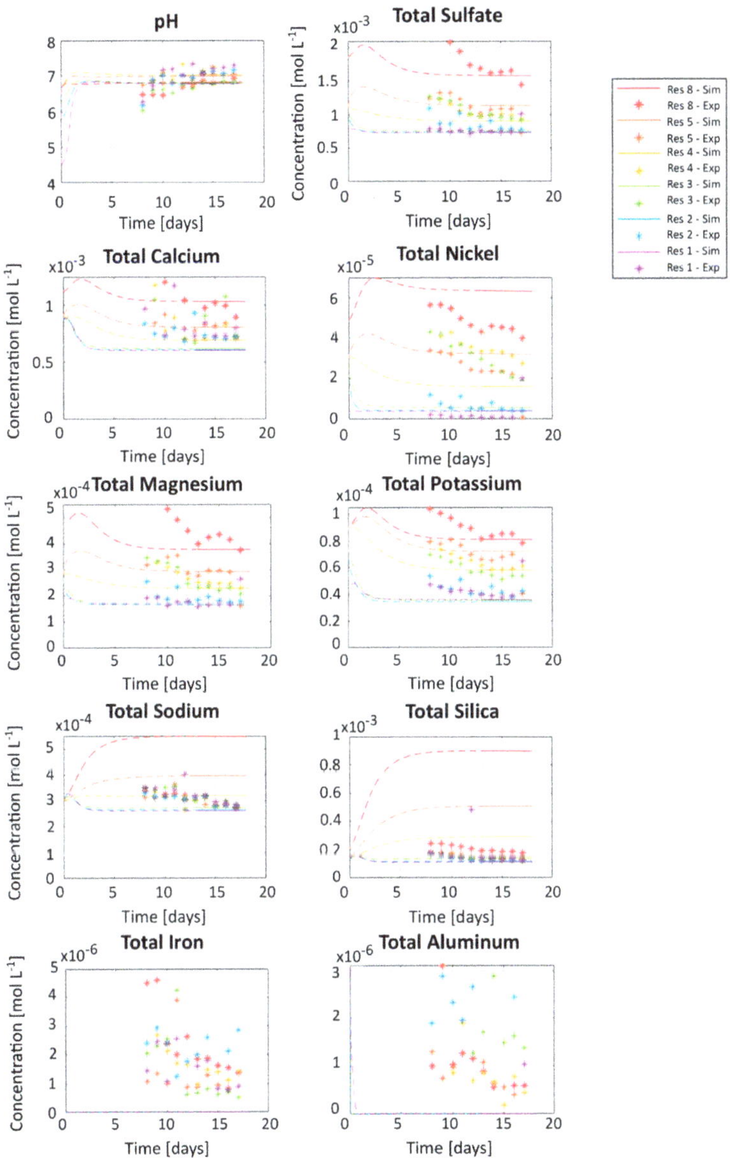

Figure 5. Results of the simulated versus observed effluent concentrations in mol L^{-1}, color coded per reservoir. Effective rate coefficients for the mineral dissolution–precipitation rates ($k_{eff,m}$) in Table 6 were calibrated through an iterative process to reproduce the observed concentrations in the effluents from the laboratory experiment.

Although high water application rates dilute the drainage chemistry, mass loading rates are expected to be similar to those under natural precipitation conditions. As mentioned previously, simulated oxygen levels did not decline throughout the domain for the conditions considered; implying that sulfide oxidation rates are not inhibited by high water application rates. However, some effect on the precipitation of secondary phases and metal attenuation is expected, since the formation of secondary minerals is more limited under high infiltration conditions. The impact is expected to most significantly affect trace metal release, which was not the focus of this study.

3.3. Reduced Infiltration Scenario Analysis

Simulation results of the flow model for a reduced infiltration rate, better aligned with real world conditions, are depicted in Figure 6. In this scenario, the infiltration rate was reduced by a factor of 10 in comparison to the conditions in the laboratory experiment. Since measured outflow rates were not available for these conditions, the boundary condition at the base of the experimental vessel had to be estimated. It was assumed that similar to the experiment with higher flow rates, a flat water table would develop, however, at a lower elevation. Simulation results indicate that the fluid flow pathways and functioning of the FCL remain very similar to the pattern observed for the higher flow rates (Figure 6), with the FCL effectively capturing most of the infiltration. Although the material zones show decreased water saturation above the water table compared to the simulations calibrated to the observational data from the laboratory experiment under higher flow conditions (compare Figure 6 to Figure 4), the simulated results suggest that water is still diverted through the sand material in the FCL (Figure 6). Furthermore, a secondary CBE between the HI material and anorthosite is also present under these reduced infiltration scenarios.

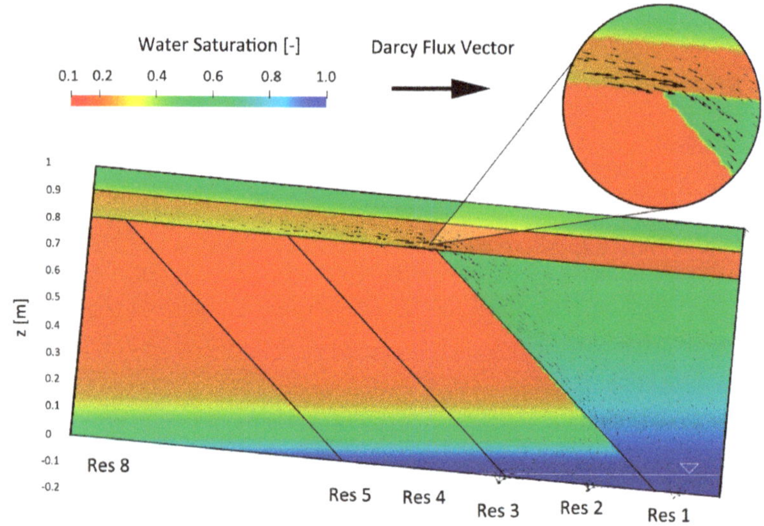

Figure 6. Simulated results of the flow model with a reduced infiltration rate and lowered water table ($z \approx -13$ cm). Simulation results show that the infiltration is diverted through the FCL towards the anorthosite material and that secondary capillary barrier effects (CBEs) between the anorthosite and HI material persist, similar to conditions in the laboratory experiment for higher infiltration rates. Darcy flux vectors are scaled the same as those shown in Figure 4, which are smaller in this case due to the reduced infiltration causing smaller fluxes. A closer look at the flow pathways for water movement from the sand into the anorthosite is included due to the reduced arrow sizes.

To evaluate the sensitivity of the results with respect to the assumed basal boundary condition, two additional simulations were conducted with the water table located at the same elevation as seen in the laboratory experiments for high inflow rates and for the conditions without a water table within the experimental vessel (suction at all reservoirs) (for results see Appendix D). These results indicate that the functioning of the FCL and internal flow pattern are not substantially impacted by the water table location and the basal boundary condition, although the drainage volume at the various reservoirs differ in response to these changes.

The low infiltration simulation analysis was only performed to investigate the flow processes with a specific focus on the performance of the FCL; an analysis on the effect of drainage chemistry was not performed due to a lack of observational data.

4. Discussion

4.1. Implications of Results from Flow Modeling

Locally sourced sand in the FCL was used to achieve a desired thickness for the engineered cover system for the full-scale piles [30]. The numerical model confirmed the development of a capillary barrier between the sand and the HI material, which results in the majority of water being diverted through the sand layer towards the uncrushed anorthosite material (Figure 4). A related study [29] using a water truck to complete a wetting test on an experimental pile for the Lac Tio site to investigate the effectiveness of the FCL at the field-scale also found that fluid diversion occurred mainly through the sand layer, which is consistent with the results forecasted by the numerical model here, as well as the previously completed modeling [61]. These findings lend support to the FCL's ability to divert most water away from the reactive HI material.

Additionally, the results from all numerical models presented here indicate the development of secondary CBEs between the HI material and uncrushed anorthosite material, which had not been previously identified. The existence of these secondary CBEs is dependent on the unsaturated soil hydraulic parameters, which were constrained by the experimental data and field observations (Table 1). The secondary CBEs are beneficial in diverting water away from the metal-rich HI material and retaining it within the uncrushed anorthosite zone, and therefore may warrant additional future investigation. The presence of secondary CBEs is also supported by the geochemical results (discussed in greater detail below).

For this study, the daily infiltration rates in the experiment to calibrate the numerical model were equivalent to one month of precipitation at the field site. In reality, water saturations within the full scale pile would be lower and infiltration processes would be transient, possibly leading to differences in flow dynamics. The effect of a reduced infiltration rate was considered through an additional scenario with an infiltration rate of 10.8 mm day^{-1} (a reduction by a factor of 10). Model results suggest that fluid flow pathways controlled by the FCL and CBEs are very similar for both infiltration rate scenarios, indicating that the FCL will also be functioning under field conditions.

Numerical modeling facilitated to account for (1) heterogeneities in the physical flow parameters between different material zones and (2) the pile geometry, owing to the unstructured grid capabilities of the MIN3P-HPC code. These factors were particularly important to investigate the influence of CBEs in the engineered cover system, since in this case the process of fluid diversion is controlled by contrast in grain size, associated hydraulic parameters and pile geometry.

4.2. Implications of Results from Reactive Transport Modeling

Both the simulated and observed effluent concentrations presented in Figure 5 support the simulated internal fluid flow pathways depicted in Figure 4. It was noted above that the highest outflow rates (both measured and simulated) were observed at reservoir 3. If the majority of this water was leaking past the FCL and percolating through the HI material zones (contrary to the simulated flow model), the effluent quality would be expected to lie closer to the observations in reservoirs 4 and 5. However, the actual measured concentrations from reservoir 3 fall somewhere in between reservoirs 4–5 and 1–2, supporting the hypothesis that only a portion of the water released at reservoir 3 moves through the HI material, while a substantial fraction percolates through the anorthosite zone and is then redirected updip towards reservoir 3. Furthermore, reservoirs 1 and 2 have very similar drainage concentrations with improved water quality relative to the other reservoirs in both the modeled and measured results. While this is expected for reservoir 1 which lies below the anorthosite material, reservoir 2 is located under the HI material. This finding further supports the results of the flow model, suggesting that water draining at reservoir 2 primarily percolates through the anorthosite material, but then moves updip and to the left towards reservoir 2 (Figure 4).

Additionally, the Darcy flux vectors shown in Figure 4 are smaller in magnitude in HI S1 towards reservoir 8 compared to the simulated flow towards reservoirs 4 and 5. Longer residence times and contact between infiltrating water and HI material are likely responsible for the increased concentrations observed in the drainage collected from reservoir 8. However, the simulations suggest that physical flow alone was insufficient to capture the differences in effluent concentrations observed between reservoirs 1 and 8. In Table 6, the differences in the rate coefficients for each material property zone are summarized. For pyrite, higher oxidative dissolution rates were required in HI S1 compared to S2 and S3 in order to reproduce the SO_4 concentrations observed at each reservoir. Consequently, the rate coefficients were also higher in zone S1 for labradorite and enstatite, although since the rate expressions for these minerals (as well as biotite) were pH dependent, the overall dissolution rates were also higher due to the greater acidity produced from faster sulfide oxidation rates.

Table 1 shows that the measured porosities from the laboratory model varied quite significantly between HI S1, S2, and S3 with values of 0.39, 0.55, and 0.58, respectively. These variations in porosity suggest heterogeneity (or potentially differences in compaction) between each zone, which may have influenced the reactive surface area and the related weathering rates. Porosity values were lower in zones from left to right following the deposition history. It is possible that materials in S1 underwent greater compaction than the other HI material zones due to the location, influencing amount (due to more overall rock), reactivity or availability of sulfide grains in this zone, although it is speculative whether this would have a greater influence over the naturally occurring chemical heterogeneities between zones. Furthermore, the grain size distribution curves (Appendix A) show that S1 displayed a smaller grain size compared to S2 and S3, which may explain the faster weathering rates due to the increased surface area and sulfide abundance [2,60,62–65].

Overall, the simulated effluent quality reproduced the measured effluent chemistries from each reservoir quite well. Similar to the physical heterogeneity being essential for capturing the influences of CBEs on flow patterns, differences in mineral weathering rates between the zones were essential for reproducing the variations in chemical loadings collected in drainage from the different reservoirs. Results suggest that for infiltration that is not diverted from the FCL, the degree of compaction and grain size may influence the residence time and the resulting effluent quality released from these zones. Simulated results would not have been able to reproduce the observational data with a homogeneous model, pointing towards the importance of considering physical and chemical heterogeneity in the numerical models going forward. In addition, the results indicate that compaction and grain size should be considered and assessed with respect to their influence on effluent quality.

5. Conclusions

RTM can be a useful tool for the quantitative interpretation of experimental data, identifying fluid migration pathways, bracketing reaction kinetics, or forecasting expected solute release rates and mass loadings. However, with many experiments aimed at explaining complex hydrogeochemical interactions, a thoughtful approach to understanding the dominant processes in the system and justifiable assumptions and limitations must be considered. In this study, RTM was useful to quantitatively assess the functioning of an engineered FCL, in the process revealing the likely development of beneficial secondary CBEs and providing insights into internal flow migration patterns. Although simplifying assumptions had to be made, the resulting model was able to reproduce the observational laboratory data including the spatially distributed drainage rates and solute concentrations, while keeping most material properties constant based on the measured parameters. The simulations were also useful to estimate the mineral weathering rates for pyrite and the aluminosilicate minerals present in the fresh HI material used in the laboratory experiment. For the simulations conducted, the consideration of physical and chemical heterogeneities between different material zones was essential to reproduce the observations made in the laboratory. The model was also useful to evaluate the effectiveness of the

FCL for the conditions of reduced infiltration rates, suggesting that the FCL will function as designed under infiltration rates more representative of field conditions.

Finally, the results demonstrate the capabilities of MIN3P-HPC as a tool for simulating the flow and reactive transport processes in complex hydrogeochemical systems such as waste rock piles. The recent implementation of unstructured grid capabilities into the model allows the user to incorporate more complex features, such as the influence of angled layers like the FCL or material zones following the angle of repose. Future advancements in computing power and improvements made to complex multi-phase interactions within RTM codes will further improve the capabilities of the models to characterize mine waste stockpiles and may be useful to contrast and improve different techniques for the design planning and storage of mine wastes.

Author Contributions: Conceptualization, K.E.R., K.U.M., B.P. (Benoît Plante), B.B.; data curation and methodology, construction of the laboratory model, experimental work, laboratory measurements by B.P. (Bissé Poaty); computer simulations by K.E.R. with guidance from N.S. and D.S.; formal analysis, investigation, K.E.R., N.S., and K.U.M.; funding acquisition, K.U.M., B.B.; resources, D.S., B.P. (Bissé Poaty), B.P. (Benoît Plante), B.B., K.U.M.; software, D.S.; validation, N.S. and K.U.M., with site-specific guidance from B.P. (Benoît Plante) and B.B.; visualization, K.E.R., N.S., D.S.; writing—original draft, K.E.R.; writing—review and editing and supervision, N.S. and K.U.M. All authors have read and agreed to the published version of the manuscript.

Funding: This research was funded by the Natural Sciences and Engineering Research Council of Canada (NSERC) through the strategic network grant "Toward Environmentally Responsible Resource Extraction Network (TERRE-NET)" (Program director D.W. Blowes), grant number 479708-2015 and an Alexander Graham Bell Canada Graduate Scholarship-Master's (CGS M), award to K.E. Raymond.

Acknowledgments: Guidance was provided by H. Modzelewski at UBC in the set up and use of high-performance computing on the UBC EOAS clusters for numerical simulations.

Conflicts of Interest: The authors declare no conflict of interest. The funders had no role in the design of the study; in the collection, analyses, or interpretation of data; in the writing of the manuscript, or in the decision to publish the results.

Appendix A. Experimental Details and Material Properties

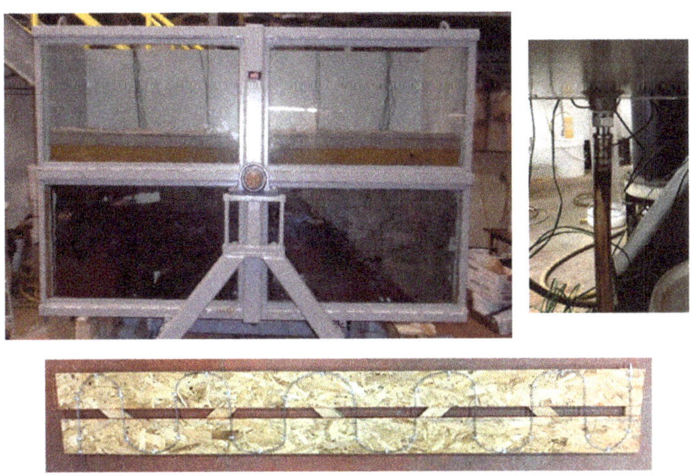

Figure A1. Photos of the laboratory experimental set-up. *Top left*: Front view of the laboratory scale model. HI rock is the dark grey material on the left which was deposited in three segregated zones; anorthosite is the light grey material (to the right of the HI); sand is orange-brown in colour along the top with crushed anorthosite (light grey) above to form the FCL. The tank was pivoted at an angle of 5° for the duration of the experiments with steady-state flow rates. *Top right*: Drainage tubes 1 cm in diameter that allow effluent to freely drain into the collection buckets below. *Bottom*: Set up of tubing with drainage holes installed along the top of the lab model to distribute the water applied at 180 L day^{-1}.

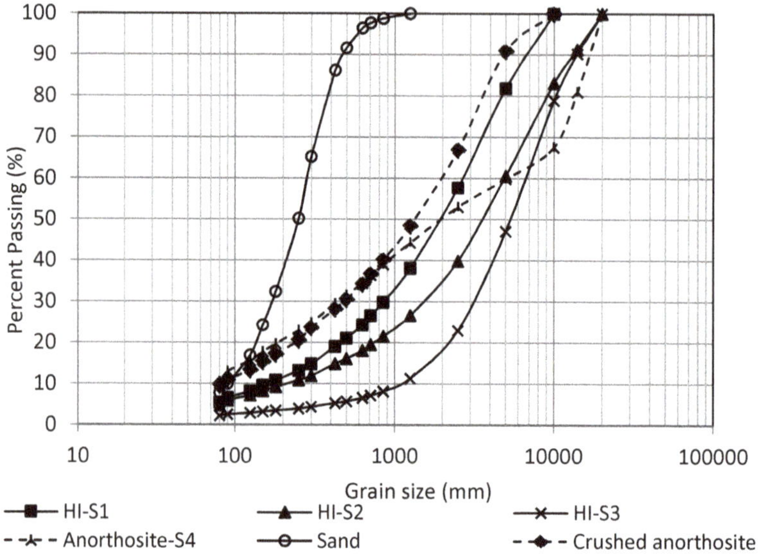

Figure A2. Grain size distribution curves for the materials in the laboratory experiment [31].

Appendix B. Model Discretization

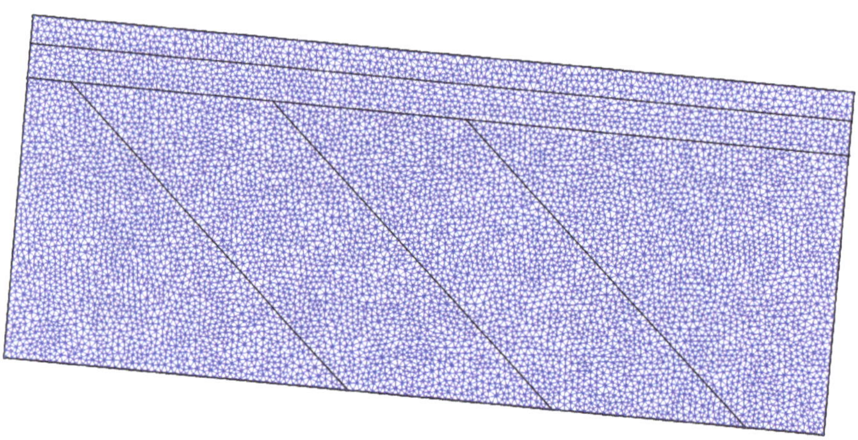

Figure A3. Triangular mesh with zone outlines used for the numerical model. The mesh file used in numerical simulations consisting of 16,602 cells sloped at a 5° angle. The cells have an average representative elemental volume of 1.4 cm^3. This unstructured grid allows for the influence of a 5° FCL slope to be investigated, as well as the effects of internal sloped boundaries following the angle of repose.

Appendix C. Statistical Analysis of Major Component Concentrations in Drainage Waters

Table A1. Statistical analysis of the major component concentrations in the drainage waters for reservoirs 1–8 for days 13–17. The averages, median, standard deviation and coefficient of variation are shown. The bolded cells indicate a coefficient of variation of CoV < 0.1.

				Res 1					
Component	Al	Ca	Fe	K	Mg	Na	Ni	S	Si
Average	0.22	30.70	0.07	1.72	4.47	6.93	0.28	25.44	3.86
Median	0.16	29.30	0.05	1.52	4.03	6.81	0.06	23.80	3.83
Standard deviation	0.16	2.09	0.03	0.45	1.09	0.46	0.50	3.79	0.16
Coefficient of variation	1.03	**0.07**	0.54	0.30	0.27	**0.07**	8.59	0.16	**0.04**
				Res 2					
Component	Al	Ca	Fe	K	Mg	Na	Ni	S	Si
Average	0.12	30.56	0.17	1.63	4.45	6.79	0.30	26.08	3.62
Median	0.10	29.00	0.15	1.61	4.39	6.52	0.28	25.20	3.59
Standard deviation	0.05	3.74	0.07	0.11	0.20	0.41	0.09	1.73	0.13
Coefficient of variation	0.46	0.13	0.50	**0.07**	**0.05**	**0.06**	0.33	**0.07**	**0.04**
				Res 3					
Component	Al	Ca	Fe	K	Mg	Na	Ni	S	Si
Average	0.05	32.23	0.04	2.13	5.56	6.69	1.46	30.68	3.41
Median	0.05	29.20	0.04	2.12	5.68	6.64	1.45	30.75	3.43
Standard deviation	0.01	6.26	0.01	0.08	0.29	0.43	0.28	0.94	0.08
Coefficient of variation	0.33	0.21	0.17	**0.04**	**0.05**	**0.06**	0.19	**0.03**	**0.02**
				Res 4					
Component	Al	Ca	Fe	K	Mg	Na	Ni	S	Si
Average	0.02	29.17	0.07	2.32	5.98	6.57	1.90	31.20	3.45
Median	0.02	28.75	0.08	2.31	6.04	6.47	1.95	31.45	3.44
Standard deviation	0.01	2.02	0.01	0.07	0.19	0.39	0.16	0.97	0.12
Coefficient of variation	0.40	**0.07**	0.15	**0.03**	**0.03**	**0.06**	**0.08**	**0.03**	**0.03**
				Res 5					
Component	Al	Ca	Fe	K	Mg	Na	Ni	S	Si
Average	0.05	31.80	0.05	2.42	6.44	6.61	1.10	31.92	3.89
Median	0.02	32.80	0.05	2.59	7.10	6.55	1.37	34.30	3.90
Standard deviation	0.07	2.29	0.01	0.47	1.42	0.15	0.58	4.97	0.10
Coefficient of variation	3.86	**0.07**	0.20	0.18	0.20	**0.02**	0.43	0.14	**0.02**
				Res 8					
Component	Al	Ca	Fe	K	Mg	Na	Ni	S	Si
Average	0.02	38.62	0.09	3.23	9.99	6.60	2.57	51.63	5.31
Median	0.02	39.05	0.09	3.25	10.10	6.59	2.60	52.40	5.33
Standard deviation	0.01	1.83	0.01	0.11	0.52	0.24	0.13	2.71	0.18
Coefficient of variation	0.39	**0.05**	0.12	**0.03**	**0.05**	**0.04**	**0.05**	**0.05**	**0.03**

Appendix D. Results of Sensitivity Analysis for Reduced Infiltration Rate

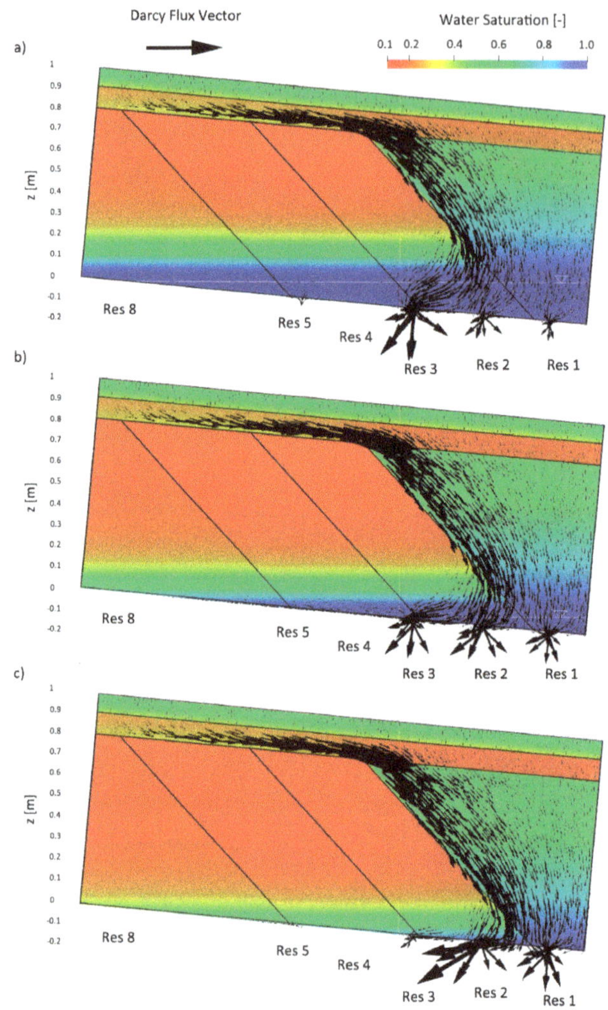

Figure A4. Simulated results of the flow models with a reduced infiltration rate: (**a**) water table as simulated for the flow model with the original infiltration rate ($z \approx -2$ cm); (**b**) water table lowered, located at $z \approx -13$ cm to represent the conditions for lower infiltration rates; and (**c**) no water table within the domain (pressure head negative at all outflow ports). Darcy flux vectors have been scaled 6-fold compared to Figures 4 and 6 for increased clarity. Flux vectors along the base demonstrate that changes in the water table location influence the outflow rates at the drainage ports depending on boundary conditions; however, the simulations also show that the results related to the functioning of the FCL and associated CBEs are not affected by the location of the water table.

References

1. Lawrence, R.W.; Wang, Y. *Determination of Neutral Potential for Acid Rock Drainage Prediction*; MEND Report 1.16.13; MEND: Smithers, BC, Canada, 1 July 1996; p. 89.
2. Blowes, D.; Ptacek, C.; Jambor, J.; Weisener, C. The Geochemistry of Acid Mine Drainage. In *Treatise on Geochemistry*; Elsevier BV: Amsterdam, The Netherlands, 2003; Volume 9, pp. 149–204.
3. MEND. *Review of Water Quality Issues in Neutral pH Drainage: Examples and Emerging Priorities for the Mining Industry in Canada*; MEND Report 10.12004; MEND: Smithers, BC, Canada, 2004; p. 58.
4. Price, W.A. *Prediction Manual for Drainage Chemistry from Sulphidic Geologic Materials*; MEND Report 1.20.1.; MEND: Smithers, BC, Canada, 2009; p. 88.
5. Vriens, B.; Arnault, M.S.; Laurenzi, L.; Smith, L.; Mayer, K.U.; Beckie, R.D. Localized Sulfide Oxidation Limited by Oxygen Supply in a Full-Scale Waste-Rock Pile. *Vadose Zone J.* **2018**, *17*, 180119. [CrossRef]
6. Sherlock, E.J.; Lawrence, R.W.; Poulin, R. On the neutralization of acid rock drainage by carbonate and silicate minerals. *Environ. Earth Sci.* **1995**, *25*, 43–54. [CrossRef]
7. Price, W.A.; Errington, J.C. *Guidelines for Metal Leaching and Acid Rock Drainage at Mine Sites in British Columbia*; Ministry of Energy Mines: Victoria, BC, Canada, 1998; pp. 1–92.
8. Tremblay, G.A.; Hogan, C.M. *Mine Environment Neutral Drainage (MEND) Manual 5.4.2d: Prevention and Control*; MEND Report 5.4.2; MEND: Smithers, BC, Canada, 2001; p. 352.
9. MEND. *Design, Construction, and Performance Monitoring of Cover Systems for Waste Rock and Tailings*; MEND Report 2.21.4a; MEND: Smithers, BC, Canada, 2004; Volume 1, p. 82.
10. Haug, M.D. *A Review of a Non-Traditional Dry Cover*; MEND Report 2.21.3b; MEND: Smithers, BC, Canada, 2002; p. 142.
11. Johnson, D.B.; Hallberg, K.B. Acid mine drainage remediation options: A review. *Sci. Total. Environ.* **2005**, *338*, 3–14. [CrossRef] [PubMed]
12. INAP GARD Guide. Available online: http://www.gardguide.com/index.php?title=Main_Page (accessed on 11 February 2020).
13. Linklater, C.M.; Sinclair, D.J.; Brown, P.L. Coupled chemistry and transport modelling of sulphidic waste rock dumps at the Aitik mine site, Sweden. *Appl. Geochem.* **2005**, *20*, 275–293. [CrossRef]
14. Molson, J.; Fala, O.; Aubertin, M.; Bussière, B. Numerical simulations of pyrite oxidation and acid mine drainage in unsaturated waste rock piles. *J. Contam. Hydrol.* **2005**, *78*, 343–371. [CrossRef] [PubMed]
15. Fala, O.; Molson, J.; Aubertin, M.; Dawood, I.; Bussière, B.; Chapuis, R. A numerical modelling approach to assess long-term unsaturated flow and geochemical transport in a waste rock pile. *Int. J. Min. Reclam. Environ.* **2012**, *27*, 38–55. [CrossRef]
16. Mayer, K.U.; Alt-Epping, P.; Jacques, D.; Arora, B.; Steefel, C.I. Benchmark problems for reactive transport modeling of the generation and attenuation of acid rock drainage. *Comput. Geosci.* **2015**, *19*, 599–611. [CrossRef]
17. Mayer, K.U.; Amos, R.T.; Blowes, D.W.; Wilson, D.; Bea, S.A.; Dipple, G.M.; Romano, C.; Su, D.; Raymond, K. Reactive Transport Modeling in Mine Waste Management—Part 1: Examination of its Role Based on Past and Present Applications. In Proceedings of the 26th Annual BC/MEND Metal Leaching/Acid Rock Drainage Workshop, Vancouver, BC, Canada, 12–16 August 2019.
18. Amos, R.; Blowes, D.W.; Bailey, B.L.; Sego, D.C.; Smith, L.; Ritchie, A.I.M. Waste-Rock hydrogeology and geochemistry. *Appl. Geochem.* **2015**, *57*, 140–156. [CrossRef]
19. Shokri, B.J.; Ardejani, F.D.; Ramazi, H.; Moradzadeh, A. Predicting pyrite oxidation and multi-component reactive transport processes from an abandoned coal waste pile by comparing 2D numerical modeling and 3D geo-electrical inversion. *Int. J. Coal Geol.* **2016**, *164*, 13–24. [CrossRef]
20. Lahmira, B.; Lefebvre, R.; Aubertin, M.; Bussière, B. Effect of material variability and compacted layers on transfer processes in heterogeneous waste rock piles. *J. Contam. Hydrol.* **2017**, *204*, 66–78. [CrossRef]
21. Blackmore, S.; Vriens, B.; Sorensen, M.; Power, I.M.; Smith, L.; Hallam, S.J.; Mayer, K.U.; Beckie, R.D. Microbial and geochemical controls on waste rock weathering and drainage quality. *Sci. Total. Environ.* **2018**, *640–641*, 1004–1014. [CrossRef] [PubMed]
22. Wilson, D.; Amos, R.T.; Blowes, D.W.; Langman, J.B.; Smith, L.; Sego, D.C. Diavik Waste Rock Project: Scale-up of a reactive transport model for temperature and sulfide-content dependent geochemical evolution of waste rock. *Appl. Geochem.* **2018**, *96*, 177–190. [CrossRef]

23. Su, D.; Mayer, K.U.; MacQuarrie, K.T.B. Numerical investigation of flow instabilities using fully unstructured discretization for variably saturated flow problems. *Adv. Water Resour.* **2020**, *143*, 103673. [CrossRef]
24. Plante, B.; Benzaazoua, M.; Bussière, B. Kinetic Testing and Sorption Studies by Modified Weathering Cells to Characterize the Potential to Generate Contaminated Neutral Drainage. *Mine Water Environ.* **2010**, *30*, 22–37. [CrossRef]
25. Plante, B.; Bussière, B.; Benzaazoua, M. Lab to field scale effects on contaminated neutral drainage prediction from the Tio mine waste rocks. *J. Geochem. Explor.* **2014**, *137*, 37–47. [CrossRef]
26. Aubertin, M.; Cifuentes, E.; Apithy, S.A.; Bussière, B.; Molson, J.; Chapuis, R.P. Analyses of water diversion along inclined covers with capillary barrier effects. *Can. Geotech. J.* **2009**, *46*, 1146–1164. [CrossRef]
27. Dubuc, J.; Pabst, T.; Aubertin, M. An Assessment of the Hydrogeological Response of the Flow Control Layer Installed on the Experimental Waste Rock Pile at the Lac Tio Mine. In Proceedings of the 70th Canadian Geotechnical Conference, Ottawa, ON, Canada, 1–4 October 2017.
28. Dimech, A.; Chouteau, M.; Martin, V.; Aubertin, M.; Bussière, B.; Plante, B. 3D Time-Lapse geoelectrical monitoring of moisture content in an experimental waste rock pile: Validation using hydrogeological data. *Symp. Appl. Geophys. Eng. Environ. Prob.* **2018**. [CrossRef]
29. Dimech, A.; Chouteau, M.; Aubertin, M.; Bussière, B.; Martin, V.; Plante, B. Three-Dimensional Time-Lapse Geoelectrical Monitoring of Water Infiltration in an Experimental Mine Waste Rock Pile. *Vadose Zone J.* **2019**, *18*, 1–19. [CrossRef]
30. Martin, V.; Bussière, B.; Plante, B.; Pabst, T.; Aubertin, M.; Medina, F.; Bréard-Lanoix, M.L.; Dimech, A.; Dubuc, J.; Poaty, B.; et al. Controlling water infiltration in waste rock piles: Design, construction, and monitoring of a large-scale in-situ pilot test pile. In Proceedings of the 70th Canadian Geotechnical Conference, Ottawa, ON, Canada, 1–4 October 2017.
31. Poaty, B. Comportement Hydrogéochimique Des Stériles D'une Halde Expérimentale Visant à Limiter la Production du Drainage Neutre Contaminé par Contrôle des Écoulements. Ph.D. Thesis, Université du Québec en Abitibi-Témiscamingue, Rouyn-Noranda, QC, Canada, 2020.
32. Bussière, B.; Plante, B.; Broda, S.; Aubertin, M.; Chen, D.; Medina, F. Water infiltration control in waste rock piles with capillary barrier effects: Design, construction and instrumentation of the experimental waste rock pile in the Lac Tio mine (in French). In Proceedings of the Symposium on Environment and Mines, Rouyn-Noranda, QC, Canada, 14–17 June 2015.
33. Plante, B.; Benzaazoua, M.; Bussière, B.; Biesinger, M.; Pratt, A. Study of Ni sorption onto Tio mine waste rock surfaces. *Appl. Geochem.* **2010**, *25*, 1830–1844. [CrossRef]
34. Plante, B.; Benzaazoua, M.; Bussière, B. Predicting Geochemical Behaviour of Waste Rock with Low Acid Generating Potential Using Laboratory Kinetic Tests. *Mine Water Environ.* **2010**, *30*, 2–21. [CrossRef]
35. Li, M.G. Acid rock drainage prediction for low-sulfide, low-neutralization potential mine wastes. In Proceedings of the 5th International Conference on Acid Rock Drainage (ICARD), Denver, CO, USA, 20–26 May 2000; pp. 567–580.
36. Jambor, J.L.; Dutrizac, J.E.; Raudsepp, M. Measured and computed neutralization potentials from static tests of diverse rock types. *Environ. Earth Sci.* **2006**, *52*, 1173–1185. [CrossRef]
37. Oelkers, E.H.; Schott, J. An experimental study of enstatite dissolution rates as a function of pH, temperature, and aqueous Mg and Si concentration, and the mechanism of pyroxene/pyroxenoid dissolution. *Geochim. Cosmochim. Acta* **2001**, *65*, 1219–1231. [CrossRef]
38. Zakaznova-Herzog, V.; Nesbitt, H.W.; Bancroft, G.; Tse, J. Characterization of leached layers on olivine and pyroxenes using high-resolution XPS and density functional calculations. *Geochim. Cosmochim. Acta* **2008**, *72*, 69–86. [CrossRef]
39. Stoch, L. Transformations of Micas in the Process of Kaolinitization of Granites and Gneisses. *Clays Clay Min.* **1976**, *24*, 156–162. [CrossRef]
40. Fala, O.; Aubertin, M.; Molson, J.W.; Bussière, B.; Wilson, W.G.; Chapuis, R.P.; Martin, V. Numerical Modelling of Unsaturated Flow in Uniform and Heterogeneous Waste Rock Piles. In Proceedings of the 6th International Conference on Acid Rock Drainage (ICARD), Cairns, QL, Australia, 14–17 July 2003; pp. 12–18.
41. Bréard-Lanoix, M.L. Characterization of Hydrogeological Properties of the Flow Control Layer Constructed on the Top of the Experimental Waste Rock Pile in the Lac Tio Mine (in French). Ph.D. Thesis, Mémoire de Maîtrise, Polytechnique de Montréal, Montréal, QC, Canada, 2017.

42. NOAA ESRL Global Monitoring Laboratory—One Year of CO2 Daily and Weekly Means at Mauna Loa. Available online: https://www.esrl.noaa.gov/gmd/webdata/ccgg/trends/co2_weekly_mlo.pdf (accessed on 18 October 2019).
43. Bea, S.A.; Mayer, K.U.; MacQuarrie, K.T.B. Reactive transport and thermo-hydro-mechanical coupling in deep sedimentary basins affected by glaciation cycles: Model development, verification, and illustrative example. *Geofluids* **2015**, *16*, 279–300. [CrossRef]
44. Su, D.; Mayer, K.U.; MacQuarrie, K.T. Parallelization of MIN3P-THCm: A high performance computational framework for subsurface flow and reactive transport simulation. *Environ. Model. Softw.* **2017**, *95*, 271–289. [CrossRef]
45. Xie, M.; Su, D.; Mayer, K.U.; Macquarrie, K. A Three-dimensional Numerical Model for Multicomponent Reactive Transport in Variably Saturated Porous Media: Verification and demonstration reports (draft). Unpublished work, 2015; 1–165.
46. Henderson, T.; Mayer, K.U.; Parker, B.L.; Al, T.A. Three-dimensional density-dependent flow and multicomponent reactive transport modeling of chlorinated solvent oxidation by potassium permanganate. *J. Contam. Hydrol.* **2009**, *106*, 195–211. [CrossRef] [PubMed]
47. Mayer, K.U.; MacQuarrie, K.T.B. Solution of the MoMaS reactive transport benchmark with MIN3P—Model formulation and simulation results. *Comput. Geosci.* **2009**, *14*, 405–419. [CrossRef]
48. Mayer, K.U.; Frind, E.O.; Blowes, D.W. Multicomponent reactive transport modeling in variably saturated porous media using a generalized formulation for kinetically controlled reactions. *Water Resour. Res.* **2002**, *38*, 13–21. [CrossRef]
49. Intissar, R. Apport des Méthodes Électriques à la Détermination de la Structure Interne d'Une Halde à Stériles. Ph.D. Thesis, Mémoire de Maîtrise, Polytechnique de Montréal, Montréal, QC, Canada, 2009.
50. Peregoedova, A. Étude Expérimentale des Propriétés Hydrogéologiques des Roches Stériles à une Échelle Intermédiaire de Laboratoire, Polytechnique, Montreal. Ph.D. Thesis, Mémoire de Maîtrise, Polytechnique de Montréal, Montréal, QC, Canada, 2012.
51. Van Genuchten, M.T. A Closed-form Equation for Predicting the Hydraulic Conductivity of Unsaturated Soils. *Soil Sci. Soc. Am. J.* **1980**, *44*, 892–898. [CrossRef]
52. Ball, J.; Nordstrom, D.K. User's manual for WATEQ4F, with revised thermodynamic data base and text cases for calculating speciation of major, trace, and redox elements in natural waters. *Open-File Rep.* **1991**. [CrossRef]
53. Demers, I.; Molson, J.; Bussiere, B.; Laflamme, D. Numerical modeling of contaminated neutral drainage from a waste-rock field test cell. *Appl. Geochem.* **2013**, *33*, 346–356. [CrossRef]
54. Williamson, M.A.; Kirby, C.S.; Rimstidt, J.D. Iron Dynamics in acid mine drainage. *J. Am. Soc. Min. Reclam.* **2006**, *2006*, 2411–2423. [CrossRef]
55. Blum, A.E.; Stillings, L.L. Chemical weathering rates of silicate minerals. *Rev. Mineral.* **1995**, *31*, 291–351.
56. Brantley, S.; Chen, Y. Chapter 4. Chemical Weathering rates of pyroxenes and amphiboles. *Chem. Weather. Rates Silic. Miner.* **1995**, *31*, 119–172. [CrossRef]
57. Nagy, K.L. Chapter 5. Dissolution and precipitation kinetics of sheet silicates. *Chem. Weather. Rates Silic. Miner.* **1995**, *31*, 173–234. [CrossRef]
58. Allison, J.D.; Brown, W.; Novo-Gradac, K.J. *MINTEQA2/PRODEFA2, A Geochemical Assessment Model for Environmental Systems: Version 3.0 User's Manual*; EPA/600/3-91/021; Environmental Research Laboratory, U.S. Environmental Protection Agency: Washington, DC, USA, 1991.
59. Lehner, S.; Savage, K. The effect of As, Co, and Ni impurities on pyrite oxidation kinetics: Batch and flow-through reactor experiments with synthetic pyrite. *Geochim. Cosmochim. Acta* **2008**, *72*, 1788–1800. [CrossRef]
60. Nicholson, R.V.; Gillham, R.W.; Reardon, E.J. Pyrite oxidation in carbonate-buffered solution: 1. Experimental kinetics. *Geochim. Cosmochim. Acta* **1988**, *52*, 1077–1085. [CrossRef]
61. Broda, S.; Aubertin, M.; Blessent, D.; Hirthe, E.; Graf, T. Improving control of contamination from waste rock piles. *Environ. Geotech.* **2017**, *4*, 274–283. [CrossRef]
62. Erguler, Z.A.; Erguler, G.K. The effect of particle size on acid mine drainage generation: Kinetic column tests. *Min. Eng.* **2015**, *76*, 154–167. [CrossRef]

63. Elghali, A.; Benzaazoua, M.; Bussière, B.; Bouzahzah, H. Determination of the available acid-generating potential of waste rock, part II: Waste management involvement. *Appl. Geochem.* **2019**, *100*, 316–325. [CrossRef]
64. Aranda, C.A. Assessment of Waste Rock Weathering Characteristics at the Antamina Mine Based on Field Cells Experiment. Master's Thesis, University of British Columbia, Vancouver, BC, Canada, 2010.
65. Komnitsas, K.; Xenidis, A.; Adam, K. Oxidation of pyrite and arsenopyrite in sulphidic spoils in Lavrion. *Min. Eng.* **1995**, *8*, 1443–1454. [CrossRef]

minerals

MDPI

Article

Implications of Sulfur Speciation on the Assessment of Acid Rock Drainage Generating Potential: A Study of South African Coal Processing Wastes

Annah Moyo [1], Juarez R. do Amaral Filho [2], Susan T.L. Harrison [1,2] and Jennifer L. Broadhurst [1,*]

[1] Minerals to Metals Initiative, Department of Chemical Engineering, University of Cape Town, Rondebosch, Cape Town 7700, South Africa; ann3mo@gmail.com (A.M.); sue.harrison@uct.ac.za (S.T.L.H.)

[2] Centre for Bioprocess Engineering Research, Department of Chemical Engineering, University of Cape Town, Rondebosch, Cape Town 7700, South Africa; juarez.amaralfilho@uct.ac.za

* Correspondence: jennifer.broadhurst@uct.ac.za

Received: 22 October 2019; Accepted: 9 December 2019; Published: 12 December 2019

Abstract: The presence of sulfur in coal processing wastes can lead to environmental impacts, such as acid rock drainage (ARD). However, not all sulfur species are acid-forming, and the implications of sulfur speciation when assessing acid rock drainage potential by means of static chemical tests are not well understood. This study set out to evaluate the implications of different sulfur forms on the assessment of acid rock drainage potential using static laboratory-scale tests and to investigate the reliability of methods for the analysis of such forms for the case of three South African coal processing wastes. Both the International Organization for Standardization (ISO) 157:1996 and Australian Coal Association Research Program (ACARP) C15034 protocols were found to be suitable tools for analyzing the different forms of sulfur. Acid-generating sulfur forms constituted between 53% and 64% of the total sulfur in the wastes evaluated, with the maximum potential acidity (MPA) and net acid-producing potential (NAPP) values calculated on the basis of acid-forming sulfur being significantly lower than those calculated on the basis of total sulfur content. Results also showed that the partial conversion of sulfur species under the relatively aggressive conditions of the acid-neutralizing capacity (ANC) and net acid generation (NAG) tests may overestimate the potential acid generating potential in the case of coal. These findings highlight the uncertainties associated with standard ARD static tests and the importance of taking sulfur speciation into account when calculating the MPA for coal processing wastes.

Keywords: sulfur forms; coal processing wastes; acid rock generating potential; acid-base accounting; net acid generation

1. Introduction

The presence of sulfur-bearing phases, particularly pyrite, in coal and coal wastes leads to environmental impacts such as acid rock drainage (ARD) upon oxidation of the sulfur in the presence of moisture and oxygen [1–3]. Although the composition of South African (SA) coal differs across seams and across the 19 coalfields, in general, the SA coals are high in ash (typically 20–40%) and low in sulfur (typically between 0.15–1.8%, although values as high as 15% have been recorded) [4]. As a result of the high extraneous mineral matter and moisture content, the run-of-mine (ROM) coal has low calorific values. Thus the coal has to be beneficiated to lower the ash (to below 15% and 21% for export and local thermal coals, respectively) and sulfur (to below 1% for both local and export thermal coals) contents, and to increase calorific value (to above 27.5 and 24.7 MJ/kg for export and local thermal coals, respectively) [5,6].

The coal preparation process to produce clean coal usually involves two major operations, namely screening and beneficiation (generally termed washing). The screened material can be sold directly or processed further by washing or beneficiation if the screening alone does not meet the required product specifications [7]. A typical South African washing plant employs dense medium separation (DMS) to beneficiate coarse (>25 mm) and intermediate/middling (1–25 mm) coal, whilst spiral concentrators are commonly used for beneficiation of fine (0.15–1 mm) coal [8]. Whilst some plants beneficiate ultra-fine (<0.15 mm) coal by means of froth flotation, many South African collieries still dispose of ultra-fines directly after dewatering. Historically, dewatering was mainly by means of thickeners, but many collieries have now installed more sophisticated dewatering equipment, such as filter presses to improve water recovery, reducing the moisture content of the slurry to roughly 25% [8]. Although the term "discards" is sometimes used in reference to all wastes arising from the coal preparation process, two types of coal processing wastes can be differentiated: discard waste from the processing of coarse, intermediate and fine coals, and ultra-fine slurry or slurry tailings [8–10]. Discards are compacted and piled into dumps which are then covered with soil and plant vegetation, whilst ultra-fine coal slurry wastes are either used as backfill material or disposed of, either in slurry dams or together with discards in the dumps [8]. Total processing waste typically accounts for between 20% and 22% of the ROM coal, 4–6% of which is in the form of an ultra-fine slurry. The typical composition of the coal wastes according to the national inventory of 2001 by the Department of Minerals and Energy (DME) is presented in Table 1.

Table 1. Typical compositions of discard and ultra-fine slurry coal waste, as adapted from the SA department of minerals and energy [9].

Parameter	Discards	Ultra-Fine Slurry
Ash (%)	30–60	10–50
Caloric value (MJ/kg)	11–20	20–27
Fixed carbon (%)	18–42	41–56
Sulfur (%)	1–5	≤2
Volatile matter (%)	18–24	17–27

1.1. The Occurrence of Sulfur in Coal and Coal Processing Wastes

Sulfur occurs in coal in three main forms: organically bound sulfur (usually <3%), inorganic sulfides and inorganic sulfates (rarely exceed 0.1% except in weathered coal samples). Sulfur in coal rarely occurs as free sulfur, which is usually produced during coal weathering [11–13]. Whilst the main inorganic sulfide sulfur occurs mainly in the form of pyrite (FeS_2), other sulfides such as galena (PbS), sphalerite (ZnS), arsenopyrite (FeAsS), and chalcopyrite ($CuFeS_2$) may occur in coal in trace amounts [11]. Organic sulfur compounds are generally grouped into thiophenes, mercaptans, disulfides and aliphatic and aryl sulfides [13]. The pyritic and organic sulfur in South African coal increases with increasing vitrinite content from west to east geologically [4]. The organic sulfur occurs embedded in the coal matrix during the formation period and is therefore hard to remove during washing, while the inorganic sulfur occurs along the cleats and cracks and is more easily removed by washing. Coal beneficiation wastes, particularly discards, thus tend to have higher inorganic sulfur/organic sulfur ratios than the ROM coal, with the absolute values depending on the coal petrography and subsequent preparation processes employed [13]. Reported values for sulfur and sulfur forms in ultra-fine slurry wastes from South African coalfields are presented in Table 2.

Table 2. Total sulfur and sulfur forms in South African ultrafine coal wastes from collieries in Waterberg, Witbank [14] and Middleburg [15,16].

Sulfur Form	Content (Mass % of Whole Coal)		
	Waterberg	Witbank	Middleburg
Total sulfur	2.04	4.18	0.80–1.10
Sulfide sulfur	0.98	2.58	0.64–1.05
Organic sulfur	0.56	0.16	0.25–0.51
Sulfate sulfur	0.50	1.44	0.05–0.50

1.2. Sulfur Speciation and ARD Formation

Previous studies have shown that the environmental risks associated with coal preparation and utilization can be attributed largely to the presence of sulfur and its speciation [16–18]. A key environmental risk is that of ARD generation, which has been mainly attributed to the presence of pyrite in coal seams and coal waste piles [17–20]. The mechanisms of the formation of ARD is documented by several researchers such as [1,21–25]. In ARD formation, pyrite (FeS_2) oxidizes in the presence of water to ferrous ions (Fe^{2+}), acidic ion (H^+) and sulfates ($SO_4{}^{2-}$) as shown in Equation (1). The ferrous iron is subsequently oxidized to ferric iron (Fe^{3+}) (Equation (2)). At pH values below 4.5, ferric iron becomes the dominant oxidant, oxidising pyrite according to the reaction shown in Equation (3) [26]. In natural disposal systems, the ARD reactions are catalysed by acidophiles such as *Acidothiobacillus ferrooxidans* (Fe and S oxidizer), *Acidothiobacillus thiooxidans* (S oxidizer) and *Leptospirillum ferrooxidans* (Fe oxidizer) [25]. The cyclic oxidation of pyrite and generation of ferric ions can continue for decades producing more acid and sulfates which can contaminate water bodies if not contained.

$$FeS_2 + \frac{7}{2}O_2 + H_2O \rightarrow Fe^{2+} + 2SO_4{}^{2-} + 2H^+ \tag{1}$$

$$Fe^{2+} + \frac{1}{4}O_2 + H^+ \rightarrow Fe^{3+} + \frac{1}{2}H_2O \tag{2}$$

$$FeS_2 + 14Fe^{3+} + 8H_2O \rightarrow 15Fe^{2+} + 2SO_4{}^{2-} + 16H^+ \tag{3}$$

Although pyrite is the acid-producing sulfide with the highest capacity in mine wastes, other sulfides such as pyrrhotite, arsenopyrite, chalcopyrite, galena and sphalerite (reported to also occur in coal [11] are also acid-producing (as shown in Equations (6)–(10)) but with a low capacity [27,28]. Reactivity is in the order pyrrhotite > galena-sphalerite > pyrite-arsenopyrite > chalcopyrite if galvanic and biological interactions are not considered [29–31].

$$Fe_{(1-x)}S + (\frac{2-x}{2})O_2 + xH_2O \rightarrow (1-x)Fe^{2+} + SO_4{}^{2-} + 2xH^+, \tag{4}$$

$$FeAsS + \frac{7}{2}O_2 + 6H_2O \rightarrow Fe(OH)_3 + SO_4{}^{2-} + H_2AsO_4{}^{-} + 3H^+, \tag{5}$$

$$2CuFeS_2 + \frac{17}{2}O_2 + 5H_2O \rightarrow 2Cu^{2+} + 2Fe(OH)_3 + 4SO_4{}^{2-} + 4H^+, \tag{6}$$

$$ZnS + 2O_2 \rightarrow Zn^{2+} + SO_4{}^{2-}, \tag{7}$$

$$PbS + 2O_2 \rightarrow Pb^{2+} + SO_4{}^{2-}. \tag{8}$$

Additional acid is generated through the formation of secondary precipitates. At pH values above 3, the hydrolysis of ferric iron, Fe^{3+}, forms hydroxide precipitates, generating acidity in accordance with the reactions in Equations (9) and (10) [25,32]. At lower pH values the formation of secondary acid-forming sulfates, such as jarosite (Equation (11)) occurs, generating further acid.

$$Fe^{3+} + 3H_2O \rightarrow Fe(OH)_3 + 3H^+, \quad (9)$$

$$Fe^{3+} + 2H_2O \rightarrow FeOOH + 3H^+, \quad (10)$$

$$3Fe^{3+} + K + 2SO_4{}^{2-} + 6H_2O \leftrightarrow KFe_3(SO_4)_2(OH)_6 + 6H^+. \quad (11)$$

Other secondary acid-forming sulfates reported to occur in weathered coal and coal wastes include alunite ($KAl_3(SO_4)_2(OH)_6$), melanterite ($FeSO_4{\cdot}7H_2O$) and schwertmannite ($Fe_8O_8(OH)_6(SO_4){\cdot}nH_2O$). These metastable sulfur species generate acid when they convert to more stable ferric hydroxides and oxides as shown in Equations (12) and (13) for jarosite and melanterite respectively [33,34]:

$$\textit{jarosite} \rightarrow \textit{geothite}: \; KFe_3(SO_4)_2(OH)_6 + 3H_2O \rightarrow 3Fe(OH)_3 + K^+ + 3H^+ 2SO_4{}^{2-}, \quad (12)$$

$$FeSO_4{\cdot}7H_2O + \frac{1}{4}O_2 \rightarrow Fe(OH)_3 + \frac{9}{2}H_2O + 2H^+ + SO_4{}^{2-}. \quad (13)$$

However, sulfates of the alkaline earth metals gypsum ($CaSO_4{\cdot}7H_2O$) and epsomite ($MgSO_4{\cdot}7H_2O$), commonly found in coal and coal wastes, are non-acid forming [33]. The organic sulfur in coal is presumed to be non-acid forming as it is mainly present in the form of heterocyclic compounds organically bonded to the coal [33,35].

The acid produced from the pyrite and other acid-forming sulfur species is neutralized by gangue material such as carbonates, silicates and clay minerals. The neutralizing potential of these minerals varies according to their reaction rates and is dependent on the environmental pH [22]. The carbonate with the highest neutralizing potential is calcite ($CaCO_3$), as it displays the most rapid dissolution. Other effective neutralizing carbonates are magnesite ($MgCO_3$), dolomite ($CaMg(CO_3)_2$) and ankerite ($CaFe(CO_3)_2$) [22]. Many of the silicate minerals, particularly quartz (SiO_2), are inert but others, such as forsterite (Mg_2SiO_4), olivine ($Mg_{1.6}Fe^{2+}{}_{0.4}SiO_4$), wollastonite ($CaSiO_3$) and serpentine ($Mg_{2.25}Fe^{2+}{}_{0.75}Si_2O_5(OH)_4$) are acid neutralizing, although they have slower dissolution rates and lower neutralizing capacities compared to carbonates [27,28]. The silicates can maintain neutral conditions if the acid formation rate is slow [22]. The net acid generation potential is, therefore, a function of both the acid-forming and neutralizing capacity, as well as the relative rates of reaction of the constituent minerals of a sample [27,28].

1.3. Assessment of Acid Rock Drainage Potential

Due to their relatively low cost and rapid turn-around times, laboratory-scale static geochemical tests are commonly used to assess the potential for mineral-bearing solids to generate ARD. Among these, acid-base accounting (ABA) and net acid generation (NAG) tests are widely applied [25,36]. In the ABA method, the net acid-producing potential (NAPP) is calculated as the difference between maximum potential acidity (MPA) and acid-neutralizing potential (ANC). The MPA is usually determined by multiplying the total sulfur content (%) by the stoichiometric factor of 30.6 giving a result in kg H_2SO_4/t [36]. However, researchers have reported that estimating MPA based on total sulfur may misclassify the test samples because not all the sulfur is present as pyritic sulfur [33,34]. This is particularly the case for coal and coal wastes, which contain sulfur in the form of non-acid generating organic sulfur and sulfates. Using total sulfur to calculate MPA thus overestimates the NAPP, prompting researchers to recommend that MPA calculations be based on pyritic sulfur, rather than total sulfur, content [33–35].

The ANC is determined empirically by reacting the solid with a known amount of HCl and back-titrating the excess acid with a standard solution of NaOH. In conducting the ANC tests, it is assumed only acid-neutralizing minerals react with the added HCl and the sulfur species will remain unchanged. However, due to the lack of selectivity, some acid-soluble sulfides or sulfates can be partially oxidized or solubilized [37,38], thus potentially under-estimating the ANC of a sample. Although a number of versions of the ANC method exist, each with slight modifications to the method [34,36,39,40], most of these aim to determine the neutralizing capacity associated with the reactive carbonate minerals (such as calcite, dolomite) and fast weathering silicates (such as anorthite, olivine and wollastonite). However, many researchers have reported that the relatively aggressive conditions of the ANC tests overestimate the neutralizing capacity by promoting the dissolution of silicates that are slow reacting or even completely inert under environmental conditions [27,28,41,42].

In NAG tests the sample is dissolved in a known quantity of hydrogen peroxide (H_2O_2) as an oxidant, to allow for both the oxidative dissolution of acid-generating forming minerals and dissolution of acid-neutralizing minerals, before back titrating the excess acid with a standard solution of NaOH. This test assumes that all the sulfide sulfur is converted into soluble sulfate while all the acid-neutralizing components simultaneously buffer the acid produced. The NAG pH and the NAG capacity at pH 4.5 and pH 7 give an indication of the net acid-generating capacity. Misleading results have been reported for samples containing high pyritic sulfur (>0.7–1%), as the H_2O_2 decomposes before complete oxidation of the sulfide occurs. In such cases, a sequence of the NAG tests is repeated until complete oxidation occurs. In samples with high organic contents (>5%), the H_2O_2 can cause partial oxidation of the carbonaceous material in coal waste forming organic acids that contribute to net acidity, causing an overestimation of the sample acidity potential [35]. The extended boil protocol developed in Australia corrects for the effects of organic acid. However, aggressive boiling can also decompose free acidity [34,35].

1.4. Analysis of Total Sulfur and Sulfur Forms in Coal

The discussions above have highlighted the importance of developing a quantitative understanding of sulfur and its forms in coal wastes in the context of characterizing the potential for ARD generation. The most commonly used methods for total sulfur determination are the Eschka method (ASTM D1377, ISO 334:1993, SANS 334:1992), bomb washing (ASTM D-3286) and high-temperature combustion (ASTM D-4239, SABS 931) [13]. Sulfur is also determined by means of instrumental techniques, including ICP-AES/OES, ICP-MS, Ultraviolet-Visible Spectroscopy (UV-VIS) and chromatography, following complete digestion of the samples.

The most commonly used method to determine sulfur forms in coal and coal wastes is the International Organization for Standardization (ISO) 157:1996 protocol [13]. This protocol involves a two-stage sequential acid leach to determine sulfate sulfur and pyritic sulfur [13]. In the first leach step, dilute HCl is used to selectively dissolve the sulfate minerals, and the leachate analyzed for sulfate using the gravimetric $BaCl_2$ method. In the second step, HNO_3 is used to selectively solubilize the pyrite in the residue from the HCl digestion step, and the leachate subsequently analyzed for iron by means of atomic absorption spectroscopy (AAS), titration or colorimetric methods. Organic sulfur is then calculated by subtracting the sulfate sulfur and pyritic sulfur from the total sulfur content. The ISO 157:1996 protocol is based on a number of assumptions, one being that sulfate minerals are dissolved in the HCl leach step, with sulfide minerals and organic sulfur remaining in the solid phase. Another key assumption is that dilute HNO_3 only dissolves pyritic iron and that pyrite is the only sulfide mineral in the sample [13]. Finally, all the unaccounted sulfur is assumed to be present as organic sulfur. Iron from the dissolution of iron silicates or iron oxides can cause overestimation of pyritic sulfur in the standard method, whilst incomplete extraction of pyritic sulfur by HNO_3 can result in underestimation [43]. Elemental sulfur in the sample causes the overestimation of the organic sulfur because it is not extracted by either HCl or HNO_3 [32]. The presence of jarosite species causes overestimation of pyritic sulfur and underestimation of sulfate sulfur, as jarosite is only partially soluble in HCl, with the majority dissolving in HNO_3 and reporting as pyrite [12]. High levels of Ca or Ba in the sample may result in the formation of insoluble $BaSO_4$ or partially soluble $CaSO_4$, thus lowering sulfate sulfur values [13].

A research project supported by the Australian Coal Association Research Program (ACARP) C15034 developed a sulfur speciation protocol that aimed to overcome most of the above-mentioned shortcomings of the ISO 157:1996 method. The ACARP C15034 protocol has been specifically developed for coal "washery" wastes. For routine determination of sulfur species in coal processing wastes, the protocol is performed in three stages, each conducted on a separate representative sub-sample. In the first stage, total sulfur is determined using the standard combustion method. In the next stage, the chromium reducible sulfur (CRS) method [32] is used to determine the sulfide sulfur content. In this method sulfide and elemental sulfur are selectively reduced by chromium to form H_2S, which can be trapped in zinc acetate solution as ZnS. Although an acetone extraction step can be included to selectively remove and determine elemental sulfur [44], this is usually not necessary as coal processing wastes usually have only trace amounts of elemental sulfur [33]. In the third stage, all sulfate minerals (except jarosite) are selectively dissolved in the KCl solution and the filtrate separated into two portions. One portion is used to determine total sulfate, using the gravimetric $BaCl_2$ method, and the other to determine soluble acid-forming sulfates (such as melanterite) through back-titration with NaOH [33,34,44]. The remaining sulfur, which includes organic sulfur and sulfur present as jarosite, is termed "low-risk" sulfur.

This study sets out to evaluate and compare these methods for analyzing sulfur and sulfur species in South African coal processing wastes, and to investigate the implications of sulfur speciation and the behavior of sulfur forms on the classification of the ARD potential when using standard static laboratory tests.

2. Experimental Work

2.1. Sample Description and Preparation

To test a range of coal processing wastes, test work was conducted on three samples from different South African coalfields and processing stages: an ultrafine (d80 < 150 μm) thickener feed slurry from a colliery in Waterberg coalfield (sample A) and an ultrafine (d80 < 150 μm) thickener underflow slurry sample (sample B) and a coarse discard (d80 > 25 mm) sample (sample C) from a colliery in the Witbank coalfield. These samples were collected as grab samples and can thus be considered typical, but not necessarily representative, of the coalfields and operations from which they were collected. For reference, a South African (Free State) coal standard reference material, SARM 19, was used to assess the reproducibility and accuracy of the total sulfur methods. The air-dried samples were pulverized to −75 μm and split to the required amounts according to the test. The experimental work consisted of the analysis of total sulfur and sulfur forms, semi-quantitative mineralogical analysis, and ARD geochemical tests.

2.2. Mineralogical Analysis

A semi-quantitative analysis of the mineralogical composition of the three coal waste samples was carried out using a Field Emission Gun (FEG) Quantitative Evaluation of Minerals by Scanning Electron Microscopy (QEMSCAN) 650F machine with two XFlash 6130 detectors (manufactured by Bruker, Billerica, MA, USA). Bruker detectors use electron dispersion scatter (EDS) detection to measure elements. These element concentration values can be seen in the iDiscover programme for each Pixel measured. The QEMSCAN then uses a mineral library system, called the species identification protocol (SIP), to determine how the elemental information measured by the detectors is classified into a mineralogical composition or species. The carbon-coated sample blocks were put in a Quorum Q150 € coater to disperse the electron charge. The results of this analysis are considered to be semi-quantitative, as the SIP file for Waterberg and Witbank coal is still under development.

2.3. Total Sulfur Analysis

The total sulfur was determined using the standard gravimetric Eschka method (SANS 334:1992/ISO 334/1992) and the standard combustion method (ASTM D-4239, sub-method C; SABS method 931) using a LECO analyser SC 632. All tests were performed in triplicate by the same person for repeatability indication. The standard error analysis gave the precision of the method, and a relative standard error (RSE) of the SARM 19 results relative to certified values allowed for assessment of the accuracy of the methods.

2.4. Sulfur Speciation

The standard ISO 157:1996 protocol was conducted by ALS laboratories to determine the sulfur forms in the coal wastes and the coal standard. Analyses were conducted on three representative sub-samples over a protracted period. The ACARP C15034 protocol [33], was applied in-house to determine sulfide sulfur, total soluble sulfate and soluble acid-forming sulfate in the three coal waste samples and the residues from the leach stages of the standard ANC and NAG tests outlined in Section 2.5. Organic/low-risk sulfur content for the two protocols was calculated by difference, using the total sulfur results obtained from the LECO analysis (Section 2.3).

2.5. Characterisation of ARD Potential

The ARD potential of the wastes was determined using the conventional ABA and NAG tests, as well as the extended boil NAG test. In the ABA test, ANC was determined using the incremental H_2O_2 Modified Sobek method [45]. The MPA was calculated on the basis of total sulfur content determined by LECO analysis (Section 2.3), pyritic sulfur as determined by the ISO 157:1996 protocol, and acid-generating sulfur (sulfide and acid-forming sulfate sulfur) as determined using the ACARP C15034 protocol (Section 2.4). The NAG pH was determined using the single addition NAG [36] and the extended boil NAG protocol [34]. All the tests were conducted in triplicate.

2.6. Deportment of Sulfur Forms under ARD Test Conditions

The three waste samples were digested according to the ANC and NAG test methods, and the residues dried and analyzed for total sulfur and sulfur forms using LECO analysis and the ACARP C15034 method, respectively. The results were compared to the content of sulfur species in the waste samples before digestion to calculate their deportment under static test leach conditions.

3. Results and Discussion

3.1. Mineralogical Composition

Generally, the coal processing wastes were found to be made up of predominantly of coal (19–45%), carbominerite (coal made of 20–60% mineral matter content and 40–80% organic content; varying from 6–30%), quartz (SiO_2; 3–21%), kaolinite ($Al_2Si_2(OH)_4$; 21–52%), pyrite (FeS_2; 1.8–2.7%) and k-feldspar ($KAlSi_3O_8$; 0.6–2%). Whilst pyrite was the major sulfide mineral in all the samples, chalcopyrite ($CuFeS_2$ < 0.5%) was also identified in the Waterberg slurry and Witbank discard wastes. Identified sulfate minerals included trace amounts of gypsum ($CaSO_4{\cdot}2H_2O$ < 0.2%) in the two slurry wastes, and jarosite ($KFe_3(SO_4)_2(OH)_6$; 0.01%) in the Waterberg slurry waste. No other sulfates, such as alunite ($KAl_3(SO_4)_2(OH)_6$) or melanterite ($FeSO_4{\cdot}7H_2O$), were detected. Only the Waterberg slurry waste contained a significant quantity of carbonate, in the form of calcite ($CaCO_3$; 1.4%), with trace amounts occurring in the Witbank coal slurry (<0.05%). Amphibole, generally considered to be a silicate of intermediate reactivity, was identified as being present in significant quantities in the Waterberg slurry waste (~0.7%) and in trace quantities (<0.05%) in the Witbank slurry wastes. Other minor-trace minerals identified in the coal wastes included the silicate mineral mica ($K_2Al_4Si_6Al_2O_{20}(OH)_4$), the non-neutralizing carbonate mineral, siderite ($FeCO_3$), and the complex oxide minerals, iron oxyhydroxides, rutile (TiO_2) and apatite ($Ca_5(PO_4)_3(OH, F, Cl)$).

3.2. Total Sulfur Content

The results in Table 3 show that the sulfur content in the samples analysed decreases in the order Witbank discards < Waterberg slurry < SARM 19 standard < Witbank slurry. Whilst all the values were within the general range reported for South African coal wastes (c.f. Table 1), the sulfur content in the

Witbank coal slurry (1.06–1.11%) was significantly lower than that reported for previous slurry wastes from a different colliery in the Witbank coalfields (4.18%) [14]. This is consistent with the reported variance in the sulfur content of different coal seams within this region [4].

Table 3. Total sulfur contents in coal standard SARM 19 and coal waste samples determined by the Eschka method and LECO analysis.

Sample	Total S Content (Mass %)	
	LECO Method	Eschka Method
SARM 19	1.54 ± 0.03	1.30 ± 0.02
Waterberg coal slurry	1.84 ± 0.01	1.89 ± 0.01
Witbank coal slurry	1.06 ± 0.06	1.11 ± 0.05
Witbank coal discards	1.94 ± 0.08	n.d.

n.d. = not determined.

A comparison of total sulfur results with the certified value for the coal standard SARM 19 (Table 4), indicates that the LECO and Eschka methods gave comparable results with low standard errors (range of 0.01–0.03%), indicating the methods were highly precise. However, whilst the LECO results are consistent with the 95% confidence limit range reported for the certified coal standard with a relative standard error of 3.13%, the relative error for the Eschka test results was over 12%.

Table 4. The relative standard error of total sulfur analyzing techniques based on coal standard SARM 19 certified values.

Certified Values (%)		Mean Measured Values (%)		Relative Error (%)	
Mean	95% Confidence Limit	LECO-ALS	Eschka	LECO-ALS	Eschka
1.49	1.42–1.55	1.54 ± 0.03	1.30 ± 0.02	3.13	−12.53

3.3. *Sulfur Speciation Results*

The sulfur speciation results obtained for the two chemical protocols, ISO 157:1996 and ACARP C15034, are shown in Table 5 and Figure 1.

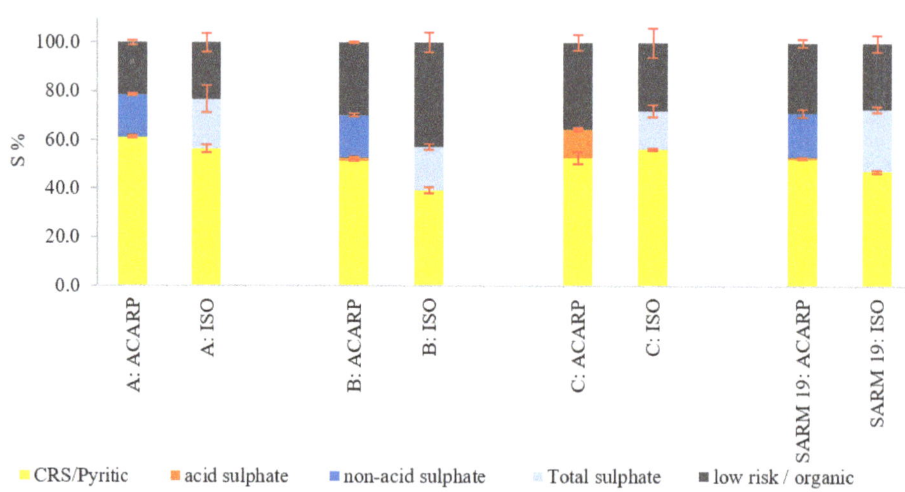

Figure 1. Distribution of sulfur forms in coal standard SARM 19 and coal waste samples A (Waterberg coal slurry), B (Witbank coal slurry) and C (Witbank coal discards) by ISO 157:1996 and ACARP C15034 protocols.

Table 5. The concentration of sulfur forms in coal standard (SARM 19) and coal waste samples determined by ISO 157:1996 and ACARP C15034 protocols.

Sulfur Form	Mass (%) of Sulfur Form in Sample							
	Waterberg Coal Slurry		Witbank Coal Slurry		Witbank Coal Discards		SARM 19	
	ACARP	ISO	ACARP	ISO	ACARP	ISO	ACARP	ISO
CRS/Pyritic [1]	1.13 ± 0.01	1.04 ± 0.02	0.55 ± 0.00	0.42 ± 0.01	1.02 ± 0.03	1.10 ± 0.01	0.80 ± 0.00	0.73 ± 0.01
Acid sulfate [2]	0.00 ± 0.00	-	0.01 ± 0.00	-	0.23 ± 0.01	-	0.01 ± 0.00	-
Non-acid sulfate [3]	0.32 ± 0.00	-	0.19 ± 0.00	-	-	-	0.28 ± 0.02	0.39 ± 0.01
Total Sulfate	0.32 ± 0.00	0.37 ± 0.06	0.20 ± 0.00	0.19 ± 0.01	0.23 ± 0.01	0.31 ± 0.02	0.29 ± 0.02	0.39 ± 0.01
Low-risk [4]**/organic**	0.39 ± 0.01	0.43 ± 0.04	0.32 ± 0.00	0.45 ± 0.04	0.69 ± 0.03	0.54 ± 0.06	0.44 ± 0.02	0.42 ± 0.03

[1] Determined as pyritic sulfur in ISO 157:1996 protocol and as chromium reducible sulfide (CRS) sulfur in ACARP C15034 protocol. [2] Not determined in ISO157:1996 protocol. [3] Not determined in ISO157:1996 protocol. [4] Calculated from the difference between total sulfur and sum of sulfate and sulfide/pyritic sulfur values.

Consistent with the findings of [16], the results in Figure 1 show that, in general, pyritic or sulfide sulfur accounts for more than 50% of the total sulfur. Sulfate sulfur accounts for 18–26% of the total sulfur for the coal standard (SARM 19) and both the coal slurry wastes, but only 12–15% of total sulfur for the Witbank coal discards. The ACARP test results indicate, furthermore, that all the soluble sulfate in Witbank coal discards is present as acid-forming soluble sulfate minerals (e.g., melanterite), whilst in the case of the Waterberg and Witbank slurry wastes and the SARM standard, the majority of the soluble sulfate sulfur (100%, 95.5% and 98.8%, respectively) is present as non-acid forming sulfates (e.g., gypsum and epsomite). The contribution of the remaining sulfur species, comprised mainly of organic sulfur, varies between 18% and 43%, with the proportion of organic sulfur being lowest in the Waterberg coal slurry waste (18–20%) and highest in Witbank coal discards (23–43%).

The sulfide or pyritic sulfur and soluble sulfate sulfur results are compared graphically in Figures 2 and 3, respectively. The results show that the standard ISO 157:1996 protocol resulted in lower sulfide sulfur values than the ACARP C15034 protocol for all samples except the Witbank discards. Conversely, in the case of sulfate sulfur, the standard ISO 157:1996 protocol resulted in slightly higher values than the ACARP C15034 protocol for all samples except the Witbank coal slurry (for which the results derived from both protocols were equivalent). The higher sulfate and lower pyritic sulfur values obtained for the ISO 157:1996 protocol in comparison to the ACARP C15034 protocol for the majority of the samples indicate that the relatively aggressive HCl leach may have resulted in some dissolution of sulfide or pyritic sulfur, thus overestimating the sulfate sulfur and underestimating the pyritic sulfur contents in the case of the ISO 157:1996 protocol. It is unclear why the ISO 157:1996 protocol resulted in higher sulfide sulfur values than the ACARP C15034 protocol in the case of the Witbank discards sample. This may possibly be due to partial dissolution of relatively stable Fe-bearing mineral phases other than pyrite (e.g., chalcopyrite and jarosite) during the HNO_3 leach step, resulting in an overestimation of the sulfide sulfur. However, whilst the mineralogical analysis indicated that this sample contains higher quantities of chalcopyrite than the other coal waste samples, there was no evidence of elevated jarosite concentrations. It should, furthermore, be noted that discrepancies between the two methods are relatively minor (mostly <10%), especially for sulfide sulfur. The results also indicate that, with the exception of one result (ISO 157:1996 total sulfate in the Waterberg slurry waste), the standard deviations obtained for both protocols were generally <10%.

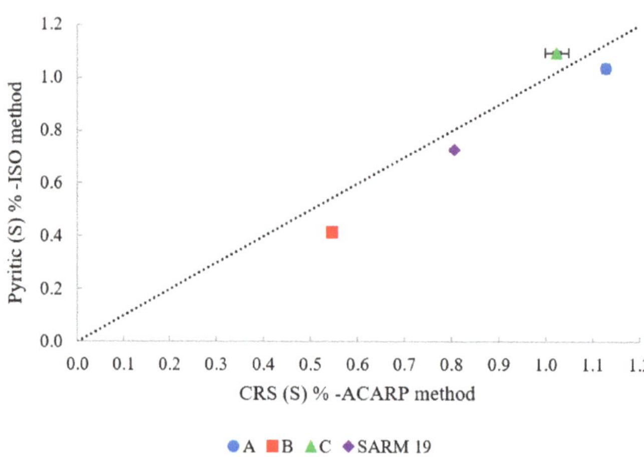

Figure 2. Comparison of ISO157:1996 and ACARP C15034 protocols in determining pyritic/sulfide sulfur on coal standard SARM 19 and coal waste samples A (Waterberg coal slurry), B (Witbank coal slurry) and C (Witbank coal discards).

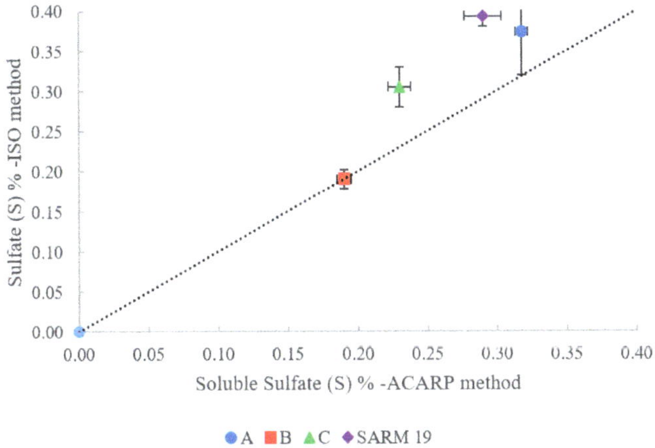

Figure 3. Comparison of ISO157:1996 and ACARP C15034 methods in determining sulfate sulfur on coal standard SARM 19 and coal waste samples A (Waterberg coal slurry), B (Witbank coal slurry) and C (Witbank coal discards).

3.4. The Implication of Sulfur Species on ARD Classification

The results for the static acid-base accounting (ABA) and net acid generation (NAG) tests for the classification of ARD potential are presented in Table 6, where the maximum potential acidity (MPA) and net acid-producing potential (NAPP) for the ABA tests have been calculated on the basis of total sulfur (S (T), as determined by means of LECO), pyritic sulfur (S^{2-}, as determined by means of the ISO 157:1996 protocol), as well as acid-forming sulfur (S (AF)), as determined by means of the ACARP C15034 protocol. The results in Table 6 indicate that the MPA and NAPP values calculated on the basis of pyritic and acid-forming sulfur are significantly lower than those calculated on the basis of total sulfur.

Table 6. Acid-base accounting and net acid generation test results for coal waste samples.

Parameter	Derivation	Waterberg Coal Slurry	Witbank Coal Slurry	Witbank Discards
Sulfur content (%)	LECO S (T)	1.84	1.06	1.94
	ISO S^{2-}	1.04	0.42	1.10
	ACARP S (AF)	1.13	0.56	1.25
MPA (kg H_2SO_4/t)	LECO S (T)	56.30	32.44	59.36
	ISO S^{2-}	31.84	12.85	33.66
	ACARP S (AF)	34.52	17.14	38.27
ANC	Modified Sobek	102.30	29.21	29.10
NAPP (kg H_2SO_4/t)	LECO S (T)	−46.79	3.22	30.26
	ISO S^{2-}	−71.27	−16.36	4.56
	ACARP S (AF)	−68.27	−12.08	9.17
NAG pH	Standard NAG	5.20	3.90	2.56
	Extended boil NAG	n/a	5.20	2.59

Where: T (S) = total sulfur, as determined by LECO; S^{2-} = pyritic sulfur as determined by the ISO 157:1996 protocol; S (AF) = acid-forming sulfur as determined by the ACARP protocol C15034 and is based on the content of CRS sulfide sulfur and the acid-forming soluble sulfate; MPA = maximum potential acidity; ANC = acid-neutralizing capacity; NAPP = net acid-producing potential; NAG = net acid generating.

A more detailed comparison (Figure 4) of the calculated MPA values indicates, furthermore, that the MPA values based on the pyritic sulfur content as derived from the ISO 157:1996 protocol are lower (between 2.7 and 4.6 kg H_2SO_4/t than those based on the acid-forming sulfur content as derived from the ACARP C15034 protocol. As discussed in the previous section, this can be largely attributed to the underestimation of the pyritic sulfur in the case of the ISO 157:1996 protocol. This difference is, furthermore, slightly larger for the Witbank discards (sample C) which contain significant quantities of acid-forming sulfate, not accounted for by the standard ISO 157:1996 protocol.

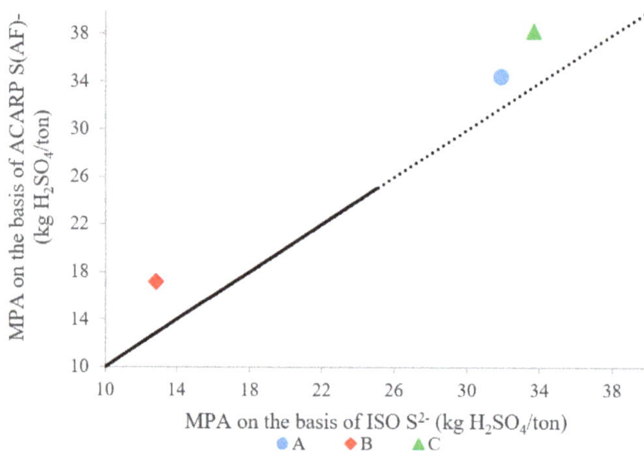

Figure 4. Comparison of maximum potential activity (MPA) values determined from ACARP C15034 acid-forming sulfur (S (AF)) and ISO157:1996 pyritic sulfur (S^{2-}) for the coal waste samples A (Waterberg coal slurry), B (Witbank coal slurry) and C (Witbank coal discards).

The combined ABA-NAG classification plot in Figure 5 indicates that the classification of the Waterberg coal slurry remained non-acid forming (NAF), with a negative NAPP and a NAG pH > 4.5, for all test cases. This can be attributed to the high experimentally-derived acid-neutralizing capacity (ANC) relative to the acid-forming sulfur content. Conversely, the classification of the Witbank discard sample remains potentially acid-forming (PAF), with a positive NAPP and a NAG pH < 4.5, for all test cases. The results in Table 6 and Figure 5 also confirm that the changes in the NAG pH on including an extended boil step were insignificant, indicating the negligible formation of organic acids under NAG test conditions.

In contrast to the Waterberg slurry and Witbank discard wastes, the classification of the Witbank coal slurry varied, depending on the test conditions and parameters. Standard NAG and ABA tests, with the NAPP calculated on the basis of total sulfur, give rise to a PAF classification. When calculating the NAPP on the basis of pyritic or acid-forming sulfur contents, the NAPP becomes negative, resulting in an "uncertain" or inconclusive classification. Furthermore, the NAG pH increased from a value of 3.9 to 5.2 during the extended boil step, changing the classification from PAF to uncertain when using total sulfur to calculate the MPA in the ABA tests, and from uncertain to NAF when using pyritic sulfur or acid-forming sulfur to calculate MPA. In contrast to the Witbank discards, the change in NAG pH during the extended boil is indicative of significant organic acid formation under the relatively aggressive NAG leach conditions. Although resulting in slightly higher NAPP values, the use of the ISO-derived sulfide sulfur, as opposed to the ACARP-derived acid-forming sulfur, did not affect the classification of the coal waste samples in any of the cases investigated in this study.

3.5. Conversion of Sulfur Species under ARD Static Tests

As indicated in Section 2.6, ACARP C15034 sulfur speciation tests were also conducted on the leach residues from the static chemical tests in order to assess the behavior of the various sulfur forms during the leach stages. The results in Figure 6 show that under ANC leach conditions, the CRS sulfide sulfur content in the solids decreased by 21%, 50% and 17% in the case of the Waterberg coal slurry, Witbank coal slurry and Witbank coal discard samples respectively. In the case of the slurry waste samples, results are indicative of partial conversion of sulfide sulfur to acid-forming sulfate (e.g., melanterite) and low-risk sulfur forms (e.g., jarosite and elemental sulfur), whilst in the case of the Witbank discard sample, a net decrease in the acid-forming soluble sulfate and low-risk sulfur can probably be attributed to both the formation and subsequent dissolution of acid-forming soluble sulfates and jarosite over the duration of the leach test. The conversion of sulfide sulfur, as well as the dissolution of acid-forming sulfates such as melanterite and jarosite, generate acid, thus resulting in an underestimation of the ANC values, in the order of 4.5–5.5 kg H_2SO_4/t in the case of the slurry wastes, and 10–12 kg H_2SO_4/t in the case of the Witbank discards.

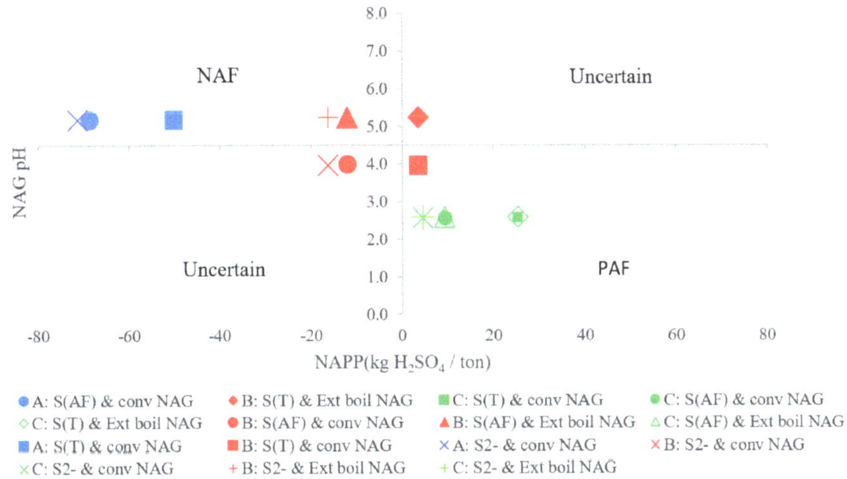

Figure 5. Classification of the acid-rock drainage (ARD) potential of coal waste samples A (Waterberg coal slurry), B (Witbank coal slurry) and C (Witbank coal discards) by conventional (conv) NAG and extended (Ext) boil net acid generation (NAG) tests in combination with acid-base accounting (ABA) tests considering LECO total sulfur S (T), ISO 157:1996 pyritic sulfur (S^{2-}) and ACARP C15034 acid-forming sulfur S (AF).

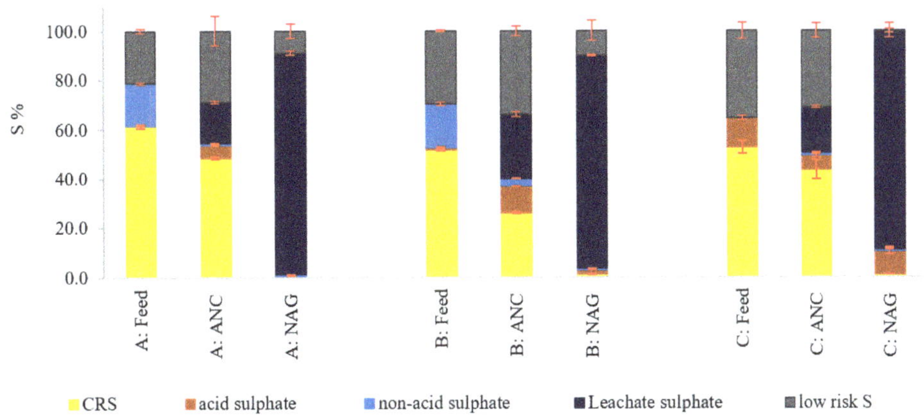

Figure 6. Conversion of sulfur species in coal waste samples A (Waterberg coal slurry), B (Witbank coal slurry) and C (Witbank coal discards) under acid-neutralizing capacity (ANC) and NAG static test conditions.

In the case of the slurry waste samples, the conversion of sulfur species during ANC leaching is relatively low and is thus unlikely to have influenced the classification of the samples. However, the relatively significant effect on ANC in the case of the Witbank discards may have resulted in a change in classification from "uncertain" to acid-forming. This relatively large effect can be largely attributed to the dissolution of acid-forming soluble sulfate, which accounted for up to 7.0 kg H_2SO_4/t (60% of the additional acid formation) for the Witbank discard waste. The possible reactions of sulfur minerals under ANC leach conditions and associated quantities of acid produced on the basis of the conversion of sulfur forms are presented in Table 7.

Table 7. Possible reactions of sulfur species under ANC test conditions and the estimated quantities of acid produced (A: Waterberg coal slurry, B: Witbank coal slurry, C: Witbank discards).

Reaction Equations	Amount of Acid Produced ($kg\ H_2SO_4/t$)		
	A	B	C
$FeS_2 + \frac{45}{4}O_2 + \frac{15}{2}H_2O + K^+ \rightarrow KFe_3(SO_4)_2(OH)_6 + 4SO_4^{2-} + 9H^+$	3.27	0.92	0.00
$FeS_2 + \frac{7}{2}O_2 + H_2O \rightarrow FeSO_4 + H_2SO_4$	1.38	1.53	1.68
$FeS_2 + \frac{15}{4}O_2 + \frac{7}{2}H_2O \rightarrow Fe(OH)_3 + 2H_2SO_4$	0.00	3.06	1.23–4.60
$KFe_3(SO_4)_2(OH)_6 + 3H_2O \rightarrow 3Fe(OH)_3 + K^+ + 2SO_4^{2-} + 3H^+$	0.00	0.00	1.83
$FeSO_4 + \frac{1}{2}O_2 + \frac{5}{2}H_2O \rightarrow Fe(OH)_3 + H_2SO_4$	0.00	0.00	3.70–7.04

The NAG leach tests resulted in the oxidative dissolution of the majority of the CRS sulfide sulfur (99.9%, 95.8% and 98.5% in the case of the Waterberg coal slurry, Witbank coal slurry and Witbank coal discards, respectively). The relatively aggressive NAG conditions, however, also resulted in a significant dissolution of low-risk sulfur, amounting to 53%, 38% and 99% in the case of the Waterberg coal slurry, Witbank coal slurry and Witbank coal discards respectively. This dissolution may result in an overestimation of the acid generating potential of the sample, particularly in the case of the Witbank discards. Organic sulfur compounds, such as thiols, sulfides and disulfides, can be oxidized by H_2O_2 to form sulfonic acid as shown by Equations (14) and (15) [46]:

$$R-SH \overset{H_2O_2}{\rightarrow} R-S-S-R \overset{H_2O_2}{\rightarrow} R-SO-SO-R \overset{H_2O_2}{\rightarrow} R-SO-SOO-R \overset{H_2O}{\rightarrow} R-SO_3H, \tag{14}$$

$$R_2S \overset{H_2O_2}{\rightarrow} R_2SO \overset{H_2O_2}{\rightarrow} R_2SO_2 \overset{H_2O}{\rightarrow} RSO_3H. \tag{15}$$

4. Conclusions

The study showed the total sulfur content in coal processing wastes can be reliably evaluated using the standard LECO combustion method. In terms of methods for analyzing sulfur forms, the ISO157:1996 protocol resulted in a slight underestimation of sulfide sulfur and the overestimation of sulfate sulfur content. However, the differences in the sulfur speciation results obtained with the ISO and the ACARP C15034 protocols were not significant, and both methods can thus be considered sufficiently reliable for analyzing the forms of sulfur.

Whilst speciation results indicated that the majority (52–61%) of the sulfur is in the form of sulfide sulfur in the samples evaluated, the presence of non-acid forming sulfates and organic sulfur can cause overestimation of acid-generating potential if calculated on the basis of total sulfur. The results of this study have shown, furthermore, that consideration of total sulfur instead of the acid-forming sulfur (such as sulfide and acid-forming soluble sulfates) can result in coals wastes being erroneously classified as acid-generating by means of static chemical tests, especially in cases where samples do not have a high acid generating potential (NAPP < 10 and NAG pH > 3).

Another source of inaccuracy in the characterization and classification of ARD potential on the basis of static chemical tests can be attributed to the undesirable reaction of sulfur species under the relatively aggressive test conditions, thus overestimating the acid generating potential.

In general, the findings of this study highlight the complex sulfur chemistry of coal processing wastes and the associated uncertainties in terms of reliably characterizing and classifying their ARD generating by means of standard chemical static tests. Further validation of the potential ARD generating potential using tests that provide information on the time-related nature of the acid-generating behavior, under conditions which more closely simulate those likely to occur in a disposal scenario, is thus recommended. It is also recommended that mineralogical analysis be used in conjunction with chemical analysis to assist in the interpretation of the results of ARD tests, particularly in terms of understanding the behavior of both the acid-forming and acid-neutralizing minerals.

Author Contributions: Conceptualization, S.T.L.H. and J.L.B.; methodology, A.M., J.R.D.A.F., S.T.L.H. and J.L.B.; formal analysis, A.M., J.R.D.A.F., S.T.L.H. and J.L.B.; investigation, A.M., J.R.D.A.F., S.T.L.H. and J.L.B.; resources, J.R.D.A.F., S.T.L.H. and J.L.B.; writing—original draft preparation, A.M., J.R.D.A.F., S.T.L.H. and J.L.B.; writing—review and editing, A.M., J.R.D.A.F., S.T.L.H. and J.L.B.; supervision, J.R.D.A.F., S.T.L.H. and J.L.B.; project administration, J.L.B.; funding acquisition, S.T.L.H. and J.L.B. All authors have read and agreed to the published version of the manuscript.

Acknowledgments: Financial support for this study was provided by the Water Research Commission of South Africa (project number K5/2580) and the University of Cape Town. The authors wish to acknowledge the contribution of Megan Becker and Gaynor Yorath in the mineralogical analysis.

Conflicts of Interest: The authors declare no conflict of interest.

References

1. Singer and Stumm. Acidic Mine Drainage—Rate-Determining Step. *Science* **1970**, *167*, 1121–1123. [CrossRef] [PubMed]
2. US Environmental Protection Agency. *Acid Mine Drainage Prediction, Acid Mine Drainage Prediction*; EPA 530-R-94-036; US Environmental Protection Agency: Washington, DC, USA, 1994.
3. Szczepanska, J.; Twardowska, I. Distribution and environmental impact of coal-mining wastes in Upper Silesia, Poland. *Environ. Geol.* **1999**, *38*, 249–258. [CrossRef]
4. Hancox, P.J.; Goetz, A.E. South Africa's coalfields—A 2014 perspective. *Int. J. Coal Geol.* **2014**, *132*, 170–254. [CrossRef]
5. Steyn, M.; Minnitt, R.C.A. Thermal coal products in South Africa. *J. South. Afr. Inst. Min. Metall.* **2010**, *110*, 593–599.
6. Eberhard, A. The future of South African coal: Market, investment, and policy challenges. *Progr. Energy Sustain. Dev.* **2011**, 1–44. [CrossRef]
7. Prevost, X. A clean future for coal. *Insid. Min.* **2010**, *2010*, 14–17.
8. SACRM (South African Coal Road Map). *Overview of the South African Coal Value Chain, South African Coal Roadmap*; Technical Report; 2011; 299p, Available online: http://www.fossilfule.co.za/initiatives/2013/SACRM-Value-Chain-Overview.pdf (accessed on 9 December 2019).
9. Department of Minerals and Energy. National Inventory Discard and Duff Coal—2001 Summary Report. *Energy* **2001**, *2001*.
10. Reddick, J.F.; Von Blottnitz, H.; Kothuis, B. A cleaner production assessment of the ultra-fine coal waste generated in South Africa. *J. South. Afr. Inst. Min. Metall.* **2007**, *107*, 811–816.
11. Gluskoter, H.J. Inorganic sulfur in coal. *IL State Geol. Surv.* **1974**, *3*, 94–98. [CrossRef]
12. European Commision. *EUR 17980-Technical Coal Research: Coal Conversion Characterisation of Sulphur in Coal and Coal Products by Standard and Non-Standard Methods*; Office for Official Publications of the European Communities: Luxembourg, 1998.
13. Speight, J.G. Handbook of Coal Analysis. In *Chemical Analysis A Series of Monographs on Analytical Chemistry and Its Applications*; John Wiley & Sons, Inc.: Hoboken, NJ, USA, 2005; Volume 166.
14. Iroala, O.J. Combining Froth Flotation with Reflux Classification to Mitigate ARD Generating Potential of the Waterberg and Witbank Coal Ultrafines via Sulfide Removal. Ph.D. Thesis, Department of Chemical Engineering, University of Cape Town, Cape Town, South Africa, 2014.
15. Kazadi Mbamba, C.; Harrison, S.T.L.; Franzidis, J.P.; Broadhurst, J.L. Mitigating acid rock drainage risks while recovering low-sulfur coal from ultrafine colliery wastes using froth flotation. *Miner. Eng.* **2012**, *29*, 13–21. [CrossRef]

16. Kotelo, L.O. Characterising the Acid Mine Drainage Potential of Fine Coal Wastes. Ph.D. Thesis, Department of Chemical Engineering, University of Cape Town, Cape Town, South Africa, 2013.
17. Bell, F.G.; Bullock, S.E.T.; Hälbich, T.F.J.; Lindsay, P. Environmental impacts associated with an abandoned mine in the Witbank Coalfield, South Africa. *Int. J. Coal Geol.* **2001**, *45*, 195–216. [CrossRef]
18. McCarthy, T.S. The impact of acid mine drainage in South Africa. *S. Afr. J. Sci.* **2011**, *107*, 1–7. [CrossRef]
19. Pinetown, K.L.; Ward, C.R.; van der Westhuizen, W.A. Quantitative evaluation of minerals in coal deposits in the Witbank and Highveld Coalfields, and the potential impact on acid mine drainage. *Int. J. Coal Geol.* **2007**, *70*, 166–183. [CrossRef]
20. Deysel, L.M.; Vermeulen, D. A Detailed ABA Study of the Coal-Bearing Formations in the Waterberg Coalfield, Limpopo Province, South Africa. In Proceedings of the 10th International Conference on Acid Rock Drainage & IMWA Annual Conference, Santiago, Chile, 21–24 April 2015; pp. 1–10.
21. Nordstrom and Southam. Geomicrobiology of sulfide mineral oxidation. In *Geomicrobiology: Interactions between Microbes and Minerals*; Banfield, J.F., Nealson, K.H., Eds.; Mineralogical Society of America: Washington, DC, USA, 1997; Volume 35, pp. 361–390.
22. Lapakko, K. *Metal Mine Rock and Waste Characterization Tools: An Overview*; The International Institute for Environment and Development (IIED): St. Paul, MN, USA, 2002.
23. Akcil, A.; Koldas, S. Acid Mine Drainage (AMD): Causes, treatment and case studies. *J. Clean. Prod.* **2006**, *14*, 1139–1145. [CrossRef]
24. Simate, G.S.; Ndlovu, S. Acid mine drainage: Challenges and opportunities. *J. Environ. Chem. Eng.* **2014**, *2*, 1785–1803. [CrossRef]
25. Parbhakar-fox, A.; Lottermoser, B.G. A critical review of acid rock drainage prediction methods and practices. *Miner. Eng.* **2015**, *82*, 107–124. [CrossRef]
26. McKibben and Barnes. Oxidation of pyrite in low temperature acidic solutions: Rate laws and surface textures. *Geochim. Cosmochim. Acta* **1986**, *50*, 1509–1520. [CrossRef]
27. Lawrence, R.W.; Scheske, M. A method to calculate the neutralization potential of mining wastes. *Environ. Geol.* **1997**, *32*, 100–106. [CrossRef]
28. Paktunc, A.D. Mineralogical constraints on the determination of neutralization potential and prediction of acid. *Environ. Geol.* **1999**, *39*, 103–112. [CrossRef]
29. Hansford, G.S.; Vargas, T. Chemical and electrochemical basis of bioleaching processes. *Hydrometallurgy* **2001**, *59*, 135–145. [CrossRef]
30. Yahya, A.; Johnson, D.B. Bioleaching of pyrite at low pH and low redox potentials by novel mesophilic Gram-positive bacteria. *Hydrometallurgy* **2002**, *63*, 181–188. [CrossRef]
31. Rohwerder, T.; Gehrke, T.; Kinzler, K.; Sand, W. Bioleaching review part A: Progress in bioleaching: Fundamentals and mechanisms of bacterial metal sulfide oxidation. *Appl. Microbiol. Biotechnol.* **2003**, *63*, 239–248. [CrossRef] [PubMed]
32. Ahern, C.; McElnea, A.; Sullivan, L. *Acid Sulfate Soils Laboratory Methods Guidelines*; Department of Natural Resources, Mines and Energy: Indooroopilly, Australia, 2004.
33. Miller, S. *ACARP Project C15034: Development of ARD Assessment for Coal Process Wastes*; Environmental Geochemistry International Pty Ltd.: Balmain, Australia, 2008.
34. Stewart, W.; Schuman, R.; Miller, S.; Smart, R. Development of Prediction Methods for ARD Assessment of Coal Process Wastes. In Proceedings of the 8th International Conference on Acid Rock Drainage (ICARD), Skelleftea, Sweden, 22–26 June 2009; pp. 1–14.
35. Schumann, R.; Stewart, W.; Miller, S.; Kawashima, N.; Li, J.; Smart, R. Acid-base accounting assessment of mine wastes using the chromium reducible sulfur method. *Sci. Total Environ.* **2012**, *424*, 289–296. [CrossRef] [PubMed]
36. Smart, R.; Skinner, B.; Levay, G.; Gerson, A.; Thomas, J.; Sobieraj, H.; Schumann, R.; Weisener, C.; Weber, P. *AMIRA International ARD Test Handbook Project P387A Prediction & Kinetic Control of Acid Mine Drainage*; Miller, S., Stewart, W., Eds.; Ian Wark Research Institute: Melbourne, Australia; Geochemistry International Pty: Balmain, Australia, 2002.
37. Descostes, M.; Vitorge, P.; Beaucaire, C. Pyrite dissolution in acidic media. *Geochim. Cosmochim. Acta* **2004**, *68*, 4559–4569. [CrossRef]
38. Dold, B. Acid rock drainage prediction: A critical review. *J. Geochem. Explor.* **2017**, *172*, 120–132. [CrossRef]

39. Lawrence, R.W.; Wang, Y. *Determination of Neutralization Potential for Acid Rock Drainage Prediction, MEND Project 1.16.3*; University of British Columbia: Vancouver, BC, Canada, 1996.
40. Sobek, A.A.; Schuller, W.A.; Freeman, J.R.; Smith, R.M. *Field and Laboratory Methods Applicable to Overburdens and Minesoils, Environmental Protection Technology Series*; EPA: Cincinnati, OH, USA, 1978.
41. Plante, B.; Bussière, B.; Benzaazoua, M. Static tests response on 5 Canadian hard rock mine tailings with low net acid-generating potentials. *J. Geochem. Explor.* **2012**, *114*, 57–69. [CrossRef]
42. Becker, M.; Dyantyi, N.; Broadhurst, J.L.; Harrison, S.T.L.; Franzidis, J.P. A mineralogical approach to evaluating laboratory scale acid rock drainage characterisation tests. *Miner. Eng.* **2015**, *80*, 33–36. [CrossRef]
43. Laban, K.L.; Atkin, B.P. The direct determination of the forms of sulphur in coal using microwave digestion and icp-aes analysis. *Fuel* **2000**, *79*, 173–180. [CrossRef]
44. Tuttle, M.; Briggs, P.; Berry, C. A Method to Separate Phases of Sulphur in Mine-Waste piles and Natural Alteration Zones, and to Use Sulphur Isotopic Compositions to Investigate Release of Metals and Acidity to the Environment. In Proceedings of the Sixth International Conference on Acid Rock Drainage, Caines, Australia, 14–17 July 2003.
45. Weber, P.A.; Thomas, J.E.; Skinner, W.M.; Smart, R.S.C. Improved acid neutralisation capacity assessment of iron carbonates by titration and theoretical calculation. *Appl. Geochem.* **2004**, *19*, 687–694. [CrossRef]
46. Borah, D.; Baruah, M.K.; Haque, I. Oxidation of high sulphur coal. Part 2. Desulphurisation of organic sulphur by hydrogen peroxide in presence of metal ions. *Fuel* **2001**, *80*, 1475–1488. [CrossRef]

minerals

MDPI

Article

Sulfur Species, Bonding Environment, and Metal Mobilization in Mining-Impacted Lake Sediments: Column Experiments Replicating Seasonal Anoxia and Deposition of Algal Detritus

Jeff B. Langman [1,*], Jaabir Duunya Ali [2], Andrew W. Child [3], Frank M. Wilhelm [4] and James G. Moberly [3]

1 Department of Geological Sciences, University of Idaho, Moscow, ID 83844, USA
2 Faculty of Geosciences, Geoengineering, and Mining, Freiberg University of Mining and Technology, 09599 Freiberg, Germany; Jaabir-Duunya.Ali@student.tu-freiberg.de
3 Department of Chemical and Biological Engineering, University of Idaho, Moscow, ID 83844, USA; awchild@uidaho.edu (A.W.C.); jgmoberly@uidaho.edu (J.G.M.)
4 Department of Fish and Wildlife Sciences, University of Idaho, Moscow, ID 83844, USA; fwilhelm@uidaho.edu
* Correspondence: jlangman@uidaho.edu

Received: 8 August 2020; Accepted: 23 September 2020; Published: 25 September 2020

Abstract: The oxidation state of sulfur [S] is a primary control on mobility of metals in sediments impacted by legacy mining practices. Coeur d'Alene Lake of northern Idaho, USA, has been impacted by upstream legacy mining practices that deposited an estimated 75 Mt of metal(loid)- and S-rich sediments into the lake. Future lake conditions are expected to include algal blooms, which may alter S and metal remobilization during the seasonal euxinic environment. Cores of the lake sediments were exposed to anoxic and anoxic + algal detritus conditions for eight weeks at 4.5 °C through introduction of a N_2 atmosphere and addition of algal detritus. At a location 2.5 cm below the sediment-water interface, anoxic conditions promoted a shift in S species to continually larger concentrations of reduced species and an associated shift in the bonding environment reflective of increased S–metal bonds. Anoxic + algal detritus conditions suppressed the increasing trend of reduced S species and induced greater release of Mn compared to the anoxic-only conditions but did not appear to enhance the release of As, Cd, or Fe. The addition of algal detritus to the sediment-water interface of these Fe- and S-rich sediments enhanced mobilization of Mn likely because of dissimilatory metal reduction where the anaerobic oxidation of the algal detritus stimulated Mn reduction. Results of the study indicate that future metal release from the lake sediments will be altered with the likely deposition of algal detritus, but the effect may not enhance the release of acutely toxic metals, such as As or Cd, or substantially impact Fe cycling in the sediments.

Keywords: mining-impacted lake; metal contamination; lakebed sediments; sulfur reduction and speciation; sulfur bonding environment; synchrotron X-ray absorption spectroscopy

1. Introduction

Legacy mining practices have resulted in extensive contamination through the release of sulfur [S] and toxic metals into aquatic ecosystems, which can negatively influence the health of human and wildlife populations for decades to centuries [1–5]. Coeur d'Alene Lake (Lake) of northern Idaho, USA, is located in the Coeur d'Alene River Basin (Basin) and downgradient of the lead-zinc-silver Coeur d'Alene Mining District (District) (Figure 1). Legacy mining practices allowed the formation of acid rock drainage (ARD) and hydrologic transport of tailings and waste rock. These processes resulted in the deposition of an estimated 75 Mt of metal(loid)- and S-rich sediments—Including 470,000 t of lead [Pb] and substantial masses of arsenic [As], cadmium [Cd], iron [Fe], manganese [Mn], zinc [Zn]—In the Lake over the past 100+ years [6–9]. The highest concentrations of metals in the Lake sediments are near the outlet of the Coeur d'Alene River (River) in the central part of the Lake (Figure 1), and lesser concentrations are present in the southern and northern parts of the Lake [7]. Seasonal changes in environmental conditions at the sediment-water interface (SWI) can remobilize S and metals that may be transported into the Lake water column [10–20]. The Lake undergoes seasonal thermal stratification [20,21] that induces changes in reduction-oxidation (redox) conditions at the SWI [22,23] where redox-sensitive As, Fe, Mn, and associated S are concentrated at an oxic-anoxic interface in this seasonally euxinic environment [24]. The strength and persistence of potential contaminant-mobilizing conditions, such as changes in redox conditions and organic matter content, determine the impact to seasonal water quality.

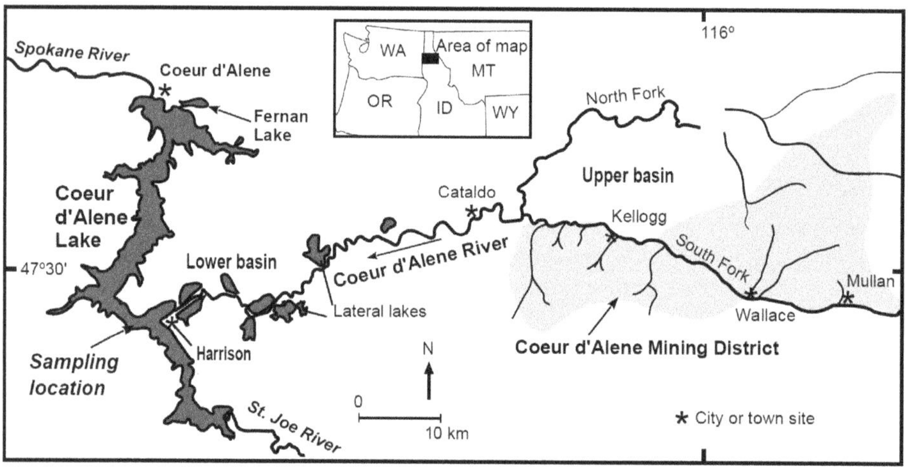

Figure 1. Coeur d'Alene Lake, River, and Mining District (Silver Valley) in northern Idaho, USA.

Historically, transport of Zn into the Lake has suppressed algal growth [25,26], which should be increasing because of an increasing input of phosphorus [P] [20,24,25,27,28]. Current seasonal conditions at the SWI have resulted in limited remobilization of S and metals [7,8,22,29–32]. Yet, there is a growing concern of increased S and metal remobilization because of an expected decrease in Zn inputs from remediation activities and increasing P inputs, which likely will produce future algal blooms similar to eutrophic and mesotrophic lakes in the region [33]. If algal blooms occur, a substantial increase in organic matter as algal detritus would be deposited at the SWI, which may alter the seasonal redox conditions that occur with thermal stratification (summer into fall) and subsequent turnover (fall). A change to the seasonal cycle of redox conditions and associated microbial communities can influence the speciation of S and influence the remobilization or retention of metals [10,14,15,34–40]. The goal of this study was to evaluate the alteration of sediment S species, changes to the S bonding environment (local bond configuration), and the influence of S redox changes on metal release under anoxic and anoxic + algal detritus conditions. These two experimental conditions are representative of current anoxia from thermal stratification and potential future conditions of anoxia and deposition of algal detritus to the SWI interface. Results of this manuscript are part of a larger study examining the Lake sediment environment, including metal release, microbial community dynamics, and nutrient inputs.

Redox-sensitive S and Fe have been identified as controls on metal mobility in the contaminated sediments of the Basin, including the lateral lakes of the lower Basin (Figure 1) [17,22,32,41,42].

The seasonal variation in the smaller, shallower, and more productive lateral lakes and select areas of the Lake cause a cycling of sediment S and metals between soluble and insoluble phases [7,11,17,27,30, 43–45]. Increased productivity of the Lake and deposition of algal detritus may strengthen and extend the seasonal shift to reducing conditions during the summer-to-fall thermal stratification. Alteration of redox conditions will influence the seasonal evolution of S species and the mobility/immobility of the associated metals, which also will be influenced by other remobilization/retention influences, such as pH, clays, Fe (oxyhydr)oxides, and low molecular weight organic acids [10,14,15,46–48]. Increases in S reduction would allow for greater scavenging of metals (authigenic metal sulfide formation), although the greater presence of organic acids may provide substrates for metal complex formation and increased metal mobility [49–53]. To understand metal mobility or retention in these mining-impacted and euxinic environments, it is necessary to understand potential changes to S speciation and the associated bonding environment.

2. Study Area, Materials, and Methods

The Basin is bounded by the Coeur d'Alene Mountains, which are part of the Bitterroot Range of the Northern Rocky Mountains. The Coeur d'Alene Mountains are composed of quartzite and argillite of the Mesoproterozoic Belt Supergroup, which contain the ore-bearing formations that have been mined since the 1880s. The District's ore primarily consists of argentiferous galena [PbS] and sphalerite [(Zn,Fe)S] with associated carbonate zones consisting primarily of siderite [$FeCO_3$] and ankerite [$Ca(Fe,Mg,Mn)(CO_3)_2$] [34,35,54]. From mining of this ore, it is estimated that 56 Mt of processed tailings containing 900,000 t of Pb and 700,000 t of Zn—along with various amounts of As, Cd, Fe, Mn, and S—were discharged into the floodplain of the Coeur d'Alene River (River) and its tributaries [55]. Additionally, an unknown amount of unprocessed waste rock was disposed of in the floodplains. Transport of the disposed tailings and waste rock by high streamflows distributed the waste rock and tailings throughout the floodplain of the Basin and into the Lake [16,54,56,57]. Furthermore, ARD formed in abandoned mines, waste rock, and contaminated sediments where carbonate minerals were insufficient to buffer acid production [9,34,58]. The Lake continues to see fluctuations of S and metal concentrations in the water column, partially from continued loading of S and metals from the River [57] and partially from release of S and metals from sediments during seasonal shifts in environmental conditions [20].

2.1. Study Design

This study was designed to replicate SWI conditions during a seasonal shift to anoxic conditions. Sediment cores were collected by boat in Aberdeen Lodge Bay (Figure 1, depth ≈ 15 m) in the central portion of the Lake across from the discharge of the River. Two sampling periods—mid-October and mid-November—were used to collect initial cores (October) for testing of sample collection and examination of sediment physical and chemical characteristics and laboratory cores (November) for replicating anoxic (control) and anoxic + algal detritus (treatment) conditions. It was assumed these fall sampling periods would be after holomixis and reoxygenation of the water column and upper sediments. The laboratory experiments were designed for control and treatment cores to be placed in anaerobic conditions (continuous N_2 atmosphere), from which a control and treatment core were removed and analyzed every two weeks over an eight-week period. Upon removal from the experiment, 1-cm sediment layers centered at 2.5 cm and 12.5 cm below the SWI were collected for X-ray absorption spectroscopy (XAS) to examine S species distribution and bonding environment. Porewater was separated from the sediment samples by centrifugation (3750 RCF (relative centrifugal force) for 15 min) and analyzed for pH and redox potential (ORP) and subsampled for analysis of dissolved (0.45-μm filtered) As, Cd, Fe, and Mn by inductively coupled plasma mass spectrometry and dissolved sulfate [SO_4] by ion chromatography at the University of Idaho Analytical Sciences Laboratory (ASL). The sediment samples were homogenized, dehydrated (lyophilized for 24 h), and preserved at −80 °C. The 2.5-cm sample location represents the near SWI environment where redox-sensitive elements have concentrated [24], and the 12.5-cm sample location represents a possible restricted zone because of low permeability in these silt-clay dominated sediments near a volcanic ash layer. This deeper zone was selected for analysis to compare depth penetration of environmental changes that could influence S speciation. Deeper sample locations were not considered because of the presence of a volcanic ash layer (Mount St. Helens 1980 [7,8]) about 15 cm below the SWI that restricts permeability and acts as a metal sorbing substrate.

2.2. Sediment Core Collection

Disinfected (70% EtOH, 30% ultrapure water), polyvinyl chloride (PVC), core barrels (5.2-cm ID × 61-cm L) were kept in a positive N_2 atmosphere for 48 h prior to sample collection. The barrels were fitted to a Kajak-Brinkhurst gravity corer to obtain sediment cores of 40 ± 1 cm with free fall from 1 m below the Lake surface (Figure 2). Immediately after core collection, overlying water was removed by siphoning to minimize disturbance of the SWI during transport. The cores were capped, vented with 0.20-μm filters, flushed with N_2, and stored upright under a N_2 atmosphere in custom-made, gas-tight containers (Figure 2) for transport to the University of Idaho Lake Social Ecological Systems (LaSES) Laboratory in Coeur d'Alene, Idaho. Lake water was collected 1 m above the SWI using a disinfected, 2-L, Van Dorn sampler. Upon arrival at the LaSES Laboratory, core containers were checked for positive pressure N_2 atmosphere and stored in the dark at 4.5 ± 0.4 °C.

Figure 2. Kajak-Brinkhurst gravity corer, core sample, and atmospherically-sealed core container with N_2 purge.

2.3. Algal Detritus

An adjoining and unimpacted eutrophic lake (Fernan Lake, Figure 1) was selected for collection of algae for simulating deposition of algal detritus to the SWI of the collected cores. Algae were collected from Fernan Lake by pumping surface water through an 80-μm mesh net. The algae were subsequently separated from remaining water by centrifugation (3750 RCF for 15 min), frozen, lyophilized, and homogenized. A slurry of algal detritus (~0.25 g of dry algae and 25 mL of lake water) was added to half of the cores along with 325 mL of Lake water (about 4 cm of head height in the core barrel) to all cores. The amount of algal detritus was based on previous sediment analyses of mesotrophic and eutrophic lakes in the region [33]. After input of the water or slurry + water, the cores were recapped, vented, flushed with N_2, and stored at 4.5 ± 0.4 °C upright in the dark under a N_2 atmosphere in the gas-tight containers to begin the 8-week incubation period.

2.4. Sediment Profile Characterization

A core from the October sample collection was sampled at the midpoint of 5-cm intervals (2.5, 7.5, 12.5, etc.) for physical and chemical characterization of the sediment profile. Core sediment characterization was performed at ASL according to quality control and assurance protocols of the North American Proficiency Testing Program. Particle size distribution was determined by hydrometer [59]. Ammonia (as NH_3–N), nitrate (as NO_3–N), phosphorus P, and organic matter (OM) concentrations were determined through standard extraction methods and analyzed by flow injection analysis (NH_3 and NO_3) or spectrophotometer (P and OM) [60,61].

2.5. X-ray Spectroscopy

The oxidation of S can be described as the reversible sequence of sulfide [S^{2-}] → disulfide [S^{1-}] → elemental S [S^0] including various polysulfides [H_2S_x or S_x^{2-}] → sulfite [SO_3^{2-}] and thiosulfate [$S_2O_3^{2-}$] → SO_4^{2-}. Schippers and Sand (1999) [62] and Druschel et al. (2003) [63] described the transformation of S^{2-}/S^{1-} to $S_x^0/SO_3^{2-}/S_2O_3^{2-}$ prior to SO_4^{2-} that has been followed by substantial exploration of the presence and development of intermediary S species/forms during sulfide oxidation and reduction, such as growth of oxic units with O_2 incorporation, possible S^{2-} to Sx^0 to S_2^{2-} evolution, Fe and S oxidation preference with H_2O and O_2 adsorption sequences, or the greater lability of the Fe–S bond than the S–S bond [64–66]. The identification of S_0 and polysulfides and $S_2O_3^{2-}$ and polythionates [$S_{x+2}O_6^{2-}$] during oxidation led to the theory of two oxidation pathways—Polysulfide or polythionate. Absorbance spectra of partially oxidized sulfides can be difficult to separate in the intermediary S range (prior to full oxidation to SO_4) because of the presence of the various S inorganic and organic forms and their reaction rate differences that cannot be viewed as a static product [16,67,68]. Although, the absorbance spectra of the range of S oxidation products can be used to view the shift in S species distribution and bonding environment reflective of temporal changes in sediment environmental conditions [17].

The primary goal of the experiments was the temporal evaluation of the shift in sediment S species and bonding environment with inducement of anoxic conditions and influence of algal detritus. The 5-cm interval samples of the October core (no anoxia or algal detritus) were analyzed to determine the metal(loid) and S characteristics of the sediment profile. An Advant'XP+ sequential X-ray fluorescence (XRF) spectrometer at Washington State University's GeoAnalytical Laboratory (Pullman, WA, USA) was used to determine element composition at each 5-cm depth (fused beads from 5 g of each sediment sample). These results provided relevant information for selection of the sampling points in the experimental cores (control and treatment cores). Additionally, each 5-cm interval sample from the October core was analyzed by synchrotron XAS for discriminating S species and bonding environment. These XAS results provided a framework for evaluating XAS results of the experiment samples (2.5-cm (upper) and 12.5-cm (lower) sediment depths) from the control and treatment cores.

Synchrotron XAS was performed at the 06B1-1 beamline (SXRMB) at the Canadian Light Source, Saskatoon, SK, Canada. Beamline 06B1-1 uses a Si (111) monochromator to deliver an incident beam energy of 1.7–10 keV, which is suitable to resolve S K-edge spectra. A broad beam (3 mm × 2 mm) was used to collect fluorescence mode spectra for X-ray absorption near edge structure (XANES) and X-ray absorption fine structure (XAFS) analysis. The spectra were processed with the XAS program ATHENA [69]. K-edge energies for the reference materials were determined as the maxima of the first derivative for the absorption spectra and were shifted to the theoretical values to account for beamline flux. For quality control purposes, each environmental sample was scanned twice to determine if additional scans were needed to reduce the signal-to-noise ratio (minimization of standard deviation near the K-edge). The final spectrum was calibrated to the energy shift apparent from the applicable standard, and the spectrum was edge-step normalized (within a maximum 15–80 eV normalization range) for comparison of environmental samples and reference materials.

The shift in predominant S oxidation states of the sediment samples were analyzed with ATHENA's linear combination fitting (LCF) capability within a K-edge range of −20 eV to +40 eV. Output of the LCF consists of the percent that each reference material contributes to the fit of the sample spectrum. The goal of the LCF analysis was not discrimination of all oxidation states present in the spectra but an evaluation of the primary oxidation states and temporal shifts with changes in redox conditions (balance of reduced and oxidized species). LCF was used to reconstruct each sample spectrum using spectra from two primary reference materials—Pyrite (FeS_2, S^{1-}) and sodium bisulfite ($NaHSO_3$, mixed state spectra of S^{4+} and S^{5+} similar to sulfone (RSO_2R, S^{4+0}) + sulfonate ($R–SO_3$, S^{5+})). This mixed state form appears as a broad double peak with K edge energies near +5.5 (S^{4+}) and +8.5 (S^{5+}) past the S^0 K edge. The FeS_2 and $NaHSO_3$ represent oxidation states that constituted the large majority of the S species present throughout the experiments. Additional oxidation states were evaluated through attempted fitting of inorganic and organic S spectra from compounds that included oxidation states of S^{2-} through S^{6+} but were not included in the LCF because of their minimal contribution to the fit. The FeS_2 reference material is representative of the reduced S endpoint. The $NaHSO_3$ represents a mixed S intermediary state that appears to be the S oxidation endpoint for solid form S. Our selection of a bisulfite with mixed state spectra is not an indication of an inorganic presence but a representation of this mixed oxidation state S form in the intermediary range (S^{4+} and S^{5+}). Prior S species analyses of Lake sediments have indicated the presence of thiols (S^{2+}) and sulfone and ester–SO_4 [42]. We did not detect the presence of S^{2+} oxidation states. Similar to Toevs et al. (2006) [42], there was minimal to no S^{2-} or S^{6+} forms in all samples collected for this study.

3. Results

The results of the Lake sediment analyses are presented in order of the October sediment characteristics and S species and bonding environment (0 to 50 cm depth) followed by results from examination of the 2.5-cm and 12.5-cm depth samples from control and treatment cores at week 0, 2, 4, 6, and 8 of the experiments. The results of the October core analyses provide the framework for examining the results of experiment (November) cores that were subjected to anoxic (control) or anoxic + algal detritus (treatment) conditions.

3.1. Sediment Characteristics

Previous coring of the Lake sediments indicated elevated levels of Ag, As, Cd, Cu, Fe, Hg, Mn, Pb, Sb, and Zn [7,8]. Total depth of contamination varied from 17 cm to 119 cm [8] indicative of the bathymetric distribution of contaminated sediments and dissolved contaminants coming from the River. Examination of the October core indicated differences in metal concentrations above and below the volcanic ash layer, which is a low permeability and sorbtive layer [70,71] that separates the older and more contaminated sediments from more recently deposited sediments (Figure 3). Substantial contaminant concentrations are present above the ash—Pb and Zn concentrations up to 2425 and 3950 ppm, respectively, along with substantial Fe (about 10 wt. % as FeO). Greater Pb concentrations are present below the ash—Up to 14,250 ppm of Pb (Zn was less than the concentrations above the ash). Sulfur concentrations have shown wide variation in Lake sediments, but Toevs et al. (2006) [42] indicated a mean concentration range of 95 to 140 mmol/kg from 0 to 30 cm at a nearby location and increasing S concentrations with depth.

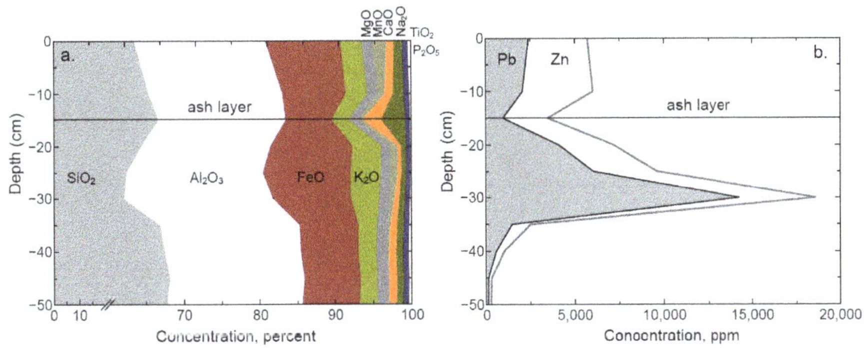

Figure 3. Cumulative area graphs of major (**a**) and trace element (**b**) concentrations for sediments collected from Aberdeen Lodge Bay in Coeur d'Alene Lake. Major elements are normalized to 100 percent. Trace element concentrations are not normalized.

The separation of the upper and lower sediments by the ash layer creates a reactive ~15 cm of contaminated sediments below the SWI. Alteration of the redox conditions at the SWI must translate to depth in the sediments (electron transfer pathways), and the permeability of the sediments must be sufficient for transport of mobilized contaminants [72]. The Lake sediments are predominantly composed of silt and clay (silt > 52%, clay > 21%) with a smaller fraction (<18%) of sand throughout the sediment profile (Table 1). The presence of substantial silt and clay indicates a sediment column of likely low permeability [73]. Additionally, there was a substantial presence of NH_3 compared to NO_3 at all depths in the October core, which is indicative of reducing conditions during this collection period. Phosphorus concentrations ranged from 15 to 27 µg/g and organic matter was low (≤5.2%) in comparison to regional lakes [33]. Benthic flux of solutes from the sediments to the Lake is considered relatively low because of a lack of bioturbation (minimal benthic community because of contamination) and limited groundwater advection [27]. Although, solute flux across the SWI does occur as indicated by previous bromide tracer tests—Flux rates of 20 and 36 mg/cm^2/yr [27].

Table 1. Results of sediment composition and nutrient analyses of samples from sediment cores collected from Aberdeen Lodge Bay in Coeur d'Alene Lake, October 2017.

Depth (cm)	Sand	Silt	Clay	Phosphorus	Nitrate (as N)	Ammonia (as N)	Organic Matter
Units	% [1]	%	%	µg/g [2]	µg/g	µg/g	%
0 to 5	12.5	65.6	21.9	24	0.80	130	5.2
5 to 10	13.6	57.6	28.8	23	<0.72	77	3.3
10 to 15	7.0	60.8	32.3	21	<0.72	56	2.8
15 to 20	6.1	59.2	34.7	20	<0.72	69	3.7
20 to 25	11.5	59.0	29.5	16	0.79	84	4.2
25 to 30	7.9	66.1	26.0	15	<0.72	130	4.8
30 to 35	17.3	56.7	26.0	25	<0.72	120	4.9
35 to 40	11.4	60.4	28.2	25	<0.72	120	4.2
40 to 45	14.7	59.1	26.2	27	0.79	120	3.8
45 to 50	16.0	52.5	31.5	24	<0.72	130	4.6

[1] %, percent; [2] µg/g; microgram per gram.

3.2. Sediment Sulfur Speciation and Bonding Environment Characteristics

For the October core, reduced and oxidized S species were present above the ash layer (0 to 12.5 cm) with a readily identifiable presence of sulfide (S^{1-} at ~2471 eV) and likely elemental S (S^0 at 2472 eV) (Figure 4), which primarily indicates a reducing environment above the ash layer during this collection period. Although, mixed oxidized S forms (likely mixed representation of S^{3+} (~2478.5), S^{4+} (~2479.5) and S^{5+} (~2481 eV)) also were present above the ash layer but at lower concentrations as indicated by the lower peak response (Figure 4). These broad spectra peaks of oxidized S are representative of the intermediary products—Inorganic sulfite/thiosulfate/polythionate along with their organic counterparts of sulfone/sulfonate. The intermediary products tend to form these broad peak portions of the S spectra that are not representative of an individual oxidation state but charge densities that reflect multiple oxidation states [74]. Compared to the sediments above the ash layer, the sediments below the ash layer (20 to 30 cm) contained greater concentrations of reduced species along with a shift towards a lower K edge ($<S^{1-}$), suggestive of non-pyritic reduced S forms. The sediments below the ash layer contained a greater variety of S species and substantially more oxidized species with increasing depth. In these deeper sediments, there was indication of substantial accumulation of intermediary products (Figure 4). The presence of these intermediary products throughout the sediment column indicates likely partial shifts in oxidation state with seasonal changes, such as onset of reducing conditions near the surface or the upwelling of oxygenated groundwater that has been identified in the nearby lateral lakes [17]. Yet, these seasonal effects are insufficient to fully reduce or oxidize the store of S in any location within the sediments. Such limited reduction and oxidation (temporally and spatially) allow for the accumulation of intermediary products between the authigenic mineral form (FeS_2, S^{1-}) and fully oxidized SO_4 form.

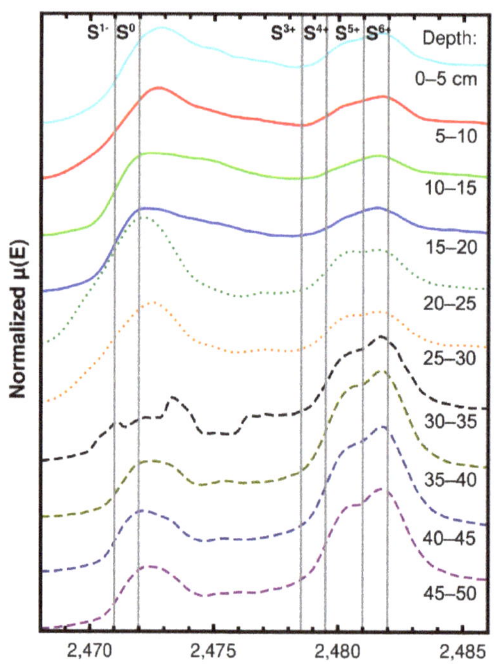

Figure 4. Sulfur absorption spectra from synchrotron XANES analysis of sediments (0 to 50 cm) collected from Aberdeen Lodge Bay in Coeur d'Alene Lake, October 2017. Theoretical sulfur oxidation states are shown to illustrate their likely K-edge location (inflection point of rising limb).

Measurement of the absorption coefficient beyond the S K-edge (XAFS region) can indicate changes in coordination number, bond length, and/or local disorder of adjacent atoms [75]. The XAFS region is a reflection of dipole transitions from core to unoccupied states of the absorbing atoms (S) that reflects characteristics of adjacent (scattering) atoms [76–78]. Due to the likely inorganic and organic

S forms and seasonal environmental changes in the Lake sediments, XAFS analysis was performed to identify relative shifts in adjacent atom characteristics (e.g., bond length) through changes in 1st sphere spectra in frequency filtered (Fourier transformation with k-weight = 1 and Hanning window (or cosine-squared taper)) for R space (Fourier transformed $\chi(R)$ in Å) and filtered R space (q space or Fourier filtered $\chi(q)$ in Å−1). The resulting complex number of the transformed spectra (R and q space) has equivalent real and imaginary numbers equal to real k and imaginary k numbers. The real portion of the transformation represents the absolute or enveloping response of the number and disorder of scatterers and the imaginary (oscillating) part represents bond length changes between the S atoms and the scatterers [79]. The real and imaginary parts complete the complex number that is reflected as a magnitude of the shell and bonding environment response to the electron wavefunction [80,81].

Non-pyritic S–S bonds are extraordinarily flexible where bond lengths can vary between 1.8 and 3.0 Å [82], but pyritic S–S (and Fe–S) bonds typically are near 2.2 Å [83,84]. Polysulfide S–S bonds tend to be shorter and near 2 Å [85]. Additionally, C–S bonds of polysulfide compounds typically are even shorter (about 1.8 Å) [86]. S–O bonds, such as SO_3 and SO_4, tend to be relatively short bonds (1.4 to 1.5 Å), but a mixed state sample tends to produce a magnitude of $\chi(R)$ near 3 Å because of constructive/deconstructive wave forms [87]. The magnitude of $\chi(R)$ for the October core is presented in Figure 5 for comparison of changes in the bonding environment with depth. The magnitude response indicates a similar bonding environment above the ash layer, a transition environment near the ash layer, and a change in the bonding environment in the deeper sediments that corresponds to the speciation differences indicated in Figure 4. The reducing conditions above the ash layer is similar in magnitude response to FeS_2 and S^0 standards (Figure 6), and the deeper sediments indicate more oxidized conditions similar to the magnitude response of sulfite, thiosulfate, and sulfate (Figure 6). The electron wave emitted by the absorbing S atom is scattered by neighboring atoms that produces a final state wavefunction from constructive and destructive responses (Figures 5b and 6b). The higher oscillation in the deeper sediments (>30 cm) likely represents constructive wave response along S–O bonds given an expected sulfite/polythionate form (sulfite equivalent shown in Figure 6). Correspondingly, this oscillation form is minimized in the more reduced S distribution, such as the lower oscillation pattern of the sediments above the ash layer where S–O bonding is less visible because of wave deconstruction with greater S–S and/or S–M bonding.

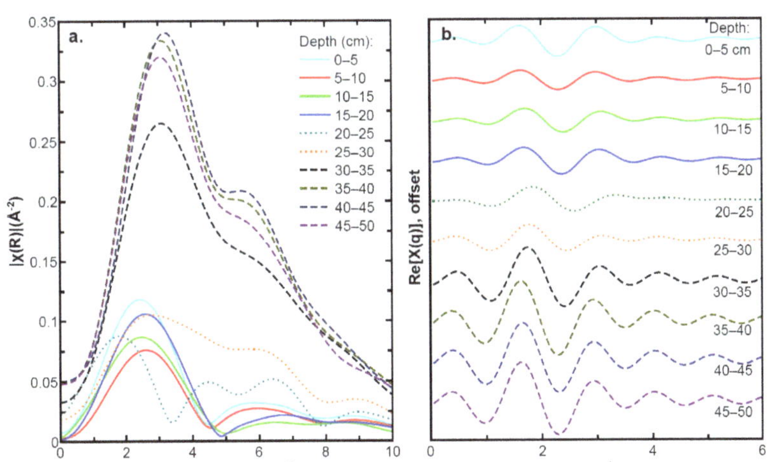

Figure 5. Sulfur XAFS spectra of 5-cm intervals of unmodified core collected from Aberdeen Lodge Bay in Coeur d'Alene Lake, October 2017: (**a**) radial bond length distribution after Fourier transformation ($|\chi(R)|$), magnitude; (**b**) Fourier filtering of $\chi(q)$, real portion.

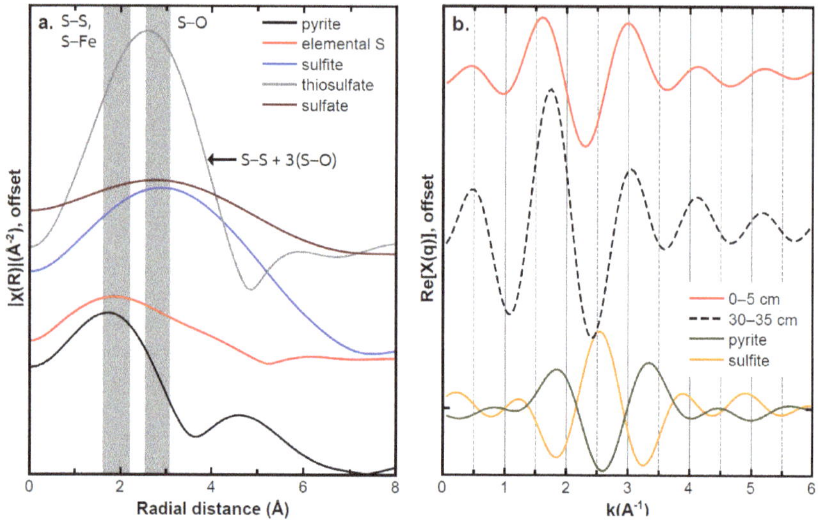

Figure 6. (**a**) Radial bond length distribution after Fourier transformation of sulfur standards ($|\chi(R)|$), magnitude, with common S–S, S–M, and S–O bond length ranges; and (**b**) Fourier filtering of $\chi(q)$, real portion, of pyrite and sulfite standards with the 0 to 5 and 30 to 35 cm spectra of the October core.

3.3. Sediment Sulfur Speciation with Induced Anoxic Conditions

There is no continuous monitoring of the Lake SWI for indication of reoxygenation of the water column with overturning of the thermocline. Therefore, core collection periods were set according to an expected reoxygenation period (October) and post-oxygenation period (November). Porewater was not extracted from the October core. The pH of porewater extracted from the November cores' upper (2.5-cm depth) and lower (12.5-cm depth) sediments over the experimental time period ranged from 6.9 to 7.1 and 6.6 to 7.6 in control and treated cores, respectively. No visible trends were apparent in pH in either core group or between depths. The ORP of porewater in upper and lower sediments ranged from −100 to +65 mV and −104 to +65 mV in control and treated cores, respectively, with primarily reducing conditions in lower sediments and mixed conditions in upper sediments. The lowest ORP values in upper and lower sediments were recorded during week 4. NO_3 and PO_4 were not detected (<0.4 mg/L) in any porewater, and SO_4 concentrations were low (<10 mg/L) with no visible trend.

A comparison of solid-phase S speciation in the upper sediments for the October (−4 week) and November (0 week) samples (Figure 7) indicate primarily reduced S in October and primarily oxidized S in November indicating that reoxygenation of the upper sediments occurred between the two sampling events. The S species of the upper sediments of the control cores (anoxia) shifted towards reduced species (predominantly S^{1-}) from week 0 to week 8 (Figure 7). The upper sediments of the treatment cores (anoxia + algae) indicated variable distributions of S species from week 0 to week 8. The input of algal detritus to the SWI appears to have minimized S reduction in the upper sediments that was apparent in the control cores. The low permeability of the sediments (Table 1) appears to restrict deeper penetration of the seasonal changes in redox conditions as indicated by predominant reduced S forms in the lower sediments for all cores (Figure 7).

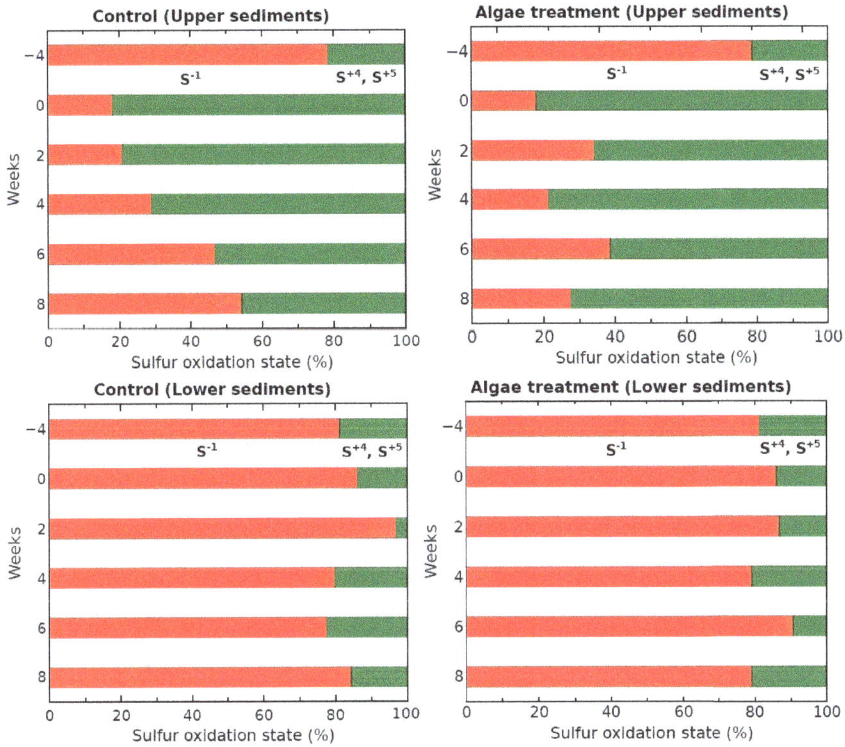

Figure 7. Distribution of the primary sulfur oxidation states (−1 and +4 to +5) in upper and lower sediment samples of the control (anoxia) and treatment cores (anoxia + algae) over the course of the 8-week experiment.

3.4. *Upper Sediment Sulfur Bonding with Induced Anoxic Conditions*

The upper sediments of the control and treatment cores were evaluated for changes in the bonding environment given the increasing trend in reduced S of the control cores and the variable redox state of S in the treatment cores over the course the experiments. The bonding environment in the upper sediments of the control cores (Figure 8a) initially (week 0) appeared to indicate mixed state bond length (radial distribution) that quickly shifted (week 2) to a polythionate-type (thiosulfate of Figure 6 or greater abundance of S–O bonds) environment that weakened in magnitude with each subsequent core (weeks 4 to 8). This decrease in constructive wavefunction is reflected in the week 2 to 8 scatterer response (Figure 8c), which corresponds to the shift in speciation from primarily oxidized forms to greater reduced S during the experiment (Figure 7). This decrease in magnitude of the radial distribution for the upper sediments reflects the reduced S bonding of the upper sediments of the October core compared to the more oxidized environment of the deeper layers in the October core. The S bonding environment of the upper sediments of the treatment cores indicated a similar initial shift in bond length from week 0 to week 2 (Figure 8b), but the bonding environment of the treatment cores did not follow a similar trend in decreasing magnitude of the bond length response. Similar to the S speciation variability (Figure 7) for the treatment cores, the bond lengths of S and its surrounding scatters varied from week to week with an increase from week 2 to 4, a decrease from week 4 to 6, followed by an increase from week 6 to 8 (Figure 8b). The scatterer response and resulting wavefunction of the treated upper sediments (Figure 8d) followed a similar pattern in constructive/deconstructive responses from week to week.

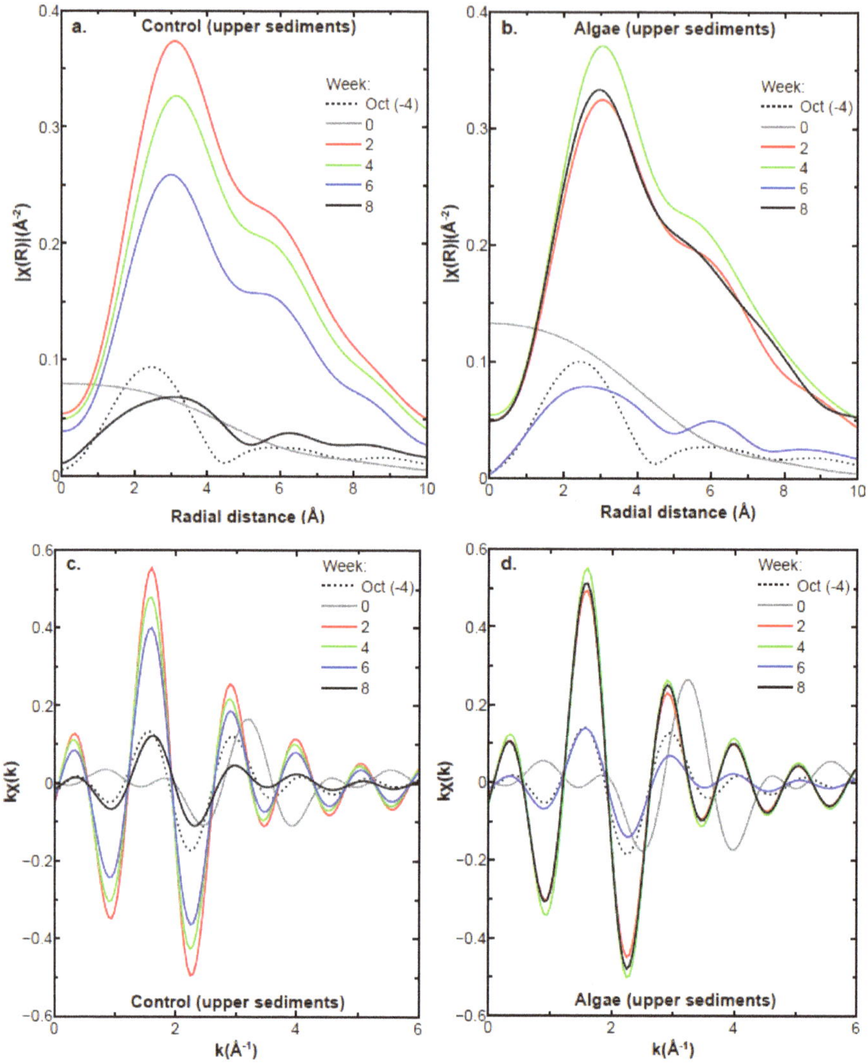

Figure 8. Sulfur XAFS spectra of control and algae-treated cores collected from Aberdeen Lodge Bay in Coeur d'Alene Lake, November 2017: (**a**,**b**) radial distribution after Fourier transformation ($|\chi(R)|$), magnitude; (**c**,**d**) Fourier filtering of $\chi(q)$, real portion.

3.5. Redox-Sensitive Elements in Porewater

The anoxic and anoxic + algae conditions produced similar trends in porewater As, Cd, and Fe concentrations for upper and lower sediments (Figure 9). The concentration trends of each contaminant were similar at each depth under the two conditions; although, the deeper sediments produced larger concentrations of As, Cd, and Fe compared to the upper sediments. This difference in upper and lower As, Cd, and Fe porewater concentrations may be a result of variability of sediment metal concentrations between the two depths. Deeper sediments released greater amounts of all three metals after week 2 then plateaued with small variations in concentrations from week 4 to week 8. Upper sediment, porewater Mn was the only element to indicate a substantial difference in concentration trend between the anoxic and anoxic + algae conditions (Figure 9). The anoxic + algae condition produced the greatest difference (+12,000 μg/L) in Mn concentrations in the upper sediments during week 2, although both conditions produced relatively similar porewater Mn concentrations by the end of the experiments.

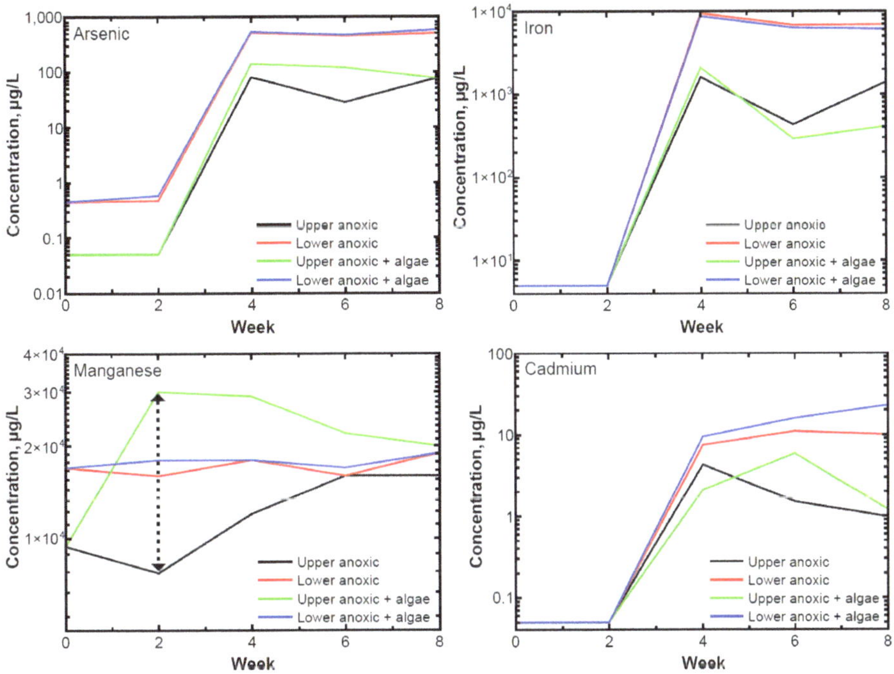

Figure 9. Metal(loid) concentrations in porewater extracted from the anoxic (control) and algae-treated (treatment) sediments analyzed for X-ray absorption spectroscopy.

4. Discussion

Anoxic conditions of the control cores produced an expected trend of increasing S reduction in the upper sediments (Figure 7). The input of algae to the SWI was expected to provide a C- and N-rich environment for further stimulation of reducing bacteria that could produce dissimilatory sulfate reduction with algae oxidation [88,89]. Dissimilatory sulfate reduction is a common process of S reduction in organic-rich environments that assists in overcoming the energy barrier of the SO_4–SO_3 redox couple [90]. Additionally, the temporal decay of the algae detritus should produce continually lower complex carbon forms that become available to the various microbial communities, which should further drive reduction of the redox-sensitive elements, such as As, Fe, Mn, and S [91,92]. Yet, the input of the algal detritus did not produce an enhancement of the S reduction trend seen with the control cores (Figure 7).

Without a stimulation of S reduction but the increased release of Mn with the addition of the algae, it appears that a dissimilatory metal reduction sequence became a dominant process in the upper sediments. Although, anoxic and anoxic + algae conditions produced similar trends in release of redox-sensitive As and Fe and a similar release of the non-redox-sensitive Cd. The substantial change to Mn release with the anoxic + algae condition suggests dissimilatory reduction of Mn with algae decomposition but not inclusion of Fe or S in this stimulated reducing effect. Commonly, organic matter decomposition and metal reduction influence the redox reactions of Fe and Mn in aqueous systems when these two metals are of substantial concentration [93–95]. Bacteria—such as *Clostridium*, *Geobacter*, *Aeromonas*, *Rhodoferax*, *Desulfobulbus*, and *Shewanella*—have shown the ability to couple the oxidation of organic matter and reduction of Fe^{3+} and Mn^{5+}. These metals become the primary electron acceptors and allow for oxidation of fatty acids and monoaromatic compounds [95,96].

The coupling of the anaerobic oxidation of algal detritus and reduction of Mn is not unusual given the identification of microbially-driven dissimilatory metal reduction in Fe- and Mn-rich, anoxic, neutral pH sediment environments [97], but the lack of significant change in Fe release (no enhancing effect) to the porewater is unexpected. Fe has been identified as a primary element participating in redox processes within the Lake sediments where Fe (oxyhydr)oxides are reduced to Fe^{2+} minerals and an increase in dissolved Fe occurs [23]. Given the substantially greater remobilization of Mn early in the treated core experiments, it would be expected that additional Fe would release to the porewater. Some of this lack of Fe mobilization can be attributed to the neutral pH of the porewater throughout the experiments (limited Fe solubility compared to Mn [98]), but the lack of increased porewater Fe concentrations with the input of algae also is apparent in the similar release of As and Cd under anoxic and anoxic + algae conditions. Fe (oxyhydr)oxides are sorbing substrates in the Lake sediments and are considered a primary control on the mobility of contaminants, such as As and Cd [23,42]. Given no corresponding change in porewater As, Cd, or Fe compared to Mn for the anoxic + algae condition, anaerobic oxidation of the algae appears to only stimulate the reduction and increased release of Mn.

5. Conclusions

Examination of the mining-contaminated sediments of Coeur d'Alene Lake indicate a metal- and sulfur-rich environment that undergoes seasonal anoxia due to thermal stratification of the water column. Replication of anoxic conditions for sediment cores collected from the lakebed produced an increasing trend in reduced sulfur species and enhanced S–M bonding in sediments near the sediment-water interface. No changes to the sulfur species distribution were observed for sediments in a deeper layer that remained a continuously reducing environment during all sampling and experimental conditions. Pairing an introduction of algal detritus to the sediment water interface with anoxic conditions suppressed any temporal trend in increased sulfur reduction and S–M bonding in sediments near the sediment-water interface and had no influence on sulfur speciation lower in the sediments. The sulfur bonding environment of the anoxic and anoxic + algal detritus conditions paralleled the S speciation results indicative of increasing reduced S–S and S–M bonds with anoxic conditions and variable S species and bonding environments under the anoxic + algae conditions. The influence of anoxia + algal detritus on sulfur speciation near the sediment-water interface produced a substantial increase in the early release of manganese to the sediment porewater compared to the anoxia-only condition. This dissimilatory metal reduction was the dominant redox process early in the experiments, which produced a substantial increase in the early release of manganese but did not enhance the release of arsenic, cadmium, and iron. This temporal effect from algal detritus appears to have altered microbial communities influence on early transformation of sulfur species and sulfur's influence on retention of manganese in the near surface sediments. Alteration of the species distribution subsequently altered the bonding environment and induced greater release of Mn compared to the anoxia-only condition. Overall, the input of the algae suppressed S reduction, enhanced the reduction of Mn, and did not appear to influence the release of other redox-sensitive metals or release of non-redox-sensitive metals.

Author Contributions: Conceptualization, J.B.L., J.G.M., and F.M.W.; methodology, A.W.C., J.B.L, J.G.M., and F.M.W.; validation, J.B.L, J.G.M., F.M.W., and A.W.C.; formal analysis, J.B.L, J.G.M., F.M.W., A.W.C., and J.D.A.; investigation, J.B.L, J.G.M., F.M.W., A.W.C., and J.D.A.; writing—original draft preparation, J.B.L. and J.D.A.; writing—review and editing, J.B.L, J.G.M., F.M.W., A.W.C., and J.D.A.; supervision, J.B.L. and J.G.M.; funding acquisition, J.B.L. and J.G.M. All authors have read and agreed to the published version of the manuscript.

Funding: This project was funded in part by the National Science Foundation Research Infrastructure Improvement award managed by Idaho EPSCoR under award number IIA-1301792 and the US Geological Survey 104b program as administered through the Idaho Water Resources Research Institute (grant #G16AP00050). Research described in this paper was performed at the Canadian Light Source (CLS), which is supported by the Canada Foundation for Innovation, Natural Sciences and Engineering Research Council of Canada, the University of Saskatchewan, the Government of Saskatchewan, Western Economic Diversification Canada, the National Research Council

Canada, and the Canadian Institutes of Health Research. We wish to thank YongFeng Hu and Qunfeng Xiao of the Soft X-Ray Microcharacterization Beamline at CLS for all their help and guidance.

Conflicts of Interest: The authors declare no conflict of interest.

References

1. Bigham, J.M. Mineralogy of ochre deposits formed by sulfide oxidation. In *Environmental Geochemistry of Sulfide Mine-Wastes*; Jambor, J.L., Blowes, D.W., Eds.; Mineralogical Association of Canada: Quebec City, QC, Canada, 1994; Volume 22, pp. 103–132. ISBN 978-0-8412-2772-9.
2. Bigham, J.M.; Nordstrom, D.K. Iron and aluminum hydroxysulfates from acid sulfate waters. *Rev. Mineral. Geochem.* **2000**, *40*, 351–403. [CrossRef]
3. Dudka, S.; Adriano, D.C. Environmental impacts of metal ore mining and processing: A review. *J. Environ. Qual.* **1997**, *26*, 590. [CrossRef]
4. Moore, J.N.; Luoma, S.N. Hazardous wastes from large-scale metal extraction. A case study. *Environ. Sci. Technol.* **1990**, *24*, 1278–1285. [CrossRef]
5. Pyatt, F.B.; Grattan, J.P. Some consequences of ancient mining activities on the health of ancient and modern human populations. *J. Public Health* **2001**, *23*, 235–236. [CrossRef] [PubMed]
6. Balistrieri, L.S.; Bookstrom, A.A.; Box, S.E.; Ikramuddin, M. *Drainage from Adits and Tailings Piles in the Coeur d'Alene Mining District, Idaho; Sampling, Analytical Methods and Results*; US Geological Survey Open-File Report; United States Geological Survey: Reston, VA, USA, 1998; p. 19.
7. Horowitz, A.J.; Elrick, K.A.; Robbins, J.A.; Cook, R.B. A summary of the effects of mining and related activities on the sediment-trace element geochemistry of Lake Coeur d'Alene, Idaho, USA. *J. Geochem. Explor.* **1995**, *52*, 135–144. [CrossRef]
8. Horowitz, A.J.; Elrick, K.A.; Robbins, J.A.; Cook, R.B. Effect of mining and related activities on the sediment trace element geochemistry of Lake Coeur d'Alene, Idaho, USA part II: Subsurface sediments. *Hydrol. Process.* **1995**, *9*, 35–54. [CrossRef]
9. Paulson, A.J. Biogeochemical removal of Zn and Cd in the Coeur d'Alene River (Idaho, USA), downstream of a mining district. *Sci. Total Environ.* **2001**, *278*, 31–44. [CrossRef]
10. Aiken, G.R.; Hsu-Kim, H.; Ryan, J.N. Influence of dissolved organic matter on the environmental fate of metals, nanoparticles, and colloids. *Environ. Sci. Technol.* **2011**, *45*, 3196–3201. [CrossRef]
11. Haus, K.L.; Hooper, R.L.; Strumness, L.A.; Mahoney, J.B. Analysis of arsenic speciation in mine contaminated lacustrine sediment using selective sequential extraction, HR-ICPMS and TEM. *Appl. Geochem.* **2008**, *23*, 692–704. [CrossRef]
12. Hochella, M.F., Jr.; Mogk, D.W.; Ranville, J.; Allen, I.C.; Luther, G.W.; Marr, L.C.; McGrail, B.P.; Murayama, M.; Qafoku, N.P.; Rosso, K.M.; et al. Natural, incidental, and engineered nanomaterials and their impacts on the Earth system. *Science* **2019**, *363*. [CrossRef]
13. Hochella Jr, M.F.; Lower, S.K.; Maurice, P.A.; Penn, R.L.; Sahai, N.; Sparks, D.L.; Twining, B.S. Nanominerals, mineral nanoparticles, and Earth systems. *Science* **2008**, *319*, 1631–1635. [CrossRef] [PubMed]
14. Hoffmann, S.R.; Shafer, M.M.; Armstrong, D.E. Strong colloidal and dissolved organic ligands binding copper and zinc in rivers. *Environ. Sci. Technol.* **2007**, *41*, 6996–7002. [CrossRef]
15. Kretzschmar, R.; Schäfer, T. Metal retention and transport on colloidal particles in the environment. *Elements* **2005**, *1*, 205–210. [CrossRef]
16. Langman, J.; Torso, K.; Moberly, J. Seasonal and basinal influences on the formation and transport of dissolved trace metal forms in a mining-impacted riverine environment. *Hydrology* **2018**, *5*, 35. [CrossRef]
17. Langman, J.B.; Behrens, D.; Moberly, J.G. Seasonal formation and stability of dissolved metal particles in mining-impacted, lacustrine sediments. *J. Contam. Hydrol.* **2020**, *232*, 103655. [CrossRef] [PubMed]
18. Plathe, K.L.; von der Kammer, F.; Hassellöv, M.; Moore, J.N.; Murayama, M.; Hofmann, T.; Hochella, M.F., Jr. The role of nanominerals and mineral nanoparticles in the transport of toxic trace metals: Field-flow fractionation and analytical TEM analyses after nanoparticle isolation and density separation. *Geochim. Cosmochim. Acta* **2013**, *102*, 213–225. [CrossRef]
19. Wigginton, N.S.; Haus, K.L.; Jr, M.F.H. Aquatic environmental nanoparticles. *J. Environ. Monit.* **2007**, *9*, 1306–1316. [CrossRef]

20. Wood, M.S.; Beckwith, M.A. *Coeur d'Alene Lake, Idaho: Insights Gained From Limnological Studies of 1991–92 and 2004–06*; United States Geological Survey: Reston, VA, USA, 2008; p. 40.
21. La Force, M.J.; Hansel, C.M.; Fendorf, S. Arsenic speciation, seasonal transformations, and Co-distribution with iron in a mine waste-influenced palustrine emergent wetland. *Environ. Sci. Technol.* **2000**, *34*, 3937–3943. [CrossRef]
22. Arora, B.; Şengör, S.S.; Spycher, N.F.; Steefel, C.I. A reactive transport benchmark on heavy metal cycling in lake sediments. *Comput. Geosci.* **2015**, *19*, 613–633. [CrossRef]
23. Cummings, D.E.; March, A.W.; Bostick, B.; Spring, S.; Caccavo, F.; Fendorf, S.; Rosenzweig, R.F. Evidence for microbial Fe(III) reduction in anoxic, mining-impacted lake sediments (Lake Coeur d'Alene, Idaho). *Appl. Environ. Microbiol.* **2000**, *66*, 154–162. [CrossRef]
24. Morra, M.J.; Carter, M.M.; Rember, W.C.; Kaste, J.M. Reconstructing the history of mining and remediation in the Coeur d'Alene, Idaho Mining District using lake sediments. *Chemosphere* **2015**, *134*, 319–327. [CrossRef] [PubMed]
25. Kuwabara, J.S.; Topping, B.R.; Woods, P.F.; Carter, J.L. Free zinc ion and dissolved orthophosphate effects on phytoplankton from Coeur d'Alene Lake, Idaho. *Environ. Sci. Technol.* **2007**, *41*, 2811–2817. [CrossRef] [PubMed]
26. Saxena, P. Harish Toxicity assessment of ZnO nanoparticles to freshwater microalgae Coelastrella terrestris. *Environ. Sci. Pollut. Res.* **2019**, *26*, 26991–27001. [CrossRef] [PubMed]
27. Kuwabara, J.S.; Woods, P.F.; Berelson, W.M.; Balistrieri, L.S.; Carter, J.L.; Topping, B.R.; Fend, S.V. Importance of sediment–water interactions in Coeur d'Alene Lake, Idaho, USA: Management implications. *Environ. Manag.* **2003**, *32*, 348–359. [CrossRef]
28. Woods, P.F.; Beckwith, M.A. *Nutrient and Trace-Element Enrichment of COEUR d'Alene Lake, Idaho*; United States Geological Survey: Reston, VA, USA, 1997; p. 93.
29. Harrington, J.M.; Fendorf, S.E.; Rosenzweig, R.F. Biotic generation of arsenic(III) in metal(loid)-contaminated freshwater lake sediments. *Environ. Sci. Technol.* **1998**, *32*, 2425–2430. [CrossRef]
30. Harrington, J.M.; LaForce, M.J.; Rember, W.C.; Fendorf, S.E.; Rosenzweig, R.F. Phase associations and mobilization of iron and trace elements in Coeur d'Alene Lake, Idaho. *Environ. Sci. Technol.* **1998**, *32*, 650–656. [CrossRef]
31. Pedersen, T.F. A comment on the future environmental status of Coeur d'Alene Lake, Idaho. *Northwest Sci.* **1996**, *70*, 179–182.
32. Şengör, S.S.; Spycher, N.F.; Ginn, T.R.; Sani, R.K.; Peyton, B. Biogeochemical reactive–diffusive transport of heavy metals in Lake Coeur d'Alene sediments. *Appl. Geochem.* **2007**, *22*, 2569–2594. [CrossRef]
33. Child, A.W.; Moore, B.C.; Vervoort, J.D.; Beutel, M.W. Bioavailability and uptake of smelter emissions in freshwater zooplankton in northeastern Washington, USA lakes using Pb isotope analysis and trace metal concentrations. *Environ. Pollut.* **2018**, *238*, 348–358. [CrossRef]
34. Balistrieri, L.S.; Box, S.E.; Tonkin, J.W. Modeling precipitation and sorption of elements during mixing of river water and porewater in the Coeur d'Alene River Basin. *Environ. Sci. Technol.* **2003**, *37*, 4694–4701. [CrossRef]
35. Balistrieri, L.S.; Blank, R.G. Dissolved and labile concentrations of Cd, Cu, Pb, and Zn in the South Fork Coeur d'Alene River, Idaho: Comparisons among chemical equilibrium models and implications for biotic ligand models. *Appl. Geochem.* **2008**, *23*, 3355–3371. [CrossRef]
36. Dockrey, J.W.; Lindsay, M.B.J.; Mayer, K.U.; Beckie, R.D.; Norlund, K.L.I.; Warren, L.A.; Southam, G. Acidic microenvironments in waste rock characterized by neutral drainage: Bacteria–mineral interactions at sulfide surfaces. *Minerals* **2014**, *4*, 170–190. [CrossRef]
37. Gao, Y.; Kan, A.T.; Tomson, M.B. Critical evaluation of desorption phenomena of heavy metals from natural sediments. *Environ. Sci. Technol.* **2003**, *37*, 5566–5573. [CrossRef]
38. Rickard, D.; Luther, G.W. Chemistry of iron sulfides. *Chem. Rev.* **2007**, *107*, 514–562. [CrossRef] [PubMed]
39. Schaider, L.A.; Senn, D.B.; Estes, E.R.; Brabander, D.J.; Shine, J.P. Sources and fates of heavy metals in a mining-impacted stream: Temporal variability and the role of iron oxides. *Sci. Total Environ.* **2014**, *490*, 456–466. [CrossRef] [PubMed]
40. Schippers, A. Biogeochemistry of metal sulfide oxidation in mining environments, sediments, and soils. *Geol. Soc. Am. Spec. Pap.* **2004**, *379*, 49–62. [CrossRef]

41. Toevs, G.; Morra, M.J.; Winowiecki, L.; Strawn, D.; Polizzotto, M.L.; Fendorf, S. Depositional influences on porewater arsenic in sediments of a mining-contaminated freshwater lake. *Environ. Sci. Technol.* **2008**, *42*, 6823–6829. [CrossRef]
42. Toevs, G.R.; Morra, M.J.; Polizzotto, M.L.; Strawn, D.G.; Bostick, B.C.; Fendorf, S. Metal(loid) diagenesis in mine-impacted sediments of Lake Coeur d'Alene, Idaho. *Environ. Sci. Technol.* **2006**, *40*, 2537–2543. [CrossRef]
43. Bostick, B.C.; Hansel, C.M.; Fendorf, S. Seasonal fluctuations in zinc speciation within a contaminated wetland. *Environ. Sci. Technol.* **2001**, *35*, 3823–3829. [CrossRef]
44. Moberly, J.; D'Imperio, S.; Parker, A.; Peyton, B. Microbial community signature in Lake Coeur d'Alene: Association of environmental variables and toxic heavy metal phases. *Appl. Geochem.* **2016**, *66*, 174–183. [CrossRef]
45. Moberly, J.G.; Borch, T.; Sani, R.K.; Spycher, N.F.; Şengör, S.S.; Ginn, T.R.; Peyton, B.M. Heavy metal–mineral associations in Coeur d'Alene River sediments: A synchrotron-based analysis. *Water. Air. Soil Pollut.* **2008**, *201*, 195–208. [CrossRef]
46. Boujelben, N.; Bouzid, J.; Elouear, Z. Adsorption of nickel and copper onto natural iron oxide-coated sand from aqueous solutions: Study in single and binary systems. *J. Hazard. Mater.* **2009**, *163*, 376–382. [CrossRef] [PubMed]
47. Cornell, R.M. Simultaneous incorporation of Mn, Ni and Co in the goethite (alpha -FeOOH) structure. *Clay Miner.* **1991**, *26*, 427–430. [CrossRef]
48. Galán, E.; Gómez-Ariza, J.L.; González, I.; Fernández-Caliani, J.C.; Morales, E.; Giráldez, I. Heavy metal partitioning in river sediments severely polluted by acid mine drainage in the Iberian Pyrite Belt. *Appl. Geochem.* **2003**, *18*, 409–421. [CrossRef]
49. Chen, K.L.; Elimelech, M. Influence of humic acid on the aggregation kinetics of fullerene (C60) nanoparticles in monovalent and divalent electrolyte solutions. *J. Colloid Interface Sci.* **2007**, *309*, 126–134. [CrossRef] [PubMed]
50. Filella, M.; Buffle, J. Factors controlling the stability of submicron colloids in natural waters. In *Colloids in the Aquatic Environment*; Tadros, T.F., Gregory, J., Eds.; Elsevier: Oxford, UK, 1993; pp. 255–273. ISBN 978-1-85861-038-2.
51. Hyung, H.; Fortner, J.D.; Hughes, J.B.; Kim, J.-H. Natural organic matter stabilizes carbon nanotubes in the aqueous phase. *Environ. Sci. Technol.* **2007**, *41*, 179–184. [CrossRef]
52. Keller, A.A.; Wang, H.; Zhou, D.; Lenihan, H.S.; Cherr, G.; Cardinale, B.J.; Miller, R.; Ji, Z. Stability and aggregation of metal oxide nanoparticles in natural aqueous matrices. *Environ. Sci. Technol.* **2010**, *44*, 1962–1967. [CrossRef]
53. Pizarro, J.; Belzile, N.; Filella, M.; Leppard, G.G.; Negre, J.-C.; Perret, D.; Buffle, J. Coagulation/sedimentation of submicron iron particles in a eutrophic lake. *Water Res.* **1995**, *29*, 617–632. [CrossRef]
54. National Research Council. *Superfund and Mining Megasites: Lessons from the Coeur d'Alene River Basin*; The National Academies of Sciences, Engineering, Medicine: Washington, DC, USA, 2005; p. 484.
55. Long, K.R. *Production and Disposal of Mill Tailings in the Coeur d'Alene Mining Region, Shoshone County, Idaho; Preliminary Estimates*; United States Geological Survey: Reston, VA, USA, 1998.
56. Balistrieri, L.S.; Box, S.E.; Bookstrom, A.A.; Hooper, R.L.; Mahoney, J.B. *Impacts of Historical Mining in the Coeur d'Alene River Basin*; Bulletin; United States Geological Survey: Reston, VA, USA, 2010.
57. Clark, G.M.; Mebane, C.A. *Sources, Transport, and Trends for Selected Trace Metals and Nutrients in the Coeur d'Alene and Spokane River Basins, Northern Idaho, 1990–2013*; United States Geological Survey: Reston, VA, USA, 2014; p. 62.
58. Tonkin, J.W.; Balistrieri, L.S.; Murray, J.W. Modeling metal removal onto natural particles formed during mixing of acid rock drainage with ambient surface water. *Environ. Sci. Technol.* **2002**, *36*, 484–492. [CrossRef]
59. Holliday, V.T. Methods of soil analysis, part 1, physical and mineralogical methods. *Geoarchaeology* **1990**, *5*, 87–89. [CrossRef]
60. Gavlak, R.; Horneck, D.; Miller, R.O. Soil, Plant, and Water Reference Methods for the Western Region. In *Proceedings of the WERA-103 Technical Committee*, 3rd ed.; Oregon State University: Corvallis, OR, USA, 2005; Volume WREP 125.
61. Sims, J.R.; Haby, V.A. Simplified colorimetric determination of soil organic matter. *Soil Sci.* **1971**, *101*, 137–141. [CrossRef]

62. Schippers, A.; Sand, W. Bacterial leaching of metal sulfides proceeds by two indirect mechanisms via thiosulfate or via polysulfides and sulfur. *Appl. Environ. Microbiol.* **1999**, *65*, 319–321. [CrossRef]
63. Druschel, G.K.; Hamers, R.J.; Banfield, J.F. Kinetics and mechanism of polythionate oxidation to sulfate at low pH by O_2 and Fe^{3+}. *Geochim. Cosmochim. Acta* **2003**, *67*, 4457–4469. [CrossRef]
64. Li, Y.; Chen, J.; Chen, Y.; Zhao, C.; Zhang, Y.; Ke, B. Interactions of oxygen and water molecules with pyrite surface: A new insight. *Langmuir* **2018**, *34*, 1941–1952. [CrossRef]
65. Terranova, U.; Mitchell, C.; Sankar, M.; Morgan, D.; de Leeuw, N.H. Initial oxygen incorporation in the prismatic surfaces of troilite FeS. *J. Phys. Chem. C* **2018**, *122*, 12810–12818. [CrossRef]
66. Zhang, T.; Wang, Y.; Hu, Y.; Wang, Z.; Chen, J.; Niu, X.; Li, Y.; Gong, X. HO selective cleavage FeS bond for FeS_2 electrolysis in alkaline solution. *Electrochimica Acta* **2019**, *306*, 327–338. [CrossRef]
67. Findlay, A.J.; Kamyshny, A. Turnover rates of intermediate sulfur species (S_x^{2-}, S^0, $S_2O_3^{2-}$, $S_4O_6^{2-}$, SO_3^{2-}) in anoxic freshwater and sediments. *Front. Microbiol.* **2017**, *8*. [CrossRef]
68. Langman, J.B.; Blowes, D.W.; Sinclair, S.A.; Krentz, A.; Amos, R.T.; Smith, L.J.D.; Pham, H.N.; Sego, D.C.; Smith, L. Early evolution of weathering and sulfide depletion of a low-sulfur, granitic, waste rock in an Arctic climate: A laboratory and field site comparison. *J. Geochem. Explor.* **2015**, *156*, 61–71. [CrossRef]
69. Ravel, B.; Newville, M. ATHENA, ARTEMIS, HEPHAESTUS: Data analysis for X-ray absorption spectroscopy using IFEFFIT. *J. Synchrotron Radiat.* **2005**, *12*, 537–541. [CrossRef]
70. Heap, M.J.; Reuschlé, T.; Farquharson, J.I.; Baud, P. Permeability of volcanic rocks to gas and water. *J. Volcanol. Geotherm. Res.* **2018**, *354*, 29–38. [CrossRef]
71. Jensen, L.C.; Becerra, J.R.; Escudey, M. Impact of physical/chemical properties of volcanic ash-derived soils on mechanisms involved during sorption of ionisable and non-ionisable herbicides. *Adv. Sorpt. Process Appl.* **2018**. [CrossRef]
72. Huettel, M.; Røy, H.; Precht, E.; Ehrenhauss, S. Hydrodynamical impact on biogeochemical processes in aquatic sediments. In *The Interactions between Sediments and Water*; Kronvang, B., Ed.; Springer: Dordrecht, The Netherlands, 2003; pp. 231–236.
73. Bryant, W.R. Permeability of clays, silty-clays and clayey-silts. *Gulf Coast Assoc. Geol. Soc. Trans.* **2002**, *52*, 1069–1077.
74. Behyan, S.; Hu, Y.; Urquhart, S.G. Sulfur 1s near-edge X-ray absorption fine structure (NEXAFS) of thiol and thioether compounds. *J. Chem. Phys.* **2011**, *134*, 244304. [CrossRef] [PubMed]
75. Sham, T.K. Nanoparticles and nanowires: Synchrotron spectroscopy studies. *Int. J. Nanotechnol.* **2008**, *5*, 1194–1246. [CrossRef]
76. Alp, E.E.; Mini, S.M.; Ramanathan, M. *X-ray Absorption Spectroscopy: EXAFS and XANES-A Versatile Tool to Study the Atomic and Electronic Structure of Materials*; Argonne National Laboratory: Lemont, IL, USA, 1990; pp. 25–36.
77. Penner-Hahn, J.E. X-ray absorption spectroscopy. *eLS* **2005**. [CrossRef]
78. Yang, Y.; Harmer, S.; Chen, M. Synchrotron-based XPS and NEXAFS study of surface chemical species during electrochemical oxidation of chalcopyrite. *Hydrometallurgy* **2015**, *156*, 89–98. [CrossRef]
79. Koningsberger, D.C.; Ramaker, D.E. Applications of X-ray Absorption Spectroscopy in Heterogeneous Catalysis: EXAFS, Atomic XAFS, and Delta XANES. In *Handbook of Heterogeneous Catalysis*; American Cancer Society: Atlanta, GA, USA, 2008; pp. 774–803. ISBN 978-3-527-61004-4.
80. Koningsberger, D.C.; Mojet, B.L.; van Dorssen, G.E.; Ramaker, D.E. XAFS spectroscopy; fundamental principles and data analysis. *Top. Catal.* **2000**, *10*, 143–155. [CrossRef]
81. Newville, M. Fundamentals of XAFS. *Rev. Mineral. Geochem.* **2014**, *78*, 33–74. [CrossRef]
82. Steudel, R. Properties of sulfur-sulfur bonds. *Angew. Chem. Int.* **1975**, *14*, 655–664. [CrossRef]
83. Dos Santos, E.C.; Lourenço, M.P.; Pettersson, L.G.M.; Duarte, H.A. Stability, structure, and electronic properties of the pyrite/arsenopyrite solid–solid interface—A DFT study. *J. Phys. Chem. C* **2017**, *121*, 8042–8051. [CrossRef]
84. Dos Santos, E.C.; de Mendonça Silva, J.C.; Duarte, H.A. Pyrite oxidation mechanism by oxygen in aqueous medium. *J. Phys. Chem. C* **2016**, *120*, 2760–2768. [CrossRef]
85. El Jaroudi, O.; Picquenard, E.; Demortier, A.; Lelieur, J.-P.; Corset, J. Polysulfide anions. 1. Structure and vibrational spectra of the S22- and S32- anions. Influence of the cations on bond length and angle. *Inorg. Chem.* **1999**, *38*, 2394–2401. [CrossRef]

86. Nikoo, S.; Meister, P.J.; Hayward, J.J.; Gauld, J.W. An assessment of computational methods for calculating accurate structures and energies of bio-relevant polysulfur/selenium-containing compounds. *Molecules* **2018**, *23*, 3323. [CrossRef] [PubMed]
87. Hawthorne, F.C.; Krivovichev, S.V.; Burns, P.C. The crystal chemistry of sulfate minerals. *Rev. Mineral. Geochem.* **2000**, *40*, 1–112. [CrossRef]
88. Bradley, A.S.; Leavitt, W.D.; Johnston, D.T. Revisiting the dissimilatory sulfate reduction pathway. *Geobiology* **2011**, *9*, 446–457. [CrossRef]
89. Plugge, C.M.; Zhang, W.; Scholten, J.C.M.; Stams, A.J.M. Metabolic flexibility of sulfate-reducing bacteria. *Front. Microbiol.* **2011**, *2*. [CrossRef]
90. Santos, A.A.; Venceslau, S.S.; Grein, F.; Leavitt, W.D.; Dahl, C.; Johnston, D.T.; Pereira, I.A.C. A protein trisulfide couples dissimilatory sulfate reduction to energy conservation. *Science* **2015**, *350*, 1541–1545. [CrossRef]
91. Canfield, D.E. Sulfate reduction and oxic respiration in marine sediments: Implications for organic carbon preservation in euxinic environments. *Deep Sea Res. Part Oceanogr. Res. Pap.* **1989**, *36*, 121–138. [CrossRef]
92. Holmer, M.; Storkholm, P. Sulphate reduction and sulphur cycling in lake sediments: A review. *Freshw. Biol.* **2001**, *46*, 431–451. [CrossRef]
93. Lovley, D.R. Fe(III) and Mn(IV) reduction. *Environ. Microbe-Met. Interact.* **2000**, 3–30. [CrossRef]
94. Lovley, D.R. Dissimilatory Fe(III) and Mn(IV) reduction. *Microbiol. Mol. Biol. Rev.* **1991**, *55*, 259–287. [CrossRef]
95. Nealson, K.H.; Myers, C.R. Microbial reduction of manganese and iron: New approaches to carbon cycling. *Appl. Environ. Microbiol.* **1992**, *58*, 439–443. [CrossRef] [PubMed]
96. Bretschger, O.; Obraztsova, A.; Sturm, C.A.; Chang, I.S.; Gorby, Y.A.; Reed, S.B.; Culley, D.E.; Reardon, C.L.; Barua, S.; Romine, M.F.; et al. Current production and metal oxide reduction by shewanella oneidensis MR-1 wild type and mutants. *Appl. Environ. Microbiol.* **2007**, *73*, 7003–7012. [CrossRef]
97. Lovley, D.R.; Giovannoni, S.J.; White, D.C.; Champine, J.E.; Phillips, E.J.P.; Gorby, Y.A.; Goodwin, S. Geobacter metallireducens gen. nov. sp. nov., a microorganism capable of coupling the complete oxidation of organic compounds to the reduction of iron and other metals. *Arch. Microbiol.* **1993**, *159*, 336–344. [CrossRef] [PubMed]
98. Hem, J.D. Chemical factors that influence the availability of iron and manganese in aqueous systems. *GSA Bull.* **1972**, *83*, 443–450. [CrossRef]

minerals

MDPI

Review

Stable Isotope Imprints during Pyrite Leaching: Implications for Acid Rock Drainage Characterization

Ágnes Ódri [1,*], Megan Becker [1,2], Jennifer Broadhurst [1], Susan T. L. Harrison [1,3] and Mansour Edraki [4]

1 Minerals to Metals Initiative (MtM), Department of Chemical Engineering, University of Cape Town, Private Bag X3, Rondebosch, Cape Town 7701, South Africa; megan.becker@uct.ac.za (M.B.); jennifer.broadhurst@uct.ac.za (J.B.); sue.harrison@uct.ac.za (S.T.L.H.)
2 Centre for Minerals Research, Department of Chemical Engineering, University of Cape Town, Private Bag X3, Rondebosch, Cape Town 7701, South Africa
3 Centre for Bioprocess Engineering Research, Department of Chemical Engineering, University of Cape Town, Private Bag X3, Rondebosch, Cape Town 7701, South Africa
4 Centre for Mined Land Rehabilitation, Sustainable Minerals Institute, The University of Queensland, St Lucia, Brisbane, QLD 4072, Australia; m.edraki@cmlr.uq.edu.au
* Correspondence: odriagnes@gmail.com

Received: 1 September 2020; Accepted: 1 November 2020; Published: 4 November 2020

Abstract: The characterization of acid rock drainage (ARD) is traditionally based on mineralogical and geochemical techniques (e.g., Acid Base Accounting tests). The complexity of ARD processes warrants contribution of methods from various disciplines. In the past decade, the increasing role of environmental isotopes in pollution monitoring has enabled the successful application of isotope methods in ARD investigations. While isotopic compositions of different pollutants can refer to their parent mineral, the degree of isotope fractionations are indicative of the mechanisms taking place during the release and transportation of ARD-related contaminants. In natural environments, however, the measured isotope fractionations are predominantly the result of several coexisting or sequential processes. Therefore, the identification and quantification of the distinct contributions of these processes to isotope variations is difficult and requires well-defined laboratory conditions, where the influence of ARD generation on different isotope systems can be assessed with greater certainty. This review provides readers with a single source of information regarding isotopic variations generated by laboratory pyrite leaching.

Keywords: stable isotopes; acid rock drainage; pyrite leaching; mine waste; tailings

1. Introduction

Acid rock drainage (ARD) is one of the major environmental problems at active and closed coal and metalliferous mining sites [1]. ARD forms naturally when sulfidic material comes in contact with atmospheric oxygen and the oxidation of sulfide minerals releases acid and contaminants. Although, pyrite is the most common sulfide in nature and is widely regarded as the main cause of ARD, other sulfides including arsenopyrite and iron-rich sphalerite also contribute to ARD generation [2–4]. The process is slow in nature, but the increase of contact surface areas at mine sites due to the excavation of mine workings and production of waste accelerates the rate of acidic drainage generation [5]. The large-scale ARD that is generated within the mining area is called acid mine drainage (AMD). Oxidative dissolution of sulfide minerals not only acidifies the environment, but also allows soluble metals and salts to enter into the solution phase and to be transported outside the mining site boundaries by surface and groundwater systems [6–8]. Metal pollution associated with acidic effluents has toxic effects on aquatic ecosystems, wildlife, surrounding vegetation and soil ecosystem [9–11]. Also, ARD affected soils are physically degraded and characterized by high concentrations of deleterious elements [12]. The high acidity increases the availability of soluble metals that affects the microbial activity, microbial community structure, biomass, and nutrient production [13,14]. Because of the potential remobilization of accumulated metals, these acidic soils can be considered as secondary sources of pollution.

The water levels in the mine workings are kept to a minimum during active mining. However, after mine closure, rising and particularly fluctuating water levels increase the risk of ARD generation [15]. For example, water that refills the underground workings due to the ceased pumping dissolves the mineral precipitates from sulfide oxidation that have accumulated on the pore space of the exposed walls and ceilings of the workings during the operation. These secondary minerals tend to store metals, metalloids, acidity, and sulfate [16,17], and their dissolution results in an initial drainage with higher acidity and metal contents (called first flush phenomena) than the subsequent ARD. On the surface, the exposure of residual sulfide minerals to water and air in waste storage facilities leads to ARD formation. The small particle sizes of mine tailings (<0.2 mm) [18] has the potential to enhance the rate of oxidation owing to increased sulfide liberation.

To characterize and predict ARD accurately, the fundamental controls on ARD generation including pH, temperature, redox conditions, type and concentration of oxidants, mineralogy, and textural characteristics need to be understood. These factors influence the oxidation and dissolution rate of pyrite. For example, the increase of the concentration of oxidants (Fe^{3+} and O_2), temperature, and Eh of solution result in accelerated pyrite dissolution rates under acidic conditions [19]. At the same time, the source and various attenuation reactions of pollutants both at the early stage of ARD formation and during their subsequent solute transport need to be identified. Formation of secondary water-soluble iron phases and minerals not only affects the ARD chemistry but also influences pyrite oxidation dynamics directly and indirectly by providing H^+ and/or Fe^{3+} and changing the mineralogy of the oxidized layers of pyrite [20,21]. The latter has significance in surface passivation that, in turn, affects the reaction rates of pyrite oxidation and dissolution. Several field and laboratory tests have been developed to characterize ARD generation potential and inform modelling studies for ARD prediction [22,23]. Static tests generate Acid Base Accounting (ABA) and Net Acid Generation (NAG) data and give an indication of the acid generating and neutralizing potential of mine wastes. These methods are considered static in nature as they do not provide information about the rate of acid production or neutralization [24–26]. Laboratory kinetic tests (e.g., humidity cells test) provide insight into ARD generation rates [27,28]. More detailed column leach tests have been developed in recent years for modelling, but due to their complex and time-consuming nature, their application remains limited [29]. Detailed mineralogy of mine waste is also suggested to inform ARD characterization [30–32] and reduce uncertainty or error in static and kinetic tests [33]. In addition, the compositional heterogeneity of pyrite can result in resistivity variations that affect its chemical reactivity and leaching behavior [34–37]. The oxidation potential of pyrite with various trace element composition is also relevant for accurate ARD characterization. Whereas the traditional ARD characterization tests involve mainly geochemical and basic mineralogical approaches, the analytical techniques for accurate measurement of a wide range of environmental isotopes have undergone continual refinement over the past few decades and present a powerful analytical approach in pollution monitoring [38,39]. In addition to the source identification by comparing isotopic composition of contaminants and their potential sources, the redistribution of stable isotopes between different phases during physicochemical and biotic processes help to understand the mechanisms that trigger the isotopic variations [40,41]. The isotope technique has the potential to complement the classic ARD characterization tests effectively because static and kinetic tests do not provide information about the source of various ARD contaminants and their fate during transportation. To apply different isotope systems as indicators of physicochemical or biotic processes, it is essential to know the degree of isotopic fractionation to relate it to different mechanisms. In nature, the isotopic variations are controlled by several coexisting or sequential processes that are difficult to distinguish based on the "sum total" of the different fractionation factors. This review, therefore, provides an overview on the application of sulfur, oxygen, and iron isotope analyses in the interpretation of different reaction mechanisms taking place during the oxidative dissolution of pyrite under controlled laboratory conditions. The aim is to highlight potential research areas where this information can be applied and where the isotope technique is not routinely used yet. A summary of the fundamentals of stable isotope geochemistry and ARD generation is provided in Section 2. In Sections 3 and 4, the behavior of sulfur and oxygen isotopes during the oxidative dissolution of pyrite is presented, followed by that of the iron isotopes in Section 5. Finally, the summary and future perspectives are presented.

2. Stable Isotopes in ARD-Related Processes

The different isotopes of a given element have distinct number of neutrons, and thus different masses. Mass differences cause variations in their physical characteristic and reactivity that ultimately results in the partitioning of heavier and lighter isotopes during physicochemical and biological processes [42]. The degree of partitioning of different isotopes between two co-existing reservoirs is mainly controlled by mass-dependent kinetic and equilibrium laws [43,44] and described by the isotopic fractionation factor:

$$\alpha_{A-B} = \frac{R_A}{R_B} \quad (1)$$

where R is the ratio of the heavy isotope to the light isotope, and A and B denote the two co-existing phases, chemical compounds, or the isotope ratio of a starting material and the product. As the value of α is very close to 1, the degree of fractionation can also be approximated by the difference in isotopic composition of two reservoirs, if delta values are <10 per mil (‰) [45]:

$$1000 \ln \alpha_{A-B} \approx \delta_A - \delta_B \approx \Delta_{A-B} \quad (2)$$

where δ notation reports the per mil differences in the isotopic ratios of a sample relative to an internationally accepted standard:

$$\delta x\ (‰) = \frac{R_{sample} - R_{standard}}{R_{standard}} \times 1000 \quad (3)$$

x refers to the type of the isotope, and R is the ratio of the heavy isotope to the light isotope in the sample (R_{sample}) and in the reference ($R_{standard}$).

The oxidation of sulfide minerals takes place through several chemical reactions, as shown here for pyrite in Reactions (4)–(6) [46]:

$$FeS_2 + 3.5O_2 + H_2O \rightarrow Fe^{2+} + 2SO_4^{2-} + 2H^+ \quad (4)$$

$$Fe^{2+} + H^+ + 0.25O_2 \rightarrow Fe^{3+} + 0.5H_2O \quad (5)$$

$$FeS_2 + 14Fe^{3+} + 8H_2O \rightarrow 15Fe^{2+} + 2SO_4^{2-} + 16H^+ \quad (6)$$

The rate of Reaction (6) is limited by the oxidation of Fe^{2+} in Reaction (5), which is greatly catalyzed by iron oxidizing microbial species such as *Acidithiobacillus ferrooxidans* by up to five orders of magnitude [47]. Other key microbial species involved include *Leptospirillum ferriphilum*, *Acidiplasma cupricumulans*, and *Ferroplasma* species [48,49]. Therefore, the rate of pyrite oxidation via Fe^{3+} can be enhanced by several orders of magnitude compared to the abiotic oxidation via O_2 (from 18 to 170 times more rapid than the Reaction (4)) [47,50].

Reactions (4)–(6) demonstrate only the overall stoichiometry of pyrite oxidation that is generally recognized as a complex, multistep chemical, electrochemical, and bacterially mediated mechanism [51]. Numerous studies have suggested that pyrite oxidation/dissolution rates in aqueous solutions are controlled by electrochemical processes that depend on the semiconducting properties of pyrite [52–58]. This type of dissolution involves the distinct anodic and cathodic sites on the pyrite surface where electron transfer reactions occur during the oxidation and reduction processes. The oxidation of pyrite under the moist-air settings typical of mine waste piles is considered both chemical and electrochemical [35,59]. The contribution of these two mechanisms to pyrite oxidation is still highly debated as well as the intermediate steps of ARD formation [60,61]. As stable isotopes are intensively involved in these reactions, they may provide a potential tool to identify the different intermediate mechanisms that control the ARD generation [52,61–65].

3. Sulfur Isotope Signatures of Pyrite Leaching

Sulfur isotope composition of dissolved sulfates in ARD relates to the source of sulfur, therefore it has the potential to be successfully used in mass balance calculations as part of the pollution monitoring [66–69].

Pyrite leaching experiments demonstrate that the quantitative conversion of pyrite to sulfate results in only minor sulfur isotope fractionation ($\Delta^{34}S_{SO4-FeS2}$) under acidic, aerobic-biotic, or abiotic conditions ranging between −1.3‰ and +0.4‰ [63,70–74]. In addition, processes like precipitation and dissolution of secondary sulfate phases were demonstrated to produce negligible S isotope fractionation (e.g., $\Delta^{34}S_{schwertmannite-sulfate}$ −0.2‰ – +0.3‰ and $\Delta^{34}S_{H-jarosite-sulfate}$ +0.3‰ – +1.1‰) [75–77]. Although these relatively small S isotope fractionation factors allow the identification of the parent sulfides in ARD-impacted water, the seasonal precipitation and dissolution cycles of sulfate minerals might cause small variations in the isotopic composition of sulfate as well. Bioleaching of pyrite at pH 2.05 by Brunner et al. [78] showed a fractionation factor of −0.2‰, in agreement with previously determined fractionation values [63,70–74]. In addition, inconsistent isotopic signatures of SO_4 were demonstrated throughout the pyrite dissolution, where the initial phase of the experiment was characterized by high $\delta^{34}S$ values compared to parent mineral and only moderate $\delta^{34}S$ enrichment was determined in the main stage (Figure 1) [78]. The weaker correlation and lower slope of the initial phase of the experiment suggests that processes other than sulfur leaching from pyrite affect the sulfur isotope composition of sulfates. The authors of Brunner et al. [78] concluded that the degassing of SO_2 at the beginning of the experiment enriches the residual aqueous SO_4 in the heavier sulfur isotope (^{34}S) due to the dominant kinetic isotope effect of this conversion. The loss of SO_2 is also in accordance with the non-stoichiometric initial pyrite dissolution findings described by others, where the molar ratio of SO_4^{2-}/Fe^t was 1.1 for biotic leaching and ranged from 1.5 to 1.6 for abiotic leaching at pH 1.2 to 2 [79,80]. Sulfur isotope fractionation values of +0.4‰ and −1.3‰ were obtained by Pisapia et al. [73] during the non-stoichiometric and stoichiometric phases of their biotic, aerobic (<pH 2) pyrite dissolution experiments. The authors of Pisapia et al. [73] attributed the opposing positive and negative $\Delta^{34}S_{SO4-FeS2}$ values to different pyrite oxidation pathways and did not relate the ^{34}S enrichment of sulfate directly to SO_2 degassing.

According to the overall stoichiometry of pyrite oxidation mechanisms, abiotic oxidation can take place only when there is enough dissolved oxygen to complete the reaction [81]. This suggests that although Reactions (4) and (6) are the most frequently used pathways to demonstrate pyrite oxidation, the actual mechanisms are much more complex and involve the occurrence of intermediate sulfur species like thiosulfate ($S_2O_3^{2-}$), elemental sulfur (S^o), and sulfite (SO_3^{2-}), depending on the availability of oxygen. The authors of Basolo and Pearson [82] argue that, according to the overall oxidation reactions of pyrite, seven electrons are transferred from the mineral to the oxidant, but because it is unlikely that more than two electrons are transferred at a time, pyrite oxidation always requires intermediate steps to produce sulfate (Figure 2). The formation of sulfur intermediates during abiotic and biotic pyrite oxidation was confirmed by several authors both under acidic and alkaline conditions, reporting systematically smaller amounts of intermediates at lower pH due to their rapid oxidation to sulfate [52,53,60,80,83–89]. The mechanism of microbial oxidation of sulfide minerals at low pH occurs via thiosulfate or polysulfide pathways [90]. The different dissolution reactions are a result of (i) differences between the crystal structures of acid-soluble and non-acid-soluble metal sulfides and (ii) the distinct oxidizing strategies used by different microorganisms [91]. Pyrite-as an acid-insoluble sulfide-is mostly oxidized by bacterially generated Fe^{3+} through the thiosulfate pathway by breaking the Fe–S bonds of the mineral. The Fe^{3+} ion is reduced to Fe^{2+} after accepting an electron from pyrite and becomes re-oxidized by microbes. The reduction-oxidation cycles continue until Fe^{2+} and thiosulfate are released from the pyrite, resulting ultimately in sulfate as the end-product via the oxidation of thiosulfate → tetrathionate → sulfite oxyanions (Figure 2). With respect to circumneutral and alkaline pH values, tetrathionate and sulfate formation takes place at pH 6–7, whereas thiosulfate and sulfite are formed at pH 9 [84]. In the polysulfide pathway, acid-soluble sulfides, e.g., sphalerite (ZnS) and chalcopyrite ($CuFeS_2$), are dissolved by both Fe^{3+} and H^+, allowing protons to break M-S bonds in the sulfides [90,92]. The main sulfur intermediates from these reactions are polysulfides (S_n^{2-}) and elemental sulfur, which may eventually oxidize to sulfate (Figure 2). Nevertheless, polysulfide formation has been detected alongside sulfate and minor iron-oxides from acidic pyrite oxidation by *A. ferrooxidans* under experimental aerobic conditions [73].

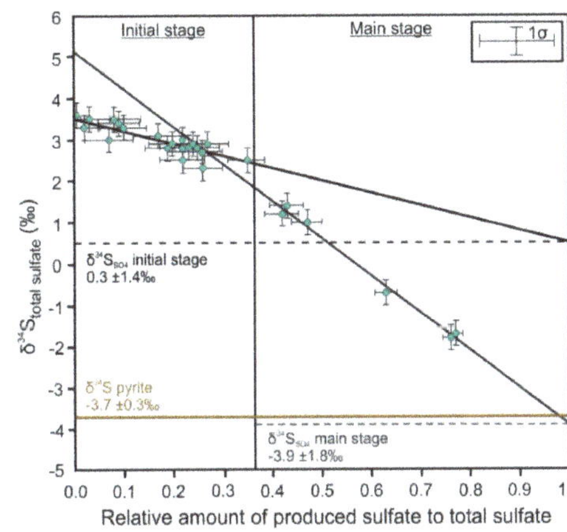

Figure 1. $\delta^{34}S$ variations of sulfates as a function of their increasing amount during the initial and main stages of pyrite leaching. The dashed lines show the $\delta^{34}S$ values of sulfate that are produced in the initial and the main stages. The vertical line demonstrates the transition from the initial to the main stage of pyrite leaching. Redrawn from Brunner et al. [78].

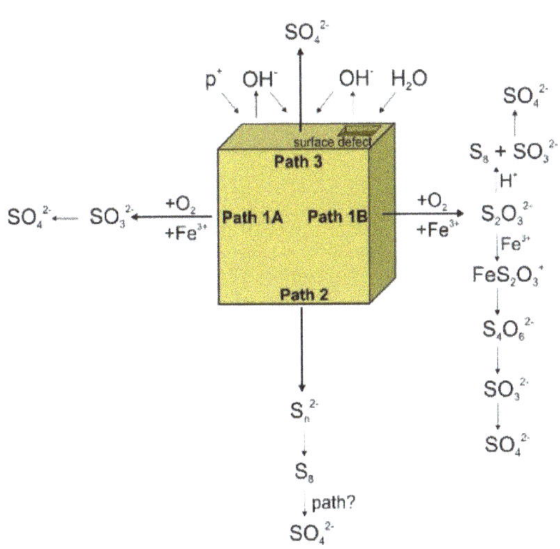

Figure 2. Schematic representation of different sulfide oxidation pathways: Paths 1A and 1B represent thiosulfate oxidation pathways where thiosulfate detaches and where the S–S bond breaks instead of

the Fe–S bond releasing sulfite directly [90]. Path 2 represents the sulfide–polysulfide-elemental sulfur pathway of monosulfides [90] and Path 3 represents the defect/photochemically driven pathway where surface electron defects can initiate pyrite oxidation by surface water and drive the subsequent sulfur oxidation by OH^- [93]. Figure modified from Druschel and Borda [94].

The incomplete oxidation of pyrite or its stepwise oxidation to sulfate is frequently accompanied by sulfur isotope variation in agreement with the general concepts of isotope fractionation, and the more oxidized form of sulfur is enriched in the heavier isotope compared to more reduced forms [52,95,96]. The redistribution of sulfur isotopes in association with the oxidation of different sulfur compounds was demonstrated by experimental and field studies under various conditions [53,72,74,97–101]. As S and O isotopes of sulfate are intensively involved in redox reactions that control the sulfur oxyanion formations, the isotope fractionation has the potential to provide insight on the governing pathways of sulfur oxidation or reduction. Due to the complexity of these redox processes along with the large number of variables that control the sulfur conversion mechanisms (e.g., residence time of sulfur intermediates, pH, crystal structure of the leached mineral, bacterial occurrence, oxygen availability, dominant oxidant), current understanding of the influence of sulfur intermediates on both the ARD sulfur cycle dynamics and its isotope geochemical characteristics is relatively limited.

4. Oxygen Isotope Signatures of Pyrite Leaching

The oxygen isotope composition of dissolved sulfate can be used to understand the oxidation reaction pathways of sulfide minerals. Sulfate-generated via pyrite oxidation-preserves its oxygen isotope composition after formation both due to the very slow kinetics of oxygen isotope equilibration between water and sulfate [102–104] and the small fractionation between dissolved sulfate and secondary sulfate minerals during precipitation (average $\Delta^{18}O_{precipitate\text{–}sulfate}$ +0.5‰ – +0.6‰) [76]. This allows for the identification of the reaction mechanisms that are responsible for sulfide oxidation. If most of the sulfate oxygen is derived from the dissolved molecular oxygen via Reaction (4), the emerging SO_4 should be enriched in ^{18}O relative to SO_4 that incorporates dominantly water-derived oxygen (WDO) following Reaction (6). The reason for this is that air is characterized by $\delta^{18}O$ = +23.8 ‰ [105–107], whereas meteoric water has variable $\delta^{18}O$ values, mainly $\delta^{18}O \leq 0$ ‰ [108]. Using Equations (7) and (8), the relative contributions of WDO and molecular oxygen to dissolved sulfate can be calculated [64].

$$\delta^{18}O_{SO4} = X(\delta^{18}O_{H2O} + \varepsilon_{H2O}) + (1 - X)\,(\delta^{18}O_{O2} + \varepsilon_{O2}) \tag{7}$$

where X is the fraction of sulfate produced by Reaction (6) and (1 − X) is the fraction of sulfate from Reaction (4). ε is the enrichment factor that describes the partition of oxygen between the sulfate and air or water. $\varepsilon^{18}O_{SO4\text{-}H2O}$ = 0.0‰ to +4.1‰ for both anaerobic biotic and abiotic experiments and $\varepsilon^{18}O_{SO4\text{-}O2}$ = −10.0‰ to −11.4‰ for biotic and −4.3‰ to −9.8‰ for abiotic reactions [63,64,72]. By measuring the $\delta^{18}O_{SO4}$ and $\delta^{18}O_{H2O}$, Equation (7) can be re-arranged according to Equation (8) [99]:

$$\delta^{18}O_{SO4} = X(\delta^{18}O_{H2O} + \varepsilon_{H2O} - \delta^{18}O_{O2} - \varepsilon_{O2}) + (\,\delta^{18}O_{O2} + \varepsilon_{O2}) \tag{8}$$

The above calculations require the assumption of both constant oxygen isotope enrichment factors and the lack of isotope exchange reactions between the water and dissolved sulfate after the oxidation.

The general concept regarding the relative percent contribution of various oxygen sources to sulfate is that: (i) if pyrite oxidation is anaerobic and Reaction (6) is the dominant, then 100% of the sulfate oxygen is derived from water, and (ii) if dissolved molecular oxygen is present, sulfate oxygen shows a combination of air and water oxygen [68,109–111]. The incorporation of WDO in sulfate is controlled by Fe oxidizing bacteria [112]. The overall pyrite oxidation reaction (Reaction (4)) suggests that 87.5% of sulfate oxygen is dissolved molecular oxygen and the remaining 12.5% is water derived. The $\delta^{18}O$ and therefore WDO values of sulfate, however, do not necessarily reflect the stoichiometry of Reaction (4).

Pyrite oxidation experiments conducted by Taylor et al. [63] showed that under water-saturated, anaerobic-abiotic and -biotic conditions at pH 2, mainly water oxygen is involved in the formation of sulfate, with WDO values of 94% and 65%, respectively. Their estimated WDO of sulfate during the alternation of wet and dry cycles in the absence of bacteria was 72%, whereas biotic wet and dry pyrite oxidation indicated the lowest value of 23%. These results demonstrate that pyrite oxidation is primarily controlled by Fe^{3+}, and biological reactions appear to incorporate more atmospheric oxygen in sulfate compared with abiotic mechanisms. The authors of Balci et al. [72], however, did not find a distinct microbial control on the pyrite oxidation pathways during the biotic and abiotic leaching experiments that were conducted under anaerobic, aerobic, and acidic pH conditions. According to their results, Fe^{3+} was the main oxidant for pyrite, and consequently, water oxygen was predominantly incorporated in sulfate. The short- and long-term experiments resulted in WDO values of 85% and 92% respectively, under aerobic-biotic conditions, and 87% for aerobic abiotic reactions. Because of the similar estimated WDO values, Balci et al. [72] considered the incubation time and pH as the main controls on the oxygen isotope composition of sulfates. This is also in accordance with the results of Qureshi [113] which suggest pH as the major factor controlling the isotopic composition of sulfate.

To understand the reason for significant water oxygen incorporation into sulfate at low pH even under oxygen-saturated conditions, it is necessary to review the mechanism of pyrite oxidation. The initial step is the oxidation of the adsorbed $Fe^{2+}(H_2O)_6$ complexes on the pyrite surface by molecular oxygen to form $Fe^{3+}(H_2O)_6$ complexes that rapidly oxidize the pyrite (Figure 3) [53,87]. During the initial release of Fe^{2+} to solution, the pyrite surface is dominated by sulfur atoms [114]. The adsorbed iron continuously experiences oxidation and reduction during the interaction with molecular oxygen and the acceptance of electrons from the pyrite (Figure 3). The latter process is governed by the electron loss of iron on the pyrite surface during the interaction via molecular oxygen [52]. The electron originates from the sulfur site and its transmission to Fe^{3+} takes place via the iron site of the mineral (Figure 3b).

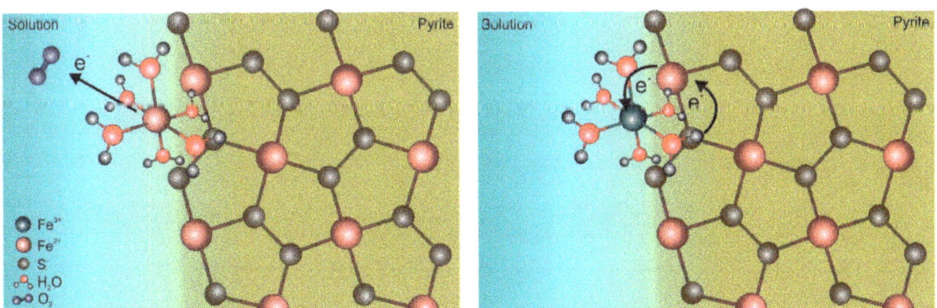

(**a**) (**b**)

Figure 3. Illustration of the abiotic oxidation mechanism of pyrite. The adsorption of released Fe^{2+} from pyrite as aqueous $Fe^{2+}(H_2O)_6$ is the first step of pyrite oxidation. (**a**) The oxidation of $Fe^{2+}(H_2O)_6$ complex by the dissolved molecular oxygen forms (**b**) Fe^{3+} that accepts electron from the pyrite during its oxidation.

An oxygen atom from the adsorbed $Fe^{3+}(H_2O)_6$ complex is added to the sulfur site of pyrite as the closing step of pyrite oxidation that produces different sulfur intermediates (Figure 4a) [85,88,94]. The reaction sequence of the oxidation-reduction of the Fe^{3+} in the adsorbed aqueous complex, and the transfer of oxygen from water molecules to the sulfur site of pyrite, continues until a sulfur oxyanion is more stable in solution, and disassociates from the pyrite surface (Figure 4b) [53].

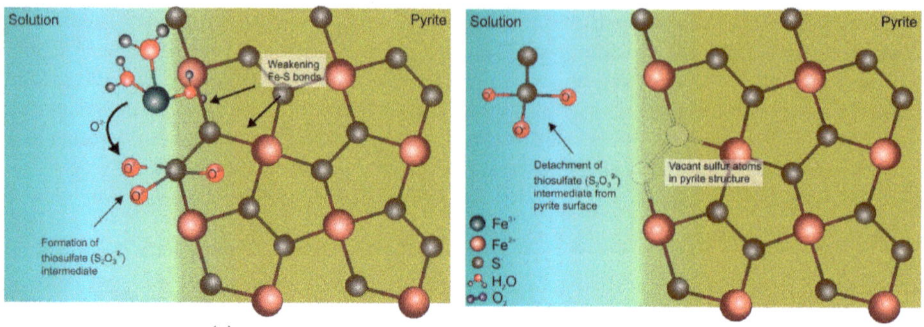

(**a**) (**b**)

Figure 4. Illustration of the formation of sulfur oxyanion during the abiotic oxidation of pyrite using the example of thiosulfate. (**a**) Addition of oxygen molecules from the aqueous $Fe^{3+}(H_2O)_6$ complex to pyrite sulfur to form thiosulfate. (**b**) The oxygen transfer continues until the detachment of the stable thiosulfate.

In accordance with the mechanism described above, oxygen plays a role in the initiation of the pyrite oxidation reaction, but the reaction pathway is mainly controlled by Fe^{3+} as the oxidant. As molecular oxygen is not directly involved in the formation of sulfur oxyanion intermediates, the incorporation of water oxygen into thiosulfate from $Fe^{3+}(H_2O)_6$ complexes is the major process that affects the oxygen isotope composition of this sulfur compound [115]. Although these oxidation steps are consistent with the high proportion of WDO in the experimentally produced sulfates, these do not explain the minor contribution of atmospheric oxygen in the sulfur oxidation reactions. The dominance of water oxygen in the formation of sulfur intermediates suggests that the participation of molecular oxygen in sulfate takes place at the more advanced stages of oxidation. Figure 5 demonstrates that the transition of sulfite to sulfate is the only conversion during the oxidation sequence when O_2 may participate in the reactions. Consequently, if anything slows down or inhibits this oxidation step (e.g., a decrease in pH reduces the amounts of sulfur intermediates) [84] that would increase the relative contribution of WDO to sulfate.

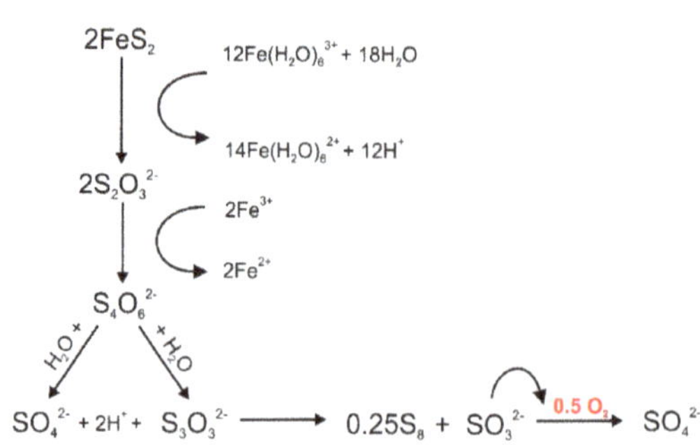

Figure 5. Pyrite oxidation pathway under oxygen-saturated conditions. Molecular oxygen-indicated in red-shows the incorporation of O_2 into sulfates. This oxidation step is considered to decrease the relative percent contribution of WDO to produced sulfate. Figure modified from Balci et al. [72].

The distribution mechanism of oxygen between water and sulfate based on their isotopic signatures was demonstrated by Brunner et al. [78]. Although the generated sulfate and water showed identical $\delta^{18}O$ value of 1.3‰ at the end of the pyrite dissolution, inconsistent oxygen isotope partitioning is presented on Figure 6. Sulfates are characterized by strong and various heavy oxygen isotope (^{18}O) enrichment at the initial stage, whereas these become isotopically lighter as a function of their increasing amount during the main stage. At the start of dissolution, sulfates show ~2.0‰ lower $\delta^{18}O$ values relative to sulfate oxygen that would be derived from water characterized by $\delta^{18}O$ = 1.3‰. According to the authors of Brunner et al. [78], sulfite intermediates formed at the early stages of dissolution are not converted to sulfate immediately. This allows a prolonged initial oxygen isotope equilibration between sulfite and water that can lower the $\delta^{18}O$ values of the initial sulfates produced.

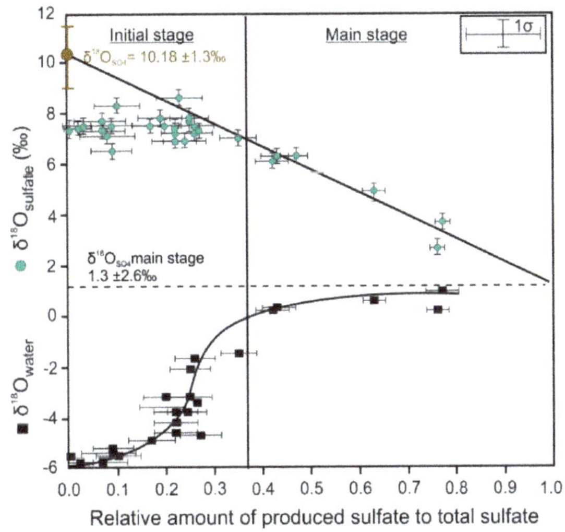

Figure 6. $\delta^{18}O$ variations of sulfates and interacting water as a function of the increasing amount of sulfate during pyrite leaching. The dashed line shows the $\delta^{18}O$ values of sulfate that are produced in the main stage of dissolution. The gold circle illustrates the supposed $\delta^{18}O$ value of initial sulfates in case of immediate conversion of sulfite to sulfate. The vertical line demonstrates the transition from the initial to the main stage of pyrite leaching. Redrawn from Brunner et al. [78].

Based on the above, the major potential factors that are considered to cause variations in sulfate oxygen isotope values and the ratio of WDO and O_2 are: (i) the contribution of different oxygen sources, (ii) degree of the oxygen isotope exchange between water and various sulfur intermediates, (iii) environmental conditions such as pH or availability of oxygen, and (iv) the reaction mechanisms of pyrite oxidation through the related sulfur intermediates [65,86,116].

5. Iron Isotope Signatures of Pyrite Leaching

The release of iron and other transition metals from sulfide minerals during oxidative dissolution can be demonstrated by the isotopic composition of both minerals and the solution [41,117]. Iron isotope variations were previously considered as biological signatures, however, Zhu et al. [118] demonstrated using the example of Cu and Fe that abiotic processes can also be responsible for the isotope fractionation of the transition metals. Fractionation of Fe isotopes have been recognized since the mid-1990s [119–121]. There is an overall agreement that changes in chemical conditions such as pH or redox potential strongly affect the mobilization of iron [122,123], but the interpretation of related Fe isotope variations is still under debate because of the relative novelty of the technique.

Laboratory experiments have shown that the dissolution reactions take place on the surface of the minerals and have demonstrated the dominance of kinetically controlled faster reaction rates of light isotopes (kinetic isotope effect) at the beginning of dissolution [124]. Wiederhold [125] found that during the reductive, abiotic leaching of goethite, the lighter Fe isotopes (^{54}Fe) were released into the solution at the early stages, which resulted in the enrichment of the heavier isotopes (^{56}Fe) at the mineral surface. However, the increase of the $\delta^{56}Fe$ value of the solution during the second stage suggested both the dominance of the equilibrium isotope effect—the forward and backward reaction rates of isotopes are identical between product and reactant—and the transient role of the initial kinetic fractionation. The preservation of the initial isotope composition depends on the extent of dissolution. The kinetic fractionation can be overwritten and erased by the subsequent isotope effects in the case of completed dissolution, which would result in the identical isotopic signature of the solution and the reacting mineral.

Similar to reductive weathering, both kinetic and equilibrium isotope fractionation between coexisting Fe(II) and Fe(III) complexes in solution was also suggested by experimental studies under oxidative conditions [126,127]. Fractionation values of 2.63 ± 0.11‰ and 2.75 ± 0.01‰ were estimated between aqueous Fe(II) and precipitated Fe(III) at pH values of 5.5 and 2.5, respectively [128]. The authors of Anbar et al. [129] suggested a slightly higher fractionation of 3.0 ‰ between aqueous Fe(II) and Fe(III) complexes. The degree of Fe isotope fractionation during Fe(III) oxides and hydroxides precipitation is significantly affected by factors such as grain size and the rate of precipitation in addition to pH conditions [130]. The enrichment of ^{56}Fe in the fluid phase at the start of oxidative dissolution of sulfide-rich rocks at pH 2 has been demonstrated [131] (Figure 7a). The initial isotopic enrichment of solution suggested a ^{56}Fe-dominated reservoir of the mineral that was washed into the leachate. The authors of Fernandez and Borrok [131] identified the ^{56}Fe-rich reservoir as a thin layer of ferric-oxide on the sulfide surface that was the result of air oxidation prior to leaching. This transient redox isotope effect was followed by the dominance of kinetic fractionation as more iron was released with time, which decreased the $\delta^{56}Fe$ value of the solution. Whilst in the acidic experiments, the influence of the redox-driven, surface-related mechanisms were significant, at pH 5, the precipitation of secondary ferric minerals from the solution controlled the redistribution of Fe isotopes (Figure 7b). Figure 7b shows that during the dissolution of sulfide-rich rocks under circumneutral conditions, the leachate is consistently characterized by lower $\delta^{56}Fe$ values compared to bulk rock due to the immediate formation of secondary precipitates [131]. Whereas the aqueous oxidation of Fe^{2+} under acidic abiotic conditions (pH2)-described by the Reaction (5) in Section 2-is limited due to its sluggish kinetics [50], the rapid oxidation of aqueous Fe(II) and the subsequent precipitation of Fe(III) oxyhydroxides is enabled at circumneutral pH values (pH 5). The system reaches the equilibrium isotope fractionation between aqueous Fe(II) and Fe(III) quickly and the precipitated Fe(III) prefers the incorporation of heavy Fe isotopes [127]. This results in a decrease of the Fe isotope ratios in the fluid phase (Figure 7b). Nevertheless, it is important to note that secondary minerals such as jarosite and Fe-(oxy)hydroxides can show lighter Fe isotopic signatures compared to their parent material or the solution under certain conditions [132,133].

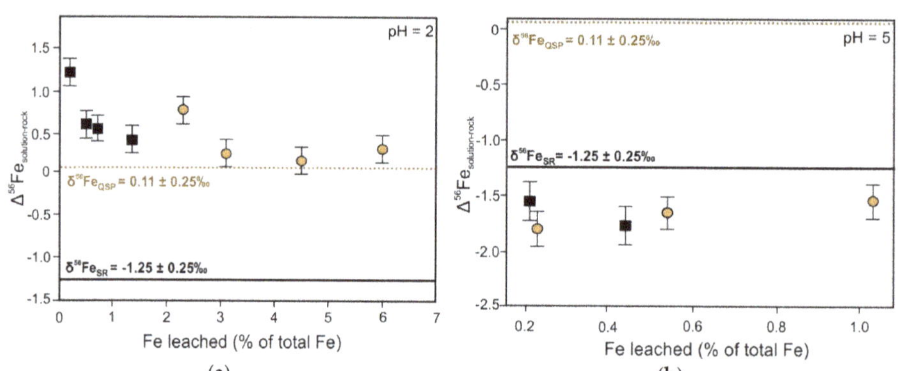

Figure 7. Fractionation of iron isotopes between solution and sulfide-rich rocks ($\Delta_{solution\text{-}rock} = \delta^{56}Fe_{solution} - \delta^{56}Fe_{rock}$) as a function of Fe recovery (**a**) under pH = 2 and (**b**) pH = 5 conditions during continuous batch leaching. Square symbols—leachate of sulfide rock (SR) group: pyrite, chalcopyrite,

galena, sericite, pyrophyllite, anglesite, sphalerite. $\delta^{56}Fe_{SR} = -1.25 \pm 0.25‰$. Circle symbols—leachate of quartz-sericite-pyrite (QSP) group: pyrite, sericite, pyrophyllite, quartz. $\delta^{56}Fe_{QSP} = 0.11 \pm 0.25‰$. Redrawn from Fernandez and Borrok [131].

Influence of Microorganisms on Iron Isotope Fractionation

Various biotic leaching experiments have been performed to investigate the effect of bacterial activity on the isotope fractionation in geological environments [72,134–138]. Microorganisms such as *Acidithiobacillus ferrooxidans* grow in a naturally acidic environment mainly on the surface of different sulfide minerals [139,140]. The *Acidithiobacillus* group of organisms are known for their ability to dissolve transient metals from sulfide minerals by both contact and indirect oxidation [141,142]. Thus, the presence of these bacteria and other microorganisms around sulfide deposits has a potential to accelerate the oxidation that facilitates the increased rates of ARD generation.

Leaching experiments of different iron minerals demonstrate both the presence and absence of bacterial control on Fe isotopes during the dissolution. Although many biological processes are kinetically controlled [28], experimental works indicate ^{56}Fe uptake by microorganisms [143,144]. The assimilation of heavy Fe isotope by bacteria decreases the $\delta^{56}Fe$ value of the solution. The authors of Croal et al. [145] demonstrated 1.5‰ microbial fractionation between aqueous Fe(II) and Fe(III). Based on this result, the biological control on redox processes and subsequent isotope fractionations were suggested. In contrast, the authors of Balci et al. [133] identified abiotically controlled Fe isotope fractionation between Fe(II) and Fe(III) under acidic conditions (pH < 3) for biotic and abiotic oxidation experiments. Bio- and electro-chemical leaching experiments of metal sulfides under consistent redox conditions found no significant difference in Fe isotope fractionation (Figure 8) [138]. This emphasizes the importance of redox potential conditions as the main factor governing the redistribution of Fe isotopes during sulfide dissolution, which seems to be a unique signature of the sulfide system.

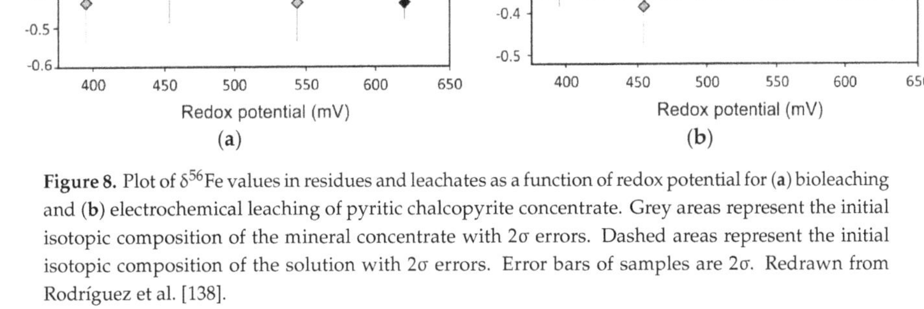

Figure 8. Plot of $\delta^{56}Fe$ values in residues and leachates as a function of redox potential for (**a**) bioleaching and (**b**) electrochemical leaching of pyritic chalcopyrite concentrate. Grey areas represent the initial isotopic composition of the mineral concentrate with 2σ errors. Dashed areas represent the initial isotopic composition of the solution with 2σ errors. Error bars of samples are 2σ. Redrawn from Rodríguez et al. [138].

Concordant Fe isotope variation mechanisms in biotic and abiotic experiments do not allow for the distinction between biological Fe fractionation and that of abiotic chemical reactions, and points to the complexity of controlling mechanisms of various isotope systems even under laboratory conditions. Further investigations that allow for the accurate characterization of abiotic and biotic processes are warranted.

6. Summary and Future Perspective

Pyrite, as the most common sulfide mineral in nature, plays a significant role in directly controlling sulfur and iron cycles and in indirectly controlling the oxygen cycle. The importance of the oxidative dissolution of pyrite relates to its environmental impact through ARD generation. Therefore, an understanding of the processes and mechanisms governing pyrite dissolution facilitates the predictive modelling of ARD, which in turn can be used to inform effective geotechnical designs and rehabilitation strategies. In summary, under controlled laboratory conditions, the use of pure mineral phases allows measurement of the discrete isotope fractionations and therefore the effects of ARD-related processes on different isotope systems can be inferred with greater certainty. Where reactions cannot be controlled due to their coexistence, the closed system in the laboratory enables the recognition of additional mechanisms by retaining the end-products, e.g., SO_2 degassing during the initial dissolution of pyrite. Interpretation of results in the field are therefore strengthened by controlled laboratory experiments that more accurately estimate the discrete fractionation factors that belong to well-defined processes. Sulfur, oxygen, and iron isotope signatures and their key roles in ARD characterization are summarized in Table 1.

Table 1. Summary of the main isotopic signatures during the oxidative leaching of pyrite

Isotope	Process	Isotopic Signature of the Process	Role in ARD Characterization	Notes
Sulfur	Quantitative conversion of pyrite to sulfate	Fractionation ranges between −1.3 ‰ and +0.4 ‰ [63,70–74]	Allows the identification of pyrite as the source of sulfur	Inconsistent sulfate $\delta^{34}S$ values during pyrite dissolution due to SO_2 degassing [78]
	Stepwise or incomplete conversion of pyrite to sulfate	Degree of fractionation varies with the oxidation state of sulfur [72,74,97]	Provide information about the pathways of sulfur oxidation or reduction	^{34}S enrichment follows the general trend: $SO_4^{2-} > SO_3^{2-} > S_2O_3^{2-} > S^0 > S^{2-}$ [96]
Oxygen	Quantitative conversion of pyrite to sulfate	Sulfate $\delta^{18}O$ mainly depends on the relative contribution of oxygen sources [68,109–112]	Indicative of the dominant reaction mechanism that is responsible for pyrite oxidation	Inconsistent sulfate $\delta^{18}O$ values during pyrite dissolution due to prolonged oxygen isotope equilibration [78]
	Stepwise or incomplete conversion of pyrite to sulfate	Sulfate $\delta^{18}O$ is influenced by the presence of sulfur intermediates [61]	Provide insights into intermediate mechanisms that control pyrite oxidation	Oxygen exchange kinetics between sulfur oxyanions and water varies with the oxidation state of sulfur [146]
Iron	Oxidative dissolution of pyrite	Fractionation between $Fe^{2+}{}_{FeS2}$ and $Fe^{3+}{}_{precipitate}$ ranges from −1.7 ‰ to 3.0 ‰ [128,129,131] [1]	Provide insights into source and (bio)geochemical cycling of iron	Inconsistent solution $\delta^{56}Fe$ values at pH 2 due to dissolution of air oxidized layer [131]

[1] Presence of bacteria might affect the iron isotope system by decreasing the $\delta^{56}Fe$ value of the solution due to the uptake of ^{56}Fe [144,145].

Based on the summaries presented here, some potential focus areas for future studies are outlined:

1. Understanding the role of pyrite type: Most laboratory experiments have used hydrothermal pyrite, even though it is well established that pyrites from different geological environments, morphologies, or electrochemical properties [37] show distinctly different oxidation rates and dissolution signatures [147,148]. Consequently, there is a need to extend the database of isotope fractionation factors to include all forms and physical parameters of pyrites in order to apply the most appropriate fractionation factors in field studies.
2. Understanding acid neutralization processes: An understudied but promising area in the application of isotope geochemistry involves the isotope signatures of neutralization processes.

ARD-related carbon isotope fractionation would not only provide information on the mechanisms of neutralization reactions, but also has the potential to determine the relative proportion of the main neutralizing carbonate phases using mixing calculations. This may provide more accurate Acid Neutralizing Capacity (ANC) calculations, as currently calcite is considered as the most dominant neutralizing mineral [3] without its quantified relative contribution to neutralization.

3. Understanding the role of different mineral processing and metallurgical activities on isotope processes: Although several isotope investigations have been undertaken in mining areas to track ARD-related processes in underground and surface water bodies, the method still has not become common practice. In addition, most of these field studies are focused on underground mine workings, pit lakes, as well as waste and water storage facilities that are collectively considered as the major environments of ARD generation. To our knowledge, no studies have examined the role of ore beneficiation processes in ARD formation using the isotope techniques in detail. The water consumption of different processing steps such as milling, dense media separation, or flotation may also provide the required conditions of sulfide oxidation, allowing the initial steps of ARD generation to take place. Tracking the sulfide oxidation pathways and sulfur transformations during various beneficiations steps would help to assess the relative contribution of mineral processing to ARD generation. Eliminating the ARD potential before waste disposal would have a significant beneficial impact on mine water quality. In addition, the quantification of mineral oxidation rates of certain processing steps via isotopes could add to process optimization by the evaluation of more accurate residence times.
4. ARD prevention, mitigation, and management: In order to minimize the potential impacts of ARD on natural water systems, it is necessary to understand (i) the mechanisms taking place during ARD formation and (ii) the mechanisms that control the mobility of contaminants. Although isotopes are successfully used in pollution source-tracking and transportation modelling, the applications of isotopes in ARD generation processes are not yet fully explored. To create more accurate prediction models and thus be able to select the most optimal pollution emission control strategies, more accurate isotope fractionation factors and the identification of causative mechanisms are required. Accurate measurements of fractionation would allow the estimation of various reaction rates with greater certainty. This would help to improve our ability to evaluate the chemical evolution of both mine waste material and the impacted water systems, potentially informing ARD control measures in a timely manner. In addition to tracking ARD generation and transportation, isotopes also can be used in evaluating the effectiveness of different remediation methods, for example, passive treatment systems [75,149].

The application of stable isotopes in environmental studies has a significant potential to facilitate the sustainability in water resources management. Precise quantification of discrete fractionation factors will extend this methodology into new areas and has implications in describing complex natural systems with greater certainty.

Author Contributions: Conceptualization, M.B., J.B., S.T.L.H. and Á.Ó.; validation, M.B., M.E. and S.T.L.H.; writing—original draft preparation, Á.Ó.; writing—review and editing, Á.Ó., M.B., M.E., J.B. and S.T.L.H.; visualization, Á.Ó.; supervision, M.B.; funding acquisition, S.T.L.H. and J.B. All authors have read and agreed to the published version of the manuscript.

Funding: National Research Foundation of South Africa (NRF)-South African Research Chairs Initiative (SARChI Chair in Minerals Beneficiation), UID 64829.

Acknowledgments: We thank Chris Harris for his helpful comments on this manuscript. This work is based on research supported by the National Research Foundation of South Africa (NRF) through their South African Research Chairs Initiative (SARChI Chair in Minerals Beneficiation, held by Jochen Petersen, UID 64829). Any opinion, finding, conclusions, or recommendation expressed in this material is that of the authors and the NRF does not accept any liability in this regard. The three anonymous reviewers and Carlito Tabelin are thanked for their constructive reviews that substantially improved this paper.

Conflicts of Interest: The authors declare no conflict of interest.

References

1. Egiebor, N.O.; Oni, B. Acid Rock Drainage Formation and Treatment: A Review. *Asia Pac. J. Chem. Eng.* **2007**, *2*, 47–62. [CrossRef]
2. Moncur, M.C.; Jambor, J.L.; Ptacek, C.J.; Blowes, D.W. Mine Drainage from the Weathering of Sulfide Minerals and Magnetite. *Appl. Geochem.* **2009**, *24*, 2362–2373. [CrossRef]
3. Dold, B. Basic Concepts in Environmental Geochemistry of Sulfidic Mine-Waste Management. In *Waste Management*; Kumar, E.S., Ed.; InTech: Rijeka, Croatia, 2010. Available online: http://www.intechopen.com/books/waste-management/basic-concepts-in-environmental-geochemistry-ofsulfidic-mine-waste-management (accessed on 22 May 2020).
4. Chopard, A.; Benzaazoua, M.; Bouzahzah, H.; Plante, B.; Marion, P. A Contribution to Improve the Calculation of the Acid Generating Potential of Mining Wastes. *Chemosphere* **2017**, *175*, 97–107. [CrossRef]
5. Akcil, A.; Koldas, S. Acid Mine Drainage (AMD): Causes, Treatment and Case Studies. *J. Clean. Prod.* **2006**, *14*, 1139–1145. [CrossRef]
6. Larsen, D.; Mann, R. Origin of High Manganese Concentrations in Coal Mine Drainage, Eastern Tennessee. *J. Geochem. Explor.* **2005**, *86*, 143–163. [CrossRef]
7. Cravotta, C.A. Dissolved Metals and Associated Constituents in Abandoned Coal-Mine Discharges, Pennsylvania, USA. Part 1: Constituent Quantities and Correlations. *Appl. Geochem.* **2008**, *23*, 166–202. [CrossRef]
8. Vyawahre, A.; Rai, S. Acid Mine Drainage: A Case Study of An Indian Coal Mine. *Int. J. Sci. Res. Sci. Eng. Technol.* **2016**, *2*, 1297–1301.
9. Bell, F.G.; Bullock, S.E.T.; Hälbich, T.F.; Lindsay, P. Environmental Impacts Associated with an Abandoned Mine in The Witbank Coalfield, Sout Africa. *Int. J. Coal Geol.* **2001**, *45*, 195–216. [CrossRef]
10. Casiot, C.; Egal, M.; Elbaz-Poulichet, F.; Bruneel, O.; Bancon-Montigny, C.; Cordier, M.-A.; Gomez, E.; Aliaume, C. Hydrological and Geochemical Control of Metals and Arsenic in a Mediterranean River Contaminated by Acid Mine Drainage (the Amous River, France); Preliminary Assessment of Impacts on Fish (*Leuciscus Cephalus*). *Appl. Geochem.* **2009**, *24*, 787–799. [CrossRef]
11. Olenici, A.; Blanco, S.; Borrego-Ramos, M.; Momeu, L.; Baciu, C. Exploring the Effects of Acid Mine Drainage on Diatom Teratology Using Geometric Morphometry. *Ecotoxicology* **2017**, *26*, 1018–1030. [CrossRef]
12. Vega, F.A.; Covelo, E.F.; Andrade, M.L. Competitive Sorption and Desorption of Heavy Metals in Mine Soils: Influence of Mine Soil Characteristics. *J. Colloid Interface Sci.* **2006**, *298*, 582–592. [CrossRef] [PubMed]
13. Bååth, E.; Anderson, T.-H. Comparison of Soil Fungal/Bacterial Ratios in a PH Gradient Using Physiological and PLFA-Based Techniques. *Soil Biol. Biochem.* **2003**, *35*, 955–963. [CrossRef]
14. Rodríguez, L.; Ruiz, E.; Alonso-Azcárate, J.; Rincón, J. Heavy Metal Distribution and Chemical Speciation in Tailings and Soils around a Pb–Zn Mine in Spain. *J. Environ. Manag.* **2009**, *90*, 1106–1116. [CrossRef] [PubMed]
15. Skousen, J.; Zipper, C.E.; Rose, A.; Ziemkiewicz, P.F.; Nairn, R.; McDonald, L.M.; Kleinmann, R.L. Review of Passive Systems for Acid Mine Drainage Treatment. *Mine Water Environ.* **2017**, *36*, 133–153. [CrossRef]
16. Clarke, L.B. *Coal Mining and Water Quality*; Technical Report; IEA Coal Research: London, UK, 1995.
17. Jambor, J.L.; Nordstrom, D.K.; Alpers, C.N. Metal-Sulfate Salts from Sulfide Mineral Oxidation. *Rev. Mineral. Geochem.* **2000**, *40*, 303–350. [CrossRef]
18. Nicholson, R.Y. Iron-Sulfide Oxidation Mechanisms; Laboratory Studies. In *The Environmental Geochemistry of Sulfide, Mine-Wastes*; Short Course Handbook; Blowes, D.W., Jambor, J.L., Eds.; Mineralogical Association of Canada: Nepean, ON, Canada, 1994; Volume 22, pp. 163–183.
19. Chandra, A.P.; Gerson, A.R. The Mechanisms of Pyrite Oxidation and Leaching: A Fundamental Perspective. *Surf. Sci. Rep.* **2010**, *65*, 293–315. [CrossRef]
20. Tabelin, C.B.; Corpuz, R.D.; Igarashi, T.; Villacorte-Tabelin, M.; Alorro, R.D.; Yoo, K.; Raval, S.; Ito, M.; Hiroyoshi, N. Acid Mine Drainage Formation and Arsenic Mobility under Strongly Acidic Conditions: Importance of Soluble Phases, Iron Oxyhydroxides/Oxides and Nature of Oxidation Layer on Pyrite. *J. Hazard. Mater.* **2020**, *399*, 122844. [CrossRef]
21. Dold, B. Evolution of Acid Mine Drainage Formation in Sulphidic Mine Tailings. *Minerals* **2014**, *4*, 621–641. [CrossRef]

22. Parbhakar-Fox, A.; Lottermoser, B.G. A Critical Review of Acid Rock Drainage Prediction Methods and Practices. *Miner. Eng.* **2015**, *82*, 107–124. [CrossRef]
23. Dold, B. Acid Rock Drainage Prediction: A Critical Review. *J. Geochem. Explor.* **2017**, *172*, 120–132. [CrossRef]
24. Brady, K.B.C.; Perry, E.F.; Beam, R.L.; Bisko, D.C.; Gardner, M.D.; Tarantino, J.M. Evaluation of Acid-Base Accounting to Predict the Quality of Drainage at Surface Coal Mines in Pennsylvania, U.S.A. *J. Am. Soc. Min. Reclam.* **1994**, *1*, 138–147. [CrossRef]
25. Paktunc, A.D. Mineralogical Constraints on the Determination of Neutralization Potential and Prediction of Acid Mine Drainage. *Environ. Geol.* **1999**, *39*, 103–112. [CrossRef]
26. Verburg, R.; Bezuidenhout, N.; Chatwin, T.; Ferguson, K. The Global Acid Rock Drainage Guide (GARD Guide). *Mine Water Environ.* **2009**, *28*, 305–310. [CrossRef]
27. Morin, K.A.; Hutt, N.M.; Ferguson, K.D. Measured Rates of Sulfide Oxidation and Neutralization in Kinetic Tests: Statistical Lessons from the Database. In Proceedings of the Sudbury '95: Mining and the Environment, Sudbury, ON, Canada, 23 May–1 June 1995; pp. 525–536.
28. Banerjee, D. Acid Drainage Potential from Coal Mine Wastes: Environmental Assessment through Static and Kinetic Tests. *Int. J. Environ. Sci. Technol.* **2014**, *11*, 1365–1378. [CrossRef]
29. Tabelin, C.B.; Igarashi, T.; Villacorte-Tabelin, M.; Park, I.; Opiso, E.M.; Ito, M.; Hiroyoshi, N. Arsenic, Selenium, Boron, Lead, Cadmium, Copper, and Zinc in Naturally Contaminated Rocks: A Review of Their Sources, Modes of Enrichment, Mechanisms of Release, and Mitigation Strategies. *Sci. Total Environ.* **2018**, *645*, 1522–1553. [CrossRef]
30. Becker, M.; Dyantyi, N.; Broadhurst, J.L.; Harrison, S.T.L.; Franzidis, J.P. A Mineralogical Approach to Evaluating Laboratory Scale Acid Rock Drainage Characterisation Tests. *Miner. Eng.* **2015**, *80*, 33–36. [CrossRef]
31. Jamieson, H.E.; Walker, S.R.; Parsons, M.B. Mineralogical Characterization of Mine Waste. *Appl. Geochem.* **2015**, *57*, 85–105. [CrossRef]
32. Guseva, O.; Opitz, A.; Broadhurst, J.; Harrison, S.; Becker, M. Characterisation and prediction of acid rock drainage in waste rock: Value of integrating quantitative mineralogical and textural measurements. *Miner. Eng.* Under review.
33. Moodley, I.; Sheridan, C.M.; Kappelmeyer, U.; Akcil, A. Environmentally Sustainable Acid Mine Drainage Remediation: Research Developments with a Focus on Waste/by-Products. *Miner. Eng.* **2018**, *126*, 207–220. [CrossRef]
34. Abraitis, P.K.; Pattrick, R.A.D.; Vaughan, D.J. Variations in the Compositional, Textural and Electrical Properties of Natural Pyrite: A Review. *Int. J. Miner. Process.* **2004**, *74*, 41–59. [CrossRef]
35. Savage, K.S.; Stefan, D.; Lehner, S.W. Impurities and Heterogeneity in Pyrite: Influences on Electrical Properties and Oxidation Products. *Appl. Geochem.* **2008**, *23*, 103–120. [CrossRef]
36. Tabelin, C.B.; Igarashi, T.; Tamoto, S.; Takahashi, R. The Roles of Pyrite and Calcite in the Mobilization of Arsenic and Lead from Hydrothermally Altered Rocks Excavated in Hokkaido, Japan. *J. Geochem. Explor.* **2012**, *119–120*, 17–31. [CrossRef]
37. Lotter, N.O.; Bradshaw, D.J.; Barnes, A.R. Classification of the Major Copper Sulphides into Semiconductor Types, and Associated Flotation Characteristics. *Miner. Eng.* **2016**, *96–97*, 177–184. [CrossRef]
38. Miljević, N.; Golobočanin, D. Potential Use of Environmental Isotopes in Pollutant Migration Studies. *Arh. Hig. Rada Toksikol.* **2007**, *58*, 251–262. [CrossRef]
39. Wiederhold, J.G. Metal Stable Isotope Signatures as Tracers in Environmental Geochemistry. *Environ. Sci. Technol.* **2015**, *49*, 2606–2624. [CrossRef]
40. Seal, R.R.; Alpers, C.N.; Rye, R.O. Stable Isotope Systematics of Sulfate Minerals. *Rev. Mineral. Geochem.* **2000**, *40*, 541–602. [CrossRef]
41. Skierszkan, E.; Mayer, U.; Beckie, R.; Weis, D. Tracking the Fate of Metals in Mining Waste Rock Using Metal Stable Isotopes. In Proceedings of the Agreeing on Solutions for more Sustainable Mine Water Management, 10th International Conference on Acid Rock Drainage and IMWA Annual Conference, Santiago, Chile, 21–24 April 2015; Brown, A., Bucknam, C., Burgess, J., Carballo, M., Castendyk, D., Figueroa, L., Kirk, L., McLemore, V., McPhee, J., O'Kane, M., et al., Eds.; p. 236.
42. Michener, R.H.; Lajtha, K. *Stable Isotopes in Ecology and Environmental Science*, 2nd ed.; Blackwell Publishing Ltd.: Oxford, UK, 2007; pp. 6–11.

43. Weiss, D.J.; Rehkämper, M.; Schoenberg, R.; McLaughlin, M.; Kirby, J.; Campbell, P.G.C.; Arnold, T.; Chapman, J.; Peel, K.; Gioia, S. Application of Non-traditional Stable-isotope Systems to the Study of Sources and Fate of Metals in the Environment. *Environ. Sci. Technol.* **2008**, *42*, 655–664. [CrossRef]
44. Young, E.D.; Galy, A.; Nagahara, H. Kinetic and Equilibrium Mass-dependent Isotope Fractionation Laws in Nature and Their Geochemical and Cosmochemical Significance. *Geochim. Cosmochim. Acta* **2002**, *66*, 1095–1104. [CrossRef]
45. Sharp, Z. *Principles of Stable Isotope Geochemistry*, 2nd ed.; Pearson Education: Upper Saddle River, NJ, USA, 2017.
46. Garrels, R.; Thompson, M. Oxidation of Pyrite by Iron Sulfate Solutions. *Am. J. Sci.* **1960**, *258*, 57–67.
47. Nordstrom, K.D.; Alpers, C.N. Geochemistry of Acid Mine Waters. In *The Environmental Geochemistry of Mineral Deposits. Part A: Processes, Techniques, and Health Issues*; Plumlee, G.D., Logsdon, M.J., Filipek, L.F., Eds.; Society of Economic Geologists: Littleton, CO, USA, 1999; Volume 6, pp. 133–160.
48. Smart, M.; Huddy, R.J.; Edward, C.J.; Fourie, C.; Shumba, T.; Iron, J.; Harrison, S.T.L. Linking Microbial Community Dynamics in BIOX® Leaching Tanks to Process Conditions: Integrating Lab and Commercial Experience. *Solid State Phenom.* **2017**, *262*, 38–42. [CrossRef]
49. Tupikina, O.V.; Minnaar, S.H.; van Hille, R.P.; van Wyk, N.; Rautenbach, G.F.; Dew, D.; Harrison, S.T.L. Determining the Effect of Acid Stress on the Persistence and Growth of Thermophilic Microbial Species after Mesophilic Colonisation of Low Grade Ore in a Heap Leach Environment. *Miner. Eng.* **2013**, *53*, 152–159. [CrossRef]
50. Nordstrom, K.D.; Southam, G. Geomicrobiology of Sulfide Mineral Oxidation. In *Geomicrobiology: Interactions between Microbes and Minerals: Reviews in Mineralogy*; Banfield, J.F., Nealson, K.H., Eds.; Mineralogical Society of America: Washington, DC, USA, 1997; Volume 35, pp. 361–390.
51. Mustin, C.; Berthelin, J.; Marion, P.; de Donato, P. Corrosion and Electrochemical Oxidation of a Pyrite by Thiobacillus ferrooxidans. *Appl. Environ. Microbiol.* **1992**, *58*, 1175–1182. [CrossRef]
52. Rimstidt, D.D.; Vaughan, D.J. Pyrite Oxidation: A State-of-the-Art Assessment of the Reaction Mechanism. *Geochim. Cosmochim. Acta* **2003**, *67*, 873–880. [CrossRef]
53. Moses, C.O.; Herman, J.S. Pyrite Oxidation at Circumneutral pH. *Geochim. Cosmochim. Acta* **1991**, *55*, 471–482. [CrossRef]
54. Kelsall, G.H.; Yin, Q.; Vaughan, D.J.; England, K.E.R.; Brandon, N.P. Electrochemical Oxidation of Pyrite (FeS2) in Aqueous Electrolytes. *J. Electroanal. Chem.* **1999**, *471*, 116–125. [CrossRef]
55. Tabelin, C.B.; Veerawattananun, S.; Ito, M.; Hiroyoshi, N.; Igarashi, T. Pyrite Oxidation in the Presence of Hematite and Alumina: II. Effects on the Cathodic and Anodic Half-Cell Reactions. *Sci. Total Environ.* **2017**, *581–582*, 126–135. [CrossRef]
56. Tu, Z.; Wan, J.; Guo, C.; Fan, C.; Zhang, T.; Lu, G.; Reinfelder, J.R.; Dang, Z. Electrochemical Oxidation of Pyrite in PH 2 Electrolyte. *Electrochim. Acta* **2017**, *239*, 25–35. [CrossRef]
57. Li, X.; Gao, M.; Hiroyoshi, N.; Tabelin, C.B.; Taketsugu, T.; Ito, M. Suppression of Pyrite Oxidation by Ferric-Catecholate Complexes: An Electrochemical Study. *Miner. Eng.* **2019**, *138*, 226–237. [CrossRef]
58. Nourmohamadi, H.; Aghazadeh, V.; Esrafili, M.D. A Comparative DFT Study of Fe^{3+} and Fe^{2+} Ions Adsorption on (100) and (110) Surfaces of Pyrite: An Electrochemical Point of View. *Surf. Interface Anal.* **2020**, *52*, 110–118. [CrossRef]
59. Jerz, J.K.; Rimstidt, J.D. Pyrite Oxidation in Moist Air. Geochim. *Cosmochim. Acta* **2004**, *68*, 701–714. [CrossRef]
60. Borda, M.J.; Strongin, D.R.; Schoonen, M.A.A. Vibrational Spectroscopic Study of Pyrite Oxidation. *Am. Mineral.* **2003**, *88*, 1318–1323. [CrossRef]
61. Nordstrom, K.D.; Wright, W.G.; Mast, A.M.; Bove, D.J.; Rye, R.O. Aqueous-Sulfate Stable Isotopes-A Study of Mining-Affected and Undisturbed Acidic Drainage. In *Integrated Investigations of Environmental Effects of Historical Mining in the Animas River Watershed, San Juan County, Colorado*; Church, S.E., von Guerard, P., Finger, S.E., Eds.; U.S. Geological Survey Professional Paper 1651; USGS: Lawrence, KS, USA, 2007; Volume 1, pp. 387–416.
62. Field, C.W. Sulfur Isotopic Method for Discriminating between Sulfates of Hypogene and Supergene Origin. *Econ. Geol.* **1966**, *61*, 1428–1435. [CrossRef]
63. Taylor, B.E.; Wheeler, M.C.; Nordstrom, D.K. Stable Isotope Geochemistry of Acid Mine Drainage: Experimental Oxidation of Pyrite. *Geochim. Cosmochim. Acta* **1984**, *48*, 2669–2678. [CrossRef]

64. Taylor, B.E.; Wheeler, M.C. Sulfur- and Oxygen-isotope Geochemistry of Acid Mine Drainage in the Western United States. In *Environmental Geochemistry of Sulfide Oxidation*; Alpers, C.N., Blowes, D.W., Eds.; American Chemical Society Symposium Series: Washington, DC, USA, 1993; Volume 550, pp. 481–514.
65. Van Stempvoort, D.R.; Krouse, H.R. Controls of δ^{18}O in Sulfate: Review of Experimental Data and Application to Specific Environments. In *Environmental Geochemistry of Sulfide Oxidation*; Alpers, C.N., Blowes, D.W., Eds.; American Chemical Society Symposium Series: Washington, DC, USA, 1993; Volume 550, pp. 446–480.
66. Krouse, H.R.; Mayer, B. Sulphur and Oxygen Isotopes in Sulphate. In *Environmental Tracers in Subsurface Hydrology*; Cook, P., Herczeg, A., Eds.; Kluwer Academic Publishers: Dordrecht, The Netherlands, 2000; pp. 195–231.
67. Otero, N.; Soler, A.; Canals, À. Controls of δ^{34}S and δ^{18}O in Dissolved Sulphate: Learning from a Detailed Survey in the Llobregat River (Spain). *Appl. Geochem.* **2008**, *23*, 1166–1185. [CrossRef]
68. Migaszewski, Z.M.; Gałuszka, A.; Hałas, S.; Dołegowska, S.; Dabek, J.; Starnawska, E. Geochemistry and Stable Sulfur and Oxygen Isotope Ratios of the Podwiśniówka Pit Pond Water Generated by Acid Mine Drainage (Holy Cross Mountains, South-Central Poland). *Appl. Geochem.* **2008**, *23*, 3620–3634. [CrossRef]
69. Migaszewski, Z.M.; Gałuszka, A.; Michalik, A.; Dołegowska, S.; Migaszewski, A.; Hałas, S.; Trembaczowski, A. The Use of Stable Sulfur, Oxygen and Hydrogen Isotope Ratios as Geochemical Tracers of Sulfates in the Podwiśniówka Acid Drainage Area (South-Central Poland). *Aquat. Geochem.* **2013**, *19*, 261–280. [CrossRef]
70. Nakai, N.; Jensen, M.L. The Kinetic Isotope Effect in the Bacterial Reduction and Oxidation of Sulfur. *Geochim. Cosmochim. Acta* **1964**, *28*, 1893–1912. [CrossRef]
71. McCready, R.G.L.; Krouse, H.R. Sulfur Isotope Fractionation during the Oxidation of Elemental Sulfur by Thiobacilli in a Solonetzic Soil. *Can. J. Soil Sci.* **1982**, *62*, 105–110. [CrossRef]
72. Balci, N.; Shanks, W.C.; Mayer, B.; Mandernack, K.W. Oxygen and Sulfur Isotope Systematics of Sulfate Produced by Bacterial and Abiotic Oxidation of Pyrite. *Geochim. Cosmochim. Acta* **2007**, *71*, 3796–3811. [CrossRef]
73. Pisapia, C.; Chaussidon, M.; Mustin, C.; Humbert, B. O and S Isotopic Composition of Dissolved and Attached Oxidation Products of Pyrite by *Acidithiobacillus ferrooxidans*: Comparison with Abiotic Oxidations. *Geochim. Cosmochim. Acta* **2007**, *71*, 2474–2490. [CrossRef]
74. Heidel, C.; Tichomirowa, M. The Isotopic Composition of Sulfate from Anaerobic and Low Oxygen Pyrite Oxidation Experiments with Ferric Iron—New Insights into Oxidation Mechanisms. *Chem. Geol.* **2011**, *281*, 305–316. [CrossRef]
75. Lefticariu, L.; Behum, P.; Bender, K.; Lefticariu, M. Sulfur Isotope Fractionation as an Indicator of Biogeochemical Processes in an AMD Passive Bioremediation System. *Minerals* **2017**, *7*, 41. [CrossRef]
76. Alvaro, A.; Roldán, F.V. Stable Isotope Composition of Acid Mine Drainage Minerals from San Miguel Massive Sulphide Mine Waste. In Proceedings of the Securing the Future: Mining, Metals and the Environment in a Sustainable Society, 8th International Conference on Acid Rock Drainage, Skellefteå, Sweden, 22–26 June 2009; pp. 166–176.
77. Gallo, A.A.; Roldán, F.V. Stable Isotope Study in Iron-Rich Sulphates from San Miguel Mine Wastes (Iberian Pyrite Belt, Spain). *Rev. Soc. Española Mineral.* **2009**, *11*, 27–28.
78. Brunner, B.; Yu, J.Y.; Mielke, R.E.; MacAskill, J.A.; Madzunkov, S.; McGenity, T.J.; Coleman, M. Different Isotope and Chemical Patterns of Pyrite Oxidation Related to Lag and Exponential Growth Phases of *Acidithiobacillus ferrooxidans* Reveal a Microbial Growth Strategy. *Earth Planet. Sci. Lett.* **2008**, *270*, 63–72. [CrossRef]
79. Yu, J.Y.; McGenity, T.J.; Coleman, M.L. Solution Chemistry during the Lag Phase and Exponential Phase of Pyrite Oxidation by *Thiobacillus ferrooxidans*. *Chem. Geol.* **2001**, *175*, 307–317. [CrossRef]
80. Descostes, M.; Vitorge, P.; Beaucaire, C. Pyrite Dissolution in Acidic Media. *Geochim. Cosmochim. Acta* **2004**, *68*, 4559–4569. [CrossRef]
81. Granger, H.C.; Warren, C.G. Unstable Sulfur Compounds and the Origin of Roll-Type Uranium Deposits. *Econ. Geol.* **1969**, *64*, 160–171. [CrossRef]
82. Basolo, F.; Pearson, R.G. *Mechanisms of Inorganic Reactions—A Study of Metal Complexes in Solution*, 2nd ed.; John Wiley and Sons Ltd.: New York, NY, USA, 1967.
83. Goldhaber, M.B. Experimental Study of Metastable Sulfur Oxyanion Formation during Pyrite Oxidation at pH 6–9 and 30 °C. *Am. J. Sci.* **1983**, *283*, 193–217. [CrossRef]
84. Schippers, A.; Jozsa, P.G.; Sand, W. Sulfur Chemistry in the Leaching of Pyrite. *Appl. Environm. Microbiol.* **1996**, *62*, 3424–3431. [CrossRef]

85. Toran, L.; Harris, R.F. Interpretation of Sulfur and Oxygen Isotopes in Biological and Abiological Sulfide Oxidation. *Geochim. Cosmochim. Acta* **1989**, *53*, 2341–2348. [CrossRef]
86. McKibben, M.A.; Barnes, H.L. Oxidation of Pyrite in Low Temperature Acidic Solutions: Rate Laws and Surface Textures. *Geochim. Cosmochim. Acta* **1986**, *50*, 1509–1520. [CrossRef]
87. Luther, G.W. Pyrite Oxidation and Reduction: Molecular Orbital Theory Considerations. *Geochim. Cosmochim. Acta* **1987**, *51*, 3193–3199. [CrossRef]
88. Moses, C.O.; Nordstrom, K.D.; Herman, J.S.; Mills, A.L. Aqueous Pyrite Oxidation by Dissolved Oxygen and by Ferric Iron. *Geochim. Cosmochim. Acta* **1987**, *51*, 1561–1571. [CrossRef]
89. Demoisson, F.; Mullet, M.; Humbert, B. Pyrite Oxidation in Acidic Medium: Overall Reaction Pathway. *Surf. Interface Anal.* **2008**, *40*, 343–348. [CrossRef]
90. Schippers, A.; Sand, W. Bacterial Leaching of Metal Sulfides Proceeds by Two Indirect Mechanisms via Thiosulfate or via Polysulfides and Sulfur. *Appl. Environ. Microbiol.* **1999**, *65*, 319–321. [CrossRef]
91. Sand, W.; Gehrke, T.; Jozsa, P.G.; Schippers, A. (Bio)Chemistry of Bacterial Leaching—Direct vs. Indirect Bioleaching. *Hydrometallurgy* **2001**, *59*, 159–175. [CrossRef]
92. Rohwerder, T.; Gehrke, T.; Kinzler, K.; Sand, W. Bioleaching Review Part A: Progress in bioleaching: Fundamentals and mechanisms of bacterial metal sulfide oxidation. *Appl. Microbiol. Biotechnol.* **2003**, *63*, 239–248. [CrossRef]
93. Borda, M.J.; Elsetinow, A.R.; Strongin, D.R.; Schoonen, M.A.A. Mechanism for the Production of Hydroxyl Radical at Surface Defect Sites on Pyrite. *Geochim. Cosmochim. Acta* **2003**, *67*, 935–939. [CrossRef]
94. Druschel, G.; Borda, M. Comment on "Pyrite Dissolution in Acidic Media" by M. Descostes, P. Vitorge, and C. Beaucaire. *Geochim. Cosmochim. Acta* **2006**, *70*, 5246–5250. [CrossRef]
95. Krouse, H.R.; McCready, R.G.L. Reductive Reactions in the Sulfur Cycle. In *Biochemical Cycling of Mineral-Forming Elements*, 1st ed.; Trudinger, P., Swaine, D., Eds.; Elsevier Science: Amsterdam, The Netherlands, 1979; Volume 3, pp. 315–368.
96. Edraki, M.; Golding, S.D.; Baublys, K.A.; Lawrence, M.G. Hydrochemistry, Mineralogy and Sulfur Isotope Geochemistry of Acid Mine Drainage at the Mt. Morgan Mine Environment, Queensland, Australia. *Appl. Geochem.* **2005**, *20*, 789–805. [CrossRef]
97. Canfield, D.; Thamdrup, B. The Production of ^{34}S-Depleted Sulfide during Bacterial Disproportionation of Elemental Sulfur. *Science* **1994**, *266*, 1973–1975. [CrossRef]
98. Böttcher, M.E.; Thamdrup, B.; Vennemann, T.W. Oxygen and Sulfur Isotope Fractionation during Anaerobic Bacterial Disproportionation of Elemental Sulfur. *Geochim. Cosmochim. Acta* **2001**, *65*, 1601–1609. [CrossRef]
99. Balci, N.; Brunner, B.; Turchyn, A.V. Tetrathionate and Elemental Sulfur Shape the Isotope Composition of Sulfate in Acid Mine Drainage. *Front. Microbiol.* **2017**, *8*, 1564. [CrossRef]
100. Müller, I.A.; Brunner, B.; Coleman, M. Isotopic Evidence of the Pivotal Role of Sulfite Oxidation in Shaping the Oxygen Isotope Signature of Sulfate. *Chem. Geol.* **2013**, *354*, 186–202. [CrossRef]
101. Kokh, M.A.; Assayag, N.; Mounic, S.; Cartigny, P.; Gurenko, A.; Pokrovski, G.S. Multiple Sulfur Isotope Fractionation in Hydrothermal Systems in the Presence of Radical Ions and Molecular Sulfur. *Geochim. Cosmochim. Acta* **2020**, *285*, 100–128. [CrossRef]
102. Lloyd, R.M. Oxygen Isotope Behavior in the Sulfate-Water System. *J. Geophys. Res.* **1968**, *73*, 6099–6110. [CrossRef]
103. Chiba, H.; Sakai, H. Oxygen Isotope Exchange Rate between Dissolved Sulfate and Water at Hydrothermal Temperatures. *Geochim. Cosmochim. Acta* **1985**, *49*, 993–1000. [CrossRef]
104. Seal, R.R.; Wandless, G.A. Sulfur Isotope Evidence for Sea-floor Mineralizing Processes at the Bald Mountain and Mount Chase Massive Sulfide Deposits, Northern Maine. In *Massive Sulfide Deposits of the Bathurst Mining Camp, New Brunswick, and Northern Maine*; Goodfellow, W.D., McCutcheon, S.R., Peter, J.M., Eds.; Society of Economic Geologists: Littleton, CO, USA, 2003; Volume 11, pp. 567–587.
105. Dole, M.; Lane, G.A.; Rudd, D.P.; Zaukelies, D.A. Isotopic Composition of Atmospheric Oxygen and Nitrogen. *Geochim. Cosmochim. Acta* **1954**, *6*, 65–78. [CrossRef]
106. Kroopnick, P.; Craig, H. Atmospheric Oxygen: Isotopic Composition and Solubility Fractionation. *Science* **1972**, *175*, 54–55. [CrossRef]
107. Van Everdingen, R.O.; Krouse, H.R. Isotope Composition of Sulphates Generated by Bacterial and Abiological Oxidation. *Nature* **1985**, *315*, 395–396. [CrossRef]

108. Epstein, S.; Mayeda, T. Variations in the ^{18}O Content of Waters from Natural Sources. *Geochim. Cosmochim. Acta* **1953**, *4*, 213–224. [CrossRef]
109. Rose, A.W.; Cravotta, C.A. Geochemistry of Coal Mine Drainage. In *Coal Mine Drainage Prediction and Pollution Prevention in Pennsylvania*; Brady, K.B.C., Smith, M.W., Schueck, J., Eds.; Pennsylvania Department of Environmental Protection: Harrisburg, PA, USA, 1998; pp. 1.1–1.22.
110. Gammons, C.H.; Duaime, T.E.; Parker, S.R.; Poulson, S.R.; Kennelly, P. Geochemistry and Stable Isotope Investigation of Acid Mine Drainage Associated with Abandoned Coal Mines in Central Montana, USA. *Chem. Geol.* **2010**, *269*, 100–112. [CrossRef]
111. Salifu, M.; Aiglsperger, T.; Mörth, C.-M.; Alakangas, L. Stable Sulphur and Oxygen Isotopes as Indicators of Sulphide Oxidation Reaction Pathways and Historical Environmental Conditions in a Cu–W–F Skarn Tailings Piles, South-Central Sweden. *Appl. Geochem.* **2019**, *110*, 104426. [CrossRef]
112. Butler, T.W. Isotope Geochemistry of Drainage from an Acid Mine Impaired Watershed, Oakland, California. *Appl. Geochem.* **2007**, *22*, 1416–1426. [CrossRef]
113. Qureshi, R.M. The Isotopic Composition of Aqueous Sulfate (A Laboratory Investigation). Ph.D. Thesis, University of Waterloo, Waterloo, ON, Canada, 1986.
114. Sasaki, K.; Tsunekawa, M.; Ohtsuka, T.; Konno, H. Confirmation of a Sulfur-rich Layer on Pyrite after Oxidative Dissolution by Fe(III) Ions around pH2. *Geochim. Cosmochim. Acta* **1995**, *59*, 3155–3158. [CrossRef]
115. Krouse, H.R.; Gould, W.D.; McCready, R.G.L.; Rajan, S. ^{18}O Incorporation into Sulphate during the Bacterial Oxidation of Sulphide Minerals and the Potential for Oxygen Isotope Exchange between O_2, H_2O and Oxidized Sulphur Intermediates. *Earth Planet. Sci. Lett.* **1991**, *107*, 90–94. [CrossRef]
116. Holt, B.D.; Kumar, R.; Cunningham, P.T. Oxygen-18 Study of the Aqueous-Phase of Sulfur Dioxide. *Atmos. Environ.* **1981**, *15*, 557–566. [CrossRef]
117. Viers, J.; Grande Gil, J.A.; Zouiten, C.; Freydier, R.; Masbou, J.; Valente, T.; de la Torre, M.L.; Destrigneville, C.; Pokrovsky, O.S. Are Cu Isotopes a Useful Tool to Trace Metal Sources and Processes in Acid Mine Drainage (AMD) Context? *Chemosphere* **2018**, *193*, 1071–1079. [CrossRef]
118. Zhu, X.K.; Guo, Y.; Williams, R.J.P.; O'Nions, R.K.; Matthews, A.; Belshaw, N.S.; Salvato, B. Mass Fractionation Processes of Transition Metal Isotopes. *Earth Planet. Sci. Lett.* **2002**, *200*, 47–62. [CrossRef]
119. Ehrlich, S.; Butler, I.; Halicz, L.; Rickard, D.; Oldroyd, A.; Matthews, A. Experimental Study of the Copper Isotope Fractionation between Aqueous Cu(II) and Covellite, CuS. *Chem. Geol.* **2004**, *209*, 259–269. [CrossRef]
120. Fantle, M.S.; De Paolo, D.J. Iron Isotopic Fractionation during Continental Weathering. *Earth Planet. Sci. Lett.* **2004**, *228*, 547–562. [CrossRef]
121. Borrok, D.M.; Nimick, D.A.; Wanty, R.B.; Ridley, W.I. Isotopic Variations of Dissolved Copper and Zinc in Stream Waters Affected by Historical Mining. *Geochim. Cosmochim. Acta* **2008**, *72*, 329–344. [CrossRef]
122. Calmano, W.; Hong, J.; Forstner, U. Binding and Mobilization of Heavy Metals in Contaminated Sediments Affected by PH and Redox Potential. *Water Sci. Technol.* **1993**, *28*, 223–235. [CrossRef]
123. Bourg, A.C.M.; Loch, J.P.G. Mobilization of Heavy Metals as Affected by pH and Redox Conditions. In *Biogeodynamics of Pollutants in Soils and Sediments: Risk Assessment of Delayed and Non-Linear Responses*; Salomons, W., Stigliani, W.M., Eds.; Springer: Berlin/Heidelberg, Germany, 1995; pp. 87–102.
124. Schott, J.; Pokrovsky, O.S.; Oelkers, E.H. The Link between Mineral Dissolution/Precipitation Kinetics and Solution Chemistry. *Rev. Mineral. Geochem.* **2009**, *70*, 207–258. [CrossRef]
125. Wiederhold, J. Iron Isotope Fractionation in Soils from Phenomena to Process Identification. Ph.D. Thesis, Eidgenössische Technische Hochschule Zürich, Zürich, Switzerland, 2006.
126. Matthews, A.; Zhu, X.K.; O'Nions, K. Kinetic iron stable isotope fractionation between iron (-II) and (-III) complexes in solution. *Earth Planet. Sci. Lett.* **2001**, *192*, 81–92. [CrossRef]
127. Welch, S.A.; Beard, B.L.; Johnson, C.M.; Braterman, P.S. Kinetic and equilibrium Fe isotope fractionation between aqueous Fe(II) and Fe(III). *Geochim. Cosmochim. Acta.* **2003**, *67*, 4231–4250. [CrossRef]
128. Johnson, C.M.; Skulan, J.L.; Beard, B.L.; Sun, H.; Nealson, K.H.; Braterman, P.S. Isotopic fractionation between Fe(III) and Fe(II) in aqueous solutions. *Earth Planet. Sci. Lett.* **2002**, *195*, 141–153. [CrossRef]
129. Anbar, A.D.; Jarzecki, A.A.; Spiro, T.G. Theoretical investigation of iron isotope fractionation between $Fe(H_2O)_6^{3+}$ and $Fe(H_2O)_6^{2+}$: Implications for iron stable isotope geochemistry. *Geochim. Cosmochim. Acta* **2005**, *69*, 825–837. [CrossRef]
130. Skulan, J.L.; Beard, B.L.; Johnson, C.M. Kinetic and equilibrium Fe isotope fractionation between aqueous Fe(III) and hematite. *Geochim. Cosmochim. Acta* **2002**, *66*, 2995–3015. [CrossRef]

131. Fernandez, A.; Borrok, D.M. Fractionation of Cu, Fe, and Zn Isotopes during the Oxidative Weathering of Sulfide-Rich Rocks. *Chem. Geol.* **2009**, *264*, 1–12. [CrossRef]
132. Rodríguez, N.P.; Engström, E.; Rodushkin, I.; Nason, P.; Alakangas, L.; Öhlander, B. Copper and Iron Isotope Fractionation in Mine Tailings at the Laver and Kristineberg Mines, Northern Sweden. *Appl. Geochem.* **2013**, *32*, 204–215. [CrossRef]
133. Balci, N.; Bullen, T.D.; Witte-Lien, K.; Shanks, W.C.; Motelica, M.; Mandernack, K.W. Iron Isotope Fractionation during Microbially Stimulated Fe(II) Oxidation and Fe(III) Precipitation. *Geochim. Cosmochim. Acta* **2006**, *70*, 622–639. [CrossRef]
134. Brantley, S.L.; Liermann, L.; Bullen, T. Fractionation of Fe Isotopes by Soil Microbes and Organic Acids. *Geology* **2001**, *29*, 535–538. [CrossRef]
135. Beard, B.L.; Johnson, C.M.; Skulan, J.L.; Nealson, K.H.; Cox, L.; Sun, H. Application of Fe Isotopes to Tracing the Geochemical and Biological Cycling of Fe. *Chem. Geol.* **2003**, *195*, 87–117. [CrossRef]
136. Mathur, R.; Ruiz, J.; Titley, S.; Liermann, L.; Buss, H.; Brantley, S. Cu Isotopic Fractionation in the Supergene Environment with and without Bacteria. *Geochim. Cosmochim. Acta* **2005**, *69*, 5233–5246. [CrossRef]
137. Johnson, C.M.; Beard, B.L.; Roden, E.E. The Iron Isotope Fingerprints of Redox and Biogeochemical Cycling in Modern and Ancient Earth. *Annu. Rev. Earth Planet. Sci.* **2008**, *36*, 457–493. [CrossRef]
138. Rodríguez, N.P.; Khoshkhoo, M.; Sandström, Å.; Rodushkin, I.; Alakangas, L.; Öhlander, B. Isotopic Signature of Cu and Fe during Bioleaching and Electrochemical Leaching of a Chalcopyrite Concentrate. *Int. J. Miner. Process.* **2015**, *134*, 58–65. [CrossRef]
139. Falco, L.; Pogliani, C.; Curutchet, G.; Donati, E. A Comparison of Bioleaching of Covellite Using Pure Cultures of *Acidithiobacillus ferrooxidans* and *Acidithiobacillus thiooxidans* or a Mixed Culture of *Leptospirillum ferrooxidans* and *Acidithiobacillus thiooxidans*. *Hydrometallurgy* **2003**, *71*, 31–36. [CrossRef]
140. Garcia, O.; Bigham, J.M.; Tuovinen, O.H. Oxidation of Isochemical FeS_2 (Marcasite-Pyrite) by *Acidithiobacillus Thiooxidans* and *Acidithiobacillus ferrooxidans*. *Miner. Eng.* **2007**, *20*, 98–101. [CrossRef]
141. Tributsch, H. Direct versus indirect bioleaching. *Hydrometallurgy* **2001**, *59*, 177–185. [CrossRef]
142. Natarajan, K.A. Microbial Aspects of Acid Mine Drainage and Its Bioremediation. *Trans. Nonferrous Met. Soc. China (Engl. Ed.)* **2008**, *18*, 1352–1360. [CrossRef]
143. Brantley, S.L.; Liermann, L.J.; Guynn, R.L.; Anbar, A.; Icopini, G.A.; Barling, J. Fe Isotopic Fractionation during Mineral Dissolution with and without Bacteria. *Geochim. Cosmochim. Acta* **2004**, *68*, 3189–3204. [CrossRef]
144. Wasylenki, L.E.; Anbar, A.D.; Liermann, L.J.; Mathur, R.; Gordon, G.W.; Brantley, S.L. Isotope Fractionation during Microbial Metal Uptake Measured by MC-ICP-MS. *J. Anal. At. Spectrom.* **2007**, *22*, 905. [CrossRef]
145. Croal, L.R.; Johnson, C.M.; Beard, B.L.; Newman, D.K. Iron isotope fractionation by Fe(II)-oxidizing photoautotrophic bacteria [1]. *Geochim. Cosmochim. Acta* **2004**, *68*, 1227–1242. [CrossRef]
146. Hubbard, C.G.; Black, S.; Coleman, M.L. Aqueous Geochemistry and Oxygen Isotope Compositions of Acid Mine Drainage from the Río Tinto, SW Spain, Highlight Inconsistencies in Current Models. *Chem. Geol.* **2009**, *265*, 321–334. [CrossRef]
147. Malmström, M.E.; Destouni, G.; Banwart, S.A.; Strömberg, B.H.E. Resolving the Scale-Dependence of Mineral Weathering Rates. *Environ. Sci. Technol.* **2000**, *34*, 1375–1378. [CrossRef]
148. Liu, R.; Wolfe, A.L.; Dzombak, D.A.; Stewart, B.W.; Capo, R.C. Comparison of Dissolution under Oxic Acid Drainage Conditions for Eight Sedimentary and Hydrothermal Pyrite Samples. *Environ. Geol.* **2008**, *56*, 171–182. [CrossRef]
149. Matthies, R.; Aplin, A.C.; Boyce, A.J.; Jarvis, A.P. Geochemical and Stable Isotopic Constraints on the Generation and Passive Treatment of Acidic, Fe–SO_4 Rich Waters. *Sci. Total Environ.* **2012**, *420*, 238–249. [CrossRef]

minerals

MDPI

Article

Mineralogical Study and Genetic Model of Efflorescent Salts and Crusts from Two Abandoned Tailings in the Taxco Mining District, Guerrero (Mexico)

Teresa Pi-Puig [1,2,*], Jesús Solé [1,2] and Adriana Gómez Cruz [3]

1 Instituto de Geología, Universidad Nacional Autónoma de México (UNAM), Cd. Universitaria, Coyoacán, Mexico City 04510, Mexico; jsole@geologia.unam.mx

2 Laboratorio Nacional de Geoquímica y Mineralogía (LANGEM), Universidad Nacional Autónoma de México (UNAM), Cd. Universitaria, Coyoacán, Mexico City 04510, Mexico

3 Posgrado en Ciencias de la Tierra, Universidad Nacional Autónoma de México (UNAM), Cd. Universitaria, Coyoacán, Mexico City 04510, Mexico; adry.gc.6116@hotmail.com

* Correspondence: tpuig@geologia.unam.mx

Received: 31 July 2020; Accepted: 4 September 2020; Published: 1 October 2020

Abstract: The mineralogy of surface crusts and efflorescent salts of two old abandoned tailings (Xochula and Remedios) of the mining district of Taxco, Guerrero, was studied by short-wave infrared spectroscopy (SWIR), X-ray diffraction (XRD), and scanning electron microscopy (SEM). The chemistry of the selected samples was studied with induced coupled plasma mass spectrometry (ICP–MS). Principal mineralogy is composed of hydrated sulfates of iron (di-trivalent), other less-soluble sulfates such as gypsum and jarosite, and oxides–hydroxides of iron, all of which are the result of the weathering of sulfides (mainly pyrite) accumulated in mining waste. Using quantitative X-ray diffraction (Rietveld method) and the spatial distribution of secondary minerals, two main zones (lateral and central) have been established in both tailings. The genetic model for their formation shows that the secondary iron minerals (sulfates, oxides, and hydroxides) in these deposits have been formed at least in three different stages, mainly by oxidation, dissolution, and precipitation processes in combination with dehydration and primary sulfides neutralization reactions. The chemical analyses of some salts confirm the presence of potentially toxic elements (PTE) in their structure and therefore indicate that the dissolution of the efflorescences in these deposits is a potential source of contamination during the wet season.

Keywords: hydrated sulfates; mining waste; X-ray diffraction; Rietveld method; SWIR

---

1. Introduction

The precipitation of sulfates in the form of efflorescent scabs or layers is a phenomenon that is characteristic of areas with acid drainage identified since antiquity [1]—because of its flower-like and concretion appearance—in the most superficial parts of the mining waste. They present a wide range of colorations (white, yellow, red, green, blue, etc.), and in general, there is a lack of vegetation in the area where they are located. Early civilizations established long ago the toxicity of the efflorescent salts [2].

In many Mexican mining districts, these efflorescences have been ignored, although the study of their mineralogy, their paragenetic sequence, and their spatial distribution is a critical aspect, since these minerals incorporate very toxic trace elements in their structures (in the substitution of iron and aluminum mainly) that can be easily released to water and soil by dissolution [3–5]. Its formation or dissolution modifies the hydrogen potential, as well as the content and transport of metals and sulfate from acid drainage waters. It is widely known that these sulfates are very susceptible to dissolve in rainy seasons or due to the action of surface waters releasing episodic pulses of highly acidic solutions. Then, they integrate toxic or contaminating elements into the drainage systems that can infiltrate the subsoil and impact the aquifers and the ecosystem in general [6–8].

The process of formation of these salts begins with the oxidation of sulfides such as pyrite in the presence of air or water, generating the necessary degree of acidity and the presence of cations. The rate of sulfide oxidation in tailings is high because a large volume of residues—characterized by a small crystal size and large reactive surface—is exposed to the action of the atmosphere. The efflorescent minerals precipitate by the evaporation of mine acid waters when the climate is warm and dry enough; they form a wide variety of crystal habits and aggregates, and their formation is determined mainly by the temperature, the degree of humidity, and the degree of oxidation on the surface of the mining waste [8–15]. Additionally, because the efflorescences are formed by fine material, they are susceptible to being transported during the rainy or windy periods [16,17].

In general, the crystals of these metastable phases are very small and contain elements with diverse oxidation states. They can be difficult to identify because they form complex mixtures and their structure can be modified very easily by dehydration [18]. The identification of these sulfates provides information about the depositional environment and the interactions of mine tailings wastes with water and air, as they are very sensitive to environmental conditions [6,8,17]. The present study aims to review the presence of hazardous mineralogical phases in small abandoned mining residues which are very old, not delimited and labeled, and even with vegetation; they are frequently found in Mexico near mining areas.

The main goals of the present work are as follows: (a) to know the nature, proportion, and distribution of the sulfate and oxide minerals in the two studied areas, (b) to understand the degree of oxidation and evolution of efflorescences and, based on the above, (c) to be able to propose an evolutionary model about their formation and (d) evaluate their potential as a source of contamination.

Historical Background and Description of the Study Area

The Taxco Mining District (Guerrero State) is located at 1700 m of altitude on the northern of the Balsas-Mezcala basin on the flank of the Paleozoic Taxco-Zitácuaro outcrops [19].

The geology of the Taxco district is characterized by a marine section with Mesozoic schists in the basement, Lower Cretaceous limestones, and Upper Cretaceous shales, which are overlaid by continental conglomerates and Cenozoic volcanic rocks. The rocks are faulted and intruded by dikes and magmatic bodies of diverse age and composition (diorite, felsitic dikes, and diabases).

The Taxco ore deposits were described as polymetallic intermediate sulfidation epithermal deposits [19–21] with fluid mineralization temperatures between 200 °C and 300 °C. The orebodies, with NW to NS orientation, are veins, stockworks, replacement mantos, and breccias, which are associated mainly with carbonates of the Mezcala and Morelos formations and less frequently with the Taxco Schist formation [22–24]. In distant parts, the skarn mineralization type is also present. Taxco is an important silver producer that is mainly associated with veins, but zinc and lead are also mined, mostly in mantos [24]. A significant amount of the mineralized bodies is constituted by galena (silver bearing), chalcopyrite, sphalerite, native silver, argentite, and silver secondary sulfosalts such as polybasite, proustite, and pyrargyrite. The gangue minerals include pyrite, pyrrhotite, stibnite, jamesonite, marcasite quartz, barite, fluorite, calcite, rhodochrosite, hematite, tremolite, epidote, and ilvaite [19,23].

Since the 16th century, metal extraction has been carried out in the Taxco region and its surroundings, releasing a large amount of mineral waste that has been deposited in the environment over time [25]. There is little evidence of colonial waste, but by 1920, the first selective flotation profit plant had been installed by American Smelting and Refining Company [19], initiating the deposit of tailings through the Taxco district [23]. For a long time, the tailings were placed on the banks of rivers and other areas that over the years have been surrounded by human settlements [23,26].

In recent years, some research has been done on this type of deposit in the Taxco region, where the mining activity takes place [23–29]. However, to our knowledge, there are no published works that focus on the detailed mineralogical study of scabs and efflorescences of sulfates that form on the surface of the tailings because of their exposure to the weather.

The Xochula tailings (Figure 1) have approximate dimensions 85 m long by 25 m wide and a height of 3 m. Only in the superficial central part is there vegetation that decays toward the ends of the deposit where crusts of violet to red coloration were identified. Xochula can be described as an oxidized tailings because its yellowish and brown colors are related with the presence of iron oxides and hydroxides or iron hydrated sulfates. At the NE of these tailings, there are small channels that seem to have been generated by the runoff of water coming from the highest parts. On the walls of these channels, white-colored salts with a fibrous habit and very irregular distribution were identified, which were probably formed by evaporation processes and forming thin crusts (≤1 cm) that only in certain points can be considered as true efflorescence (Figure 2).

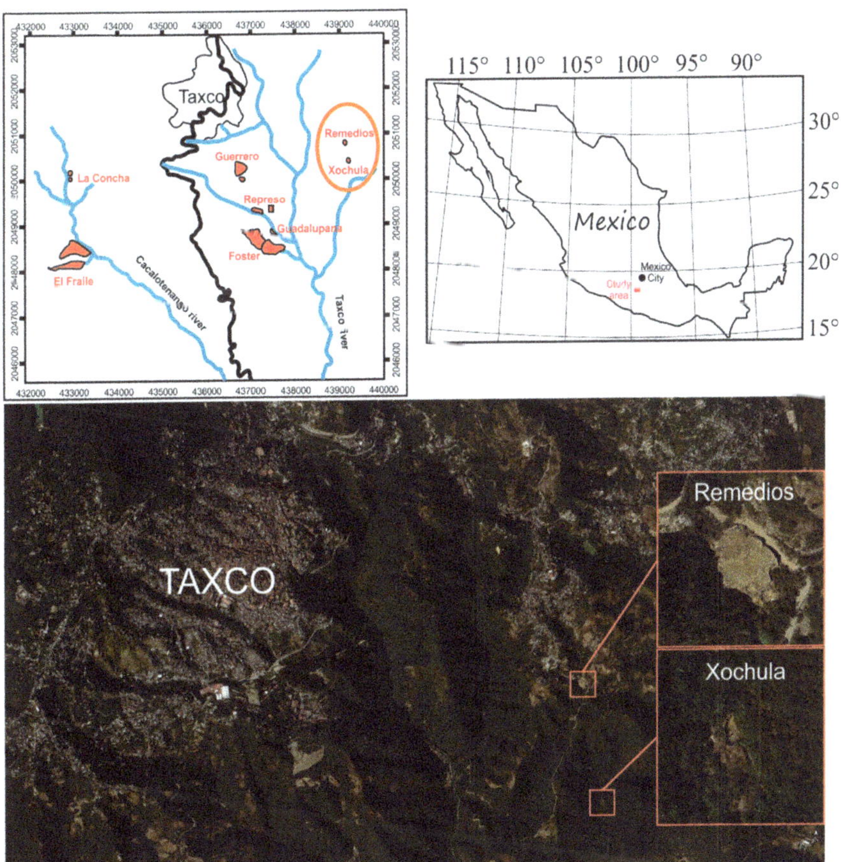

Figure 1. Location of the abandoned tailings studied in this work, produced by the mining activity in the Taxco district. Due to their size, the Fraile tailings are those that have been most studied [23,27], but little work has been made up to this date to characterize the smaller tailings as Remedios and Xochula. Orange oval in the upper left-hand figure delimits the studied area.

Figure 2. Field photographs of the Xochula tailings. (**A**) Appearance of the JH-38 sample with hydrated iron sulfates. (**B**) Sample JH-03 shows the field characteristics that generally present the materials without hydrated iron sulfates. (**C**) The gutters made at one end of the deposits are highlighted by the draining of the water from the surface toward the lower parts. The formation of structures of fibrous appearance in the walls of the channels. (**D**) Superficial view of the Xochula deposit, the presence of vegetation, and the formation of soils in some portions can be appreciated. (**E**,**F**) Views of two walls of the tailings highlight the presence of whitish colorations with efflorescent salts on both sides.

The Remedios tailings (Figure 1) have dimensions 55 m long by 45 m wide with a height from the ground of 10 m. It is also oxidized tailings with low pyrite content, a stacked accumulated mining waste forming at least two independent terraces. Unlike the Xochula tailings deposit, it does not present vegetation, and there is no formation of soil on the surface. The colorations observed are very varied, although there is no pattern that defines their distribution. In this deposit, efflorescence is observed especially on the side walls, which are characterized by bright orange, yellowish, greenish, and whitish tones that are generally mixed. In the superficial part, different colorations also appear, although the efflorescence is less clear and with a more irregular distribution (Figure 3).

Figure 3. Field photographs of the Remedios tailings. (**A–C**) Photographs of samples containing iron hydrated sulfates; the macroscopic aspect of salts and efflorescences are highlighted. (**D**) Image of sample JT-33 located on the surface of the tailings shows the general macroscopic characteristics of materials that do not contain hydrated iron sulfates. (**E**,**F**) are panoramic views of the deposit both on the surface and on the wall.

2. Materials and Methods

2.1. Sampling

Prior to the collection of samples, the tailings were delimited with the help of images from Google Earth; a sampling square grid with a side length of 10 m was established. A total of 90 samples were taken, 41 corresponding to the Xochula tailings and 49 corresponding to the Remedios tailings. In addition to the 10 m grid sampling, changes in the colorations of the materials were also considered as sampling criteria. The samples chosen were mainly those with a fine-grained (sand and silt) and salt-like appearance. The samples were collected on the surface (up to 2 cm depth). The samples were collected during the dry season (May) with an aluminum spoon and stored in small airtight plastic bottles; then, they were stored in a dark room at 25 °C, where they remained until they were analyzed. The basic characteristics of each of the samples are detailed in Table 1 (Xochula tailings) and Table 2 (Remedios tailings). The sampling pattern was included as Figure S1.

Table 1. Physical description of samples from the superficial (JH-01 to JH-36) and lateral (JH-37 to JH-41) parts of Xochula Tailings.

Sample	Origin	Coloration	Description
JH-01	Tailings	Yellowish orange	Very fine grain efflorescent salts
JH-02	Tailings	Pale yellow	Soft crust without vegetation
JH-03	Tailings	Yellowish orange	Material with fibrous texture
JH-04	Soil	Yellowish brown and bluish	Fine grain with vegetation
JH-05	Tailings	Brownish-yellowish orange	Very fine grain without efflorescence
JH-5-1	Tailings and crust	Intense orange and violet red	Consolidated crust
JH-06	Soil	Orange brown	Fine grain with vegetation
JH-07	Soil	Yellowish brown	Fine grain with vegetation
JH-08	Soil	Brown	Consolidated material without vegetation
JH-09	Soil	Yellowish brown	Fine grain with vegetation
JH-10	Soil	Yellowish brown	Fine grain with vegetation
JH-11	Soil	Brown	Fine grain with vegetation
JH-12	Soil	Yellowish brown	Fine grain with vegetation
JH-13	Soil	Yellowish reddish brown	Fine grain with vegetation
JH-14	Soil	Orange brown	Fine grain with vegetation
JH-15	Soil	Yellowish brown	Fine grain with vegetation
JH-16	Soil	Dark yellowish brown	Fine grain with vegetation
JH-17	Tailings	Intense orange	Crust without vegetation
JH-18	Soil	Yellowish brown	Fine grain with vegetation
JH-19	Soil	Yellowish brown	Fine grain with vegetation
JH-20	Tailings	Yellowish white	Very fine grain efflorescent salts
JH-21	Tailings	Bright yellow orange	Loose material without vegetation
JH-22	Tailings	Dark Yellowish brown	Consolidated without vegetation
JH-23	Soil	Light yellowish brown	Fine grain with vegetation
JH-24	Soil	Yellowish brown	Fine grain with vegetation
JH-25	Soil	Orange-reddish brown	Fine grain with vegetation
JH-26	Soil	Yellowish brown	Fine grain with vegetation
JH-27	Soil	Yellowish brown	Fine grain with vegetation
JH-28	Soil	Yellowish-orange brown	Fine grain with vegetation
JH-29	Soil	Yellowish brown	Fine grain with vegetation
JH-30	Tailings	Intense orange	Fine grain without vegetation
JH-31	Tailings	Yellowish white	Very fine grain efflorescent salts
JH-32	Tailings	Yellowish white	Very fine grain efflorescent salts
JH-33	Tailings	Intense orange	Loose sandy material
JH-34	Tailings	Intense orange and yellow	Loose sandy material
JH-35	Tailings	Intense orange	Loose sandy material
JH-36	Tailings	Reddish violet and orange	Slightly hardened scab
JH-37	Tailings	Yellowish white	Very fine grain efflorescent salts
JH-38	Tailings	Whitish	Very fine grain efflorescent salts
JH-39	Tailings	White orange	Massive, fine-grained efflorescent salts
JH-40	Tailings	Pink white	Very fine grain efflorescent salts
JH-41	Tailings	Gray bluish lead	Consolidated material with fibrous texture

2.2. *Short-Wave Infrared Spectroscopy*

The analysis by short-wave infrared spectroscopy (SWIR) was performed for all 90 samples collected; the measurements were made with a portable Pro LabSpec spectrophotometer (Analytical Spectral Devices, Inc., Boulder, CO, USA). Before performing the measurement of the samples, a barite blank was measured to calibrate the equipment; this calibration was repeated every 5 measured samples. Before being analyzed, the samples were gently disintegrated with a spatula and placed in a small container. Each sample was measured in duplicate to verify the response of the spectrophotometer. At the end of the measurement, each sample was returned to their original container and stored in the room at 25 °C for later analysis.

Table 2. Physical description of samples from the superficial (JT-01 to JT-47) and lateral (JT-03, JT-04, JT-05, JT-06, JT-16, JT-44, JT-45, JT-46, JT-48 and JT-49) parts of Remedios Tailings.

Sample	Origin	Coloration	Description
JT-01	Tailings	Loose sandy material	Loose sandy material
JT-02	Tailings	Yellowish brown	Material with fibrous texture
JT-07	Tailings	Violet gray	Loose sandy material
JT-08	Tailings	Orange brown	Very fine grain efflorescent salts
JT-09	Tailings	Greenish yellow	Very fine grain efflorescent salts
JT-10	Tailings	Orange brown	Loose sandy material
JT-11	Tailings	Orange red	Loose sandy material
JT-12	Tailings	Intense orange and yellow	Loose sandy material
JT-13	Tailings	Yellowish orange	Weakly consolidated
JT-14	Tailings	Pink violet	Slightly hardened crust
JT-15	Tailings	Yellowish brown	Loose sandy material
JT-17	Tailings	Red brown	Slightly hardened scab
JT-18	Tailings	Red brown	Slightly hardened scab
JT-19	Tailings	Pink violet	Slightly hardened scab
JT-20	Tailings	Dark brown violet	Slightly hardened scab
JT-21	Tailings	Yellowish brown	Crust with disease cracks
JT-22	Tailings	Yellowish orange	Very fine grain efflorescent salts
JT-23	Tailings	Orange-red brown	Crust with disease cracks
JT-24	Tailings	Intense orange brown	Soft crust with efflorescence of fine particle size
JT-25	Tailings	Yellowish brown	Loose sandy material
JT-26	Tailings	Orange and greenish yellow	Loose sandy material
JT-27	Tailings	Amarillo greenish	Very fine grain efflorescent salts
JT-28	Tailings	Intense violet	Loose sandy material
JT-29	Tailings	Yellowish orange	Slightly hardened crust
JT-30	Tailings	Intense orange brown	Loose sandy material
JT-31	Tailings	Intense orange	Crust lightly hardened
JT-32	Tailings	Bright yellow	Loose sandy material
JT-33	Tailings	Bright orange	Very fine grain efflorescent salts
JT-34	Soil	Black	Crust lightly hardened
JT-35	Tailings	White orange	Very fine grain efflorescent salts
JT-36	Tailings	Bright violet	Very fine grain efflorescent salts
JT-37	Tailings	Reddish brown pink	Crust lightly hardened
JT-38	Tailings	Yellowish orange	Loose sandy material
JT-39	Tailings	Bright yellow	Stratified layers and crusts
JT-40	Tailings	Reddish brown	Loose sandy material
JT-41	Tailings	Yellowish orange	Stratified layers and crusts
JT-42	Tailings	Bright brown	Crust lightly hardened
JT-43	Tailings	Yellowish orange	Crust lightly hardened
JT-47	Tailings	Orange white	Very fine grain and massive efflorescent salts
JT-03	Tailings	Bright orange and blue	Very fine grain and massive efflorescent salts
JT-04	Tailings	Whitish	Very fine grain and massive efflorescent salts
JT-05	Tailings	Bright greenish yellow	Very fine grain and massive efflorescent salts
JT-06	Tailings	Bright orange	Very fine grain and massive efflorescent salts
JT-16	Tailings	Yellowish white	Stratified layers and crusts
JT-44	Tailings	Yellowish white	Very fine grain and massive efflorescent salts
JT-45	Tailings	White	Very fine grain and massive efflorescent salts
JT-46	Tailings	Bright orange	Very fine grain and massive efflorescent salts
JT-48	Tailings	Bright brown	Weakly hardened scab
JT-49	Tailings	Bright white brown	Very fine grain and massive efflorescent salts

2.3. X-ray Diffraction Analysis

Bulk rock analysis was performed for all samples. Approximately two grams of each sample were taken, manually ground in an agate mortar, sieved to less than 45 microns, and loaded onto back side aluminum sample holders. The mineralogical analysis of the clay fraction was performed for 10 selected samples, which were gently disaggregated to avoid artificial grain size reduction [30]. Each sample was broken into small chips (<2 mm) using a porcelain crusher and subsequently dispersed in deionized water. The clay separation was done using the most non-aggressive method [31,32]. Grain size fractions of <2 μm were separated in distilled water according to Stoke's law with a Rotanta (Hettich) centrifuge operating at 6000 rpm for 5 min. From the <2 μm fractions, air-dried oriented preparations were obtained by pipetting some drops of the Mg-saturated suspensions onto a glass slide, which was then dried at 30 °C for a few hours [32]. Ethylene glycol solvation of the slides was achieved by exposing them to ethylene glycol vapor at 70 °C for a minimum of 12 h.

For X-ray diffraction analysis, an Empyrean diffractometer (Malvern Panalytical, Malvern, UK) equipped with a long fine tube (copper and cobalt), as well nickel and iron filters was used, operating at 40 mA and 45 kV, in a range of 4° to 70° (2 theta), with a step scan of 0.003° (2θ) and 30 s of integration time. Phase identification and quantification by the Rietveld method was made using HighScore v4.5 software, making use of both International Center of Diffraction Database (ICDD) and Inorganic Crystal Structure Database (ICSD). Relevant data from the diffractometer used are shown in Table 3. The specimen-dependent parameters that were refined were the zero error, displacement error, polynomial fitting for the background with six coefficients, cell parameters, and atomic coordinates.

With the quantitative XRD results, a hierarchical clustering was performed with Statgraphics software (version 5.1) to recognize the genetic relation between the different mineral groups.

Table 3. Relevant data from the diffractometer used during measurement and Rietveld refinement.

Parameter	Value	Parameter	Value
Geometry	Bragg-Brentano	Divergence slit	$\frac{1}{2}$°
Goniometer radius	240 mm	Incident antiscatter slit	1°
Radiation source	CuKα and CoKα	Reflected antiscatter slit	7.5°
Generator	45 kV, 40 mA	Soller slits	0.04 rad
Tube	Long Fine Focus	Detector	Solid state (PIXcel3D)
Monochromator	None	Step size	0.003° (2θ)
Filter	Nickel (Cu) and Iron (Co)	Integration time	30 s
Incident beam optics	Parallel mirror (Göbel)	Spinning	15 rpm

2.4. Scanning Electron Microscopy

A low vacuum scanning electron microscope Hitachi TM-1000 (Hitachi, Tokyo, Japan) was used. The preparation of the samples was not necessary; however, small fragments of efflorescence were selected, since the use of sprayed samples is not possible due to the vacuum system. The identification of the minerals observed was based on both the results of the Energy-Dispersive Spectroscopy (EDS) analysis and the observation of the morphologies and textures of the crystals and aggregates [12].

2.5. Chemistry of Whole Rock and Trace Elements by Inductively Coupled Plasma

In order to verify if the identified hydrated sulfates contain potentially toxic heavy elements (PTE), three samples of efflorescences from the Remedios tailings were analyzed by ICP–MS (inductively coupled plasma) at Activation Laboratories Ltd. (Ancaster, ON, Canada). A combination of packages Code 4B (lithium metaborate/tetraborate fusion ICP–OES whole rock) and Code 4B2 (trace element ICP–MS) was used. Fused samples were continuously mixed with a 5% nitric acid solution containing an internal standard until the sample was completely dissolved and analyzed with a Perkin Elmer Sciex ELAN ICP–MS. Three blanks and five controls (three before each sample group and two after) were analyzed with the samples.

3. Results

3.1. Short-Wave Infrared Spectroscopy

The data obtained were corrected and plotted in Grapher (Golden Software, version 16, Golden, CO, USA), thus generating the spectra for each measured sample; it was plotted in the ranges of the near infrared (900 to 1400 nm) and short-wave infrared (1400 to 2550 nm). Mineral identification by this technique was based on the position of the absorption features of the representative spectrum of each group, as well as its general shape. In this study, a visual and numerical comparison of these characteristics was made with spectral libraries [33] and with spectra from [34,35]. The SWIR technique has been mainly used to discriminate samples with soluble sulfates from samples that do not contain these minerals (JH-I_{SWIR} and JT_{SWIR} groups) in a way that is quick and qualitative (and in some cases during field work). This discrimination is important because soluble sulfates are potentially more dangerous since they dissolve easily in rainy seasons.

To diagnose whether sulfates were present, the existence of a very distinct absorption band between 1700 and 1850 nm (approximately 1740 nm for gypsum and approximately 1850 nm for jarosite) was used (ν + $\delta OH/H_2O$ and rotational fundamentals or δS-O overtones [35]). The bands located between 2100 and 2300 nm are related to the vibrations of the SO bonds, while those related to the metal–OH bond are found between 1480 and 2250 nm and can seldom be discriminated from the SO bands. The bands related with H_2O + OH bonds, those between 1440 and 1476 nm, and those related with the H_2O molecule alone are found between 1947 and 1969 nm [35].

Based on the absorption characteristics of the spectra obtained, and due to the great similarity that was found among many of them, a grouping was made and six clusters were established: three belonging to the Xochula tailings (JH-I_{Swir}, JH-II_{Swir}, JH-III_{Swir}; Figure 4) and three belonging to the Remedios one (JT-I_{Swir}, JT-II_{Swir}, JT-III_{Swir}; Figure 4). It is necessary to mention that there are small spectral variations within each group, since two samples with strictly equal absorption characteristics were not found. The first group was characterized in both tailings (JH-I_{Swir} and JT-I_{Swir}) by presenting a very flat spectrum in which only a small peak near 1850 nm stands out, which coincides with a characteristic band of jarosite mineral. This group is the most abundant and is characterized by the absence of soluble iron sulfates. Samples of the second group (JH-II_{Swir} and JT-II_{Swir}) were characterized by the presence of two important bands at 1850 nm and at 1400 nm (H_2O + OH). For these spectra—also in the zone ranging from 2100 to 2300 nm—a single broad band appears, which corresponds mainly to the Fe–OH bond. Samples of the third group (JH-III_{Swir}, JT-III_{Swir}) were defined by the same absorption bands as the samples of the second group, but in the zone between 2100 and 2400 nm, two different bands were discriminated, indicating the presence of two types of bonds between OH and metals.

A graphical comparison was also made of the spectra described above against the spectra of the main iron sulfates found in the literature [34,35]. From this comparison, it was possible to determine the presence of some characteristic absorption bands (bending and stretching, mainly of H_2O, OH, SO_4, and metal–OH) and relate them with the presence of rozenite, copiapite, jarosite, and hexahydrite (Figure 5). The SWIR equipment used is not state-of-the-art, but it can be used with confidence. Nevertheless, SWIR has not allowed us to discriminate the identity of all the soluble sulfates present in the samples, since these are generally complex mixtures with many phases and the absorption bands of the hydrated sulfates are often very similar to each other, as shown in Figure 4. To discriminate the different sulfates and quantify their relative abundance, the infrared data were complemented with the X-ray diffraction data.

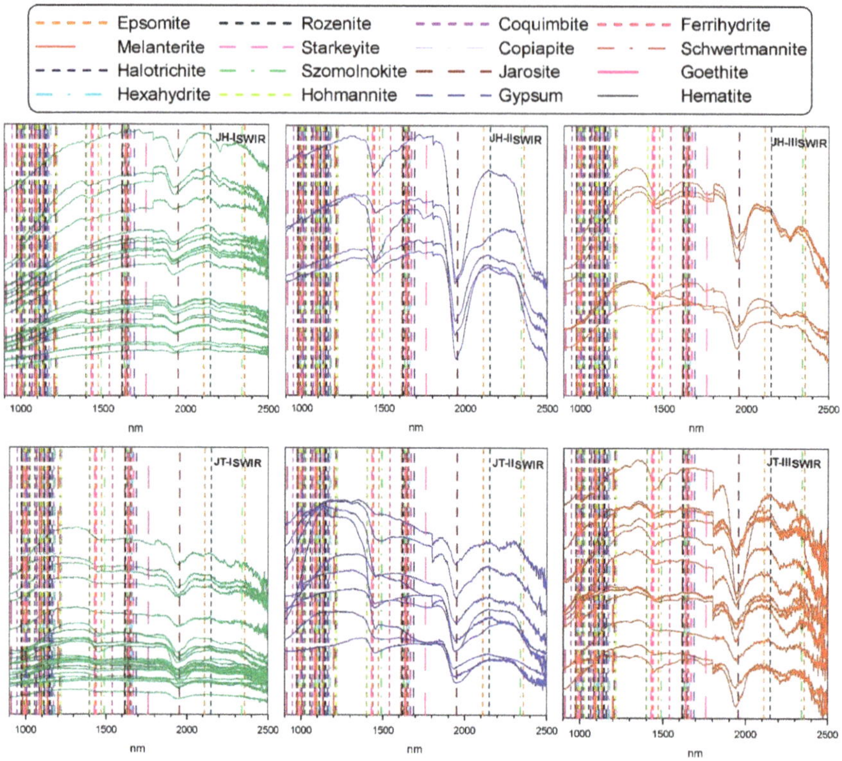

Figure 4. Groups established during the application of the analysis by infrared spectroscopy in the samples of the Xochula (JH) tailings and Remedios (JT) tailings. To facilitate the comparison with the minerals reported by other authors [33,35], the different bands (vertical lines) of the most important phases identified by X-ray diffraction were introduced. The first group of samples (green) is characterized by the presence of a very flat spectrum, in which only a small peak near 1850 nm stands out, coinciding with a characteristic band of jarosite mineral. This group is the most abundant and is also characterized by the absence of soluble iron sulfates. The second group (blue) is characterized by the presence of two important bands at 1850 nm and 1400 nm (H_2O + OH). For these spectra, in the zone ranging from 2100 to 2300 nm, a single broad band appears, which corresponds mainly to the Fe–OH bond. The third group (brown) displays the same absorption bands as the samples of the second group, but it is characterized by the presence of two bands in the zone between 2100 and 2400 nm, indicating the presence of two types of bonds between OH and metal.

3.2. X-ray Diffraction

Most of the minerals identified are of secondary (predominance of Fe^{2+}) and tertiary (predominance of Fe^{3+}) origin, and they correspond (Table 4) mainly to iron sulfates with different degrees of hydration, demonstrating that the main primary iron mineral accumulated in both mining residues was probably pyrite. The hydrated sulfates of iron predominate, but it is worth noting the presence, in low proportion, of hydrated magnesium sulfates with different hydration status (7 water molecules for epsomite, 6 water molecules for hexahydrite, 4 water molecules for starkeyite, and one water molecule for kieserite) in both tailings. For each of the tailings, a sample (JH20 and JT47) with gunningite was identified; this is a hydrated zinc and manganese sulfate. These data are congruent with the high zinc content of the samples analyzed by ICP–MS in this mining waste, as well as with the presence of sphalerite in the Taxco mineral deposit. Furthermore, a sample of Remedios tailings (JT47) with botryogen (magnesium and iron hydroxysulfate) was identified.

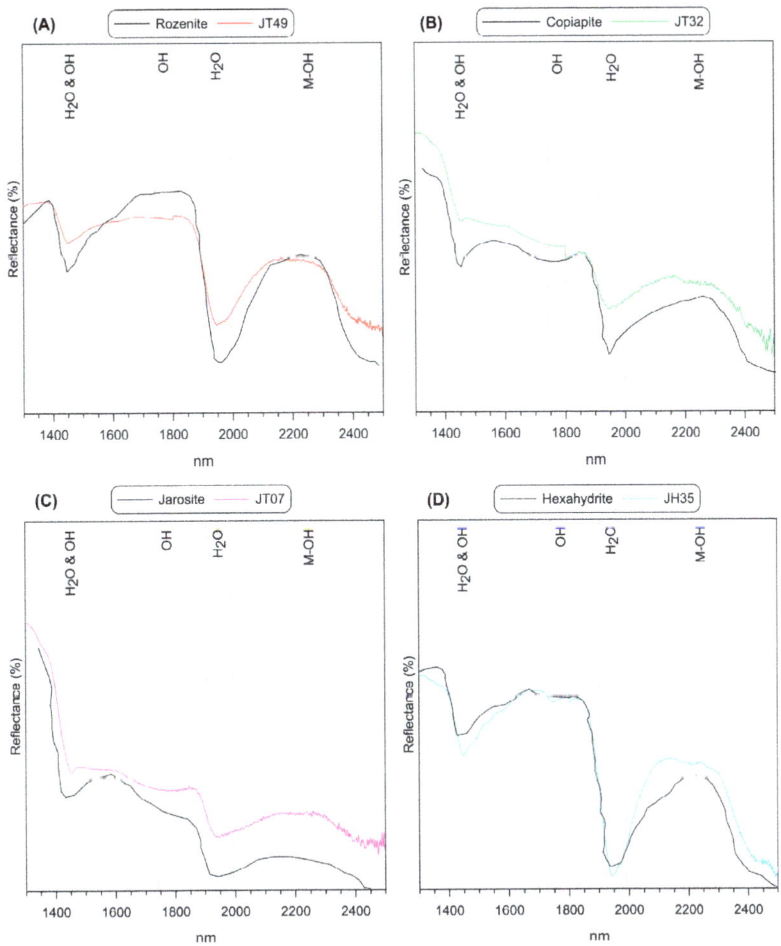

Figure 5. Characteristic short-wave infrared spectroscopy (SWIR) absorption spectra of some iron-bearing sulfates found at Taxco tailings. Reflectance spectra of (**A**) rozenite, (**B**) copiapite, (**C**) jarosite, and (**D**) hexahydrite, compared with digitalized literature spectra (Velasco et al., 2005; Cloutis et al., 2006). In general, only when the sample has more than 25% of the phase can a good correlation be observed.

3.2.1. Xochula Tailings

The primary minerals identified were quartz, feldspars, pyrite, fluorite, calcite, and clay minerals. For example, quartz was detected in 39 samples (17–82%); feldspars, mainly plagioclase, were detected in 19 samples (4–15%); pyrite was found in 13 samples (1–11%); fluorite was found in six samples (2–8%); carbonates were found in three samples (with a maximum of 7%), and zeolites were found only in one sample (5% in sample JH22). Phyllosilicates are more abundant than in the Remedios tailings and were identified in clay fraction (<2 μm), mica-illite was identified in 27 samples (2–27%), kaolinite was identified in 26 samples (3–17%), smectite was identified in four samples (6–12%), and chlorite was identified in only two samples, with a maximum ratio of 17% in sample JH14. These minerals are quite crystalline and were defined as primary minerals. The most important hydrated sulfate mineral in the Xochula tailings are coquimbite (in six samples with a maximum ratio of 30% in sample JH32), copiapite (in three samples in a range of 7% to 48%), rozenite (in three samples with a range of 15% to 38%), szomolnokite (in two samples with a maximum of 13% in sample JH38), and halotrichite (in two samples with a range of 5% to 16%). Other minerals of this group are starkeyite (approximately 69% in sample JH20), hexahydrite (58% in sample JH35), gunningite (9% in sample JH20), and melanterite (4% in sample JH04); all of them were only found in one sample each. As for insoluble sulfates, jarosite was identified in 18 samples (1% to 19%), alunite was identified in two samples (3% to 5%), and gypsum was identified in 22 samples (4% to 46%). Regarding iron oxides and hydroxides, goethite was found in 10 samples (1% to 28%), hematite was found in seven samples (2–10%, always in a lower ratio than goethite), ferrihydrite was found in two samples (8 and 9%), and schwertmannite was found only in one sample JH37 (8%). Jarosite and gypsum were found both in samples rich in hydrated iron sulfates and in those that are characterized by the presence of oxides of this element, such as goethite and hematite. Table S1 shows the mineral phases identified and their relative abundance estimated by the Rietveld refinement method.

Table 4. Table with the non-primary minerals found in the studied efflorescences of Xochula (JH) and Remedios (JT) tailings.

Mineral Name	Abbr.	Chemical Formula	Group	XRD Pattern Ref.	Characteristic XRD Peaks (Å)		
Botryogen	Bot	$MgFe^{3+}(SO_4)_2(OH){\cdot}7(H_2O)$	Iron and Mg hydroxysulfate	ICSD 980034682	6.36	8.96	5.52
Epsomite	Ep	$MgSO_4{\cdot}7(H_2O)$	Hydrous Mg sulfate	ICSD 980029384	4.22	4.21	5.36
Halotrichite	Hal	$Fe^{2+}Al_2(SO_4)_4{\cdot}22(H_2O)$	Hydrous Fe and Al sulfate	ICSD 980096598	4.80	15.84	4.95
Hexahydrite	Hex	$MgSO_4{\cdot}6(H_2O)$	Hydrous Mg sulfate	ICSD 980016546	4.36	5.45	4.03
Hohmannite	Hoh	$Fe^{3+}{}_2(SO_4)_2(OH)_2{\cdot}7(H_2O)$	Iron hydroxysulfate	ICSD 980037328	7.96	8.73	10.42
Fibroferrite	Fi	$Fe^{3+(}SO_4)(OH){\cdot}5(H_2O)$	Iron hydroxysulfate	ICSD 980100721	12.08	4.07	3.45
Gunningite	Gun	$(Zn, Mn^{2+})SO_4{\cdot}H_2O$	Hydrous Zn and Mn sulfate	ICSD 980071348	4.40	4.76	3.05
Kieserite	Kie	$MgSO_4{\cdot}H_2O$	Hydrous Mg sulfate	ICSD 980015924	4.79	3.40	3.36
Melanterite	Me	$FeSO_4{\cdot}7H_2O$	Hydrous Fe sulfate	ICSD 980016589	4.90	3.77	4.87
Rostite	Ros	$Al(SO_4)(OH){\cdot}5H_2O$	Hydroxysulfate	ICDD 000421427	4.25	4.18	3.90
Rozenite	Roz	$FeSO_4{\cdot}4H_2O$	Hydrous Fe sulfate	ICSD 980023914	5.47	4.50	3.98
Starkeyite	Stk	$MgSO_4{\cdot}4(H_2O)$	Hydrous Mg sulfate	ICDD 010721096	4.47	5.43	3.95
Szomolnokite	Szo	$Fe^{2+}SO_4{\cdot}(H_2O)$	Hydrous Fe sulfate	ICSD 980027098	3.44	3.11	4.86
Copiapite	Cop	$Fe^{2+}Fe^{3+}{}_4(SO_4)_6(OH)_2\ 20(H_2O)$	Hydroxysulfate	ICDD 000110395	10.5	6.87	3.43
Coquimbite	Coq	$Fe^{3+}{}_2(SO_4)_3{\cdot}9(H_2O)$	Hydrous Fe sulfate	ICDD 010802836	8.27	9.45	3.36
Alunite	Al	$KAl_3(SO_4)_2(OH)_6$	Hydrous sulfate of K and Al	ICSD 980012106	5.40	7.37	10.81
Jarosite	Jar	$KFe^{3+}{}_3(SO_4)_2(OH)_6$	Hydrous sulfate of K and Fe	ICSD 980157717	3.08	5.10	3.11
Beudantite	Beu	$PbFe_3(AsO_4)(SO_4)(OH)_6$	Hydrous arsenate of Pb and Fe	ICSD 980067455	3.07	5.94	3.66
Pb-Jarosite	PbJar	$Pb_{0.5}Fe^{3+}{}_3(SO_4)_2(OH)_6$	Hydrous sulfate of Pb and Fe	ICSD 980169961	3.07	5.93	2.25
Gypsum	Gy	$CaSO_4{\cdot}2H_2O$	Calcium sulfate dihydrate	ICSD 980161622	7.59	4.28	2.87
Ferrihydrite	Ferri	$Fe^{3+}{}_2O_3{\cdot}0.5(H_2O)$	Hydrous ferric oxyhydroxide	ICSD 980056287	1.15	1.02	0.90
Goethite	Goe	$Fe^{+3}O(OH)$	Iron oxyhydroxide	ICDD 010755065	4.18	2.45	3.88
Hematite	Hem	Fe_2O_3	Iron oxide	ICDD 010715088	2.70	2.52	3.68
Schwertmann.	Schw	$Fe^{3+}{}_{16}O_{16}(OH)_{12}(SO_4)_2$	Iron oxyhydroxysulfate	ICSD 980169971	2.55	3.39	4.86

3.2.2. Remedios Tailings

The primary minerals identified were quartz, pyrite, anglesite (only in sample JT36), fluorite, zeolites, and clay minerals; they were found mainly in the samples impoverished in hydrous iron sulfates. These phases are distributed irregularly throughout the sampled area. For example, quartz was detected in almost all samples (5–82%), pyrite was found only in seven samples in a proportion of 2% to 7%, fluorite was found only in two samples (6–7%), and carbonates and feldspars were not detected in the analyzed samples. Phyllosilicates were identified in clay fraction (<2 μm) as mica-illite in 16 samples (1–16%), kaolinite was identified in two samples (9–17%), smectite was identified only as traces (<5% in several samples), and chlorite was identified in only three samples with a maximum percentage of 8%. These minerals are always primary minerals. The most important hydrated sulfate mineral in the Remedios tailings is copiapite, which is present in 14 samples (1% to 38%). Other abundant minerals of this group are rozenite (in seven samples with percentages that vary from 2% to 33%), coquimbite (in five samples with percentages that vary from 1% to 30%), szomolnokite (in four samples with a maximum of 21% in sample JT06), and epsomite (in four samples with percentages from 2% to 24%). Minerals such as hohmannite (45% in sample JT46), metahohmannite (26% in sample JT03), fibroferrite (3% in sample JT06), starkeyite (81% in sample JT04), and kieserite (20% in sample JT47) were identified only in one or two samples. Minerals of the alunite group such as hydronium or K-jarosite (identified in 42 samples with percentages that vary from 2% to 30%), beudantite (identified in five samples with a maximum ratio of 5% in sample JT39), and plumbojarosite (identified in two samples with a maximum ratio of 11% in sample JT36) are present both in samples taken from the side walls of the deposit and in most of those taken from the surface. In the JT-04 sample, the presence of ferrihydrite (4%) was confirmed. As a product of the oxidation and subsequent dehydration of primary sulfides [36], hematite was identified in 24 samples (2–12%), and goethite was identified in 30 samples (2–33%). Both minerals were found both in samples taken on the surface and in those taken on the side walls of the deposit. Gypsum, present in 46 samples, is a very ubiquitous mineral (2–35%). Anhydrite was identified only as traces in the case of sample JT 46. Table S1 shows the mineral phases identified and their relative abundance estimated by the Rietveld refinement method [37].

3.2.3. X-ray Diffraction General Grouping

In order to compare the mineralogical results more easily and efficiently, seven genetic categories of minerals were established (primary (including gangue and ore minerals), phyllosilicates, secondary sulfates (with Fe^{2+}), tertiary sulfates (with Fe^{3+}), alunite group (alunite and all types of jarosite), iron oxides and hydroxides, and gypsum). Making these groupings was essential, since various minerals were discretely identified only in a few samples. Based on the XRD results and these seven mineralogical associations, three main groups of samples could be discriminated in each of the tailings (Figure 6).

Group I (JH-I_{DRX} & JT-I_{DRX}): samples with primary minerals (mainly quartz) and phyllosilicates (mainly mica), gypsum, jarosite and, in a variable proportion, iron oxides and hydroxides. This group is characterized by the absence of soluble hydrated sulfates. Small amounts of pyrite were identified in some of these samples, which is more frequent in JH tailings (13 samples) than in JT ones (7 samples). Phyllosilicates are more abundant for JH samples, while oxides and hydroxides, gypsum, and jarosite are more abundant for JT samples (Figure 7).

Group II (JH-II_{DRX} and JT-II_{DRX}): samples with a predominance of secondary hydrated sulfates with reduced iron (Fe^{2+}). The predominant mineral of this group in both tailings is rozenite. Samples from that group may also contain jarosite, gypsum, quartz, or oxides and hydroxides in varying proportions.

Group III (JH-III_{DRX} and JT-III_{DRX}): samples with a predominance of hydrated tertiary iron sulfates (mainly coquimbite, as well as hohmannite and fibroferrite in specific samples), but which may also present a mixture of secondary and tertiary sulfates (Fe^{2+} + Fe^{3+}) or minerals such as coquimbite, which has both oxidized and reduced iron. In the case of JH tailings, the predominant mineral of this type is coquimbite, while for JT samples, it is copiapite. Samples from that group may also contain quartz, jarosite, gypsum, iron oxides, and hydroxides in varying proportions.

The first group of samples established by XRD coincides for both mining tailings in more than 90% of the cases with group I of samples established by SWIR. The second and third groups established by XRD only partially coincide (63%) with groups II and II established by SWIR. This is because the diffraction groups were established based on the presence of Fe^{2+} or Fe^{3+} in the minerals, but it is probable that groups II and III established by SWIR should have a more direct relationship with the structure of the different types of sulfate minerals [38]. However, since each sample is generally a complex mineral mixture, it has not been possible to dig deeper into this issue.

3.3. Scanning Electron Microscopy

By means of this technique, we were able to verify that the habit of iron-hydrated sulfates is mainly fibrous and that the crystals are generally of small size (clay and silt) (Figure 8). In many of the samples, the mixture of different hydrated phases was observed, evidencing that they were formed quickly by evaporation of the acid drainage mainly in the walls of the tailings, where the runoff processes are observed. Since many of the secondary minerals are iron sulfates with different oxidation and hydration states, it was not possible to discriminate among them by means of EDS measurements.

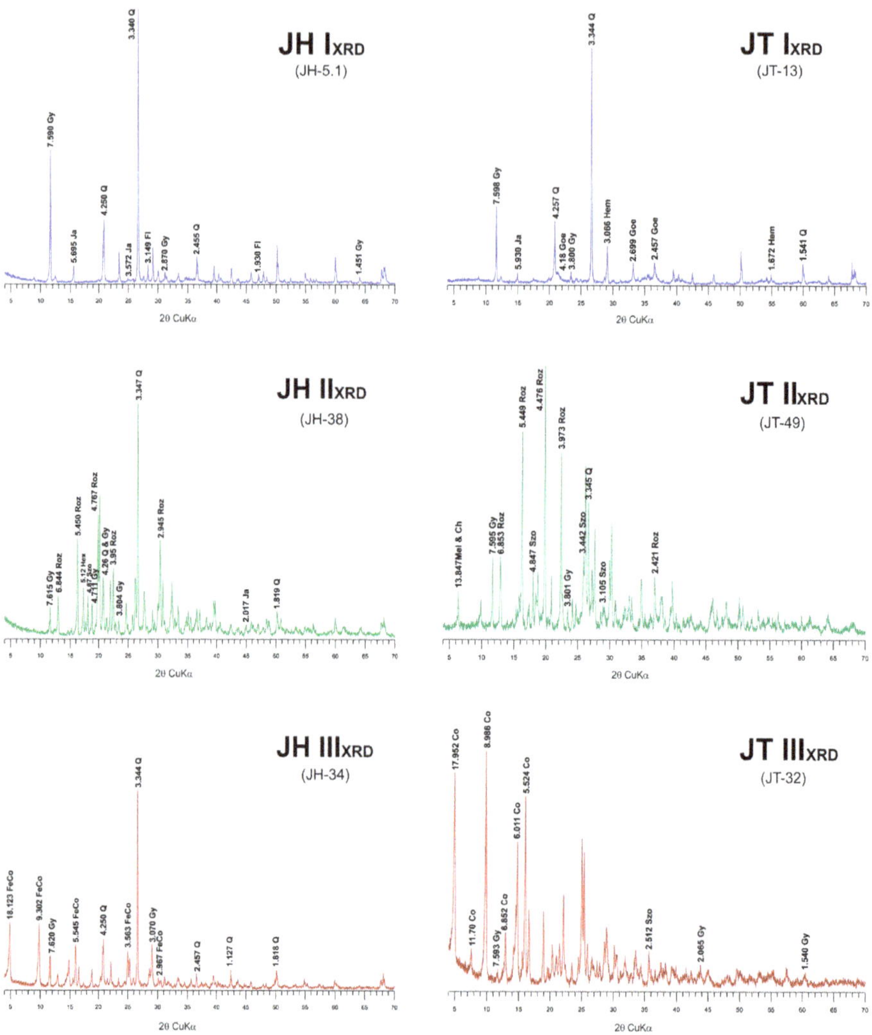

Figure 6. Representative XRD patterns from each of the three sample groups for Xochula (JH) tailings and Remedios (JT) tailings. The first group (blue) is characterized by the presence of primary minerals (mainly quartz) and phyllosilicates (mainly mica), gypsum, jarosite, and a variable proportion of iron oxides and hydroxides. The second group of samples (green) has a predominance of secondary hydrated sulfates with reduced iron (Fe^{2+}). Samples from that group may also contain jarosite, gypsum, iron, and/or quartz oxides and hydroxides in varying proportions. The third group of samples (brown) has a predominance of hydrated tertiary iron sulfates or a mixture of secondary and tertiary sulfates (Fe^{2+} + Fe^{3+}). Samples from that group may also contain quartz, jarosite, gypsum, iron oxides, and hydroxides in varying proportions. Co: copiapite; Ch: chlorite; Hex: hexahydrite; FeCo: ferrocopiapite; Fl: fluorite; Goe: goethite; Gy: gypsum; Hem: hematite; Ja: jarosite; Q: quartz; Mel: melanterite; Roz: rozenite; Szo: szomolnokite.

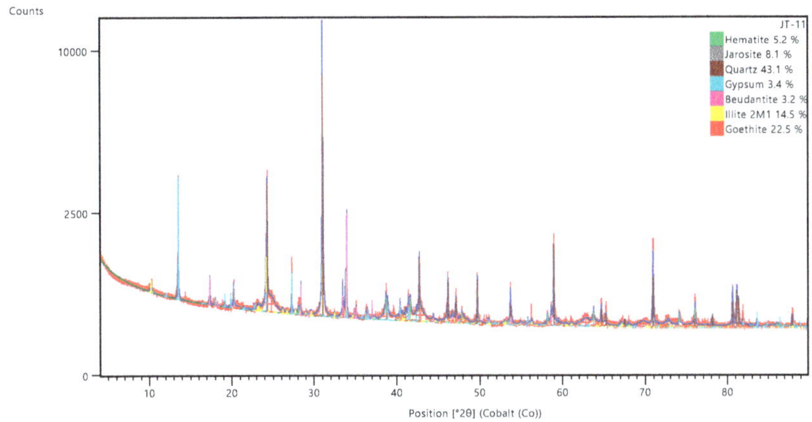

Figure 7. Example of quantification of crystalline phases using XRD and the Rietveld refinement method for sample JT11 (JT-I_{DRX}).

3.4. Geochemistry

To verify if the hydrated sulfates identified contain in their structure potentially toxic heavy elements (PTE), three samples of efflorescences from the Remedios tailings were analyzed for major (Table 5) and trace elements (Table 6).

Regarding the major elements, all the samples analyzed have approximately the same amount of silica (21% to 28%), which coincides with the proportion of quartz (close to 20%) that these samples present. The values of loss by ignition are very high (29% to 43%) in all samples, demonstrating that these are mainly constituted by sulfates (loss of SO_3) with different degrees of hydration (loss of H_2O). The iron percentage is also very high and very close to 25% Fe_2O_3 in the three samples. It can be related to the presence of iron oxides, hydrated iron sulfates, and traces of pyrite, the latter only for sample JT48. The Al_2O_3 content is always low and related to the presence of small amounts of phyllosilicates. The content of CaO (4% to 7%) is related with the presence of gypsum in all samples (7% to 18% by weight).

Table 5. Major element analysis from three samples of Remedios tailings. Concentrations in wt %. All detection limits 0.01%, except Mn, 0.001%. LOI: loss on ignition.

Sample	SiO_2	Al_2O_3	Fe_2O_3(T)	MnO	MgO	CaO	Na_2O	K_2O	TiO_2	P_2O_5	LOI	Total
JT-05	20.47	1.92	26.23	0.222	0.24	3.46	0.02	0.34	0.071	0.03	43.25	96.25
JT-03	20.80	1.33	25.69	0.585	0.26	4.70	0.01	0.24	0.040	0.02	38.11	91.78
JT-48	27.51	1.43	25.89	0.486	0.23	6.47	0.02	0.38	0.103	0.03	29.38	91.92

Regarding the trace elements (Table 6), the samples are enriched in the following potentially toxic elements (PTE): Cu (850 to 980 ppm), Zn (2980 to 9000 ppm), As (453 to 569 ppm), Sn (573 to >1000 ppm), Pb (1260 to 4730 ppm), and, to a lesser extent, Sb (38 to 60 ppm) and W (34 to 59 ppm). Sample JT03 is the one that is more enriched in Pb and Zn, while sample JT05 is the one with the highest content of As. The highest contents of Cu, Sn, Sb, and W were found in sample JT48. Zn and Cu can be hosted in copiapites (soluble sulfate); on the other hand, Pb and As are hosted in minerals from the jarosite group (less soluble sulfate). By comparison of these results with those obtained in other abandoned tailings in Mexico—such as those of San Felipe [39] or those of Nacozari de García [40]—the studied tailings of the Taxco mining district are those with the highest Zn and Cu enrichment; they are less enriched in As and Pb than those in San Felipe (>6000 ppm As, >8000 ppm Pb) [39] but much more enriched in these elements than the tailings of Nacozari de García (29 ppm As, 39 ppm Pb) [40]. Based on the XRD mineralogical data, we can confirm that these potentially toxic elements are clearly associated with the hydrated iron sulfates. The concentrations of Cu, Zn, As, Pb, Sn, and Sb exceed for all the samples the maximum permissible limits of toxicity for soils reported by the World Health Organization [41] and the regulations of the Secretary of Environment and Natural Resources (SEMARNAT) of the United States of Mexico [42].

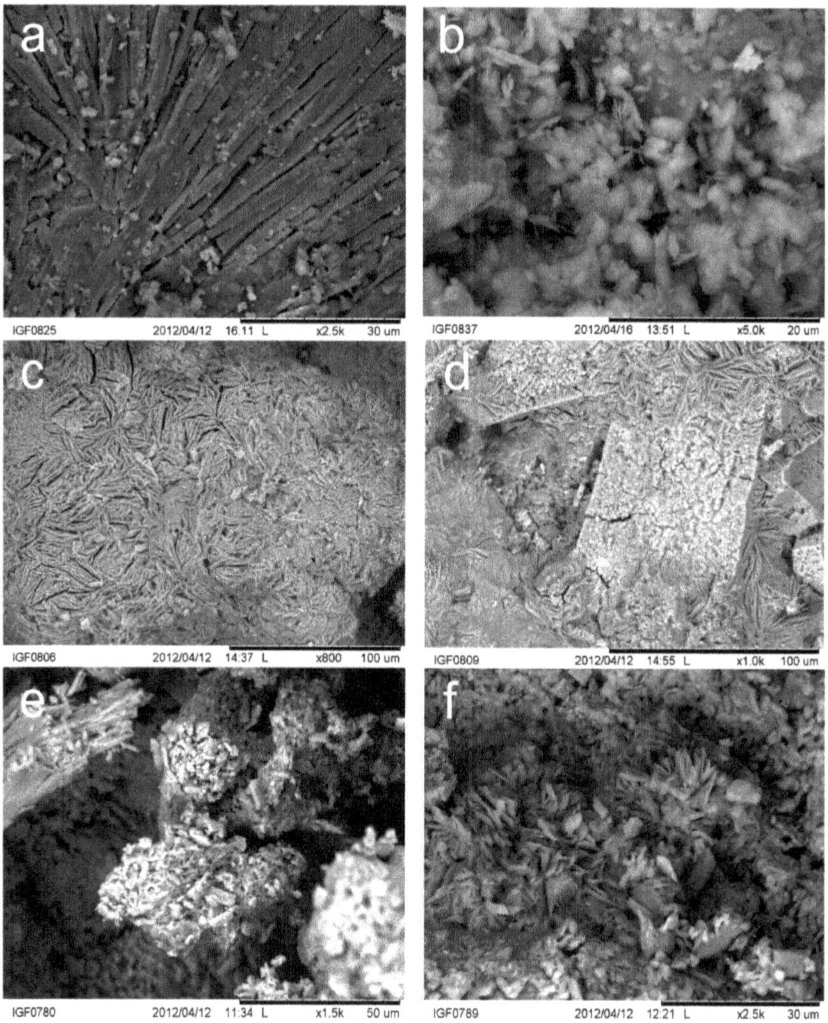

Figure 8. Microphotographs of samples with efflorescence. JH: Xochula; JT: Remedios. (**a**) Sample JH41, prismatic gypsum crystals; (**b**) Sample JH02, general aspect of the sample with different types of sulfates (mainly jarosite with anhedral crystals and tabular copiapite); (**c**) Sample JH48, tabular copiapite crystals; (**d**) Sample JH48, gypsum, jarosite, and clay minerals (central part), with tabular copiapite crystals at the external part; (**e**) Sample JT03, general appearance of the sample. It is observed that the sample is formed by an irregular mixture of different types of sulfates mixed with residues of the primary material; (**f**) Sample JT03, aggregate of copiapite crystals. In scale bar um = μm.

Table 6. Induced coupled plasma mass spectrometry (ICP–MS) analysis of some important trace elements from three samples of Remedios tailings. Concentrations in μg g^{-1} (ppm). Detection limits: Cu (10), Zn (30), As (5), Sn (1), Sb (0.2), W (10), Pb (30). RBV: Regional background values [23]; NOM-141: NOM-141-SEMARNAT-2003 [42].

Sample	Cu	Zn	As	Sn	Sb	W	Pb
JT-05	850	2980	569	815	37.6	45.8	1260
JT-03	880	9000	453	573	48.6	33.6	4730
JT-48	980	6920	519	>1000	59.5	58.8	2830
RBV	25	64	30	-	9.0	-	26
NOM-141	100	100	20	-	-	-	100

4. Discussion

4.1. Mineral Zoning of the Mining Waste

To establish the mineral zoning, the quantitative XRD results of the seven defined mineralogical groups (primary minerals, phyllosilicates, secondary hydrated sulfates, tertiary hydrated sulfates, alunite group, gypsum, and iron oxides and hydroxides) have been used as variables to (a) draw a contour map for Xochula (Figure 9) and Remedios Tailings (Figure 10) and (b) calculate a dendrogram using hierarchical clustering (Figure 11) that illustrates the statistic association between these groups in each of the tailings.

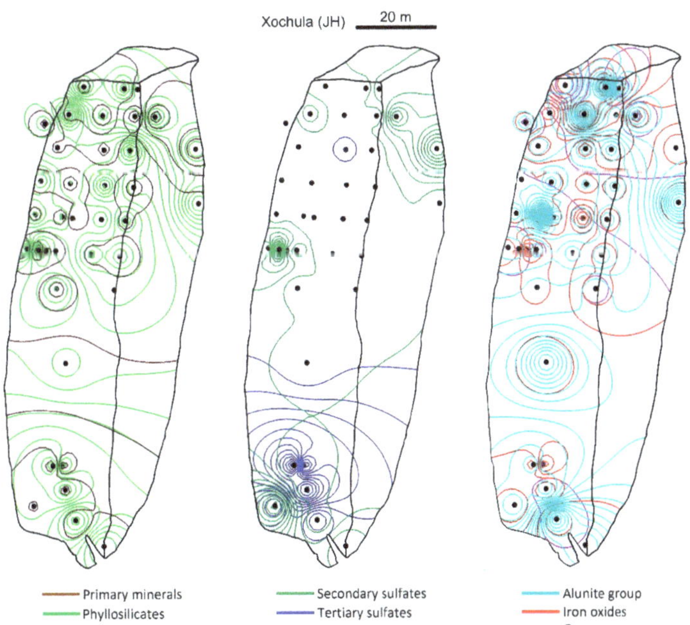

Figure 9. Contour map calculated from the quantitative data obtained using X-ray diffraction and the Rietveld method for Xochula (JH)) tailings. The groups considered are: primary minerals (pyrite, fluorite, calcite, and feldspars) in brown color, phyllosilicates (micas, smectites, kaolinite, chlorite) in light green color, secondary hydrated sulfates (Fe^{2+}) in dark green color, tertiary hydrated sulfates (Fe^{3+}) in blue color, alunite group minerals (mainly jarosite) in cyan color, gypsum in purple color, and iron oxides and hydroxides (schwertmannite, ferrihydrite, hematite, and goethite) in red color. The black line represents the outline of the tailings (superficial and lateral views).

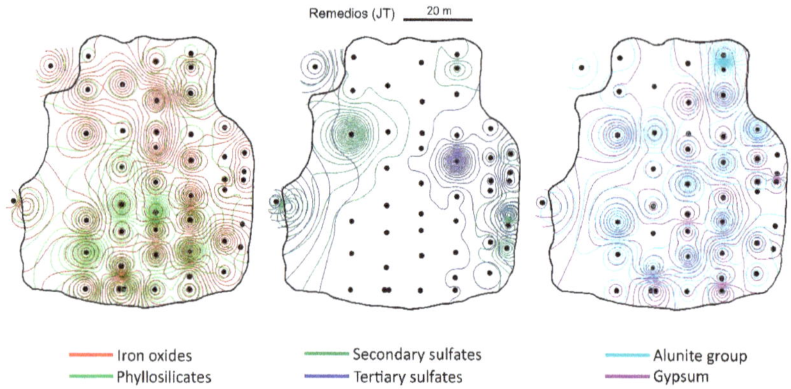

Figure 10. Contour map calculated from the quantitative data obtained using X-ray diffraction and the Rietveld method for Remedios (JT) tailings. The groups considered are: iron oxides and hydroxides (schwertmannite, ferrihydrite, hematite, and goethite) in red color, phyllosilicates (micas, smectites, kaolinite, and chlorite) in light green color, secondary hydrated sulfates (Fe^{2+}) in dark green color, tertiary hydrated sulfates (Fe^{3+}) in blue color, alunite group minerals (mainly jarosite) in cyan color, gypsum in purple color. The black line represents the outline of the tailings (superficial view).

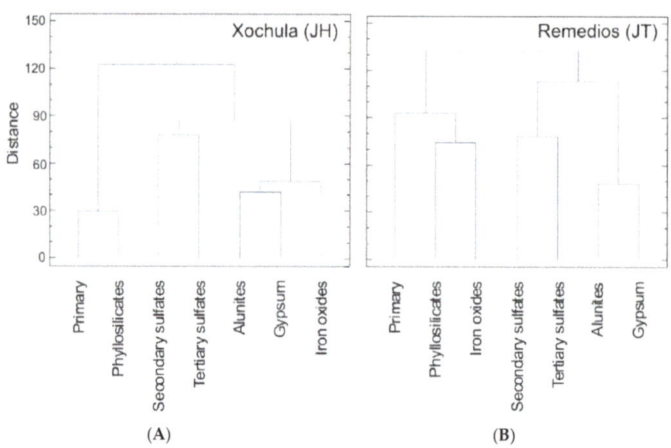

Figure 11. Dendrogram based on the quantitative X-ray diffraction data calculated with the Rietveld method. The groups considered are: primary minerals (pyrite, fluorite, calcite, and feldspars), secondary hydrated sulfates (Fe^{2+}), tertiary hydrated sulfates (Fe^{3+}), alunite group minerals (mainly jarosite), phyllosilicates (micas, smectites, kaolinite, and chlorite), gypsum, iron oxides and hydroxides (schwertmannite, ferrihydrite, hematite, and goethite). (**A**) Xochula (JH) and (**B**) Remedios (JT).

Based on these graphs, two different mineralogical zones have been distinguished for both tailings, which are the central, more horizontal area and the lateral or edge zone of the tailings. Secondary and tertiary hydrated soluble sulfates clearly predominate in the lateral zones and are practically non-existent in the central part of the tailings (Figures 9 and 10). Comparing the lateral areas of both tailings, tertiary (with Fe^{3+}) hydrated soluble sulfates predominate in Remedios, while secondary ones (Fe^{2+}) are more abundant in the Remedios tailings. In both tailings, there is a good correlation between poorly soluble sulfates (jarosites and gypsum). However, while in the Xochula the distribution of alunite mineral group is similar to Remedios, the distribution of gypsum is more uneven. The iron oxides are distributed throughout all the Remedios tailings and correlate well with the distribution of primary minerals. In Xochula, the presence of iron oxides is much more reduced, and its distribution is directly related to the presence of jarosite (Figure 11A). The relationship between phyllosilicates and primary minerals is very clear in Xochula and less in Remedios, where the phyllosilicates are only found in the central part. These results are concordant with the color (yellowish to intense orange, violet, blue in the lateral part) distribution in the tailings.

The described zonation was governed mainly by the water runoff. In the areas through which the deposits drain, a greater presence of efflorescent salts was observed. In places where drainage channels were absent, few agglomerations of the efflorescent salts were observed. Runoff water infiltrates from the surface to the internal part and the sides of the tailings, precipitating by evaporation many of the elements that it carries in solution. The formation of the efflorescence occurs when the water saturated in different elements evaporates, depositing sulfates in the lower and lateral parts of the deposit. In the superficial part of the waste deposits, secondary minerals were mainly non-soluble sulfates such as gypsum and jarosite, as well as more evolved minerals such as iron oxides.

4.2. Comparison of the Xochula Tailings with the Remedios Tailings

Xochula (JH samples) is an old tailing (probably more than 100 years old) with the development of soil and vegetation on its surface. Remedios (JT samples) is a younger and bigger deposit without evidence of vegetation or soil formation, which is characterized by a greater diversity and abundance of soluble hydrated sulfates. In the case of Xochula, the predominant tertiary hydrated sulfate (with Fe^{3+}) mineral is coquimbite, while for Remedios (JT samples) it is copiapite. Both tailings have a lot of quartz, but other primary minerals such as potassium feldspar or small amounts of calcite were only identified in the Xochula tailings. Pyrite was identified in 13 JH samples and only in seven JT samples. Phyllosilicates are also more abundant for JH samples, whereas oxides and hydroxides, gypsum, and jarosite are at a higher abundance in JT samples. The presence of these clays on the surface of the tailings is related with the settlement of the soil and vegetation.

Almost all the identified sulfates are of iron, indicating that the tailings were originally deposits of residual pyrite. Being ancient accumulations and based on the obtained mineralogical results, it is interpreted that both have undergone several alteration cycles over the time that they have been exposed to weathering. The mineralogical evolution in both tailings is quite similar and can be classified as advanced. However, the irregular distribution of non-oxidized pyrite in tailings as old as these can only be related to waste dumping in more recent times (different episodes). The great abundance of gypsum without residues of calcite in the Remedios tailings could be an indication of the use of carbonates as a remediation mechanism in this bigger deposit.

4.3. Genetic Model

Different authors [3,9,10,36] have demonstrated that the first hydrated iron sulfate that is formed during the pyrite alteration sequence is melanterite [$FeSO_4{\cdot}7(H_2O)$], and that the process that allows its formation is evaporation. In the case of the tailings studied in Taxco, this mineral was found only in two (JT03, JH04) of all analyzed samples, which was attributed to the fact that melanterite is a mineral highly sensitive to changes in humidity and temperature, which means it could have been transformed easily to another phase. The blue-green colorations observed in the field (Table 1) suggest the presence of this mineral. From the melanterite, other iron (divalent) sulfates were formed by a dehydration process, such as rozenite [$FeSO_4{\cdot}4(H_2O)$], which was present in the samples JH37, JH38, JH39, JT04, JT16, JT20, JT35, JT44, JT46, JT47 and JT49 and szomolnokite [$FeSO_4{\cdot}H_2O$], which is a mineral that was found in samples JH38, JH39, JT03, JT06, JT47, and JT49. It is known that rozenite precipitates are associated with very acidic (pH $\approx$ 1) drainage waters [33]. Once the szomolnokite had been formed, the oxidation process began, and the dehydration continued to form di-trivalent iron sulfates, such as copiapite [$Fe^{2+}Fe^{3+}{}_4(SO_4)_6(OH)_2{\cdot}20(H_2O)$] (14 samples of JT and three samples of JH) and coquimbite [$Fe_{2-x}Al_x(SO_4)_3{\cdot}9(H_2O)$, x~0.5], which were found in 11 samples (JT09, JT26, JT28, JT32, JT39, JH01, JH05, JH08, JH32, JH34, and JH37). It is known that ferrocopiapite precipitates from extremely Fe^{3+}-rich solutions [43].

In many of the samples, these minerals originated by oxidation were mixed with other phases formed by dehydration; this mixture of processes acting on the same sample gives rise to the mineralogical variability that was identified. In a more evolved stage, schwertmannite [$(Fe_8O_8(OH)_6(SO_4)\cdot n(H_2O)$] was formed by the rapid oxidation of Fe^{2+} in acidic conditions [36]. This mineral was only identified in sample JH37, but due to its low crystallinity, it could also be present in other samples. It would be necessary to use other techniques such as Raman or X-ray absorption spectroscopy [44,45] to confirm its identification.

Iron oxides and hydroxides identified during the study were ferrihydrite, goethite, and hematite. These minerals can be formed by the oxidation and hydrolysis of primary sulfides, especially pyrite. During the iron oxidation processes, the first phase that was precipitated by hydrolysis was ferrihydrite, but since it is a metastable material, it later changed to goethite or hematite. Another possible option for the formation of ferrihydrite in the studied area is through the dissolution–oxidation of hydrated sulfates such as rozenite, szomolnokite, copiapite, and coquimbite. Once formed, ferrihydrite can be decomposed by dehydration to form goethite. If dehydration continues, hematite is formed. Ferrihydrite was only found in three samples (JT04, JH22, and JH36), which indicates that if it was formed at an earlier stage, it may have already transformed to form goethite or hematite. This mineral was not found mixed with other oxides. Hematite and goethite were observed as mixed phases in many samples.

Oxides were formed by the two aforementioned processes: the uninterrupted process of alteration of pyrite (during the wet or rainy season), and the dissolution–dehydration of ferrous sulfates, which would have been formed prior to the interruption of the oxidation process of pyrite during the dry period. Their possible formation by bacterial activity was not demonstrated in this work, but it is important to mention that, as presented by [36] in his training model, the biological environment plays a decisive role in their formation. For example, the bacterium *Acidithiobacillus ferrooxidans* is an acidophilic bacterium with a physiology based on the fixation of carbon from atmospheric CO_2, obtaining its energy from the oxidation of iron.

Jarosite is represented in half of the Xochula tailings samples and in almost all the samples from Remedios tailings (the exception of six samples). Oxides were probably formed from the alteration of the jarosite; therefore, associations of jarosite–goethite–hematite were observed in several samples.

The stability of the jarosite depends on the pH; with the increase of pH values, jarosite tends to transform into goethite. If the sulfuric acid generated by the primary oxidation of pyrite is not neutralized, the pH decreases; under these conditions, jarosite is formed from ferric iron oxides. An important role in the production of jarosite during the oxidation of ferrous iron is the bacterial activity of *Acidithiobacillus ferrooxidans* [46]. Based on this, three evolutionary stages are proposed to explain the formation of these mining wastes (Figure 12).

Stage 1: Rozenite, szomolnokite, hexahydrite, and halotriquite (secondary sulfates), which were formed by the progress of oxidation and dehydration, were identified in both tailings [36]. The scarcity of melanterite and the presence of minerals that—according to the model of genetic evolution—are subsequent to this indicates that melanterite was probably formed, but it disappeared by continuous action of the same processes forming rozenite, which is the secondary hydrated sulfate that is more common in efflorescences of Xochula and Remedios tailings.

Stage 2: The second stage of minerals in the analyzed tailings are represented by the presence of copiapite and coquimbite (tertiary sulfate minerals), which are generated by the oxidation–dehydration of the minerals of the first stage. However, a mixture of minerals of both stages is observed in general; this may be the result of only a part of the minerals in the first stage having been altered, or that—within the same sample—these processes were acting simultaneously, leading to a stepped sequence of alteration.

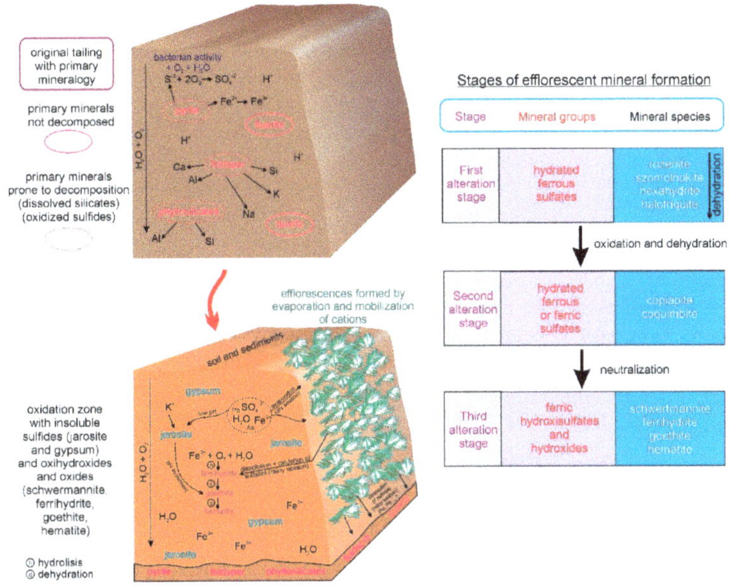

Figure 12. Stages of mineral (sulfates and oxides) precipitation formed mainly by oxidation, hydration, and evaporation.

Stage 3: The third stage of pyrite alteration in these tailings is represented by the presence of schwertmannite and ferrihydrite. Although they were identified in the studied sites, these minerals are found only in small amounts. This shortage is because they represent the last stage of alteration in which more stable minerals (goethite and hematite) are generated and there is still the generation of acidity in the deposits. The appearance of schwertmannite and ferrihydrite showcases the depletion of sources of acidity, according to [34].

According to [8], the efflorescences are a result of the consumption of the neutralization potential of the gangue minerals. The leached that determined the formation of efflorescent minerals is considered the final step in the evolution of the acid drainage system; at this point, it is too late for preventive actions, and only remediation tools can be applied (soil compactation, seal with clay, covers with sludge, dry covers, etc.) to avoid more affectations in the ecosystems [47]. For the Remedios and Xochula tailings, since they are small waste deposits, harvesting the efflorescences before the rainy season would reduce the risk of surface water contamination.

5. Conclusions

Xochula and Remedios tailings were both exposed to weathering for more than 50 years, developing a thick oxidation zone with jarosite as a predominant mineral and abundant efflorescent salts (secondary and tertiary soluble hydrated iron sulfates) in the borderline surface and lateral parts, where runoff water flows. As a result of this distribution, the removal of the efflorescences before the rainy season starts is a good alternative to avoid contamination of the water or air. Similarly to other efflorescences studied around the world, three paragenetic stages of the formation of efflorescent minerals were observed in both tailings. In the first stage, ferrous sulfates such as rozenite were formed; in the second one, ferrous and ferric sulfates (mainly copiapite and coquimbite) predominate; and in the final stage, iron hydroxides and oxides appear. The irregular distribution of non-oxidized pyrite in tailings as old as these can only be related to waste dumping in more recent times. The Remedios tailing is a younger and bigger deposit than Xochula, presenting greater mineralogical diversity but similar advanced mineralogical evolution.

The SWIR technique proved to be very useful to quickly discriminate the samples with soluble sulfate. However, this technique did not allow for the individual sulfate minerals to be discriminated from each other, since the samples are mineralogically complex mixtures of minerals with very similar SWIR patterns. Only for the most abundant minerals (>25% in weight of the sample) was it possible to conduct their discrimination by SWIR by comparison with the literature spectra. The X-ray diffraction technique using the Rietveld refinement method turned out to be an ideal tool to characterize and quantify all the individual phases. Furthermore, statistical tools applied to the quantitative XRD results allowed us to establish the genetic relationship between the different mineralogical groups.

The geochemical data showed that efflorescent salts present concentrations of potentially toxic elements (especially Zn, Cu, As and Pb) well above the values recommended by the World Health Organization and the country's legislation, demonstrating that the Xochula and Remedios tailings are still potentially harmful long after they were dumped. Tailings derived from ores similar to the ones in Taxco will not be mineralogically stabilized even after many decades of weathering, but their morphological exposure will progressively be confused with the landscape, making them a permanent source of contamination.

Supplementary Materials: The following are available online at http://www.mdpi.com/2075-163X/10/10/871/s1, Figure S1: Location of samples in Xochula (JH) and Remedios (JT) tailings., Table S1: Mineral phases identified and their relative abundance (% weight) calculated with the Rietveld refinement method.

Author Contributions: Conceptualization, methodology, software, formal analysis, investigation, resources, and writing T.P.-P.; software, validation, writing, review, and editing J.S.; methodology, formal analysis A.G.C. All authors have read and agreed to the published version of the manuscript.

Funding: This research was partially funded by CONACYT: Laboratorios Nacionales, No. 299087.

Acknowledgments: We especially thank Augusto Rodríguez for his support during the SWIR analysis measurements and Mireia Lilit Solé Pi by the English revision of the manuscript. We also acknowledge the invitation of Rosa Maria Prol Ledesma to participate in the field geological work.

Conflicts of Interest: The authors declare no conflict of interest.

References

1. Agricola, G. *De Re-Metallica; On Metallic Substances*; Hoover, H.C.; Hoover, L.H., Translators; Dover Publications: Mineola, NY, USA, 1556; p. 638.
2. Karpenko, V.; Norris, J.A. Vitriol in the history of chemistry. *Chem. Listy* **2002**, *96*, 997–1005.
3. Nordstrom, D.K.; Alpers, C.N. Negative pH, efflorescent mineralogy, and consequences for environmental restoration at the Iron Mountain Superfund site. California. *Proc. Natl. Acad. Sci. USA* **1999**, *96*, 3455–3462. [CrossRef] [PubMed]
4. Chicoș, M.M.; Damian, G.; Stumbea, D.; Buzgar, N.; Ungureanu, T.; Nica, V.; Iepure, G. Mineralogy and geochemistry of the tailings pond from Straja Valley (Suceava County, Romania). Factors affecting the mobility of the elements on the surface of the waste deposit. *Carpathian J. Earth Environ. Sci.* **2016**, *11*, 265–280.
5. Grover, B.P.C.; Johnson, R.H.; Billing, D.G.; Weiersbye, I.M.G.; Tutu, H. Mineralogy and geochemistry of efflorescent minerals on mine tailings and their potential impact on water chemistry. *Environ. Sci. Pollut. Res.* **2016**, *23*, 7338–7348. [CrossRef] [PubMed]
6. Hudson-Edwards, K.A.; Jamieson, H.E.; Lottermoser, B.G. Mine wastes: Past, present, future. *Elements* **2011**, *7*, 375–380. [CrossRef]
7. Hayes, S.M.; Root, R.A.; Perdrial, N.; Maier, R.M.; Chorover, J. Surficial weathering of iron sulfide mine tailings under semi-arid climate. *Geochim. Cosmochim. Acta* **2014**, *141*, 240–257. [CrossRef]
8. Dold, B. Evolution of Acid Mine Drainage Formation in Sulphidic Mine Tailings. *Minerals* **2014**, *4*, 621–641. [CrossRef]
9. Alpers, C.N.; Blowes, D.W.; Nordstrom, D.K.; Jambor, J.L. Secondary minerals and acid mine-water chemistry. In *Environmental Geochemistry of Sulfide Mine-Wastes (Mineralogical Association of Canada Short Course Volumes)*; Jambor, J.L., Blowes, D.W., Eds.; Mineralogical Association of Canada: Quebec, QC, Canada, 1994; Volume 22, pp. 247–270. [CrossRef]

10. Dold, B.; Fontboté, L. Element cycling and secondary mineralogy in porphyry copper tailings as a function of climate, primary mineralogy, and mineral processing. *J. Geochem. Explor.* **2001**, *74*, 3–55. [CrossRef]
11. Dold, B.; Fontboté, L. A mineralogical and geochemical study of element mobility in sulfide mine tailings of Fe oxide Cu-Au deposits from the Punta del Cobre belt, northern Chile. *Chem. Geol.* **2002**, *189*, 135–163. [CrossRef]
12. Buckby, T.; Black, S.; Coleman, M.L.; Hodson, M.E. Fe sulphate-rich evaporative mineral precipitates from Río Tinto, southwest Spain. *Mineral. Mag.* **2003**, *67*, 263–278. [CrossRef]
13. Hammarstrom, J.M.; Seal Ii, R.R.; Meier, A.L.; Kornfeld, J.M. Secondary sulfate minerals associated with acid drainage in the eastern US: Recycling of metals and acidity in surficial environments. *Chem. Geol.* **2005**, *215*, 407–431. [CrossRef]
14. Romero, A.; González, I.; Galán, E. The role of efflorescent sulfates in the storage of trace elements in stream waters polluted by acid mine-drainage: The case of Peña Del Hierro, southwestern Spain. *Can. Mineral.* **2006**, *44*, 1431–1446. [CrossRef]
15. Arranz-González, J.C.; Cala-Rivero, V.; Iribarren Campaña, I. Geochemistry and mineralogy of surface pyritic tailings impoundments at two mining sites of the Iberian Pyrite Belt (SW Spain). *Environ. Earth Sci.* **2011**, 1–13. [CrossRef]
16. Jambor, J.L.; Nordstrom, D.K.; Alpers, C.N. Metal-sulfate salts from sulfide mineral oxidation. In *Sulfate Minerals: Crystallography, Geochemistry, and Environmental Significance (Reviews in Mineralogy and Geochemistry)*; Alpers, C.N., Jambor, J.L., Nordstrom, D.K., Eds.; Mineralogical Society of America: Chantilly, VA, USA, 2000; Volume 40, pp. 303–350.
17. Nordstrom, D.K. Acid rock drainage and climate change. *J. Geochem. Explor.* **2009**, *100*, 97–104. [CrossRef]
18. Hyde, B.C.; King, P.L.; Dyar, M.D.; Spilde, M.N.; Ali, A.M.S. Methods to analyze metastable and microparticulate hydrated and hydrous iron sulfate minerals. *Am. Mineral.* **2011**, *96*, 1856–1869. [CrossRef]
19. Salas, G.P. *Economic Geology, Mexico*; Geological Society of America: Boulder, CO, USA, 1991; p. 453. ISBN 0-8137-5213-2.
20. Cambrubí, A.; Albinson, T. Epithermal deposits in México—Update of current knowledge, and an empirical reclassification. *Geol. Soc. Am. Spec. Pap.* **2007**, *422*, 1–40. [CrossRef]
21. Farfán-Panamá, J.L.; González-Partida, E.; Camprubí, A. Geología y mineralización del depósito epitermal polimetálico del distrito minero de Taxco, Guerrero, México, extended abstract. In Proceedings of the AIMMGM, XXX Convención Internacional de Minería, Acapulco, México, 16–19 October 2013.
22. IMSA. *Geology and Ore Deposits of the Taxco Mining District*; Inter. Report; IMSA: Guerrero, México, 1978; 40p.
23. Talavera-Mendoza, O.; Yta, M.; Moreno-Tovar, R.; Dótor-Almazán, A.; Flores-Mundo, N.; Duarte-Gutiérrez, C. Mineralogy and geochemistry of sulfide–bearing tailings from silver mines in the Taxco, Mexico area to evaluate their potential environmental impact. *Geofis. Int.* **2005**, *44*, 46–64.
24. López-Jiménez, R. Geological and Metallogenetic Features of the Manto Esperanza Vieja: Ag-Base-metal Epithermal Deposit of Taxco, Mexico. Master's Thesis, Universidad Complutense de Madrid, Madrid, España & Universidad Nacional Autónoma de México, Querétaro, México, 2008; 98p.
25. Barrera, M. Geoquímica de metales pesados en los jales del Fraile municipio de Taxco, Guerrero, México. Master's Thesis, Instituto de Geofísica, UNAM, Querétaro, México, February 2001; 59p.
26. Armienta, M.A.; Talavera, O.; Morton, O.; Barrera, M. Geochemistry of Metals from Mine Tailings in Taxco, Mexico. *Bull. Environ. Contam. Toxicol.* **2003**, *71*, 387–393. [CrossRef]
27. Romero, F.M. Procesos Geoquímicos Que Controlan la Movilidad de Metales y Metaloides en Jales de Sulfuros Metálicos "El Fraile", Taxco Guerrero. Ph.D. Thesis, Instituto de Geofísica, UNAM, Querétaro, México, March 2004; 155p.
28. Talavera-Mendoza, O.; Armienta-Hernández, M.A.; García-Abundis, J.; Flores-Mundo, N. Geochemistry of leachates from the El Fraile sulfide tailings piles in Taxco, Guerrero, southern Mexico. *Environ. Geochem. Health* **2006**, *28*, 243–255. [CrossRef]
29. Romero, F.M.; Armienta, M.A.; González-Hernández, G. The solid-phase control on the mobility of potentially toxic elements in an abandoned lead/zinc mine tailings impoundment, Taxco, Mexico. *Appl. Geochem.* **2007**, *22*, 109–127. [CrossRef]
30. Liewig, N.; Clauer, N.; Sommer, F. Rb-Sr and K-Ar dating of clay diagenesis in Jurassic sandstone oil reservoirs, North Sea. *Am. Assoc. Pet. Geol. Bull.* **1987**, *71*, 1467–1474.
31. Jackson, M.L. *Soil Chemical Analysis*; Prentice Hall: Englewood Cliffs, NJ, USA, 1958; p. 498.

32. Moore, D.M.; Reynolds, R.C., Jr. *X-ray Diffraction and the Identification and Analysis of Clay Minerals*, 2nd ed.; Oxford University Press: Oxford, UK, 1997; p. 378. [CrossRef]
33. Chukanov, N.V.; Chervonnyi, A.D. *Infrared Spectroscopy of Minerals and Related Compounds*; Springer: Berlin, Germany, 2016; p. 1116. ISBN 978-3-319-25347-3. [CrossRef]
34. Velasco, F.; Alvaro, A.; Suarez, S.; Herrero, J.M.; Yusta, I. Mapping Fe-bearing hydrated sulphate minerals with short wave infrared (SWIR) spectral analysis at San Miguel mine environment, Iberian Pyrite Belt (SW Spain). *J. Geochem. Explor.* **2005**, *87*, 45–72. [CrossRef]
35. Cloutis, E.A.; Hawthorne, F.C.; Mertzman, S.A.; Krenn, K.; Craig, M.A.; Marcino, D.; Methot, M.; Strong, J.; Mustard, J.F.; Blaney, D.L.; et al. Detection and discrimination of sulfate minerals using reflectance spectroscopy. *Icarus* **2006**, *184*, 121–157. [CrossRef]
36. Nordstrom, D.K. Aqueous pyrite oxidation and the consequent formation of secondary iron minerals. In *Acid Sulfate Weathering*; Kittrick, J.A., Fanning, D.S., Hossner, L.R., Eds.; Soil Science Society of America: Madison, WI, USA, 1982; Volume 10, pp. 37–56.
37. Rietveld, H. A profile refinement method for nuclear and magnetic structures. *J. Appl. Crystallogr.* **1969**, *2*, 65–71. [CrossRef]
38. Hawthorne, F.C.; Krivovichev, S.V.; Burns, P.C. The crystal chemistry of sulfate minerals. In *Reviews in Mineralogy and Geochemistry*; Sulfate Minerals: Crystallography, Geochemistry, and Environmental Significance; Alpers, C.N., Jambor, J.L., Nordstrom, D.K., Eds.; Mineralogical Society of America: Chantilly, VA, USA, 2000; Volume 40, pp. 1–101. [CrossRef]
39. Del Rio-Salas, R.; Ayala-Ramírez, Y.; Loredo-Portales, R.; Romero, F.; Molina-Freaner, F.; Minjarez-Osorio, C.; Pi-Puig, T.; Ochoa–Landín, L.; Moreno-Rodríguez, V. Mineralogy and Geochemistry of Rural Road Dust and Nearby Mine Tailings: A Case of Ignored Pollution Hazard from an Abandoned Mining Site in Semi-arid Zone. *Nat. Resour. Res.* **2019**. [CrossRef]
40. Meza-Figueroa, D.; Maier, R.M.; de la O-Villanueva, M.; Gómez-Alvarez, A.; Moreno-Zazueta, A.; Rivera, J.; Campillo, A.; Grandlic, C.H.J.; Anaya, R.; Palafox-Reyes, J. The impact of unconfined mine tailings in residential areas from a mining town in a semi-arid environment: Nacozari, Sonora, Mexico. *Chemosphere* **2009**, *77*, 140–147. [CrossRef]
41. WHO (World Health Organization). Global Health Observatory Data Repository: International Health Regulations (2005) Monitoring Framework. Available online: https://apps.who.int/gho/data/node.main-eu.IHR?lang=en (accessed on 17 June 2020).
42. NORMA Oficial Mexicana, NOM. 141-SEMARNAT-2003. In *Que Establece El Procedimiento Para Caracterizar Los Jales, así Como Las Especificaciones Y Criterios Para La Caracterización Y Preparación Del Sitio, Proyecto, Construcción, Operación Y Postoperación de Presas de Jales, Diario Oficial de La Federación*; Estados Unidos Mexicanos-Secretaria de Medio Ambiente y Recursos Naturales: Mexico City, Mexico, 2004.
43. Jamieson, H.E.; Robinson, C.; Alpers, C.N.; McCleskey, R.B.; Nordstrom, D.K.; Peterson, R.C. Major and trace element composition of copiapite-group minerals and coexisting water from the Richmond mine, Iron Mountain, California. *Chem. Geol.* **2005**, *215*, 387–405. [CrossRef]
44. Escobar-Quiroz, I.N.; Villalobos-Peñalosa, M.; Pi-Puig, T.; Martín Romero, F.; Aguilar-Carrillo de Albornoz, J. Identification of jarosite and other major mineral Fe phases in acidic environments affected by mining-metallurgy using X-ray Absorption Spectroscopy: With special emphasis on the August 2014 Cananea acid spill. *Rev. Mex. Cienc. Geol.* **2019**, *36*, 229–241. [CrossRef]
45. Jamieson, H.E.; Walker, S.R.; Parsons, M.B. Mineralogical characterization of mine waste. *Appl. Geochem.* **2015**, *57*, 85–105. [CrossRef]
46. Shahroz Khan, S.; Haq, F.; Hasan, F.; Saeed, K.; Ullah, R. Growth and Biochemical Activities of Acidithiobacillus thiooxidans Collected from Black Shale. *J. Microbiol. Res.* **2012**, *2*, 78–83. [CrossRef]
47. Pozo-Antonio, S. Techniques to correct and prevent acid mining drainage. *DYNA (Medellín, Colombia)* **2014**, *81*, 73–80. [CrossRef]

minerals

MDPI

Article

Mineral Formation under the Influence of Mine Waters (The Kizel Coal Basin, Russia)

Elena Menshikova, Boris Osovetsky, Sergey Blinov and Pavel Belkin *

Department of Geology, Perm State University, Bukireva st. 15, Perm 614990, Russia; menshikova_e@list.ru (E.M.); opal@psu.ru (B.O.); blinov_s@mail.ru (S.B.)
* Correspondence: pashabelkin@mail.ru

Received: 21 February 2020; Accepted: 15 April 2020; Published: 17 April 2020

Abstract: The development of coal deposits is accompanied by negative environmental changes. In the territory of the Kizel coal basin (Perm Region, Russia), the problem of contamination of water sources by acid mine waters and runoff from rock dumps is particularly acute. Mine waters are acidic (pH 2–3), with high mineralization (up to 25 g/L) and significant content of sulfate ions, iron, aluminum, manganese, toxic trace elements (As, Co, Ni, Pb and Zn). They are formed as a result of the interaction of underground waters from flooded mines of the Kizel basin with coal and rocks of dumps with high sulfur content (15%). Uncontrolled inflow of mine water into rivers (about 22 million m^3 annually) leads to significant amounts of iron and aluminum hydroxide precipitation. These precipitations are in active interaction with river water, polluting the rivers tens of kilometers downstream and are entering the Kama reservoir. Studies of alluvial precipitation can be considered as a method of control and predictors of technogenic water pollution. The mineral composition of river sediments was studied with the application of different methods, including studies of sand-gravel and silty-clayey sediments. The sandy-gravel grains in the bottom load are mainly composed by natural minerals and are represented by a significant number of particles of coal dumps, slags and magnetic spherules. The silty-clayey material, mixed with natural minerals, contains a significant number of amorphous phases with a predominance of iron-rich substances, which may actively concentrate toxic elements. The presence of jarosite, goethite, basaluminite, lepidorocite and copiapite in silty-clayey sediments are indicators of the influence of mine waters.

Keywords: AMD-precipitates; AMD-minerals; mine water; waste rock dumps; river systems; sedimentation; migration; Kizel coal basin; the Western Urals

1. Introduction

Modern alluvial sediments as one of the components within the upper part of the lithosphere are a traditional object of study of geological sciences, and an important element of the environment. Alluvial sediments integrate geochemical and mineral features of catchment rocks, and rivers are a link between areas of sediment mobilization in catchments and large sediment accumulation basins [1]. The alluvium composition makes it possible to identify the migration flows of matter, the zone of anthropogenic influence, and thus is used as an indicator of anthropogenic impact on water bodies. The authors have conducted studies of the impact of coal mining on the current state of the river ecosystems in Kizel coal basin (Perm Region). In this work the mineralogy of the river deposits in a polluted region is considered.

Coal mining, which has been carried out in the Kizel basin for over 200 years, has now been stopped. The cessation of mining activities and the inundation of mines did not lead to the solution of environmental problems. The main factors of the territory contamination are the self-diffusion of acidic mine waters and the runoff of rock dumps and industrial sites [2–4]. Mine waters flow into 19 rivers of the territory. In modern conditions, some rivers are polluted even more intensely and over a longer period of time than during the operation of coal mines. The scale of this process is demonstrated by the analysis of multiyear series of space images of this territory [5]. The entry of spontaneous mine water into rivers resulted in formation of considerable volumes of technogenic sediments.

The study area is located within the Western Urals folding zone adjacent to the Pre-Urals foredeep, and is composed of Paleozoic rocks (from the Middle Devonian to Lower Perm inclusive) and covering them with Quaternary deposits. The carbonaceous sequence (the Visean stage of Lower Carboniferous) is represented by sandstones, siltstones, argillites and clay shales with limestone interlayers. The rocks of the carbonaceous sequence include finely dispersed pyrite and organic sulfur. The content of sulfide and organic sulfur reaches 12%–15%. Sulfur is one of the main active elements of the coal-bearing rocks, determining acid-alkaline and redox conditions of sediment accumulation, concentration of elements in the reducing geochemical environment (with the presence of H_2S), their leaching in the oxidation zone with the formation of products affecting the ecological condition of the territory.

The purpose of the present research is an estimation of the migration and sedimentary processes of formation of modern alluvial deposits in the territory of the Kizel coal basin with consideration of applied ecological aspects. The study of the mineral occurrence forms of the substances in river sediments for areas affected by mine water discharge is of great importance, as supergenic formation of secondary minerals is the most important natural process for removing a number of elements from such solutions. These minerals represent a dynamic system in which mineralogical and chemical transformations can be caused by sudden changes in chemical and physical conditions [6–8]. Evolution of acidic sulphate waters results in a number of rather insoluble hydroxysulfate and oxyhydroxide Fe and Al minerals, which are precipitated during oxidation, hydrolysis and neutralization. These minerals, as a rule, have low crystallinity, coexist with intermediate and high crystalline phases, and are difficult to describe [9,10]. Secondary natural iron minerals have a higher retention capacity than their synthetic counterparts [11].

Although the study of secondary minerals in mine water is the subject of several studies, the role of these precipitates, their properties and associations remain a relevant topic of research, namely for the still unknown effects they may have on the ecosystems. Therefore, the present work contributes with novel results about the processes of natural and anthropogenic mineral formation in an abandoned coal basin. This research shows the specificity of mineral associations in bottom sediments and reflects the scales of display of processes of transformation in modern alluvium.

2. Methods

The study areas are located at the Vilva, Usva (the Chusovaya River basin), Kizel, Vilva (the Yaiva River basin), and Kosva Rivers and their tributaries, into which the mine water is discharged or has ever been discharged. The sampling scheme takes into account background areas that were not affected by anthropogenic factors. Suspended load material (aleurite clayey sediments) and bottom load material (sandy and sandy-gravel) were sampled. A total of 72 samples were taken at 36 sites. The locations of some sampling point are shown in the scheme (Figure 1).

The mineral composition of sandy and sandy-gravel sediments was characterized on the basis of optical analysis of 1000 grains in 0.25–0.1 mm fraction (Nikon SMZ 745, Nikon, Tokyo, Japan) as the most representative in terms of mineral diversity, and the composition of silty-clay sediments was analyzed by X-ray diffraction analysis (D2 Phaser, Bruker, Billerica, MA, USA).

The movable forms of pollutants have been estimated on the basis of general chemical analysis of water extraction from alluvial deposits (ratio of 100 g of ground per 500 mL of water, method of analysis—titrometry). The sum of water-soluble salts is the sum of all ions found by chemical analysis in the bottom sediments (in mg/kg of ground).

Figure 1. Scheme of sampling places.

X-ray fluorescent (S8 Tiger, Bruker) and inductively coupled plasma mass spectrometry (ICP-MS) (Aurora M90, Bruker) analyses were used to determine the chemical composition of individual alluvial sediments. The particle size distribution of sediments was determined by the sieve method and by using a particle size analyzer (Analysette 22 Micro Tec plus, Fritsch). A scanning electron microscope (JSM 6390 LV, Jeol, Tokyo, Japan) with microprobe analysis (EDS INCA 350 Energy, Oxford Instruments, Abingdon, UK) was used for study of certain types of technogenic products in the bottom sediments.

3. Results and Discussion

According to the conducted researches the mineral composition of sand alluvial deposits of the investigated rivers is rather diverse. In uncontaminated sediments on a background of dominant quartz there are feldspars, talc, carbonates (the content is 3%–11%), chlorite, amphibole, epidote, pyroxenes, chromite, magnetite, hematite, leucoxene, hydrogoethite, ilmenite (as a rule, in the quantity of 0.1%–2%) (Table 1). In quantities less than 0.1% there are garnets, actinolite, rutile. The mineral composition of silty-clayey sediments is presented by quartz, plagioclases and clay minerals (kaolinite, illite, hydromica, smectite and chlorite) (Table 2).

Table 1. Mineral composition of sandy and sandy-gravel sediments in the rivers.

Place of Sampling (Number in Figure 1)	Character of Influence	Natural Mineral Association *, %	Other Natural Minerals **, %	Technogenic Particles, %
		Basin of the Yaiva River		
The Kizel River, mouth (1)	Rock dumps on the catchment area and coastal zone, mine drainage	quartz (56), feldspars (13)	iron hydroxides *** (3.5), pyroxenes (1.3), magnetite (1.1), epidote (0.3)	coal-rock particles (22), slag (2.7), magnetic spherules (0.3)
		Basin of the Kosva River		
The Kosva River, upstream (2)	Background area	quartz (69), carbonates (11)	chlorite (2), amphibole (1.8), epidote (1.4), pyroxenes (0.9), chromite (0.9), magnetite (0.5), hematite (0.5), iron hydroxides (0.2) and others	coal-rock particles (2)
The Kosva River, 40 km downstream (3)	Rock dumps on the catchment area and coastal zone, mine drainage	quartz (58)	carbonates (2.6), talc (2.1), feldspar (1.9), chlorite (1.9); amphibole (1.5), epidote (1.2), pyroxenes (0.6), iron hydroxides (0.4), magnetite (0.4), garnets (0.3), chromite (0.3) and others	coal-rock particles (19.8), slag (6.5), magnetic spherules (1.1)
		Basin of the Chusovaya River		
The Usva River (4)	Background area	quartz (85), talc (4)	feldspars (1.9), chlorite (1.9), amphibole (0.8), epidote(0.6), pyroxenes (0.3), iron hydroxides (0.1), magnetite (0.1), hematite(0.1) and others	_ ****
The Usva River, below the mouth Rudyanka River (5)	Supply of contaminated river water	quartz (68), feldspars (4), talc (4)	chlorite (2.8), epidote (1.3), amphibole (1.0), magnetite (1.0), iron hydroxides (0.7), pyroxenes (0.5), actinolite (0.3), hematite (0.2), ilmenite (0.2), garnets (0.1), rutile (0.1) and others	coal-rock particles (0.5), slag (0.2), magnetic spherules (0.2)
The Vilva River (6)	Background area	quartz (68), talc (9)	feldspars (2.9), pyroxenes (1.2), epidote (1.0), magnetite (0.7), amphibole (0.6), hematite (0.5), iron hydroxides (0.2), ilmenite (0.2), chromite (0.1)	_ ****
The Vilva River, downstream (7)	Supply of contaminated river water	quartz (82), talc (4)	feldspars (2.0), chlorite (2.0), epidote (0.5), amphiboles (0.5), pyroxenes (0.2), iron hydroxides (0.2), actinolite (0.1), magnetite (0.1) and others	coal-rock particles (0.1)
The Vilva River, mouth (8)		quartz (75), epidote (4)	talc (3.5), chlorite (1.8), amphibole (1.4), pyroxenes (0.5), iron hydroxides (0.5), actinolite (0.4), magnetite (0.4), ilmenite (0.3), haematite (0.1), chromite (0.1) and others	magnetic spherules (0.3)

* The natural association of minerals includes minerals with the content of more than 4%; ** except for the specified minerals in the majority of samples where there were found fragments of rocks of natural origin with the content of 0–16%; *** Iron hydroxides in the mouth of the Kizel River were mainly of anthropogenic origin; **** components are not detected by analysis.

In the river sections located in the zone of influence of the coal deposit development and downstream, there is a change in the mineral composition of bottom sediments, which is clearly recorded not only by the results of laboratory mineralogical studies, but also directly in the field conditions. In particular, detritus material in river sediments, which are currently receptors of mine spills, is covered with a red layer of iron hydroxides. Iron hydroxides are also present in the form of individual particles, among which lamellar individuals predominate. They are often formed by replacing plant residues and often preserve the fibrous or layered structure of plants. In addition, crusts, deposits and aggregates are also present. They come in various sizes, including finely dispersed aggregates. Most noticeable is their presence in the class of 1–0.5 mm particles. Lemon-yellow differences prevail in color, but bright red particles are also observed in some places. Under the electron microscope, the inhomogeneity of the internal structure of these glandular products is detected, which is manifested by the presence of a highly porous structure and a cellular surface of the particles. Among them there are sometimes fragments with sharp edges, which testify to their extreme fragility. Probably when transported in a water stream, such particles are easily crushed, frayed and partially suspended.

In the silty-clayey sediments composition, along with natural mineral grains, in significant quantities there are components due to the specifics of man-made impact-coal-bearing particles (up to 22%), slags (up to 6.5%) and magnetic spherules (up to 1.1%) [12]. The main source of such technogenic products is coal mine dumps.

Table 2. Mineral composition of silty-clayey sediments of the rivers.

Place of Sampling (Number in Figure 1)	Character of Influence	Natural Mineral Association *, %	Secondary Minerals, %
Basin of the Yaiva River			
The North Kizel River (9)	Background area	quartz (52), plagioclases (23), clay minerals (9) **	– ***
The Sukhoi Kizel River, mouth (10)	Mine drainage in the past and wastewaters from rock dumps	quartz (36), plagioclases (10), potassium feldspars (4)	goethite (7)
The Viejasher River, mouth (11)		quartz (25), clay minerals (12), plagioclases (7)	jarosite (3), goethite (2)
The East Kizel River, mouth (12)	Mine drainage	quartz (13), clay minerals (4)	goethite (15)
The Poludenyi Kizel River, mouth (13)		quartz (15), clay minerals (3)	lepidocrocite (3), goethite (2)
The Kizel River, mouth (14)		quartz (5)	jarosite (4)
The North Vilva River, 500 m below the mouth of the Kizel River (15)	Contaminated river water runoff	quartz (15), clay minerals (4)	jarosite (11), goethite (9), lepidocrocite (4), diaspore (2)
Basin of the Kosva River			
The Ladeynaya Log River (16)	Background area	quartz (68), plagioclases (17), clay minerals (7)	goethite (3)
The Ladeynaya Log River, mouth (17)	Mine drainage	quartz (30), plagioclases (12), clay minerals (7)	goethite (3)
The Shumikha River, mouth (18)	Runoff from waste dumps, impact of sludge collector	quartz (27), clay minerals (21)	basaluminite (8), jarosite (1), goethite (1)
The Gubashka River, mouth (19)	Mine drainage in the past and wastewaters from rock dumps	quartz (36), clay minerals (9), plagioclases (9)	goethite (2)
The Kamenka River, mouth (20)		quartz (40), clay minerals (13)	goethite (1)
The Kosva River, 40 km downstream of the mine site (21)	Mine drainage and wastewaters from rock dumps	quartz (20), clay minerals (8), plagioclases (4)	goethite (<1)
The Kosva River, 70 km downstream of the mine site (22)		quartz (10)	– ***
Basin of the Chusovaya River			
The Rudyanka River (23)	Background area	quartz (77), clay minerals (9), plagioclases (4)	goethite (1)
The Rudyanka River, mouth (24)	Mine drainage	quartz (27), clay minerals (11)	goethite (1)
The Usva River (25)	Background area	quartz (66), clay minerals (7), plagioclases (5)	goethite (1)
The Usva River, below the mouth of the River Rudyanka (26)	Mine drainage, runoffs from rock dumps, contaminated river water runoff	quartz (15), clay minerals (12)	goethite (1.5), jarosite (1)
The Bolshaya Gremyachaya River (27)	Mine drainage	quartz (34)	jarosite (45)
The Bolshaya Gremyachaya River, 5 km below discharge mine water (28)		quartz (15), clay minerals (6)	copiapite (2)
The Vilva River (29)	Background area	quartz (70), plagioclases (18), clay minerals (10)	goethite (<1)
The Vilva River, mouth (30)	Supply of contaminated river water	quartz (4)	lepidocrocite (2)

* The natural mineral association includes minerals whose content is at least 4%; ** Under the term "clay minerals" are combined (hereinafter referred to as to the increase of the content in silty-clayey sediments of the rivers of the investigated territory): kaolinite, hydromica, smectite and chlorite; *** components are not detected by analysis.

Among the objects studied the highest content of technogenic material is found in alluvial deposits of the Kosva River, the whole right bank of which on territory of the town of Gubakha is covered with rocks of mine dumps. Downstream of the river silty-clayey sediments are painted black due to the high content of coal-bearing and slag particles. There are nearshore shoals, more than 90% of which are formed by anthropogenic material of sandy size.

Carbonaceous particles have a high migration capacity. They migrate as a part of the bottom, and particularly as suspended load, as well as by air at considerable distances. The territory of the Kizel basin is characterized by pyrite-containing carbonaceous particles. Although the latter are noticeably inferior to pure coal in terms of migration capacity, they are also actively transported by the Kosva River system. Calculations have shown that their migration capacity approximately corresponds to that of quartz particles [13]. Morphologically carbonaceous particles with pyrite are noticeably different from pure coal. They are characterized by a variety of forms (flattened, angular, elongated, etc.), relatively weak roundness, uneven surface, on which numerous pyrite inclusions can be seen (Figure 2). Particles of pyrite in coal are very diverse. Among them there are crystals up to 0.2 mm in size in the form of octahedron and cube, which often form tightly fused aggregates. According to microprobe analysis pyrite from coal particles contains a variable complex of toxic elements (As, Co, Ni, Cu, Zn, Sb, and Se). Arsenic, the content of which is determined with the greatest accuracy using a wave spectrometer, is noted in each pyrite-containing sample. Its content varies from 0.045% to 0.51 %, but values from 0.15% to 0.25 % prevail. There is no doubt that a considerable portion of toxic elements in the process of pyrite decomposition passes into the water medium [14]. At the same time, the forms of their occurrence can be different: colloidal formations (micelles), suspended iron hydroxide, arsenic acid, etc.

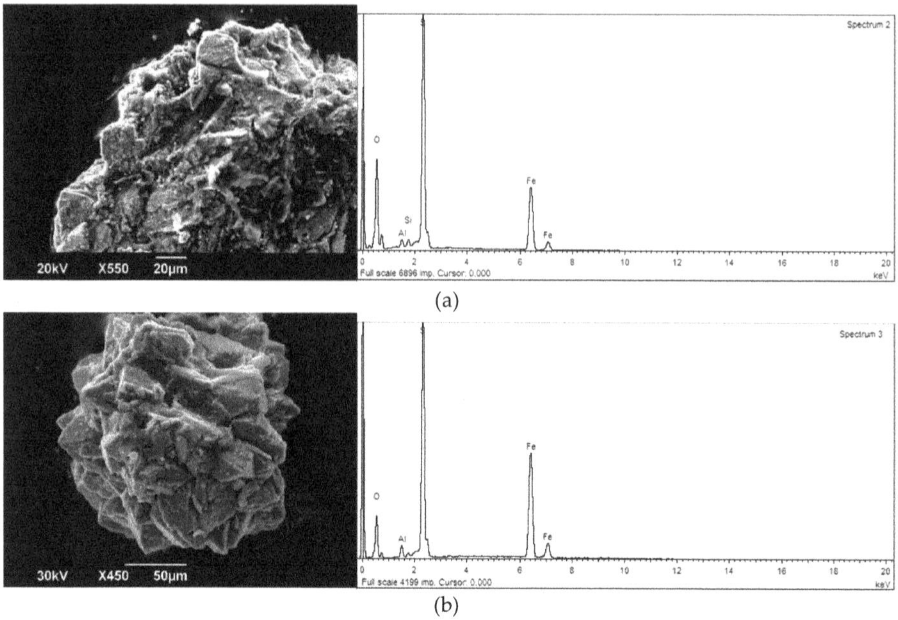

Figure 2. Clayey particles with pyrite from the Kosva River alluvium: (**a**) a fragment of a coal particle with pyrite crystals on the surface, (**b**) an aggregate of cubic and octahedral pyrite crystals isolated from the coal particles.

Magnetic spherules in alluvial sediments of the Kizel coal basin (Figure 3), according to the microprobe analysis, belong to iron (magnetite) and silicate-iron (modified magnetite aggregates) groups [15]. Iron spherules are characterized by correct spherical shape, strong semi-metallic luster, black color, smooth surface and small size (usually less than 100 μm). They have the highest magnetic susceptibility values. The share of iron oxides in their chemical composition exceeds 80%. Characteristic components reflecting the degree of change in the chemical composition of spherules are SiO_2, Al_2O_3, CaO and K_2O. The spherules, which are more affected by the changes (presence of caverns, cavities and cracks filled with secondary substances), are characterized by lower iron oxide content and higher silica and alumina content.

Silicate-iron spherules are grey or brownish in color, have a matte luster, rough surface and larger sizes. They have a lower magnetic susceptibility compared to the previous group. In their chemical composition the content of iron oxides decreases to 78%–55%, respectively of the content of Si, Al, Ti, Ca and Mg; while sometimes Na, K and Mn increases. Typical indicators of the impact of environmental chemical agents on spherules are increased contents of phosphorus and sulfur compounds in their composition. The maximum concentration of spherules is found in sediments of the Kosva River at a distance of 22 km below of the town of Gubakha, where the coke production is located, which is considered as the main source of their supply.

In the overwhelming majority of silty-clayey material of the investigated rivers, the zones of anthropogenic influence have a bright orange and rusty color. It is deposited on shallows, and covers coarse-grained deposits. X-ray amorphous iron phases may account for a significant proportion of its composition, which demonstrates a high percentage of the revealed crystalline phases for the background areas (assuming a total of 80%–90%) and their significantly lower content for river areas in the zone of influence of mine water and waste water from rock dumps. It is this finely dispersed phase of sediment—sorbing the toxic elements As, Co, Ni, Pb and Zn, on the one hand—that provides the processes of self-purification of water, on the other hand, and is a source of secondary pollution of water at the sites of active accumulation, which is confirmed for a number of other mining areas [8,16,17]. The study of the granulometry of such sediments showed a one-modal distribution of these particles by size with a distinct mode, usually in a narrow class of 16–8 μm.

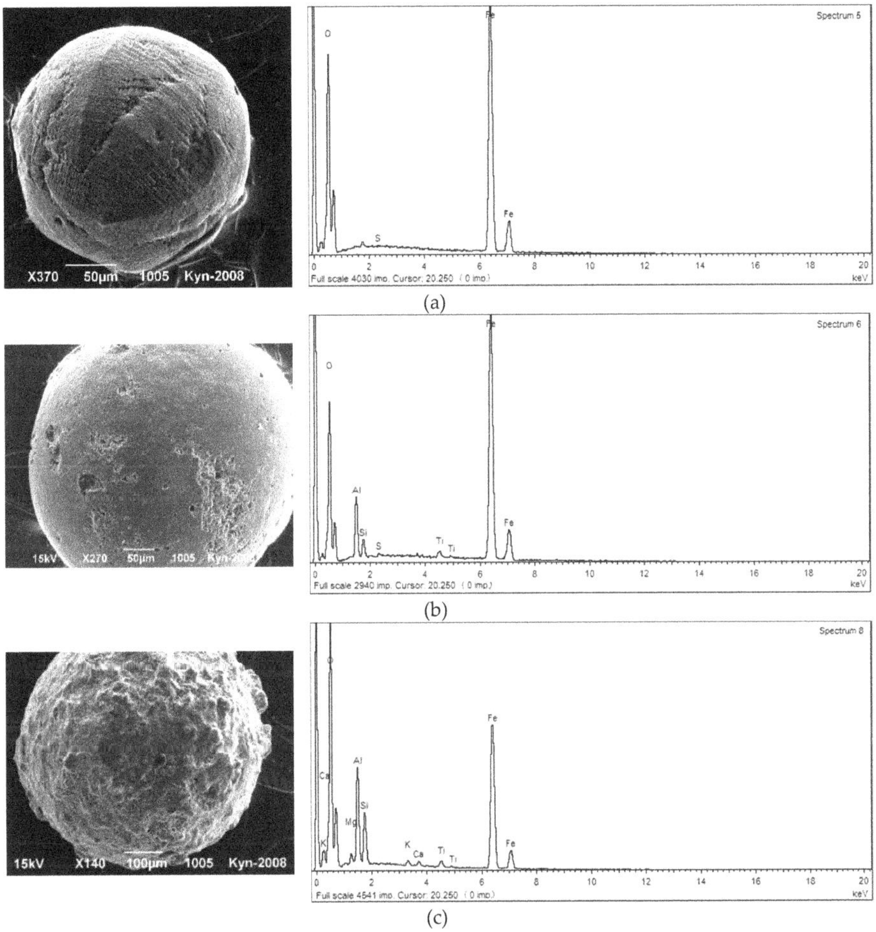

(a) (b) (c)

Figure 3. Morphology of typical magnetic spherules: (**a**) tiled microsculpture of the surface with a distinct separation of octants (the Kosva River), (**b**) initial stage of spherula change with separate corroded areas of the surface (the Kizel River), (**c**) stage of a strong spherula change (the Kizel River).

Secondary minerals (%) are present in the crystallized part of silty-clayey sediments: goethite—up to 15%, jarosite—up to 45%, basaluminite—up to 8%, lepidocrocite—up to 4%, diaspore and copiapite—up to 2%.

An example of the maximum manifestation of secondary mineral formation processes is the area of mixing neutral waters of the Vilva River with waters of the Kizel River saturated with suspended, colloidal and dissolved substances. Here in silty-clayey sediments are found (%): jarosite (up to 11%), goethite (up to 9%), lepidocrocite (up to 4%), diaspore (up to 2%), which are absent in the upstream of the Vilva River. These mineral phases, formed by sulfide oxidation processes, are among the most common in other mining areas as well [6,18–21]. Goethite is the most stable in hypergenic conditions in this group (pH range 2.5–7) [22], and therefore it is widely distributed in the group of secondary river sediment minerals for this area.

According to the analysis of water extraction silty-clayey sediments in natural conditions have close to neutral reaction of medium (pH = 6.3–7.6) and low content of water-soluble substances—200–600 mg/kg (Table 3). The extractor hood has a hydrogen carbonate-calcium composition and low concentrations of pollutants, which are typical for mine waters.

Table 3. Chemical composition of water extraction from bottom sediments in the Kizel coal basin rivers, mg/kg ground.

Components	Background Areas (n = 9)	Plots of Modern Supply of Mine Water, Distances from the Spillage			Rivers That Were Previously Discharged by Mine Water (n = 8)
		Up to 1 km (n = 7)	1 to 10 km (n = 6)	Several Tens of Kilometers (Downstream Areas) (n = 6)	
pH	6.3–7.6	3.0–3.6	5.4–6.7	6.1–7.3	6.0–7.3
HCO_3^-	76.3–369.2	n.a.	125.1–750.5	756.6–1882.4	88.5–286.8
SO_4^{2-}	57.7–257.0	2367.9–16702.7	790.1–4135.5	963.0–4404.4	64.9–1491.4
Cl^-	8.9–31.9	143.6–443.1	93.9–544.2	83.3–533.6	10.7–58.5
Ca^{2+}	13.0–133.3	164.4–796.6	82.2–1109.2	654.3–1762.5	88.2–503.0
Mg^{2+}	4.3–27.4	60.2–724.3	12.8–148.3	75.4–198.1	16.4–132.5
Na^++K^+	12.7–62.1	26.5–790.9	20.7–152.9	85.1–403.5	6.9–150.6
Fe	n.a.–6.6	688.3–5272.1	93.8–810.2	5.7–74.1	n.a.–15.0
Al^{3+}	n.a.–0.6	29.4–152.7	0.6–8.5	n.a.–0.6	n.a.–0.8
Sum of water-soluble salts	198.5–606.1	3711.8–24768.2	2048.9–7407.8	3958.3–9333.4	493.6–2528.8

n—number of samples, n.a.—component was not detected by the analysis, CO_3^{2-}—component was not detected by the analysis.

Rivers below the inflow of mine spills can be divided into three zones by the intensity of change in the chemical composition of alluvial deposits (Figure 4). The largest change is observed in the areas up to 1 km below the inflow of mine waters. The content of water-soluble substances here increases by 20–40 times in comparison with the initial one. The extractor hood has an acidic reaction of medium (pH = 3–3.6) and sulfate-iron composition. For some rivers the concentration of mobile forms of iron is 800, aluminum 25, sulfates 65 times higher than the background values. Absence of hydrogen carbonate ion in the water extraction of bottom sediments at river sections adjacent to mine spills is explained by low pH values. From amorphous iron hydroxides in river sediments here occurs the formation of goethite and jarosite (in favorable conditions for crystallization the content of jarosite may be 45%).

Downstream, at a distance of 1 to 10 km from mine spills, the content of water-soluble substances is reduced by 2–3.5 times. The content of water-soluble substances is changed to sulfate-iron-calcium and sulfate-calcium within pH 5.4–6.7. The content of sulfates decreases by 3–4 times, iron—by 6.5–7 times and aluminum—by 18–49 times. In these areas, the influence of mine waters is associated with the formation of secondary minerals such as goethite, lepidocrocite, copiapite in river sediments, and with the additional influence of runoff from mine dumps in the sediments are present basaluminite and diaspore.

In the downstream sections of the rivers located several tens of kilometers downstream of the sources of pollution, the composition of the sulfate-calcium extract is comparable to that of the background. However, the content of water-soluble substances in bottom sediments increases by a factor of 1.3–1.9 compared to the previous zone, which is due to an increase in the concentrations of hydrocarbonate and calcium ions. This phenomenon may be due to the fact that the studied rivers in their middle course cut through the carbonate rocks. In addition, the presence of significant amounts of sulfate ions in river water increases the solubility of calcium carbonate. Experimental studies [23] have shown that with pH increasing the desorption processes of sulfates from sediments increase, which naturally increases their content in water. Active transition to aqueous media of SO_4^{2-} for these sites can also be associated with the evolution of crystal mineral phases of ferrous sediments, which is described in the following studies [11,24]. For this area, changes in alluvial sediments are also due to organic matter and finely dispersed coal particles, a significant amount of which is typical for sediments from the Kizel basin rivers. The growth of the total amount of water-soluble substances is rather intensive, but the iron content in the water extraction decreases by 11–16 times and still more exceeds the background in order. The content of moving aluminum forms is reduced to background values. For these river sections, secondary river sediment minerals are represented by goethite, but most iron migrates and accumulates in an amorphous form. Experimental data [6] have confirmed that it is amorphous Fe compounds that can provide deposition of toxic metals from river waters. When physicochemical conditions change, they due to their high reactivity can be a source of toxic metals in river waters. According to the conducted researches [25,26] modern receipts of iron compounds from catchments of the Yaiva and Kosva Rivers in the Kama reservoir exceed volumes of discharge during active mining.

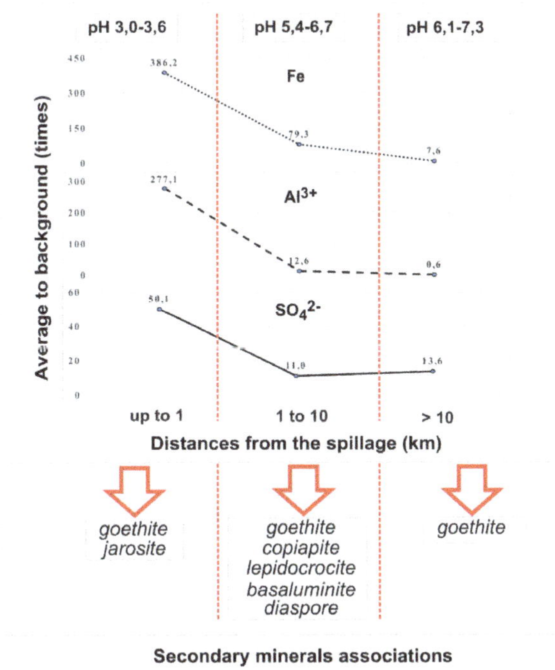

Figure 4. Variation in the content of mobile forms of basic mineral-forming ions in river sediments of the Kizel coal basin.

A study of bottom sediments in the rivers, in which discharging of mine water was stopped after the liquidation of the mines, showed that the pH values were close to the background (Table 3). The content of water-soluble substances is 2.5–4 times higher than the background values due to increased concentrations in the extraction of sulfate and calcium ions. The content of mobile forms of iron and aluminum slightly exceeds the background values. This demonstrates the deposition role of river sediments in relation to SO_4^{2-}, which have good solubility and can easily transfer to river waters.

4. Conclusions

The conducted researches have shown that in zones of intensive technogenic influence modern alluvial deposits are the complexes formed under the influence of both natural and technogenic factors. The study of their composition and occurrence forms of chemical elements allows us to identify the migration flows of matter and areas with different intensity of man-made impact on water bodies.

The river sediments up to 1 km downstream from mine spills are the most polluted. They are characterized by an acidic reaction of the environment and the highest contents of mobile forms of sulfate, iron and aluminum. However, even in the upper parts of the rivers located several tens of kilometers below the self-discharges, the sulfate and iron content is one order of magnitude higher than the background values. The regularity of changes in the chemical composition of water extraction from bottom sediments has been established: from hydrogen carbonate–calcium in the background areas to sulfate–calcium–iron in the areas adjacent to spills, and sulfate–calcium in the near-surface areas.

The anthropogenic sediments generated by the mixing of mine and river water migrate downstream in the form of suspended load. Washing out occurs particularly intensively during the flood period. Bottom sediments are the most polluted in the areas of rivers below the spillway: the Vilva River—below the town of Gremyachinsk to the mouth (further polluted by the Usva and Chusovaya Rivers), the Kosva River—below the town of Gubakha to the mouth, and the Kizel and Vilva Rivers—below the town of Kizel to the mouth (further polluted by the Yayva River). Under current conditions, the impact of development has been observed in longer sections of rivers than in the period of active coal field development. Accumulated in the backwater zone of the Kama reservoir (downstream sections of the studied rivers), technogenic sediments are a long-term source of secondary pollution of river water. The processes observed actually worsen water quality in the Kama reservoir and the Chusovaya River, which is observed at the water intakes of a number of settlements. Contamination of river water and bottom sediments has a negative impact on the existence and development of aquatic plants and animals.

Further study of mineral forms of occurrence for elements, especially toxic ones, can become the basis for predicting their behavior during migration, on geochemical barriers, in the process of diagenetic transformation. Special attention should be paid to the assessment of emerging sediments as a possible source of secondary water pollution.

Author Contributions: Conceptualization, E.M.; methodology, B.O. and S.B.; investigation, E.M., B.O. and S.B.; writing—review and editing, B.O. and P.B.; visualization, P.B.; project administration, E.M.; funding acquisition, E.M. All authors have read and agreed to the published version of the manuscript.

Funding: The reported study was funded by RFBR, project number 19-05-50071.

Acknowledgments: The authors would like to thank the staff of the nano-mineralogy sector of the "Center for Collective Use of Perm State National Research University" for their assistance in conducting research.

Conflicts of Interest: The authors declare no conflicts of interest.

References

1. Leeder, M.R. *Sedimentology and Sedimentary Basins*, 2nd ed.; Wiley-Blackwell: Hoboken, NJ, USA, 2011; pp. 245–294.
2. Blinov, S.; Krasilnikova, S. Using the industrial wastes for remediation of sites affected by acid mine water dumping. In Proceedings of the International Mine Water Association Conference, Perm, Russia, 15–19 July 2019; Khayrulina, E., Wolkersdorfer, C., Polyakova, S., Bogus, A., Eds.; Perm State University, Perm Federal Research Center of the Ural Branch of RAS: Perm, Russia, 2019; pp. 731–735.
3. Fetisova, N.; Fetisov, V.; Bachurin, B.; Imaikin, A. Natural Neutralization of Acid Mine Water in Carbonate Deposits of Kizel Coal Basin. In Proceedings of the International Mine Water Association Conference, Perm, Russia, 15–19 July 2019; Khayrulina, E., Wolkersdorfer, C., Polyakova, S., Bogus, A., Eds.; Perm State University, Perm Federal Research Center of the Ural Branch of RAS: Perm, Russia, 2019; pp. 719–724.

4. Imaikin, A. Mine waters of Kosva field of Kizel coal basin during and after its operation, forecast of hydrochemical regime of mine waters that are discharged on the surface. In Proceedings of the International Multidisciplinary Scientific GeoConference Surveying Geology and Mining Ecology Management (SGEM2014), Albena, Bulgaria, 17–26 June 2014; pp. 605–612.
5. Berezina, O.A.; Shikhov, A.N.; Abdullin, R.K. Application of multiyear series of space imagery data to assess the environmental situation in coal-mining areas (on the example of the liquidated Kizel coal basin). *Curr. Probl. Earth Remote Sens. Space* **2018**, *15*, 144–158. (In Russian) [CrossRef]
6. Consani, S.; Ianni, M.C.; Dinelli, E.; Capello, M.; Cutroneo, L.; Carbone, C. Assessment of metal distribution in different Fe precipitates related to Acid Mine Drainage through two sequential extraction procedures. *J. Geochem. Explor.* **2019**, *196*, 247–258. [CrossRef]
7. España, J.S. The Behavior of Iron and Aluminum in Acid Mine Drainage: Speciation, Mineralogy, and Environmental Significance. In *Book Thermodynamics, Solubility and Environmental Issues*; Letcher, T.M., Ed.; Elsevier: Amsterdam, The Netherlands, 2007; pp. 137–150. [CrossRef]
8. Carbone, C.; Dinelli, E.; Marescotti, P.; Gasparotto, G.; Lucchetti, G. The role of AMD secondary minerals in controlling environmental pollution: Indications from bulk leaching tests. *J. Geochem. Explor.* **2013**, *132*, 188–200. [CrossRef]
9. Bigham, J.M.; Nordstrom, D.K. Iron and aluminum hydroxysulfates from acid sulfate waters. *Rev. Mineral. Geochem.* **2000**, *40*, 351–403. [CrossRef]
10. Caraballo, M.A.; Rötting, T.S.; Nieto, J.M.; Ayora, C. Sequential extraction and DXRD applicability to poorly crystalline Fe- and Al-phase characterization from an acid mine water passive remediation system. *Am. Mineral.* **2009**, *94*, 1029–1038. [CrossRef]
11. Baleeiro, A.; Fiol, S.; Otero-Fariña, A.; Antelo, J. Surface chemistry of iron oxides formed by neutralization of acidic mine waters: Removal of trace metals. *App. Geochem.* **2018**, *89*, 129–137. [CrossRef]
12. Maximovich, N.G.; Menshikova, E.A.; Osovetskiy, B.M. Hard technogenic components in alluvium and environment. In Proceedings of the 8th International Congress International Association of Engineering Geology, Vanconver, Canada, 21–25 September 1998; Brookfield: Rotterdam, The Netherlands, 2000; pp. 4579–4582.
13. Menshikova, E.A.; Osovetsky, B.M. Coal particles in the Environment. *Probl. Reg. Ecol.* **2014**, *5*, 83–88. (In Russian)
14. Siegel, F.R. *Environmental Geochemistry of Potentially Toxic Metals*; Springer: Berlin, Germany, 2002; p. 218. [CrossRef]
15. Menshikova, E.A.; Osovetsky, B.M. Technogenic magnetic spherules. In Proceedings of the International Multidisciplinary Scientific GeoConference (SGEM2016), Albena, Bulgaria, 30 June–6 July 2016; Book 3, Water Resources, Forest, Marine and Ocean Ecosystems. Hydrology and Water Resources, Forest Ecosystems;. Curran Associates, Inc.: Makati, Philippines, 2016; Volume 3, pp. 295–302. [CrossRef]
16. Valente, T.M.; Gomes, P.; Sequeira Braga, M.A.; Pamplona, J.; Antunes, I.M.; Ríos Reyes, C.A.; Moreno, F. Mineralogical Attenuation Processes Associated With The Evolution Of Acid Mine Drainage In Sulfide-Rich Mine Wastes. In Proceedings of the International Mine Water Association Conference, Perm, Russia, 15–19 July 2019; Khayrulina, E., Wolkersdorfer, C., Polyakova, S., Bogus, A., Eds.; Perm State University, Perm Federal Research Center of the Ural Branch of RAS: Perm, Russia, 2019; pp. 146–154.
17. Valente, T.M.; Antunes, I.M.; Sequeira Braga, M.A.; Neiva, A.; Santos, A.; Moreno, F. Mobility Control Of Uranium And Other Potentially Toxic Elements In Mine Waters By Ochre-Precipitates. In Proceedings of the International Mine Water Association Conference, Perm, Russia, 15–19 July 2019; Khayrulina, E., Wolkersdorfer, C., Polyakova, S., Bogus, A., Eds.; Perm State University, Perm Federal Research Center of the Ural Branch of RAS: Perm, Russia, 2019; pp. 457–463.
18. Cravotta, C.A. Dissolved metals and associated constituents in abandoned coal-mine discharges, Pennsylvania, USA. Part 1: Constituent quantities and correlations. *Appl. Geochem.* **2007**, *23*, 166–202. [CrossRef]
19. Espana, J.S.; Pamo, E.L.; Santofimia, E.; Aduvire, O.; Reyes, J.; Barettino, D. Acid mine drainage in the Iberian Pyrite Belt (Odiel river watershed, Huelva, SW Spain): Geochemistry, mineralogy and environmental implications. *Appl. Geochem.* **2005**, *20*, 1320–1356. [CrossRef]
20. Nordstrom, D.K. Hydrogeochemical processes governing the origin, transport and fate of major and trace elements from mine wastes and mineralized rock to surface waters. *Appl. Geochem.* **2011**, *26*, 1777–1791. [CrossRef]
21. Valente, T.; Grande, J.A.; de la Torre, M.L.; Gomes, P.; Santisteban, M.; Borrego, J.; Sequeira Braga, M.A. Mineralogy and geochemistry of a clogged mining reservoir affected by historical acid mine drainage in an abandoned mining area. *J. Geochem. Explor.* **2015**, *157*, 66–76. [CrossRef]

22. Nordstrom, D.K. The effect of sulfate on aluminum concentrations in natural waters: Some stability relations in the system $A1_2O_3$-SO_3-H_2O at 298 K. *Geochim. Cosmochim. Acta* **1982**, *46*, 681–692. [CrossRef]
23. Rose, S.; Ghazi, A.M. Release of sorbed sulfate from iron oxyhydroxides precipitated from acid mine drainage associated with coal mining. *Environ. Sci. Technol.* **1997**, *31*, 2136–2140. [CrossRef]
24. Kumpulainen, S.; Carlson, L.; Räisänen, M.-L. Seasonal variations of ochreous precipitates in mine effluents in Finland. *App. Geochem.* **2007**, *22*, 760–777. [CrossRef]
25. Dvinskikh, S.A.; Vostroknutova, Y.O.; Kitaev, A.B. The role of technogenic iron in the formation of its content in the water of the Kamskoye and Votkinskoye reservoirs. *Geogr. Bull.* **2015**, *4*, 18–25. (In Russian)
26. Lyubimova, T.; Parshakova, Y.; Lepikhin, A.; Tiunov, A. The risk of river pollution due to washout from contaminated floodplain water bodies during periods of high magnitude floods. *J. Hydrol.* **2016**, *534*, 579–589. [CrossRef]

minerals

MDPI

Article

Geochemical Behavior of Potentially Toxic Elements in Riverbank-Deposited Weathered Tailings and Their Environmental Effects: Weathering of Pyrite and Manganese Pyroxene

Yeongkyoo Kim

School of Earth System Sciences, Kyungpook National University, Daegu 41566, Korea; ygkim@knu.ac.kr; Tel.: +82 53 950 5360

Received: 26 March 2020; Accepted: 2 May 2020; Published: 7 May 2020

Abstract: Uncontrolled management of mine tailings can cause serious environmental problems. Red and black deposits of weathered mine tailings are observed in the upstream of the Nakdong River in Korea, washed away from abandoned mines during floods. Herein, the geochemical and mineralogical changes that occur during weathering of these deposited mine tailings and the mobility of resulting potentially toxic elements were investigated. Primarily, johansennite (manganese pyroxene) was identified in the deposits. Goethite and jarosite were identified as secondary minerals in the red or brown layers. Manganese oxide (MnO) formed by the weathering of johansennite in the black layers and schwertmannite in the red and brown layers were also identified via energy-dispersive X-ray spectroscopy. The three most abundant potentially toxic elements in the residual and iron and manganese oxide fractions were Zn, Pb and As. The high percentage of potentially toxic elements in these oxide fractions indicated that the weathering products of pyrite and manganese pyroxene were crucial in fixing these elements, and MnO likely played an important role in controlling the behavior of heavy metals. In addition, metals were detected in significant concentrations in the exchangeable and carbonate-bound fractions, which can be toxic to the river's ecological system.

Keywords: tailing weathering; potentially toxic element; goethite; manganese oxide; johansennite

---

1. Introduction

Mining and subsequent processing can generate large quantities of tailings comprising sulfides and toxic elements; such tailings can cause acid mine drainage (AMD). Finely crushed mill tailings have a large surface area, rendering them particularly reactive, thereby causing elevated concentrations of potentially toxic elements in the soil and hydrological systems around the mine; sometimes, this can even affect distant areas through the river's flow [1,2]. Uncontrolled management of tailings can be a severe environmental threat including over one hundred cases of tailings dam failures recorded worldwide [3]. For example, mine tailings spilled owing to a dam rupture in 1998 in Aznalcóllar (South West Spain), wherein 4.5×10^5 m^3 of acidic waters and toxic tailings spilled into the rivers, causing an adverse environmental effect on the floodplains [4–6]. Residual pollution could be detected even 20 years later, posing a continued high risk of pollution dispersion to the remaining ecosystem [7].

Over the past 40 years, approximately 1000 metal mines and 330 coal mines have been decommissioned in Korea [8]. However, the waste rocks and tailings of these mines have not been properly managed, causing severe adverse environmental impacts on the nearby soil and hydrological systems, particularly near rivers [9].

Red/brown and black/gray layers are observed at several sites along the upstream riverbank of the Nakdong River in Bonghwa, Korea. Owing to the severe weathering process caused by river water, original mineral textures of the tailings were not observed; rather, these layers are the weathering products of the deposited tailings, washed away from the abandoned mines by past floods. The secondary minerals formed by the weathering of sulfides, such as ferrihydrite, schwertmannite, jarosite, and goethite, generally have a large surface area and high adsorption capacities, greatly reducing the mobility of the potentially toxic elements in the tailings [10–15]. Understanding the weathering process of tailings in a specific environment would thus allow the estimation of resultant mobility changes of the potentially toxic elements in that system because the tailings are vulnerable to weathering and the possibility of the leaching of these elements is high [16,17].

Effective mine waste management and restoration requires scientific knowledge of the behaviors of potentially toxic elements in that location. Various factors may affect the mobility of these elements in heavily weathered tailings located at a riverside. This work, therefore, aims to investigate the mineralogical and geochemical changes of the weathered tailings on the riverbank of the Nakdong River. This paper presents X-ray diffraction (XRD), field emission scanning electron microscopy (FE-SEM), determination of the total concentration of potentially toxic element including five-step sequential extraction data for the tailings samples, and discussing the effect of the weathering process on the behavior of these elements.

2. Materials and Methods

2.1. Sample Collection

Thirty-three samples were collected based on color, particle size, and texture of eight weathered tailings located on the riverbank of the upstream Nakdong River shown in Figure 1. Detailed descriptions of each location are listed in Table 1. Herein, eleven samples were collected from point 1; three samples were collected from point 2; four samples were collected from point 3; four samples were collected from point 4; two samples were collected from point 5; four samples were collected from point 6; two samples were collected from point 7; and three samples were collected from point 8. To preserve their original conditions as much as possible, the samples were collected using a shovel and transported to the laboratory in a plastic bag.

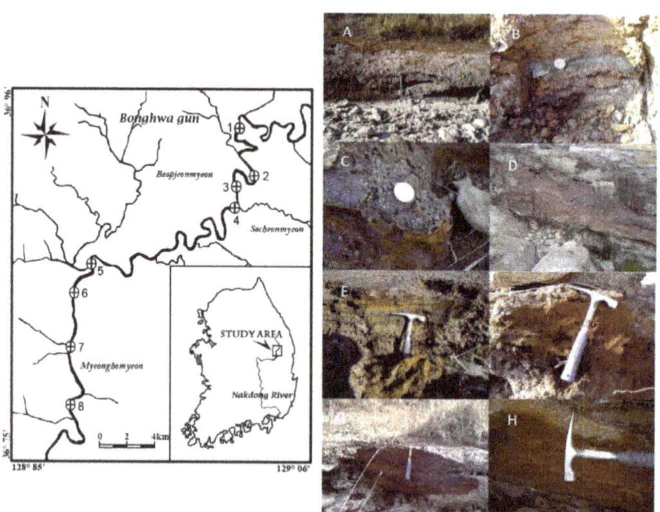

Figure 1. Sampling points (left) and weathered tailings layers (right) observed on the riverbank of the Nakdong River: (**A**,**B**) Point 1, (**C**) Point 2, (**D**) Point 3, (**E**) Point 4, (**F**) Point 5, (**G**) Point 6, and (**H**) Point 7.

Table 1. Description of each layer of weathered tailings observed on the riverbank of the Nakdong River.

Sampling Point	Number of Samples	Average Thickness	Color	pH (Average Value)
1	11	55 cm	Varied, mostly black	3.83–7.93 (6.68)
2	3	25 cm	Brown, interlayered with gray	3.22–4.05 (3.57)
3	4	45 cm	Red, gray, and black	4.15–4.87 (4.50)
4	4	15 cm	Gray, brown, black, and yellow	2.59–3.35 (3.10)
5	2	3 cm	Red with a thin black layer	6.08–6.26 (6.17)
6	4	56 cm	Red and black	4.07–5.82 (4.79)
7	2	8 cm	Brown	6.57–6.66 (6.62)
8	3	28 cm	Brown, gray, and black	6.62–6.68 (6.64)

2.2. Sample Analysis

The pH of each sample was measured using a pH meter (model 420A, ORION, Beverly, MA, USA) 24 h after mixing 2 g of the sample with 20 mL of deionized water with a shaker for 30 min.

To investigate their mineral compositions, the samples were dried at 60 °C for 24 h in an oven. The samples were ground using an agate mortar and analyzed via X-ray diffraction (XRD, Phillips X'pert APD, Hilversum, The Netherlands). The data were collected from 2θ = 5°–60° with steps of 2θ = 0.016° and a counting time of 1 s per step with Ni-filtered Cu-Kα radiation at 40 KV and 30 mA at the Daegu Center of Korea Basic Science Institute. To more precisely identify secondary iron minerals, six representative samples from sampling points 1, 4, 5, 6-1, 6-2 and 7 were mixed with distilled water and dispersed using an ultrasonic cleaner for 20 min. The relatively large quartz and feldspar particles were separated using a 0.5 mm mesh sieve. The solid samples were separated via centrifugation for 5 min and then dried at 60 °C for 24 h. These samples were then analyzed via powder XRD (Bruker D5005, Billerica, MA, USA) with Fe-filtered Co-Kα radiation at 40 kV and 30 mA at the Industrial Mineral Bank of Kangwon National University at a 2θ of 5°–60° with steps of 0.04° and a counting time of 8 s per step.

The morphology and texture of the primary and secondary minerals of the tailings were observed via field-emission scanning electronic microscopy (FE-SEM, Hitachi model S-4200, Hitachi, Tokyo, Japan) operating at 15 kV at the Daegu Center of the Basic Science Institute (KBSI). The chemical compositions of the minerals were additionally analyzed via energy-dispersive X-ray spectroscopy (EDS).

The concentrations of Al_2O_3, CaO, Fe_2O_3, K_2O, MgO, MnO, Na_2O, P_2O_5, SiO_2 and TiO_2 in the tailings were determined via X-ray fluorescence (XRF, Phillips PW1404, Phillips, Eindhoven, The Netherlands) operating in an ultrathin window and with a Sc–Mo X-ray tube at 4 kW at Kyungpook National University using the fused-glass method. Loss on ignition (LOI) was determined after heating the tailings to 900 °C.

The concentrations of eight potentially toxic elements (As, Cd, Co, Cr, Cu, Ni, Pb and Zn) were analyzed using the total digestion method (modified EPA3052) suggested by the Environmental Protection Agency of the United States. The samples were digested in aqua regia (HCl:HNO_3 = 3:1) containing HF and analyzed via an inductively coupled plasma atomic emission spectroscopy (ICP-AES, Horiba Jobin Yvon Ultima2C, Longjumeau, France) operating at 1.15 kW at a plasma gas flow rate of 12 L/min at the Seoul branch of the Korea Basic Science Institute. The spectral range was 160–800 mm with the resolution of 0.005 nm in UV. Calibration for each element confirmed that the correlation coefficient was >0.995. Sample digestion for the ICP analysis was performed using a closed-type Teflon Digestion Vessel 60 mL (Savillex, Eden Prairie, MN, USA), which is designed for rapid sample digestion via microwave or block digestion techniques. Standard samples, such as estuarine sediments

(USGS SRM 1646a) and NIST Montana Soil (NIST SRM 2711), were used to verify the analysis. The measurements of the standard samples were performed in triplicate and the mean recovery errors were within ±5%. The detection limits were 0.05 ppm for As and 0.005 ppm for other elements.

The sequential extraction method modified by Li et al. [18] based on the scheme proposed by Tessier et al. [19] was used to determine the chemical fractions of the potentially toxic elements. This scheme involves five steps for each fraction: Exchangeable metals (F1), metals bound to carbonate and specially adsorbed (F2), metals bound to Fe/Mn oxides (F3), metals bound to organic matter and sulfides (F4), and metals in the residual fraction (F5). Eight elements were analyzed via ICP-AES (Jobin Yvon Ultima2C, Longjumeau, France) and ICP-mass spectroscopy (ICP-MS, Elan DRC II, Perkin Elmer, Boston, MA, USA) at 1.5 kW of radio frequency (RF) generator power with an RF frequency of 40 MHz at the Seoul branch of the Korea Basic Science Institute. The Ar nebulizer gas flow rate was 0.84 L/min. Abundance sensitivity (at ^{23}Na) was $<1.0 \times 10^{-6}$ for the low-mass side and $<1.0 \times 10^{-7}$ for the high-mass side. The detection limits of ICP-MS were 25 ppb for As and 2.5 ppb for other elements.

Pearson's correlation coefficient analysis was performed using SPSS (version 25.0) to correlate the concentrations of metals and pH values.

3. Results

3.1. Sample pH

The resulting pH values ranged from approximately 3 to 8; the corresponding results are listed in Table 1. The sample with the lowest pH (3.2) was collected from point 4, whereas that with the highest pH (7.93) was collected from point 1.

3.2. Mineral Compositions

The results of XRD analysis using a Cu target showed that the samples comprised quartz, feldspar, johansennite (Mn–pyroxene, $CaMnSi_2O_6$), mica, amphibole, talc, and chlorite. At some sampling points, other minerals, such as gypsum (point 1), kaolinite (point 8), bassanite (point 1), and calcite (point 1) were also identified. Johansennite was primarily detected in the black samples collected from point 1.

Additional XRD analysis using a Co target was performed to identify the iron minerals of the fine particles. Goethite collected from point 6-2 and jarosite collected from point 3 were identified (Figure 2). The largest amount of johansennite was observed in the black portion of the sample collected from point 1, although it was also identified in the red/brown portions collected from other samples. Overall, the color of the samples varied with the mineral composition. The red and brown samples usually contained goethite or jarosite, whereas the black samples contained johansennite and its weathering product, manganese oxide (MnO).

SEM/EDS analysis was performed for the sample (collected from point 1-3) rich in Fe and Mn to reveal the different mineralogical features. In sample 1-3, the Fe(III) oxide and MnO coating had both smooth and irregular-shaped surfaces, as shown in Figure 3. The black sample collected from point 1-3 exhibited high concentrations of Fe and Mn: 49.7 wt% Fe and 36.67 wt% Mn, as detailed in Table 2. The Fe-containing mineral was identified as goethite via XRD. The mineral phase containing a high Mn concentration was presumed to be noncrystalline MnO because it was not identified via XRD. Therefore, goethite and MnO were the main minerals containing Fe and Mn detected in this sample. Johansennite typically has varying colors such as blue–green, gray–white, dark brown, or colorless [20]; therefore, the black color of the sample was primarily caused by Mn oxide. In the goethite and MnO, relatively high concentrations of Zn (2.82 and 17.2 wt%, respectively) and Pb (8.2 wt% in both samples) were observed (Table 2), indicating that these elements were surface-adsorbed and coprecipitated with goethite and MnO. Pb and Zn have been demonstrated to coprecipitate with iron oxide [21–23]; furthermore, Zhao et al. [24] has described the importance of the coprecipitation of Zn with MnO.

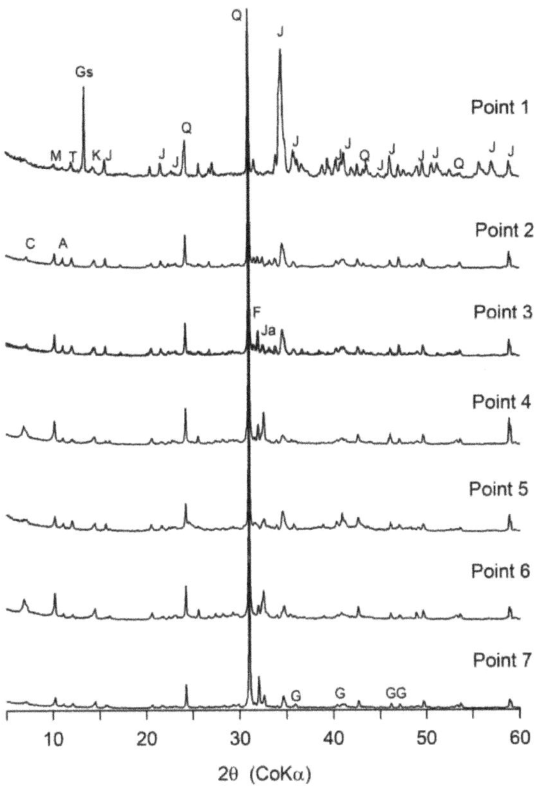

Figure 2. X-ray diffraction patterns of the representative weathered tailings samples with a cobalt target, where the A, C, F, G, Gs, J, K, M and T peaks represent amphibole, chlorite, feldspar, goethite, gypsum, johansennite, kaolinite, mica and talc, respectively.

Table 2. Chemical compositions of the samples collected from point 1-3 and 2 analyzed via EDS at each point shown in Figure 3.

B		D		E	
Element	**wt%**	**Element**	**wt%**	**Element**	**wt%**
O	39.29	O	36.67	O	56.80
Fe	49.72	Mn	36.04	Mg	1.10
Zn	2.82	Zn	17.21	S	16.11
Pb	8.18	As	1.82	Ca	1.95
		Pb	8.25	Mn	2.82
				Fe	22.22
Total	100		100		100

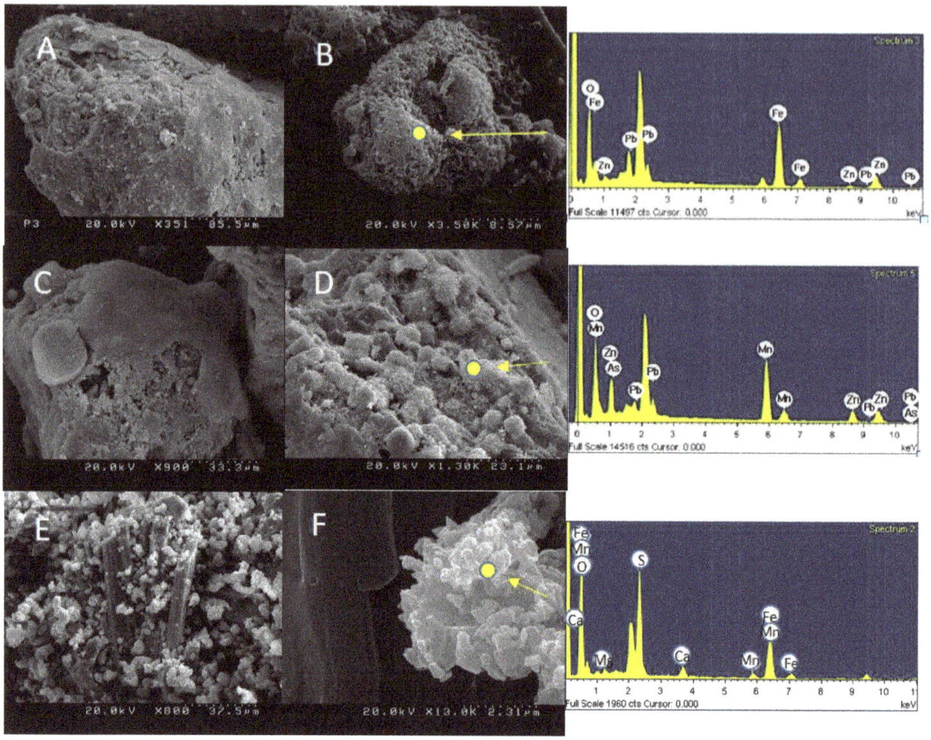

Figure 3. Scanning electronic microscope images and energy-dispersive X-ray spectroscopy results of the sample collected from point 1 and 2: (**A**) and (**B**): Goethite, (**C**) and (**D**): Mn oxide, (**E**) and (**F**): Schwertmannite formed on the gypsum crystal.

Minerals with fine white particles were observed in the sample collected from point 2. SEM imaging showed that these were needle-shaped gypsum particles comprising Ca and S (Figure 3). Considering the Fe, S and O composition as determined via EDS, the irregular-shaped minerals on the surface of the gypsum is likely schwertmannite ($Fe_8O_8(OH)_6SO_4$), which can be easily transformed to goethite [25,26].

3.3. Chemical Compositions of Major Elements

The chemical compositions of major elements analyzed via XRF for selected samples are listed in Table 3. Three samples were collected from point 1 to examine brown, gray, and black samples (samples 1-1, 1-2, and 1-3, respectively). The concentration of calcium was high for the samples collected from points 1 and 6, likely because of the presence of calcite in the tailings. The concentration of iron was high (3.83–7.93 wt%) for most samples owing to the weathering of pyrite. Additionally, a relatively high concentration of Mn was detected in samples collected from point 1 (4.07–5.27 wt%, as opposed to 0.74–2.28 wt% in other samples), in which johansennite and its weathering product were observed via XRD and SEM.

Table 3. Chemical composition of the mine tailings (wt%).

	1-1	1-2	1-3	2	3	4	5	6	7	8
Al_2O_3	6.84	5.79	6.37	11.16	9.42	9.26	7.13	5.94	11.43	11.4
CaO	6.52	15.28	6.28	2.31	2.85	2.55	5.67	9.74	3.58	3.94
Fe_2O_3	20.85	14.37	13.15	7.44	8.3	6.36	21.33	15.93	14.93	18.89
K_2O	1.10	0.83	1.58	2.87	3.02	3.4	1.45	1.36	2.10	1.93
MgO	1.49	1.55	1.33	1.14	0.95	0.84	1.45	1.08	1.45	1.20
MnO	4.78	4.07	5.27	0.80	1.63	0.74	2.28	2.72	1.61	1.61
Na_2O	0.37	0.22	0.55	1.31	1.19	1.45	0.56	0.61	1.34	0.97
P_2O_5	0.14	0.14	0.11	0.09	0.09	0.11	0.13	0.1	0.14	0.11
SiO_2	50.00	38.72	57.33	64.98	67.62	71.17	51.18	48.32	54.82	49.84
TiO	0.34	0.36	0.33	0.49	0.36	0.44	0.43	0.32	0.52	0.43
LOI	7.18	18.14	7.95	7.88	5.23	3.85	9.07	13.48	8.13	9.29
Total	99.61	99.47	100.25	100.47	100.66	100.17	100.68	99.60	100.05	99.61

3.4. Chemical Compositions of Potentially Toxic Elements

The total concentration and mobility of eight potentially toxic elements were investigated for each weathered sample and each fraction of sequentially extracted samples. Results showed that the weathered tailings contained very high concentrations of these elements: 117.2–4.191 mg/kg of As, <5–71.29 mg/kg of Cd, <5–24.53 mg/kg of Co, 11.00–131.3 mg/kg of Cr, 20.03–173.2 mg/kg of Cu, 8.446–34.40 mg/kg of Ni, 822.49–3,720 mg/kg of Pb, and 246.3–10,996 mg/kg of Zn (Table 4). Typically, the order of potentially toxic element concentrations in the tailings samples was Zn > Pb > As > Cu > Cr > Ni > Cd > Co; in all samples, the most abundant elements were Zn, Pb, and As.

Table 4. Potentially toxic element concentrations of mine tailings (mg/kg).

	1-1	1-2	1-3	2	3	4	5	6	7	8
As	2178	1281	1109	117.2	132.9	499.9	4191	2303	1746	2510
Cd	15.43	18.65	71.29	11.93	7.077	<5	6.759	19.75	5.52	5.454
Co	20.38	24.53	12.29	7.432	12.39	<5	6.56	13.7	14.59	6.233
Cr	62.49	22.5	12.69	30.12	17.89	11	27.43	19.55	131.3	25.91
Cu	173.2	135.4	135.7	50.27	70.58	20.03	105.6	95.33	66.61	60.38
Ni	17.53	34.46	24.97	11.34	12.39	8.446	12.92	14.51	14.98	12.27
Pb	3270	2261	1904	208.3	836.1	82.49	1355	1481	1406	2047
Zn	5364	3201	10996	762.8	1075	246.3	1010	1567	889.8	833.1

Although the potentially toxic element concentrations provide useful information regarding the behavior of elements, different chemical forms of the elements have different potential effects on the ecosystem. Therefore, sequential extraction was performed to obtain the chemical fraction and thereby more detailed information on the toxicity of these elements. The extracted fractions of the elements in each step are shown in Figure 4 as a percentage of the sum of all fractions. The weight percentages of Cr, Ni, and Cd in the cation-exchangeable fraction (step 1) were generally larger than other elements and in the range of 0.20–42.26%, 1.58–32.28%, and 0.52–68.66%, respectively. Ni and Cr were the most predominant in the fraction of metal bound to carbonate (step 2) and were in the range of 13.57–58.92% and 15.05–39.47%, respectively. The percentages of Pb, Cr, and Zn absorbed on Fe and Mn oxides in step 3 were 25.05–89.73%, 8.84–81.85%, and 7.33–74.93%, respectively, indicating that these three elements likely coprecipitate with or strongly adsorb on Fe and Mn oxides; this is also supported by the EDS observations. The samples collected from point 1 had the highest percentage of Pb and Zn in step 3, which may be related to the high concentration of MnO. During organic or sulfide adsorption, i.e., step 4, Cu was the most predominant at 5.7–48.68%. Most of the As was found in the residual fraction, i.e., in step 5: 86.01–99.96 wt%.

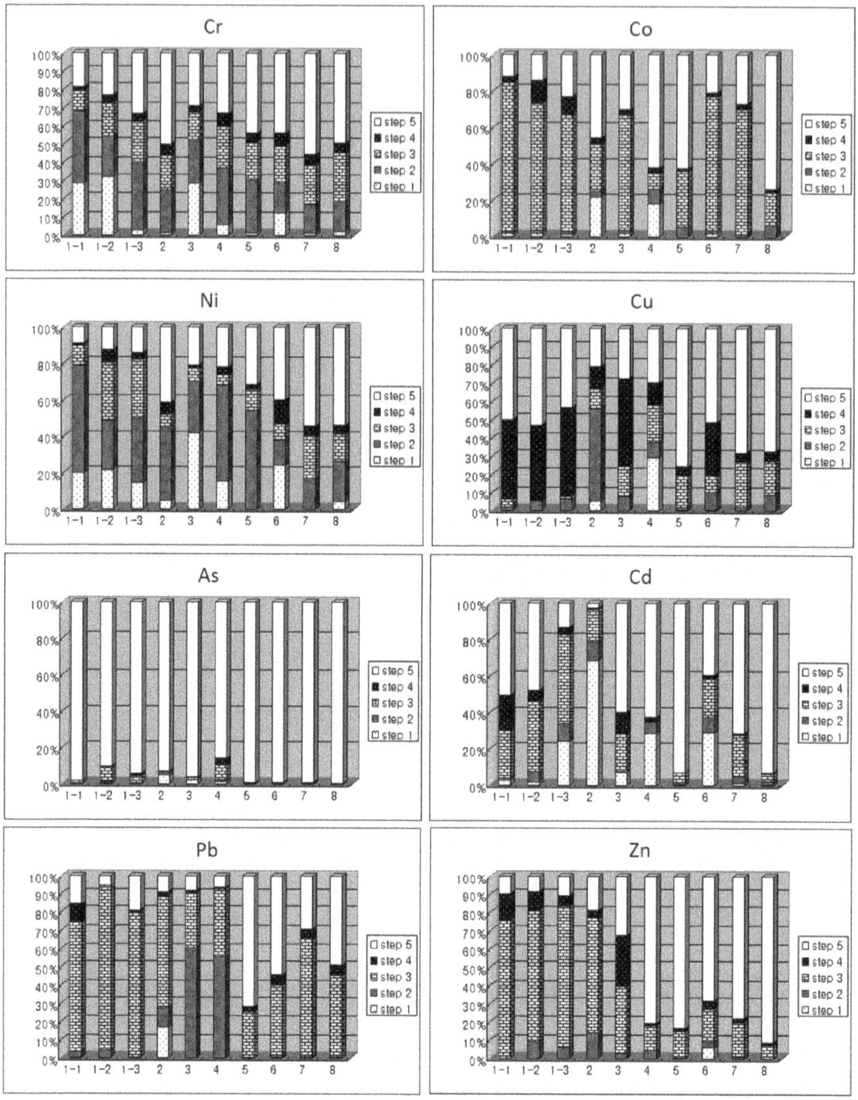

Figure 4. Relative abundance of the sequentially extracted potentially toxic elements for the weathered tailings samples. Step 1: Exchangeable metals, Step 2: Metals bound to carbonate and specially adsorbed, Step 3: Metals bound to Fe/Mn oxides, Step 4: Metals bound to organic matter and sulfides, Step 5: Metals in the residual fraction.

4. Discussion

The various minerals identified on the tailings that had been deposited and weathered on the bank of the Nakdong River comprised primary minerals that had been moved from other places and secondary minerals that had been formed by the weathering of the tailings. The primary minerals, as observed via XRD analysis, mainly contained silicates, particularly johansennite. Secondary minerals were primarily formed during the oxidation of pyrite and weathering of silicate minerals. The elements released by the weathering of pyrite (S) and calcite and pyroxene (Ca) formed gypsum and basanite, which are sulfate minerals. Pyrite also provided dissolved Fe for goethite formation and both dissolved Fe and S for jarosite precipitation. The K ions in jarosite can be provided by the weathering of K feldspar. MnO, which is important in fixing potentially toxic elements through coprecipitation and adsorption, directly precipitated by the weathering of johansennite. Thus, the color of the weathered tailings was mainly influenced by the weathering of pyrite and johansennite.

Schwertmannite, which is metastable and commonly found in AMD, was observed via SEM/EDS analysis and was considered to be transformed to goethite in the tailings layers. Therefore, most goethite found in the weathered tailings was probably transformed from schwertmannite. The pH, which plays an important role in the formation of iron minerals, ranged between 2.59 and 7.93. Jarosite is a common mineral in the weathered zone of sulfide ore deposits that precipitates with a high concentration of SO_4 at pH < 3 [14]. Jarosite was thus only observed in the sample collected from point 3, in which the pH was relatively low.

The pH values of tailings depend on the weathering of pyrite and buffering by the dissolution of calcite [27–29]. Therefore, the increase in pH observed in the samples collected from point 1 was due to the dissolution of carbonate minerals present in the mine tailings. The high concentrations of Ca detected at points 1, 5, 6, and 7 were closely related to the existence of calcite, whereas the samples collected from points 2, 3, and 4 had both a lower pH and a lower concentration of Ca. Additionally, the concentration of potentially toxic elements increased with increasing pH, showing a positive correlation with these elements and indicating that mineral compositions including pyrite and calcite were vital in controlling the behaviors of potentially toxic elements (Table 5) [28,29]. At a low pH, the mobility of these elements increases, enabling the leaching of a large portion of the elements. The samples collected from points 1, 5, 6, and 7, which had relatively high pH values and high Ca concentrations, generally showed elevated concentrations of potentially toxic elements. The correlation coefficients between the element concentration and pH for As and Cr were slightly lower than those of others, indicating that their concentrations were less correlated with pH. This is likely because they exist as oxyanions in the natural environment, unlike the other elements studied. The pH of the samples examined herein was lower than the point of zero charge of goethite and MnO; consequently, these secondary minerals have positive charges [28–31]. Therefore, As and Cr as oxyanions are dissolved less at low pH. The concentrations of As and Cr also exhibited low or negative correlation coefficients with other elements owing to their negative charges of oxyanions, compared with the high correlation coefficients among other elements.

Table 5. Pearson's correlation coefficients calculated for the potentially toxic elements and pH.

	As	Cd	Co	Cr	Cu	Ni	Pb	Zn	pH
As	1								
Cd	0.130	1							
Co	0.044	0.195	1						
Cr	0.157	−0.268	0.269	1					
Cu	0.358	0.492	0.734 *	0.046	1				
Ni	0.017	0.537	0.787 **	−0.057	0.686 *	1			
Pb	0.470	0.270	0.676 *	0.220	0.860 **	0.568	1		
Zn	−0.050	0.934 **	0.366	−0.136	0.695 *	0.592	0.518	1	
pH	0.460	0.344	0.555	0.345	0.720 *	0.670 *	0.853 **	0.546	1

*: significant at 0.05 level; **: significant at 0.01 level.

SEM/EDS studies confirmed that the weathered tailings samples contained high concentrations of Pb and Zn in the Fe oxide and Zn in the MnO, particularly those collected from point 1, which generally had higher concentrations of Fe and Mn. Goethite is highly reactive and can adsorb toxic elements dissolved during weathering, and toxicity is expected to be reduced through coprecipitation and sorption [15,32]. Schwertmannite, which is supposed to precipitate first and be transformed to goethite later, also has the capacity to adsorb potentially toxic elements [13,24,26].

In many cases, potentially toxic elements can be naturally stabilized through sorption or coprecipitation with insoluble secondary minerals that are stable in an oxidizing environment such as goethite [12,15,32]. However, no detailed studies have been performed on the role of MnO in controlling these elements in natural environments. This study showed that MnO, precipitated by the weathering of Mn–pyroxene, likely plays a more important role in fixing potentially toxic elements than Fe oxide minerals.

The sequential extraction results provided more details on the mobility of each toxic element, enabling better estimation of their potential toxicity. With slight geochemical changes, the cation-exchangeable fraction (step 1) and metal fractions bound to carbonate and specially adsorbed (step 2) are more mobile, can be easily released to the water system, and are therefore, more toxic to nearby ecological systems; metals extracted in the later steps are more stable [33–36]. Cd, Cr, and Ni were present in these first two steps, albeit in low concentrations, indicating their greater potential toxicity.

Large percentages of the potentially toxic elements Co, Cd, Ni, Pb, and Zn were bound to the Fe and Mn oxides (i.e., extracted during step 3), particularly for the samples collected from point 1. As supported by the SEM/EDS analysis for Pb and Zn, these elements were supposed to be coprecipitated with or adsorbed on the surfaces of the goethite or MnO. Considering the black color and high Mn concentration of the samples from point 1, the amorphous MnO likely played a more significant role in fixing these elements; thus, weathering manganese silicate with pyrite can reduce the potential toxicity in the tailings and related systems. The most abundant potentially toxic elements in the weathered tailings were Zn, Pb, and As. Despite their elevated concentration, their potential toxicity was determined to be relatively low, as they were extracted during later steps, unlike Cd, Cr, and Ni, which, as discussed, were extracted during the first two steps. The third most abundant element, As, was mostly found in the stable residual form (i.e., extracted during step 5), providing the lowest level of toxicity in this system.

5. Conclusions

In this study, on an upstream riverbank of the Nakdong River, red and black layers of weathered mine tailings washed down from abandoned mines during floods were sampled and analyzed. As most ore minerals in the resulting tailings were severely weathered, the samples contained high concentrations of potentially toxic elements. The XRD analysis revealed that the samples comprised silicates, such as quartz, feldspar, mica, chlorite, amphibole, talc, kaolinite, johansennite (manganese pyroxene), and other secondary minerals, including goethite, gypsum, bassanite, and jarosite. Additionally, pyrite and calcite were identified; the pH of the samples was heavily influenced by the weathering of pyrite and buffering by the dissolution of calcite. Furthermore, the pH and the weathering of primary minerals contributed to the formation of secondary minerals. Pyrite, calcite, and johansennite were the main minerals found to influence the formation of secondary minerals. The weathering of johansennite produced amorphous MnO coated on the surface of primary minerals in black layers. Goethite was transformed from schwertmannite, which was identified via SEM analysis, and jarosite was observed in the samples with low pH values.

The overall concentrations of potentially toxic elements were in the order of Zn > Pb > As > Cu > Cr > Ni > Cd > Co. Zn, Pb and As, were the most abundant elements in all samples. Although the concentrations of Cr and Ni were relatively low, they mainly existed in cation-exchangeable forms and forms easily bound to carbonate (i.e., extracted during steps 1 and 2, respectively, of a sequential extraction method); they can, therefore, easily dissolve, causing environmental problems. Co, Cd, Ni, Pb, and Zn were more easily bound to Fe and Mn oxides (i.e., extracted during step 3), suggesting coprecipitation with and sorption on these oxides. This was particularly true in the samples collected from point 1, where the concentration of Mn was high because of the presence of johansennite and amorphous MnO. This strongly indicates that both the weathering of pyrite, producing goethite, and the weathering of johansennite are important in fixing potentially toxic elements. The third most abundant element, As, was predominantly present in the final residue, which can be very stable even with the geochemical changes.

Funding: This work was supported by the National Research Foundation of Korea (NRF) grant funded by the Korean government (MSIT) (no. 2019R1A2C1002254).

Acknowledgments: I thank Ji-Hwan Shin for the preparation of a map and statistical analysis, and two anonymous reviewers for their constructive comments.

Conflicts of Interest: The authors declare no conflict of interest.

References

1. Pagnanelli, F.; Moscardini, E.; Giuliano, V.; Toro, L. Sequential extraction of heavy metals in river sediments of an abandoned pyrite mining area: Pollution detection and affinity series. *Environ. Pollut.* **2004**, *132*, 189–201. [CrossRef] [PubMed]
2. Kossoff, D.; Hudson-Edwards, K.A.; Dubbin, W.E.; Alfredsson, M. Major and trace metal mobility during weathering of mine tailings: Implications for flood plain soils. *Appl. Geochem.* **2012**, *27*, 562–576. [CrossRef]
3. Rico, M.; Benito, G.; Salgueiro, A.R.; Díez-Herrero, A.; Pereira, H.G. Reported tailings dam failures A review of the European incidents in the worldwide context. *J. Hazard. Mater.* **2008**, *152*, 846–852. [CrossRef] [PubMed]
4. Grimalt, J.O.; Ferrer, M.; Macpherson, E. The mine tailing accident in Aznalcollar. *Sci. Total Environ.* **1999**, *242*, 3–11. [CrossRef]
5. Benito, G.; Benito-Calvo, A.; Gallart, F.; Martín-Vide, J.P.; Regües, D.; Bladé, E. Hydrological and geomorphological criteria to evaluate the dispersion risk of waste sludge generated by the Aznalcollar mine spill (SW Spain). *Environ. Geol.* **2001**, *40*, 417–428. [CrossRef]
6. Kossoff, D.; Dubbin, W.E.; Alfredsson, M.; Edwards, S.J.; Macklin, M.G.; Hudson-Edwards, K.A. Mine tailings dams: Characteristics, failure, environmental impacts, and remediation. *Appl. Geochem.* **2014**, *51*, 229–245. [CrossRef]
7. García-Carmona, M.; García-Robles, H.; Turpín Torrano, C.; Fernández Ondoño, E.; Lorite Moreno, J.; Sierra Aragón, M.; Martín Peinado, F.J. Residual pollution and vegetation distribution in amended soils 20 years after a pyrute mine tailings spill (Aznalcóllar, Spain). *Sci. Total Environ.* **2019**, *650*, 933–940. [CrossRef]
8. Ji, S.; Kim, S.; Ko, J. The status of the passive treatment systems for acid mine drainage in South Korea. *Environ. Geol.* **2008**, *55*, 1181–1194. [CrossRef]
9. Lee, J.E.; Kim, Y. A quantitative estimation of factors affecting pH changes using simple geochemical data from acid mine drainage. *Environ. Geol.* **2008**, *55*, 65–75. [CrossRef]
10. Fukushi, K.; Sato, T.; Yanase, N. Solid-solution reaction in As(V) sorption by schwertmannite. *Environ. Sci. Technol.* **2003**, *37*, 3581–3586. [CrossRef]
11. Fukushi, K.; Sato, T.; Yanase, N.; Minato, J.; Yamada, H. Arsenate sorption on schwertmannite. *Am. Mineral.* **2004**, *89*, 1728–1734. [CrossRef]
12. Burgos, W.D.; Borch, T.; Troyer, L.D.; Luan, F.; Larson, L.N.; Brown, J.F.; Lambson, J.; Shimizu, M. Schwermannite and Fe oxides formed by biological lo-pH Fe(II) oxidation versus abiotic neutralization: Impact on trace metal sequestration. *Geochim. Cosmochim. Acta* **2012**, *76*, 29–44. [CrossRef]
13. Baleeiro, A.; Fiol, S.; Otero-Fariña, A.; Antelo, J. Surface chemistry of iron oxides formed by neutralization of acidic mine waters: Removal of trace metals. *Appl. Geochem.* **2018**, *89*, 129–137. [CrossRef]
14. Kim, Y. Effects of different oxyanions in solution on the precipitation of jarosite at room temperature. *J. Hazard. Mater.* **2018**, *353*, 118–126.
15. Hajji, S.; Montes-Hernandez, G.; Sarret, G.; Tordo, A.; Morin, G.; Ona-Nguema, G.; Bureau, S.; Turki, T.; Mzoughi, N. Arsenite and chromate sequestration onto ferrihydrite, siderite and goethite nanostructured minerals: Isotherms from flow-through Reactor Experiments and XAS measurements. *J. Hazard. Mater.* **2019**, *362*, 358–367. [CrossRef]
16. Mascaro, I.; Benvenuti, B.; Corsini, F.; Costagliola, P.; Lattanzi, P.; Parrini, P.; Tanelli, G. Mine wastes at the polymetallic deposit of Fenice Capanne (southern Tuscany, Italy), Mineralogy, geochemistry, and environmental impact. *Environ. Geol.* **2001**, *41*, 417–429.

17. Meza-Figueroa, D.; Maier, R.M.; de la O-Villanueva, M.; Gómez-Alvarez, A.; Morenzo-Zazueta, A.; Rivera, J.; Campillo, A.; Grandlic, C.J.; Anaya, R.; Palafox-Reyes, J. The impact of unconfined mine tailings in residential areas from a mining town in a semi-arid environment: Nacozari, Sonora, Mexico. *Chemosphere* **2009**, *77*, 140–147. [CrossRef]
18. Li, X.D.; Coles, B.J.; Ramsey, M.H.; Thornton, I. Sequential extraction of soils for multi-element analysis by ICP-AES. *Chem. Geol.* **1995**, *124*, 109–123. [CrossRef]
19. Tessier, A.; Campbell, P.G.C.; Bisson, M. Sequential extraction procedure for the speciation of particulate trace metals. *Anal. Chem.* **1979**, *51*, 844–851. [CrossRef]
20. Schaller, W.T. Johansennite, a new manganese pyroxene. *Am. Mineral.* **1938**, *23*, 575–582.
21. Martínez, C.E.; McBride, M.B. Coprecipitates of Cd, Cu, Pb and Zn in iron oxides: Solid phase transformation and metal solubility after aging and thermal treatment. *Clays Clay Miner.* **1998**, *46*, 537–545. [CrossRef]
22. Martínez, C.E.; McBride, M.B. Cd, Cu, Pb, and Zn coprecipitates in Fe oxide formed at different pH: Aging effects on metal solubility and extractability by citrate. *Environ. Toxicol. Chem.* **2001**, *20*, 122–126. [CrossRef] [PubMed]
23. Lee, G.H.; Bigham, J.M.; Faure, G. Removal of trace metals by coprecioitation with Fe, Al and Mn from natural eaters contaminated with acid mine drainage in the Bucktown mining district, Tennessee. *Appl. Geochem.* **2002**, *17*, 569–581. [CrossRef]
24. Zhao, S.; Wang, Q.; Sun, J.; Borkiewicz, O.J.; Huang, R.; Saad, E.M.; Fields, B.; Chen, S.; Zhu, M.; Tang, Y. Effect of Zn coprecipitation on the structure of layererd Mn oxides. *Chem. Geol.* **2018**, *493*, 234–245. [CrossRef]
25. Jönsson, J.; Persson, P.; Sjöberg, S.; Lövgren, L. Schwertmannite precipitated from acid mine drainage: Phase transformation, sulphate release and surface properties. *Appl. Geochem.* **2005**, *20*, 179–191. [CrossRef]
26. Schroth, A.W.; Parnell, R.A. Trace metal retention through the schwertmannite to goethite transformation as observed in a field setting, Alta Mine, MT. *Appl. Geochem.* **2005**, *20*, 907–917. [CrossRef]
27. Lee, P.K.; Touray, J.C. Characteristics of a Polluted Artificial Soil located along a Motorway and Effects of Acidification on the leaching Behavior of heavy metals (Pb, Zn, Cd). *Water Res.* **1998**, *32*, 3425–3435. [CrossRef]
28. Kim, H.J.; Kim, Y.; Choo, C.O. The effect of mineralogy on the mobility of heavy metals in mine tailings: A case study in the Samsanjeil mine, Korea. *Environ. Earth Sci.* **2014**, *71*, 3429–3441. [CrossRef]
29. Park, S.; Kim, Y. Mineralogical changes and distribution of heavy metals caused by the weathering of hydrothermally altered, pyrite-rich andesite. *Environ. Earth Sci.* **2016**, *75*, 1125. [CrossRef]
30. Jurjovec, J.; Ptacek, C.J.; Blowes, D.W. Acid neutralization mechanisms and metal release in mine tailings: A laboratory column experiment. *Geochim. Cosmochim. Acta* **2002**, *66*, 1511–1523. [CrossRef]
31. Heikkinen, P.M.; Räisänen, M.L. Mineralogical and geochemical alteration of Hitura sulphide mine tailings with emphasis on nickel mobility and retention. *J. Geochem. Explor.* **2008**, *97*, 1–20. [CrossRef]
32. Jiang, W.; Lv, J.; Luo, L.; Yang, K.; Lin, Y.; Hu, F.; Zhang, J.; Zhang, S. Arsenate and cadmium co-adsorption and co-precipitation on goethite. *J. Hazard. Mater.* **2013**, *262*, 55–63. [CrossRef] [PubMed]
33. Dold, B.; Fontboté, L. A mineralogical and geochemical study of element mobility in sulfide mine tailings of Fe oxide Cu-Au deposits from the Punta del Cobre belt, northern Chile. *Chem. Geol.* **2002**, *189*, 135–163. [CrossRef]
34. Carlsson, E.; Thunberg, J.; Ohlander, B.; Holmström, H. Sequential extraction of sulfide-rich tailings remediated by the application of till cover, Kristineberg mine, northern Sweden. *Sci. Total Environ.* **2002**, *299*, 207–226. [CrossRef]
35. Caraballo, M.A.; Rotting, T.S.; Nieto, J.M.; Ayora, C. Sequential extraction and DXRD applicability to poorly crystalline Fe- and Al-phase characterization from an acid mine water passive remediation system. *Am. Mineral.* **2009**, *94*, 1029–1038. [CrossRef]
36. Kim, Y. Mineral phases and mobilities of trace metals in white precipitates found in acid mine drainage. *Chemosphere* **2015**, *119*, 803–811. [CrossRef]

minerals

MDPI

Article

Selective Removal of As(V) Ions from Acid Mine Drainage Using Polymer Inclusion Membranes

Iwona Zawierucha [1,*], Anna Nowik-Zajac [1] and Grzegorz Malina [2]

1 Institute of Chemistry, Jan Dlugosz University in Czestochowa, 42-200 Czestochowa, Poland; a.zajac@ajd.czest.pl

2 Department of Hydrogeology and Engineering Geology, AGH University of Science and Technology, Mickiewicza 30, 30-059 Cracow, Poland; gmalina@agh.edu.pl

* Correspondence: i.zawierucha@ajd.czest.pl; Tel.: +48-343614918-(251)

Received: 20 August 2020; Accepted: 12 October 2020; Published: 14 October 2020

Abstract: Acid mine drainage (AMD) is globally recognized as one of the environmental pollutants of the priority concern due to high concentrations of toxic metals and sulfates. More rigorous environmental legislation requires exploitation of effective technologies to remove toxic metals from contaminated streams. In view of high selectivity, effectiveness, durability, and low energy demands, the separation of toxic metal ions using immobilized membranes with admixed extractants could ameliorate water quality. Cellulose triacetate based polymer inclusion membranes (PIMs), with extractant and plasticizer, were studied for their ability to transport of As(V) ions from synthetic aqueous leachates. The effects of the type and concentration of extractant, plasticizer content, and sulfuric acid concentration in source phase on the arsenic removal efficiency have been assessed. Under the best of applied conditions, PIM with Cyanex 921 as extractant and *o*-nitrophenyl octyl ether (*o*-NPOE) as plasticizer showed high repeatability and excellent transport activity for selective removal of As(V) from AMD.

Keywords: arsenic; acid mine drainage; polymer inclusion membranes

1. Introduction

Mining of minerals is associated with acid drainage problems that can give rise to long-term impairment to aquatic environment and biodiversity. Acid mine drainage (AMD) is produced when sulfide-bearing material is comes into contact with oxygen and water [1]. It can be characterized as acidic water (pH < 2) with high concentrations of sulfate and various metals and metalloids [2–4]. Penetration of AMD into surface and ground waters is a critical environmental issue and often results in biotic degradation of water bodies around the abandoned mines through, e.g., direct toxicity, changes of habitats, and in the nutrient cycle, leading to unsuitability of water resources for use in housing, agriculture, and industry [5–7]. The acidification and leaching of toxic elements like metals and arsenic are mainly reasons for the environmental degradation. The region of AMD sites includes polluted soil and groundwater that threatens the flora and fauna [8]. Acidic (ground)water can be detrimental to both aquatic and terrestrial organisms [9], and it may trigger the mobilization of diverse metals and metalloids [10]. As a result, the AMD pose an environmental and health threats and a major challenge to clean up hazardous pollutants [11].

Arsenic (As) being a common AMD pollutant is a naturally occurring metalloid very mobile in the environment. Its mobility largely depends on the parent mineral form, oxidation state, and mobilization mechanisms [12]. Arsenic exists in the −3, 0, +3 and +5 oxidation states; however, arsenite (As(III) and arsenate (As(V)) are the most frequently found in water [13–16]. Due to slow redox transformations, they are present in both reduced and oxidized environments. In the case of anaerobic (reduced) conditions, e.g., groundwater and sediment, arsenic is present mainly as arsenite, while arsenate is dominant in oxidative environments (e.g., surface waters) [17]. The pH also plays an important role in determining the state of arsenic in aqueous solutions [18]. Arsenic is well known for its toxic effects [8,19,20]. It may lead to the noxious impacts on human health like skin defects, disturbances of blood circulation, neurological problems, diabetes, difficulties with breathing, and hepatic and renal abnormalities [21].

Typical processes for treating AMD rely on arsenic removal via oxidation, precipitation, and separation [22,23]. Techniques applied conventionally to remove arsenic species from water include, oxidation, coagulation–flocculation, ion exchange adsorption, and membrane techniques [24]. Recently, advances in nanoscience and nanotechnology have triggered the development of various nanomaterials as novel adsorbents of heavy metals and arsenic from aqueous solutions, including AMD [25,26]. Bioremediation of arsenic contaminated soil and (ground)water is advantageous related to its environmental affinity and possible cost benefits. It is based on microbial activity to remove, mobilize, and hold arsenic through sorption, biomethylation–demethylation, complexation, coprecipitation, and oxidation–reduction processes [27].

The current development of AMD passive treatment systems has recently been reviewed by Skousen et al. (2017) [7]. Long-term, effective passive AMD treatment systems based on geogenic (geochemical and biological) processes allow for acidity neutralization, oxidation, or metals and arsenic reduction and precipitation, and are suitable for small and medium AMD discharges. Constructed wetlands, including vertical flow wetlands and bioreactors as examples of passive biological treatment technologies, are generally related to microbial activity and can use organic matter to stimulate microbial sulfate reduction and contaminants adsorption. In the case of passive geochemical systems, on the other hand, alkalinity-generating materials, e.g., limestone, come into contact with AMD or with uncontaminated water above the AMD discharge [7].

Sulfate reducing bacteria (SRB) are used for in situ AMD treatment; however, the choice of organic substrate as a carbon source affects the process efficiency [28]. Permeable reactive barriers (PRB) are considered innovative to treat AMD [29]. To treat AMD with metals contamination the PRB is generally composed of solid organic matter, such as, municipal/leaf compost and wood chips/sawdust. Organic matter enhances the proliferation of sulfate-reducing bacteria that may reduce sulfate to sulfide, and the subsequent formation of insoluble metal sulfides [30].

The effective removal of metal ions from aqueous waste streams can be obtained by applying a polymer inclusion membranes (PIMs) technique. Separation of metal ions using this membrane system is a multi-stage process including extraction of the substance from the source phase, diffusion of the resulting complex through the membrane and re-extraction into the receiving phase [31]. PIM is created in the form of thin, flexible, and stable film by molding a solution containing an extractant (an ion carrier), a plasticizer and a base polymer, such as cellulose triacetate (CTA) or polyvinyl chloride (PVC). The obtained membrane provides the barrier dividing source and receiving aqueous phases [31–35].

The PIM separation properties are advantageous due to high membrane selectivity with its increased durability. The membrane stability is due to the immobilization of extractant in the solid polymer matrix, while its high permeability is provided by addition of the plasticizer [31]. In addition, PIMs are characterized by mechanical strength comparable to that of filtration membranes. An important feature of PIM-based system is the possibility of modifying membrane composition to gain the appropriate selectivity and separation efficiency. The other advantages of PIM technique include ease of operation and minimum use of hazardous chemicals [36,37].

Guell et al. (2011) [38] studied the transport of arsenate from water samples across the PIM composed of CTA and Aliquat 336, and using 0.1 M NaCl as a receiving phase. They found that As(V) transport efficiency was not affected by presence of co-existing anions in natural waters. Vera et al. (2018) [39] using PIM made of 52% CTA, 48% Aliquat 336 (*w*/*w*), and 2 M NaCl as a receiving phase obtained As(V) removal efficiency within the range of 53–81% in different water samples. Moreover, they noted that transport was not dependent on the kind of polymer and the membrane thickness [39]. The transport of arsenate ions has been also investigated by De Lourdes Ballinas et al. (2004) [40] using a PIM based on CTA and dibutyl

butyl phosphonate (DBBP) as the carrier. High arsenic recovery factor (90% in 800 min) was obtained for an initial As(V) concentration of 3000 mg/L and the receiving solution contained 2 M LiCl.

In this work the removal of As(V) ions from synthetic aqueous leachates (sulfuric acid solutions) by transport through PIM has been studied. The membrane consisted of cellulose triacetate (CTA) as the polymeric support, *o*-nitrophenyl octyl ether (*o*-NPOE) as the plasticizer, and solvating extractants as ion carriers. The effects of various parameters on the removal efficiency of arsenic were studied, including the plasticizer content, type, and concentration of a carrier in the membrane, and the sulfuric acid concentration in the source phase. Moreover, in order to extend the applicability of the separation technique, we have evaluated reusability of PIM, and tested the PIM-system for selective removal of arsenic from a real AMD sample.

2. Materials and Methods

2.1. Reagents

Inorganic chemicals, i.e., arsenate(V) sodium ($Na_2HAsO_4{\cdot}7H_2O$), sodium chloride (NaCl), and sulfuric acid (H_2SO_4) of analytical grade were purchased from POCh (Gliwice, Poland). Organic reagents, i.e., cellulose triacetate (CTA), *o*-nitrophenyl octyl ether (*o*-NPOE), and dichloromethane (DCM) were also of analytical grade, purchased from Fluka (Buchs, Switzerland) and used without further purification. Aqueous solutions were prepared with double distillation water, with a conductivity of 0.1 µS/cm. The carriers: DBBP ($CH_3(CH_2)_3P(O)[O(CH_2)_3CH_3]_2$) and Cyanex 921 ($[CH_3(CH_2)_7]_3PO$) were purchased from Sigma-Aldrich (St. Louis, MO, USA).

2.2. Preparation Of Polymer Inclusion Membranes and Stability Test

The preparation pf polymer inclusion membranes was carried out based on the procedure previously described [31]. To prepare PIMs, the following solutions were used (in DCM as an organic solvent): cellulose triacetate (CTA) as a support, *o*-nitrophenyl octyl ether as a plasticizer, and DBBP and Cyanex 921 as ionic carriers. The defined volumes of the CTA solution, plasticizer, and carrier were mixed, and this mixture was poured in to a 5.0 cm glass ring attached to a plate glass with CTA—dichloromethane glue. The glass ring was left to rest overnight to DCM evaporate in room temperature. The obtained membrane was disconnected from the glass plate by wetting in cold water. The effective surface area of the membrane was of 4.9 cm^2. The membrane thickness was measured with an accuracy of 1.0 µm standard deviation over four readings by A2002M type digital ultrameter from Inco-Veritas (Warsaw, Poland). The average CTA membrane thickness was of 24 µm.

The stability of PIMs was investigated in terms of mass loss, which is related to the leaching of carrier and/or plasticizer. For that, PIMs were immersed in 100 mL of ultrapure water and was shaken for 24 h. Membranes were weighed before and after this procedure, and the mass loss was calculated. The membrane masses were also examined before and after arsenic ions transport.

2.3. Transport Studies

Transport experiments were carried out in a permeation cell described in our earlier paper [31] using a two-compartment cell where the membrane film was tightly clamped between the source and receiving phases. The source phase was of an aqueous solution of As(V) in sulfuric acid media (50 cm^3) whereas the aqueous receiving phase was 1.0 M sodium chloride (50 cm^3). The experiments were performed at a room temperature (23–25 °C) and both source and receiving aqueous phases were agitated at 600 rpm with synchronous stirrers. Samples of aqueous phases were removed periodically via a sampling port with a syringe and analyzed to determine the As(V) concentration. The acidity of both aqueous phases was controlled by pH-meter (multifunctional pH-meter, CX-731 Elmetron, with a combine pH electrode, ERH-136, Hydromet, Poland) [31,35].

Kinetics of the transport process through PIM is described by a first-order reaction in respect to a metal ion concentration [41]:

$$\ln\left(\frac{c}{c_0}\right) = -kt \quad (1)$$

where c is the metal ion concentration (mol/dm^3) in the source phase at a given time, c_0 is the initial As(V) concentration in the source phase (mol/dm^3), k is the rate constant (s^{-1}), and t is the transport time (s).

To calculate the k value, a graph of $\ln(c/c_0)$ versus time was plotted. The relationship of $\ln(c/c_0)$ versus time was linear, which was confirmed by high values of determination coefficients (r^2), i.e., ≥ 0.99. (Figure 1). The permeability coefficient *(P)* was calculated as follows [31,35,42]:

$$P = -\frac{V}{A}k \quad (2)$$

where V is the volume of aqueous source phase (m^3), and A is the surface area of membrane (m^2). The initial flux (J_0) in (μmol/m^{2}s) was determined as equal to

$$J_0 = Pc_0 \quad (3)$$

To describe the efficiency of metal ion removal from the source phase, the recovery factor (*RF*) was calculated as

$$RF = \frac{c_0 - c}{c_0} \cdot 100\% \quad (4)$$

The metal ions concentrations were measured by flame atomic absorption spectrometry (Solar 939, Unicam, Munich, Germany) and the sulfate concentration was determined by ion chromatography (861 Advanced Compact IC, Metrohm, Herrisau, Switzerland). The reported values correspond to the average of duplicates, and the standard deviation observed was within 2%.

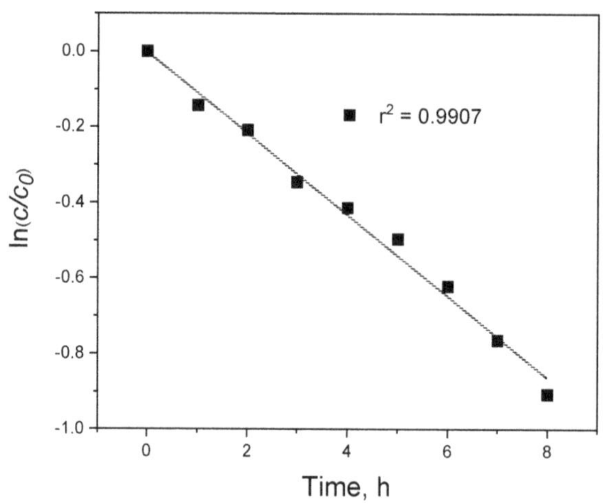

Figure 1. The relationship of $\ln(c/c_0)$ vs. time for arsenic transport across polymer inclusion membranes (PIM). Source phase: $1.0{\cdot}10^{-2}$ M As(V), 1.0 M H_2SO_4; membrane: 19 wt.% of cellulose triacetate (CTA), 3 wt.% of Cyanex 921, 78 wt.% of *o*-NPOE; receiving phase: 1.0 M NaCl.

3. Results and Discussion

3.1. Kinetics of As(V) Ions Transport through Polymer Inclusion Membranes

The changes of concentration of As(V) ions in the source, receiving and the membrane phases were analyzed to determine their transport through PIM containing Cyanex 921 by creating a concentration profile of the metal as a function of time (Figure 2). The obtained exponential curves of c/c_0 to time corroborates the kinetic model of metal ions transport suggested by Danesi et al. (1984) [41]. Figure 2 indicates that the transport of As(V) ions across membrane was speedy and parallel. The lack of the metal ions aggregation in the membrane phase was also noted.

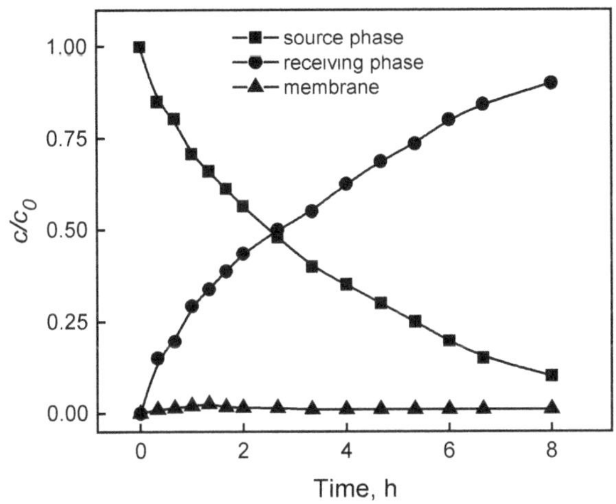

Figure 2. The profile of As(V) concentrations in the source phase, membrane and receiving phase during transport through PIM. Source phase: $1.0 \cdot 10^{-2}$ M As(V), 1.0 M H_2SO_4; membrane: 19 wt.% of CTA, 3 wt.% of Cyanex 921, 78 wt.% of *o*-NPOE; receiving phase: 1.0 M NaCl.

3.2. Effect of Membrane Composition

The transport through PIMs is mainly affected by the following factors: the composition of the membrane and selection of appropriate receiving phase [42]. The effect of the type and content of ion carriers in PIM on As(V) ions transport was studied. Membranes with a predetermined content of CTA (25 mg-19 wt.%) and plasticizer (77 wt.% of *o*-NPOE) were synthesized, while the concentration of carrier (DBBP or Cyanex 921) in the membrane was ranged from 1 to 8 wt.% (from 0.01 to 0.30 M with respect to the plasticizer volume). The *o*-NPPE was chosen as the plasticizer due to its low viscosity and a high dielectric constant. Plasticizer is a solvent for an ion carrier (extractant) and enhances the flexibility and permeability of the membrane. In turn the mechanical strength of the membrane is given by the CTA polymer [43]. The presence of carrier is essential for the facilitated transport of arsenic ions through the membrane. This fact was confirmed by results of blank experiments (in the absence of carrier), which showed no significant flux across PIM containing only the support and plasticizer.

Figure 3 shows that the transport of As(V) increased as the carrier concentration increased. However, this effect stopped at a concentration of 0.15 M (4 wt.%). The fluxes tended to a constant value at higher carriers concentrations, which was probably due to saturation of the membrane pores with metal complex species. These results indicate that the arsenic removal in the PIM system occurred in the following steps. Firstly, the metal ions moved from the source phase to the boundary layer of the membrane. Then, the complex As(V)-extractant was formed at the source solution–PIM interface. Finally, the As(V)-carrier complex transferred through PIM and As(V) was released at the PIM–receiving solution interface [44]. The arsenic transport process occurred in the presence of the carrier (extractant) contained within the membrane phase, so the coupled facilitated transport took place [45]. The carrier reacted with arsenic ions producing a complex which was transported through the membrane. The pH gradient between source and receiving phases was the driving force for the transport of metal ions across PIM [46]. The highest flux values were of 0.96 for DBBP and 1.30 μmol/m^{2}s for Cyanex 921. The highest RF values (i.e., 89.0% and 95.8%) were obtained for membranes with the 0.15 M carrier concentration (DBBP and Cyanex 921, respectively) (Figure 4). The lower RF values were for carrier concentrations greater than 0.15 M, which can be explained by the effect of increasing membrane viscosity; a higher viscosity of the membrane limited the diffusion of the As(V)−carrier complex within the membrane phase. Based on these results, for further experiments, the Cyanex 921 as ion carrier was used.

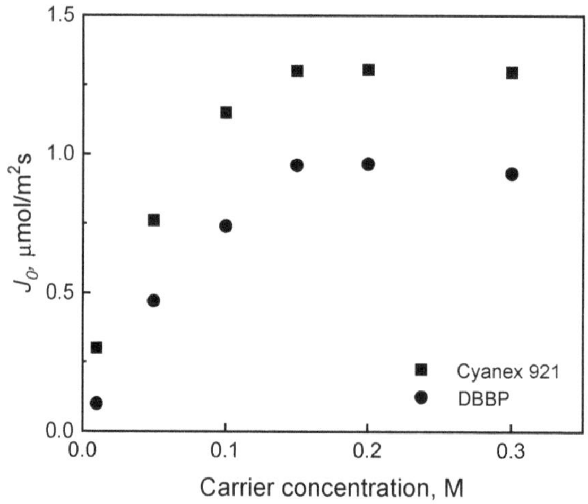

Figure 3. The As(V) fluxes vs. ions carriers concentration in the PIM. Source phase: $1.0 \cdot 10^{-2}$ M As(V), 1.0 M H_2SO_4; membrane: 19 wt.% of CTA, 77 wt.% of *o*-NPOE; receiving phase: 1.0 M NaCl.

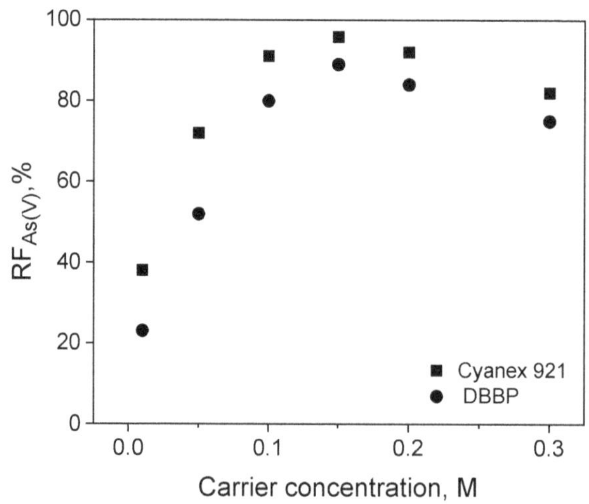

Figure 4. Recovery factor (RF) of As(V) ions vs. carrier concentration. Source phase: $1.0 \cdot 10^{-2}$ M As(V), 1.0 M H_2SO_4; membrane: 19 wt.% of CTA, 77 wt.% of *o*-NPOE; receiving phase: 1.0 M NaCl.

The Cyanex 921 extractant is composed of tri-n-octylphosphine oxide with a molecular weight of 386 [45]. In a liquid membrane system, Cyanex 921 can extract both acids and metal complexes [34,47]. The decrease in pH of the receiving phase over time indicated that protons H^+ from sulfuric acid were transferring to NaCl solution, acidifying it. However, this coextraction did not affect the As(V) stripping.

The effect of the plasticizer amount in PIM on As(V) ions transport was also examined. For this purpose, the membranes with constant carrier concentration of 0.15 M and different contents (0.5–2.0 cm^3) of *o*-NPOE/1.0 g CTA (63–87 wt.%) were prepared. Figure 5 shows the relationship between fluxes and amounts of *o*-NPOE. With the increasing amount of plasticizer, the fluxes increased. This trend had been kept until reaching a specific value of plasticizer in the membrane (1.0 cm^3 *o*-NPOE/1.0 g CTA, 77 wt.%). A further increase of the plasticizer amount was related to an increase of the thickness of PIM which probably reduced the permeability of the membrane. This resulted in a limiting the diffusion rate of metal complexes through membrane and a visible decrease in transport efficiency.

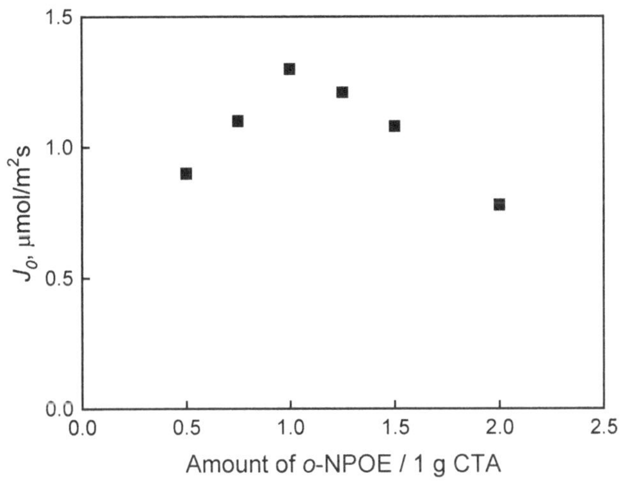

Figure 5. The effect of *o*-NPOE plasticizer on As(V) transport through PIM containing Cyanex 921. Source phase: $1.0{\cdot}10^{-2}$ M As(V), 1.0 M H_2SO_4; membrane: 19 wt.% of CTA, 4 wt.% of Cyanex 921; receiving phase: 1.0 M NaCl.

The highest RF factor of As(V) (95.8%) was achieved by transport through the PIM composed as follows: 19 wt.% of cellulose triacetate as the support, 4 wt.% of Cyanex 921, and 77 wt.% of *o*-nitrophenyl octyl ether as the plasticizer. Therefore, such PIM composition has been referred to as "optimal" and used in our further tests.

The image (3-D scan) obtained by atomic force microscopy (AFM) for membrane with an "optimal" composition is shown in Figure 6b. The colour intensity represents a vertical profile of the sample; the light areas indicate the highest spikes while dark areas the pores (organic inclusion in CTA support). The pores are noticeable as small well-defined dark regions. On the other hand, the membrane from CTA itself (Figure 6a) is non-porous and only slightly creased. The middle sized pores estimated for the optimal PIM were equal to 0.05 μm.

The stability of "optimal" PIM was investigated in terms of the extent of leaching of its components (the carrier and plasticizer) from the membrane to ultrapure water. Membrane mass loss was found to be 7 ± 1% (n = 2), which indicated good resistance of membrane components to migration from the base polymer. Plasticizer (*o*-nitrophenyl octyl ether), as a member of phthalate esters group, hydrolyzes below pH 5 and above pH 7. The degree of hydrolysis increases with increasing or decreasing pH and with longer contact times [48]. In our study (at acidic source phase), membrane mass loss after 8 h of As(V) transport was of 9 ± 0.4% (n = 2), which demonstrated that the plasticizer was properly bound to the polymer. Otherwise, in the case of significant leaching, due to toxicity of the phthalates [49], it is not necessary to use the plasticizer at the pilot or industrial scales.

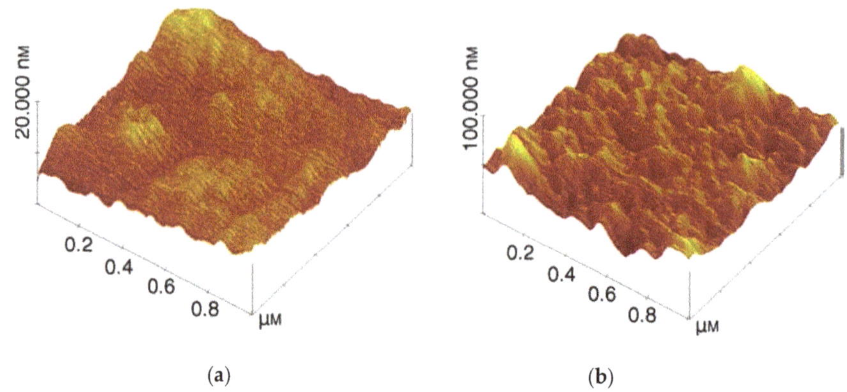

Figure 6. The 3-D image of atomic force microscopy (AFM) for CTA membrane (**a**) and "optimal" polymer inclusion membrane (**b**).

3.3. Modification of the Source Phase Acidity

The effect of sulfuric acid concentration in the source phase within the range of 0.01–2.0 M on As(V) ions transport through PIM was examined. As is seen in Figure 7 the arsenic removal efficiency increased with rising acidity of the source phase, reaching a maximum of 95.8% at the H_2SO_4 concentration of 1.0 M. This can be explained by the fact that in acidic conditions (pH < 2), As(V) is found in the form of H_3AsO_4—a neutral particle that can be removed by Cyanex 921 (solvating extractant) [50]. At low concentrations of H_2SO_4, arsenic acid is deprotonated, and thus the removal percentage is lower [47]. High efficiency of arsenic removal using PIM at 1.0 M concentration of H_2SO_4 in the source phase was also because the source and receiving phases differed significantly in the pH value. The pH gradient stimulated diffusion of the As(V)–carrier complex through the membrane and, consequently, the efficiency of arsenic separation by the PIM.

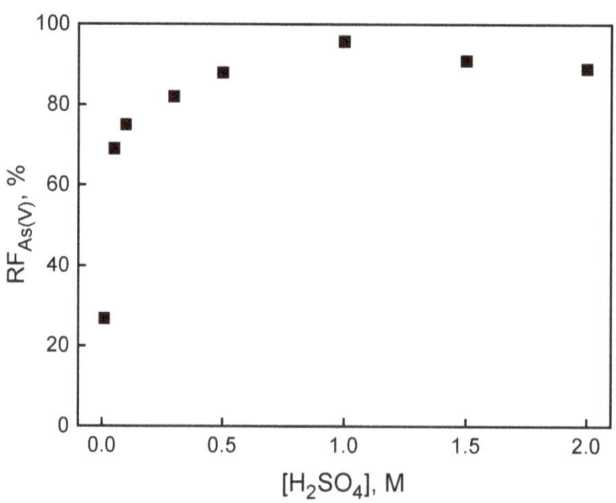

Figure 7. Recovery factor (RF) of As(V) ions vs. H_2SO_4 concentration in the source phase. Source phase: $1.0{\cdot}10^{-2}$ M As(V); membrane: 19 wt.% of CTA, 4 wt.% of Cyanex 921, 77 wt.% of *o*-NPOE; receiving phase: 1.0 M NaCl.

3.4. Membrane Reusability

The reusability, among different properties of PIMs, provides the greatest advantage, ensuring its industrial applicability [35]. In the repeated As(V) transport (eight cycles), the same PIM membrane was used while both aqueous phases were replaced for each cycle. The results indicate (Figure 8) that the efficiency of the PIM is repeatable. The RF values were above 95% in the first four cycles of the PIM (each cycle—8 h), while then slightly decreased (the RF values were ca. 92%). The optimal PIM seems to be effective for multiple use for arsenic separation process.

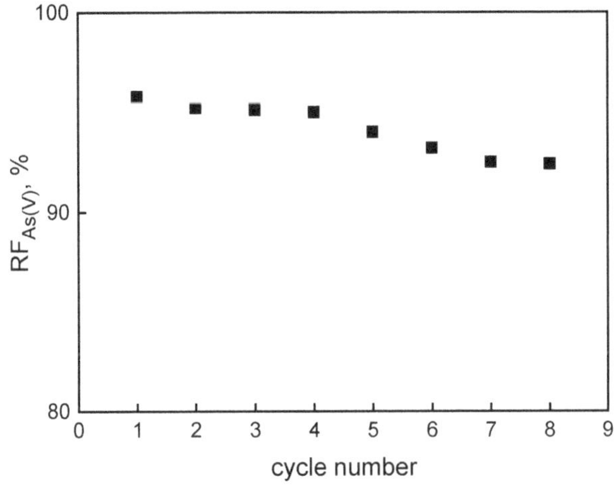

Figure 8. Recovery factor (RF) of As(V) vs. cycle number. Source phase: $1.0 \cdot 10^{-2}$ M As(V), 1.0 M H_2SO_4; membrane: 19 wt.% of CTA, 4 wt.% of Cyanex 921, 77 wt.% of *o*-NPOE; receiving phase: 1.0 M NaCl.

3.5. Selective Removal of As(V) from AMD

The optimal PIM was tested for its applicability for the removal of arsenic from an AMD sample originated from the Wisniowka mining area (south-central Poland) [51]. The AMD was acidic (pH 1.8) with As(V) and sulfate concentrations of 782 and 93,900 mg/dm^3, respectively. Concentrations of other metals like: Fe_{total}, Mn, Cu, Ni, and Cr were of 34,200, 50, 46, 27, and 20 mg/dm^3, respectively. The As(V) transport studies were carried out using the optimal PIM and 1.0 M sodium chloride as the receiving phase. After 8 h of continuous operating time the removal efficiency of arsenic was of 90% (Figure 9). Moreover, the RF values found for sulfate and other metals were below 5%, which prove the selectivity of the system for arsenic.

Microbial treatment based on the use of SRB could be considered as an approach for subsequent remediation of AMD to reducing sulfate, increase pH, and precipitate other metals as metal sulfides. For remediation to occur, SRB require, besides the presence of sulfate as electron acceptors, anaerobic conditions and an outer carbon source as electron donors, typically supplied from a simple organic compound. The process gives hydrogen sulfide and bicarbonate that treat the AMD by creating insoluble metal sulfides [52]. The produced metal sulfide sludge is more dense, faster settling, and less subject to dissolution. In addition, the process induces enough alkalinity to increase the pH without the necessity of any addition of alkaline agents [53].

In addition to microbial treatment, number of methods developed to treat metals contaminated water streams can potentially be applied as post-treatment after removing As from AMD using PIMs and prior to its discharge to the environment or further use. They include: chemical precipitation [54], ion exchange [55], adsorption [56], membrane filtration [57], reverse osmosis (RO) [58], solvent extraction [59], and electrochemical treatment [60]. For example, hydroxides or sulfur-containing precipitation agents commonly used for effective removal of metals from water and wastewater comprise [61]: potassium/sodium thiocarbonate (STC) and

2,4,6-trimercaptotiazine (TMT), and dithiocarbamate (DTC) compounds, e.g., sodium dimethyldithiocarbamate (SDTC), and ligands for permanent metal binding, e.g., 1,3-benzenediamidoethanethiol (BDETH2), 2,6-pyridinediamidoethanethiol (PyDET), a pyridine-based thiol ligand (DTPY), or ligands with open chains containing many sulfur atoms, using of a tetrahedral bonding arrangement around a central metal atom. High removal efficiencies of Cu and Cd ions of 98 and 99%, respectively, using RO were reported by Qdais and Moussa (2004) [62], whereas nanofiltration was capable of removing more than 90% of Cu ions [62]. Comparative sorption studies on the potential of indigenously synthesized carbon nanomaterials for removal of diverse metals from contaminated water streams showed higher sorption of Cd, Pb, Ni, and Zn on nanoporous carbon compared to nanocarbon [63]. The toxic metals like Hg, Pb, Cd, and Cu were also effectively removed from water using Fe_3O_4 magnetic nanoparticles coated with humic acid [64].

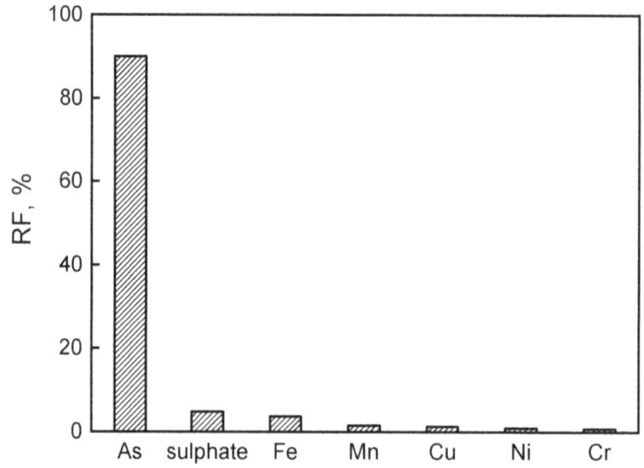

Figure 9. The RF values obtained in the competitive PIM transport of arsenic from the AMD sample. Membrane: 19 wt.% of CTA, 4 wt.% of Cyanex 921, 77 wt.% of *o*-NPOE; receiving phase: 1.0 M NaCl.

4. Conclusions

The study of As(V) ions transport through polymer inclusion membranes (PIMs) containing solvating extractant as the carrier allowed to specify the factors determining the removal efficiency. The membrane saturation obtained with 4 wt.% of Cyanex 921 for transport of As(V) through PIM gives the highest efficiency (95.8%) when the concentration of H_2SO_4 in the source phase was equal to 1.0 M, and 1.0 M sodium chloride was used as the stripping agent. The *o*-nitrophenyl octyl ether proved to be the suitable plasticizer in PIM for the effective arsenic removal, reaching the maximum recovery factor (RF) value for membranes containing 77 wt.% of *o*-NPOE. The optimal PIM composition was found to be 19 wt.% of cellulose triacetate as the support, 4 wt.% of Cyanex 921 as the ion carrier, and 77 wt.% of *o*-nitrophenyl octyl ether as the plasticizer. The transport efficiency of the PIM can be reproduced; thus, the membrane could be multiple used for arsenic separation. Moreover, the obtained results show the good performance of the PIM respected to the selective removal of arsenic from AMD with a separation efficiency of 90%.

Author Contributions: Conceptualization, I.Z. and A.N.-Z.; investigation, I.Z. and A.N.-Z.; supervision, G.M.; writing—original draft, I.Z.; writing—review and editing, A.N.-Z. All authors have read and agreed to the published version of the manuscript.

Funding: This research was funded by the National Science Centre, Poland (Grant number UMO-2015/19/D/ST10/03214).

Acknowledgments: The authors gratefully acknowledge Jan Dlugosz University in Czestochowa for the support provided for the publication of this work.

Conflicts of Interest: The authors declare no conflict of interest.

References

1. Akcil, A.; Koldas, S. Acid Mine Drainage (AMD): Causes, treatment and case studies. *J. Clean. Prod.* **2006**, *14*, 1139–1145. [CrossRef]
2. Othman, A.; Sulaiman, A.; Ibrahim, I. The use of low grade limestone in acid mine drainage treatment. *Malays. J. Anal. Sci.* **2019**, *23*, 472–478.
3. Burillo, J.C.; Castro-Larragoitia, J.; Burillo, G.; Ortega, A.; Medellin-Castillo, N. Removal of arsenic and iron from mine-tailing leachate using chitosan hydrogels synthesized by gamma radiation. *Environ. Earth Sci.* **2017**, *76*, 450. [CrossRef]
4. Serrano, J.; Leiva, E. Removal of arsenic using acid/metal-tolerant sulfate reducing bacteria: A new approach for bioremediation of high-arsenic acid mine waters. *Water* **2017**, *9*, 994. [CrossRef]
5. Choi, H.-J. Biosorption of heavy metals from acid mine drainage by modified sericite and microalgae hybrid system. *Water Air Soil Pollut.* **2015**, *226*, 185. [CrossRef]
6. Kato, T.; Kawasaki, Y.; Kadokura, M.; Suzuki, K.; Tawara, Y.; Ohara, Y.; Tokoro, C. Application of GETFLOWS coupled with chemical reactions to arsenic removal through ferrihydrite coprecipitation in an artificial wetland of a Japanese closed mine. *Minerals* **2020**, *10*, 475. [CrossRef]
7. Skousen, J.; Zipper, C.E.; Rose, A.; Ziemkiewicz, P.F.; Nairn, R.; McDonald, L.M.; Kleinmann, R.L. Review of passive systems for acid mine drainage treatment. *Mine Water Environ.* **2017**, *36*, 133–153. [CrossRef]
8. Luo, C.; Routh, J.; Dario, M.; Sarkar, S.; Wei, L.; Luo, D.; Liu, Y. Distribution and mobilization of heavy metals at an acid mine drainage affected region in South China, a post-remediation study. *Sci. Total Environ.* **2020**, *724*, 138122. [CrossRef]
9. Obiri-Nyarko, F.; Kwiatkowska-Malina, J.; Malina, G.; Wołowiec, K. Assessment of zeolite and compost-zeolite mixture as permeable reactive materials for the removal of lead from a model acidic groundwater. *J. Contam. Hydrol.* **2020**, *229*, 103597. [CrossRef]
10. Kjoller, C.; Postma, D.; Larsen, F. Groundwater acidification and the mobilization of trace metals in a sandy aquifer. *Environ. Sci. Technol.* **2004**, *38*, 2829–2835. [CrossRef]
11. Simate, G.S.; Ndlovu, S. Acid mine drainage: Challenges and opportunities. *J. Environ. Chem. Eng.* **2014**, *2*, 1785–1803. [CrossRef]
12. Al Abed, S.; Jegadeesan, G.; Purandare, J.; Allen, D. Arsenic release from iron rich mineral processing waste: Influence of pH and redox potential. *Chemosphere* **2007**, *66*, 775–782. [CrossRef] [PubMed]
13. Tokoro, C.; Kadokura, M.; Kato, T. Mechanism of arsenate coprecipitation at the solid/liquid interface of ferrihydrite: A perspective review. *Adv. Powder Technol.* **2020**, *31*, 859–866. [CrossRef]
14. Mohan, D.; Pittman, C.U., Jr. Arsenic removal from water/wastewater using adsorbents—A critical review. *J. Hazard. Mater.* **2007**, *142*, 1–53. [CrossRef]
15. Malina, G. Ecotoxicological and environmental problems associated with the former chemical plant in Tarnowskie Góry, Poland. *Toxicology* **2004**, *205*, 157–172. [CrossRef]
16. Pous, N.; Casentini, B.; Rossetti, S.; Fazi, S.; Puig, S.; Aulenta, F. Anaerobic arsenite oxidation with an electrode serving as the sole electron acceptor: A novel approach to the bioremediation of arsenic-polluted groundwater. *J. Hazard. Mater.* **2015**, *283*, 617–622. [CrossRef]
17. WHO 2011. *Arsenic in Drinking Water*; WHO: Geneva, Switzerland, 2011.
18. Henke, K.; Hutchison, A. Arsenic Chemistry. In *Arsenic: Environmental Chemistry, Health Threats and Waste Treatment*; Henke, K.R., Ed.; John Wiley & Sons Ltd.: West Sussex, UK, 2009; pp. 154–196.
19. Majzlan, J.; Plášil, J.; Škoda, R.; Gescher, J.; Kögler, F.; Rusznyak, A.; Küsel, K.; Neu, T.R.; Mangold, S.; Rothe, J. Arsenic-rich acid mine water with extreme arsenic concentration: Mineralogy, geochemistry, microbiology, and environmental implications. *Environ. Sci. Technol.* **2014**, *48*, 13685–13693. [CrossRef]
20. Tardy, V.; Casiot, C.; Fernandez-Rojo, L.; Resongles, E.; Desoeuvre, A.; Joulian, C.; Battaglia-Brunet, F.; Héry, M. Temperature and nutrients as drivers of microbially mediated arsenic oxidation and removal from acid mine drainage. *Appl. Microbiol. Biotechnol.* **2018**, *102*, 2413–2424. [CrossRef]
21. Abdul, K.S.M.; Jayasinghe, S.S.; Chandana, E.P.S.; Jayasumana, C.; Silva, P.M.C.S.D. Arsenic and human health effects: A review. *Environ. Toxicol. Pharmacol.* **2015**, *40*, 828–846. [CrossRef]
22. Abrosimova, N.A.; Saeva, O.P.; Bortnikova, S.B.; Edelev, A.V.; Korneeva, T.V.; Yurkevich, N.V. Metals and metalloids removal from mine water using natural and modified heulandite. *Int. J. Environ. Sci. Dev.* **2019**, *10*, 202–205. [CrossRef]

23. Cui, M.; Jang, M.; Cho, S.-H.; Khim, J.; Cannon, F.S. A continuous pilot-scale system using coal-mine drainage sludge to treat acid mine drainage contaminated with high concentrations of Pb, Zn, and other heavy metals. *J. Hazard. Mater.* **2012**, *215–216*, 122–128. [CrossRef] [PubMed]
24. Nicomel, N.R.; Leus, K.; Folens, K.; Van Der Voort, P.; Du Laing, G. Technologies for arsenic removal from water: Current status and future perspectives. *Int. J. Environ. Res. Public Health* **2016**, *13*, 62. [CrossRef]
25. Mondal, P.; Bhowmick, S.; Chatterjee, D.; Figoli, A.; Van der Bruggen, B. Remediation of inorganic arsenic in groundwater for safe water supply: A critical assessment of technological solutions. *Chemosphere* **2013**, *92*, 157–170. [CrossRef] [PubMed]
26. Hristovski, K.; Baumgardner, A.; Westerhoff, P. Selecting metal oxide nanomaterials for arsenic removal in fixed bed columns: From nanopowders to aggregated nanoparticle media. *J. Hazard. Mater.* **2007**, *147*, 265–274. [CrossRef] [PubMed]
27. Wang, S.; Zhao, X. On the potential of biological treatment for arsenic contaminated soils and groundwater. *J. Environ. Manag.* **2009**, *90*, 2367–2376. [CrossRef]
28. Zagury, G.J.; Kulnieks, V.I.; Neculita, C.M. Characterization and reactivity assessment of organic substrates for sulphate-reducing bacteria in acid mine drainage treatment. *Chemosphere* **2006**, *64*, 944–954. [CrossRef]
29. Costello, C. *Acid Mine Drainage: Innovative Treatment Technologies*; U.S. EPA Office of Solid Waste and Emergency Response Technology Innovation Office: Washington, DC, USA, 2003.
30. Blowes, D.W.; Ptacek, C.J.; Benner, S.G.; McRae, C.W.T.; Bennett, T.A.; Puls, R. Treatment of inorganic contaminants using permeable reactive barriers. *J. Contam. Hydrol.* **2000**, *45*, 123–137. [CrossRef]
31. Zawierucha, I.; Nowik-Zajac, A.; Kozlowski, C. Removal of Pb(II) ions using polymer inclusion membranes containing calix[4]resorcinarene derivative as ion carrier. *Polymers* **2019**, *11*, 2111. [CrossRef]
32. Almeida, M.I.G.S.; Cattrall, R.W.; Kolev, S.D. Recent trends in extraction and transport of metal ions using polymer inclusion membranes (PIMs). *J. Membr. Sci.* **2012**, *415–416*, 9–23. [CrossRef]
33. Cho, Y.; Cattrall, R.W.; Kolev, S.D. A novel polymer inclusion membrane based method for continuous clean-up of thiocyanate from gold mine tailings water. *J. Hazard. Mater.* **2018**, *341*, 297–303. [CrossRef]
34. Wieszczycka, K.; Staszak, K. Polymers in separation processes. In *Polymer Engineering*, 1st ed.; Tylkowski, B., Wieszczycka, K., Jastrzab, R., Eds.; De Gruyter: Poznan, Poland, 2017; pp. 235–276.
35. Zawierucha, I.; Nowik-Zajac, A.; Kozlowski, C. Application of Cr(VI) transport across the polymer inclusion membrane with calixresorcin[4]arene derivative as ion carrier. *Sep. Sci. Technol.* **2020**, *55*, 2204–2210. [CrossRef]
36. Nghiem, L.D.; Mornane, P.; Potter, I.D.; Perera, J.M.; Cattrall, R.W.; Kolev, S.D. Extraction and transport of metal ions and small organic compounds using polymer inclusion membranes (PIMs). *J. Membr. Sci.* **2006**, *281*, 7–41. [CrossRef]
37. Zawierucha, I.; Kozlowski, C.; Malina, G. Immobilized materials for removal of toxic metal ions from surface/groundwaters and aqueous waste streams. *Environ. Sci. Proc. Impacts* **2016**, *18*, 429–444. [CrossRef]
38. Guell, R.; Anticó, E.; Kolev, S.D.; Benavente, J.; Salvadó, V.; Fontàs, C. Development and characterization of polymer inclusion membranes for the separation and speciation of inorganic As species. *J. Membr. Sci.* **2011**, *383*, 88–95. [CrossRef]
39. Vera, R.; Antico, E.; Fontas, C. The use of a polymer inclusion membrane for arsenate determination in groundwater. *Water* **2018**, *10*, 1093. [CrossRef]
40. De Lourdes Ballinas, M.; De San Miguel, E.R.; De Jesus Rodriguez, M.T.; Silva, O.; Munoz, M.; De Gyves, J. Arsenic(V) removal with polymer inclusion membranes from sulfuric acid media using DBBP as carrier. *Environ. Sci. Technol.* **2004**, *38*, 886–891. [CrossRef] [PubMed]
41. Danesi, P.R. Separation of metal species by supported liquid membranes. *Sep. Sci. Technol.* **1984**, *19*, 857–894. [CrossRef]
42. Nowik-Zajac, A.; Zawierucha, I.; Kozlowski, C. Selective removal of silver(I) using polymer inclusion membranes containing calixpyrolles. *RSC Adv.* **2019**, *9*, 31122–31132. [CrossRef]
43. Pospiech, B.; Kujawski, W. Ionic liquids as selective extractants and ion carriers of heavy metal ions from aqueous solutions utilized in extraction and membrane separation. *Rev. Chem. Eng.* **2015**, *31*, 179–191. [CrossRef]
44. Marino, T.; Figoli, A. Arsenic removal by liquid membranes. *Membranes* **2015**, *5*, 150–167. [CrossRef]
45. Alguacil, F.J.; López-Delgado, A.; Alonso, M.; Sastre, A.M. The phosphine oxides Cyanex 921 and Cyanex 923 as carriers for facilitated transport of chromium (VI)-chloride aqueous solutions. *Chemosphere* **2004**, *57*, 813–819. [CrossRef] [PubMed]

46. Kaya, A.; Alpoguz, H.K.; Yilmaz, A. Application of Cr(VI) transport through the polymer inclusion membrane with a new synthesized calix[4]arene derivative. *Ind. Eng. Chem. Res.* **2013**, *52*, 5428–5436. [CrossRef]
47. Martinez Perez, M.E.; Reyes-Aguilera, J.A.; Saucedo, T.I.; Gonzalez, M.P.; Navarro, R.; Avila-Rodriguez, M. Study of As(V) transfer through a supported liquid membrane impregnated with trioctylphosphine oxide (Cyanex 921). *J. Membr. Sci.* **2007**, *302*, 119–126. [CrossRef]
48. US Environmental Protection Agency (EPA). *Method 8061A: Phthalate Esters by Gas Chromatography with Electron Capture Detection (GC/ECD)*; US Environmental Protection Agency: Cincinnati, OH, USA, 1996.
49. Chen, X.; Xu, S.; Tan, T.; Lee, S.T.; Cheng, S.H.; Lee, F.W.F.; Xu, S.J.L.; Ho, K.C. Toxicity and estrogenic endocrine disrupting activity of phthalates and their mixtures. *Int. J. Environ. Res. Public Health* **2014**, *11*, 3156–3168. [CrossRef] [PubMed]
50. Iberhan, L.; Wisniewski, M. Extraction of arsenic(III) and arsenic(V) with Cyanex 925, Cyanex 301 and their mixtures. *Hydrometallurgy* **2002**, *63*, 23–30. [CrossRef]
51. Migaszewski, Z.M.; Gałuszka, A.; Dołegowska, S. Extreme enrichment of arsenic and rare earth elements in acid mine drainage: Case study of Wisniowka mining area (south-central Poland). *Environ. Pollut.* **2019**, *244*, 898–906. [CrossRef]
52. Ruehl, M.D.; Hiibel, S.R. Evaluation of organic carbon and microbial inoculum for bioremediation of acid mine drainage. *Miner. Eng.* **2020**, *157*, 106554. [CrossRef]
53. Costa, M.C.; Duarte, J.C. Bioremediation of acid mine drainage using acidic soil and organic wastes for promoting sulphate-reducing bacteria activity on a column reactor. *Water Air Soil Pollut.* **2005**, *165*, 325–345. [CrossRef]
54. Fu, F.; Wang, Q. Removal of heavy metal ions from wastewaters: A review. *J. Environ. Manag.* **2011**, *92*, 407–418. [CrossRef]
55. Verma, V.; Tewari, S.; Rai, J. Ion exchange during heavy metal bio-sorption from aqueous solution by dried biomass of macrophytes. *Bioresour. Technol.* **2008**, *99*, 1932–1938. [CrossRef]
56. Cochrane, E.; Lu, S.; Gibb, S.; Villaescusa, I.; Gibb, S. A comparison of low-cost biosorbents and commercial sorbents for the removal of copper from aqueous media. *J. Hazard. Mater.* **2006**, *137*, 198–206. [CrossRef] [PubMed]
57. Landaburu-Aguirre, J.; Pongracz, E.; Peramaki, P.; Keiski, R.L. Micellar-enhanced ultrafiltration for the removal of cadmium and zinc: Use of response surface methodology to improve understanding of process performance and optimisation. *J. Hazard. Mater.* **2010**, *180*, 524–534. [CrossRef] [PubMed]
58. Yoon, J.; Amy, G.; Chung, J.; Sohn, J.; Yoon, Y. Removal of toxic ions (chromate, arsenate, and perchlorate) using reverse osmosis, nanofiltration, and ultrafiltration membranes. *Chemosphere* **2009**, *77*, 228–235. [CrossRef]
59. Lertlapwasin, R.; Bhawawet, N.; Imyim, A.; Fuangswasdi, S. Ionic liquid extraction of heavy metal ions by 2-aminothiophenol in 1-butyl-3-methylimidazolium hexafluorophosphate and their association constants. *Sep. Purif. Technol.* **2010**, *72*, 70–76. [CrossRef]
60. Akbal, F.; Camci, S. Copper, chromium and nickel removal from metal plating wastewater by electrocoagulation. *Desalination* **2011**, *269*, 214–222. [CrossRef]
61. Pohl, A. Removal of heavy metal ions from water and wastewaters by sulfur-containing precipitation agents. *Water Air Soil Pollut.* **2020**, *231*, 503. [CrossRef]
62. Qdais, H.A.; Moussa, H. Removal of heavy metals from wastewater by membrane processes: A comparative study. *Desalination* **2004**, *164*, 105–110. [CrossRef]
63. Ruparelia, J.P.; Duttagupta, S.P.; Chatterjee, A.K.; Mukherji, S. Potential of carbon nanomaterials for removal of heavy metals from water. *Desalination* **2008**, *232*, 145–156. [CrossRef]
64. Liu, J.; Zhao, Z.; Jiang, G. Coating Fe_3O_4 magnetic nanoparticles with humic acid for high efficient removal of heavy metals in water. *Environ. Sci. Technol.* **2008**, *42*, 6949–6954. [CrossRef] [PubMed]

Publisher's Note: MDPI stays neutral with regard to jurisdictional claims in published maps and institutional affiliations.

minerals

Article

Application of GETFLOWS Coupled with Chemical Reactions to Arsenic Removal through Ferrihydrite Coprecipitation in an Artificial Wetland of a Japanese Closed Mine

Tatsuya Kato [1], Yohei Kawasaki [1], Masakazu Kadokura [1], Kohei Suzuki [1], Yasuhiro Tawara [2], Yoshiyuki Ohara [3] and Chiharu Tokoro [4,*]

1. Graduate School of Creative Science and Engineering, Waseda University, 3-4-1 Okubo, Shinjuku-ku, Tokyo 169-8555, Japan; tatsuya.kato@aoni.waseda.jp (T.K.); far-flat@suou.waseda.jp (Y.K.); masa-kado0413@akane.waseda.jp (M.K.); kouhei-0403-su@ruri.waseda.jp (K.S.)
2. Geosphere Environmental Technology Corporation, 2-1 Kanda-Awaji-cho, Chiyoda-ku, Tokyo 101-0063, Japan; tawara@getc.co.jp
3. Ningyo-toge Environmental Engineering Center, Japan Atomic Energy Agency, 1550 Kamisaibara, Kagamino-cho, Tomata-gun, Okayama 708-0698, Japan; ohara.yoshiyuki@jaea.go.jp
4. Faculty of Science and Engineering, Waseda University, 3-4-1 Okubo, Shinjuku-ku, Tokyo 169-8555, Japan

* Correspondence: tokoro@waseda.jp; Tel.: +81-3-5286-3320

Received: 5 April 2020; Accepted: 20 May 2020; Published: 23 May 2020

Abstract: Passive systems that utilize a natural power such as a pond, plant, or microorganisms, is expected to be a cost-effective method for acid mine drainage (AMD) treatment. The Ningyo-toge mine, a non-operational uranium mine located in Okayama Prefecture, Japan, generates AMD containing arsenic and iron. To quantitatively study arsenic and iron ion removal in an artificial wetland and pond, chemical reactions were modeled and incorporated into the GETFLOWS (general purpose terrestrial fluid-flow simulator) software. The chemical reaction models consisted of arsenite and ferrous oxidation equations and arsenic adsorption on ferrihydrite. The X-ray diffraction analysis of sediment samples showed ferrihydrite patterns. These results were consistent with the model for arsenite/ferrous oxidation and arsenic adsorption on ferrihydrite. Geofluid simulation was conducted to simulate mass transfer with the utilized topographic model, inlet flow rate, precipitation, and evaporation. The measured arsenic and iron ions concentrations in solution samples from the wetland and pond, fitted well with the model. This indicated that the main removal mechanism was the oxidation of arsenite/ferrous ions and that arsenic was removed by adsorption rather than dilution.

Keywords: GETFLOWS; acid mine drainage; the three-dimensional topographic model; quantitative modeling; surface complexation

1. Introduction

Acid mine drainage (AMD) from abandoned or closed mines is an environmental problem associated with mining [1–3]. AMD is produced when sulfide minerals (e.g., pyrite, sphalerite, and chalcopyrite) react with oxygen and groundwater. Thus, it often contains toxic ions, such as iron (Fe(II) and Fe(III)), zinc (Zn(II)), lead (Pb(II)), and arsenic (As(III) and As(V)). To avoid contamination around mines, AMD has mainly been treated by a neutralization process to add a neutralizer, such as calcite ($CaCO_3$) and slaked lime ($Ca(OH)_2$) [4]. This process can effectively remove toxic ions from AMD through precipitation as hydroxide or adsorption of hydroxides, such as ferrihydrite ($Fe(OH)_3$) and aluminum hydroxide ($Al(OH)_3$) [5–7].

In Japan, it is necessary to treat AMD in about 100 abandoned or closed mines. To this end, the Japanese government provides financial support for AMD treatment and spends billions of yen each year [8]. Since it was reported in our previous studies that an AMD treatment has been a necessity for a long time [9,10], it is predicted that large funds will be needed to address this. To achieve a sustainable AMD treatment for the future, cost reduction of the neutralization process is required.

Many research work on passive treatment has been attracting attention recently, in a bid to reduce the cost of AMD treatment. The passive treatment utilizes natural resources such as geographical features, microorganisms, and plants, for AMD treatment [11–13]. AMD treatment by passive methods incurs lower management costs for materials, power, and labor than an active treatment, which mainly consists of the neutralization process. AMD treatment using an artificial wetland and pond is one of the methods of passive treatment. In previous studies [14–16], many researchers reported that AMD treatment using an artificial wetland and pond was effective for removing toxic ions. According to previous studies, it was suggested that not only chemical but also biological reactions, such as sulfate reduction, iron, and manganese oxidation using bacteria, occurred in an artificial wetland and pond. However, quantitative modeling based on removal mechanisms has not been conducted, due to the complexity of the reaction. To achieve a more efficient AMD treatment using an artificial wetland and pond, it is important to develop quantitative modeling based on removal mechanisms.

The objective of this study was to elucidate the mechanism of toxic ion removal, especially arsenic, using an artificial wetland and pond, and to quantify it using geochemical modeling software. In this study, the target field was the Yotsugi mill tailings pond in the Ningyo-toge mine [17–19], which is partially an artificial wetland near the inlet. In an artificial wetland and pond, arsenic and iron were removed satisfactorily enough to meet the strict local additional standards, but it is currently unknown whether the mechanism was due to dilution or chemical reaction. To identify the removal mechanisms, we performed a water quality survey and field sampling. Furthermore, a surface model was constructed using a contour map and the flow field of the water in the mineral deposit field was modeled using GETFLOWS software [20–22]. GETFLOWS is a simulation software originally developed by one of the authors and used to calculate the mass transport processes [20–27]. In this study, the chemical reactions were originally coupled with it, according to the authors' previous expertise on ferrous and arsenite ion oxidation, ferric ion precipitation, and coprecipitation of arsenite and arsenate to iron precipitates [5–7]. More specifically, based on our previous findings and the field survey results for concentration of coexisting ions in the pond, it was assumed that ferrihydrite should be the main iron precipitate, and surface complexation of arsenite and arsenate to ferrihydrite should be the dominant reaction in the coprecipitation process in the pond and the wetland, due to the ratio between iron and arsenic. A chemical reaction model was constructed based on these assumptions and we confirmed whether our assumptions were correct by comparing the results of the field survey and the analysis results. Through these procedures, we elucidated the removal mechanism of iron and arsenic ions in the artificial wetland and pond in an actual closed mining site in Japan, based on the consistency between the field survey results minimally required and the calculation results that aggregated the findings in previous mechanism investigations.

2. Materials and Methods

2.1. Field Survey and Analysis

The Ningyo-toge mine is a closed uranium mine located in the Okayama Prefecture, Japan (35°18′47.58″ N, 133°55′52.21″ E) [17–19]. The Yotsugi mill tailings pond, an artificial pond, was constructed to receive slag and other materials generated by the smelter during operation, and now has partially become an artificial wetland near the inlet. After the closure of the smelting facility in 1982, this tailing pond was utilized as a remediation pond [23]. The pond has four main inlets as sources of mine drainage from the surrounding area that flow through underground pipes towards the pond—one comes from an opencast mine, two are from underground mines, and one is from the upstream part of the mill tailings pond (Figure 1). The AMD from the four inlets flow towards a dam located at the western end of the pond. The depth of the Yotsugi mill tailings pond in summer and winter was 1.4 m and 1.2 m at point H, respectively. This decreased towards the upstream area of the Yotsugi mill tailing pond from point H to A. The depth at point A was 0.0 m.

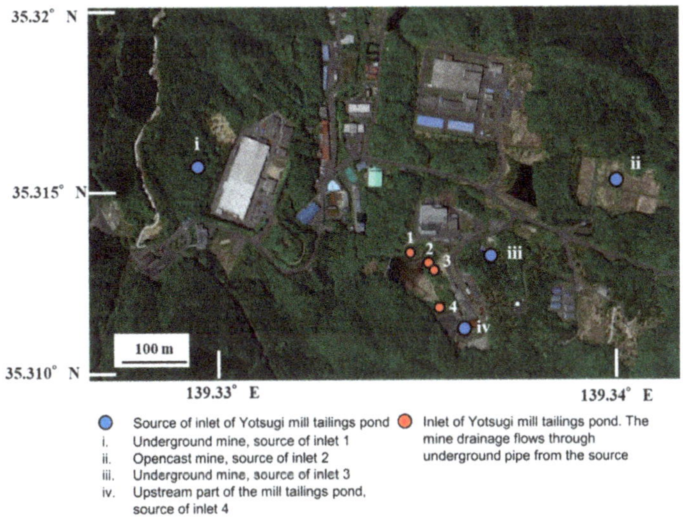

Figure 1. Inlet points of acid mine drainage in the Yotsugi mill tailings pond.

Field surveys were conducted three times in the summer and winter of 2016, respectively. For each survey, three water samples were collected from the source of each of the four inlets (points i–iv, Figure 1). In addition, three water and sediment samples at each point of the surface were collected along the Yotsugi mill tailings pond in the direction of water flow (Figure 2 and Table 1). Using a thin tube with a diameter of 20 mm, a 100 mm thick sedimentary layer was collected from the bottom of the Yotsugi tailings mill pond.

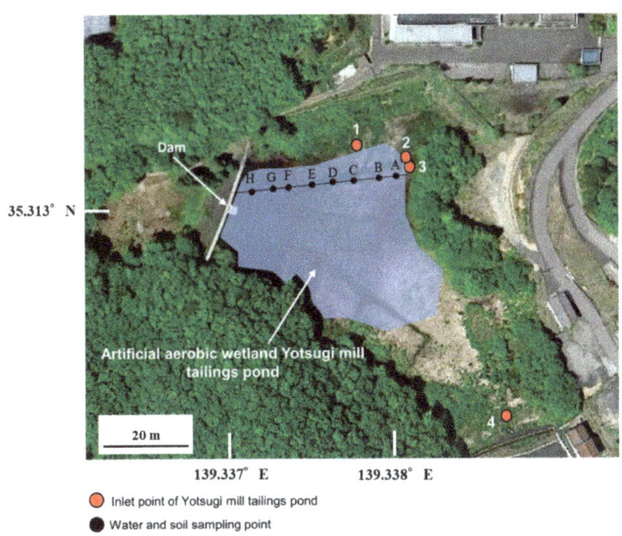

Figure 2. Sampling points along the Yotsugi mill tailings pond.

Table 1. The distance of the sampling points from A in the Yotsugi mill tailings pond (m).

Point	A	B	C	D	E	F	G	H
Distance	0	10	27	44	60	70	85	100

The pH, temperature, oxidation–reduction potential (ORP), electric conductivity (EC), and concentration of dissolved oxygen (DO) of the water samples were measured on-site, using a pH/DO meter (OM-70, HORIBA, Kyoto, Japan). After the measurement of pH, temperature, ORP, EC, and DO concentration, the water samples were filtered using a 0.45-μm membrane filter. We also confirmed that no colloidal particles were present in the sample water after extra filtration. The concentration of all cations in the filtrates was analyzed by an inductively coupled plasma-mass spectrometer (ICP-MS, 7700X, Agilent, Santa Clara, CA, USA), using the mode of semi-quantitative analysis. The concentration of a ferrous ion in the inlet was measured beforehand by the phenanthroline absorption spectrophotometric method, and the total iron ion was almost the same as the concentration of the ferrous ion. The concentration of the arsenite ion in the inlet was measured beforehand through separation, using high selectivity inorganic separating bulk (MetaSEP AnaLig TE-01, GL Sciences, Tokyo, Japan) [24], and the total arsenic ion was almost the same as the concentration of the arsenite ion. Therefore, the valency of iron and arsenic ions in the inlet was treated as ferrous and arsenite ions, respectively. Each sample was measured three times, and it was confirmed that their error was within 5%. Other cation ions of iron and arsenic obtained from the ICP–MS analysis resulted in less than several mg/L of sodium, potassium, calcium, magnesium, and aluminum. Sulfate and chloride ions were also measured by ion chromatography (IC, IC850, Metrohm, Tokyo, Japan) and were below the 50 mg/L. From the chemical equilibrium calculation using PHREEQC [4], we neglected the effect of these ions for the removal reaction of iron and arsenic ions in this study.

The sediment samples were vacuum dried and the mineral phases were ascertained using X-ray diffraction analysis (XRD, RINT Ultima III, Rigaku, Tokyo, Japan), with Cu-Kα radiation at 40 kV and 30 mA. The scan range was from 2°–80°, with a scan rate of 2°/min. The XRD specimens on the plane glass holder were prepared using a front-loading technique. The objective of the XRD analysis in this study was to detect low crystalline materials of an iron hydroxide, such as 2-line ferrihydrite. Therefore, to observe the peak of the low crystalline materials, background subtraction was not conducted on the obtained XRD patterns.

2.2. Simulation Model Using GETFLOWS

The GETFLOWS simulation software was used to simulate the mass transport processes in the Yotsugi mill tailings pond. GETFLOWS is a fully distributed and integrated watershed modeling tool that treats the transport processes of water, air, various dissolved and volatilized materials, suspended sediments in water, and heat from the surface to the underground, on arbitrary temporal-spatial scales. The surface water flow and dissolved materials transport in surface water functions were applied in this study.

The governing equation of surface water was based on the mass conservation law, as described in Equation (1):

$$-\nabla \cdot M_w + \rho_w q_w = \frac{\partial(\rho_w S_w)}{\partial t} \tag{1}$$

where ∇ is the differential operator, M_w is the mass flux of surface water (kg/m^2/s), ρ_w is the density of water (kg/m^3), q_w is the volumetric flux of the sink and source (m^3/m^3/s), S_w is the saturation of water calculated from the surface water depth and the grid height (m^3/m^3), and t is the time (s).

The simulator solved the Manning's law for estimating the surface water flow velocity, and it also solved the depth-averaged diffusion wave approximation of open channel shallow water Saint-Venant equations in two dimensions, for modeling the flows of water on land, in streams, and on slopes. Instead of solving the momentum conservation equations following Tosaka et al. [25], we embedded the velocity field in the flow term in Equation (1) (i.e., the first term on the left-hand side). The mass flux of surface water per unit area M_w in flow terms for surface flow could be expressed by adapting Manning's law as Equation (2):

$$M_w = -\frac{\rho_w R^{\frac{2}{3}}}{n_l}\sqrt{\left|\frac{\partial h}{\partial l}+\frac{\partial z}{\partial l}\right|}sgn\left(\frac{\partial h}{\partial l}+\frac{\partial z}{\partial l}\right),\ (l = x,\ y) \tag{2}$$

where n_l is Manning's roughness coefficient ($m^{-1/3}$ s), R is the hydraulic radius (m), h is the surface water depth (m), l is the distance of surface water flow direction (m), and z is the elevation from the datum level (m). The surface water depth h was computed from the water saturation S_w and the height of the grid block in the surface environment.

The mass balance equation could be modified as Equation (3), when considering the behavior of the reactive substance i dissolved in water [26]:

$$-\nabla\cdot\left(M_w C_{w,i}\right)+\nabla\cdot D_{w,i}\nabla\left(\rho_w C_{w,i}\right)+\rho_w q_w C_{w,i}-\sum_{\substack{ir=1\\ ir\neq i}}^{NC} m_{w,i\to ir}+\sum_{\substack{ir=1\\ ir\neq i}}^{NC} m_{w,ir\to i}-m_{w\to s,i}=\frac{\partial\left(\rho_w S_w C_{w,i}\right)}{\partial t} \tag{3}$$

where $C_{w,i}$ is the mass fraction of dissolved material i in water (kg/kg), $D_{w,i}$ is the diffusion coefficient of dissolved material i (m^2/s), $m_{w,i1\to i2}$ is the rate of generation/decomposition of a chemical reaction from dissolved material $i1$ to $i2$ in water (kg/m^3/s), and $m_{w\to s,i}$ is the adsorption rate of the dissolved material i (kg/m^3/s). The first term on the left-hand side of Equation (3) is the advection term, the second term is the diffusion term, and the third term is the sink/source term for drainage and precipitation. The fourth and fifth terms are the decomposition and generation terms for chemical reactions in water, respectively. The sixth term represents a transfer from the water phase to the solid phase by adsorption.

These governing equations were discretized spatially using the integral finite difference method and temporally by the fully implicit method. Due to their nonlinearity, the governing equations were iteratively converged using the Newton–Raphson method. Matrix factorization was performed using the conjugate residual method, with preconditioning by nested factorization, which targeted the nested structure of the structured mesh. To speed up the numerous calculations, we used the successive locking process [27] to remove the grids where sufficient convergence was achieved in the nonlinear iterative process. The governing equations, required data, and solution procedures were explained in detail in a previous paper [26]. To ensure its accuracy and applicability, the GETFLOWS simulator was applied to many verification test problems involving theoretical and analytical solutions. In addition, GETFLOWS was applied to more than 500 sets of field and laboratory data in Japan and overseas to validate its utility [20–22,26,28].

A three-dimensional topographic model of the Yotsugi mill tailings pond was developed using the Geographic Information System software MapInfo®, with field elevation data provided by the Japan Atomic Energy Agency [23]. The three-dimensional topographic model was used to perform geosphere fluid analysis. The model domain was set as the catchment area of the pond to consider all water flowing into the pond. The ridge surrounding the pond was set as the boundary of the analysis area (Figure 3). No flow boundary condition was applied except for a pond outlet. All water inflow from outside of the pond was considered as surface water.

Advection–diffusion analysis was performed using GETFLOWS. The topographic model, inlet flow rate, precipitation, evaporation, and water quality were utilized as input data in this simulation. Percolation and recharge in the ground did not need to be considered. The Manning's roughness coefficient was set to 0.03 as a parameter related to the surface flow. Precipitation and evaporation data were measured and provided by the Japan Atomic Energy Agency and the inlet flow rates were provided by the Ningyo-toge Environmental Technology Center [23].

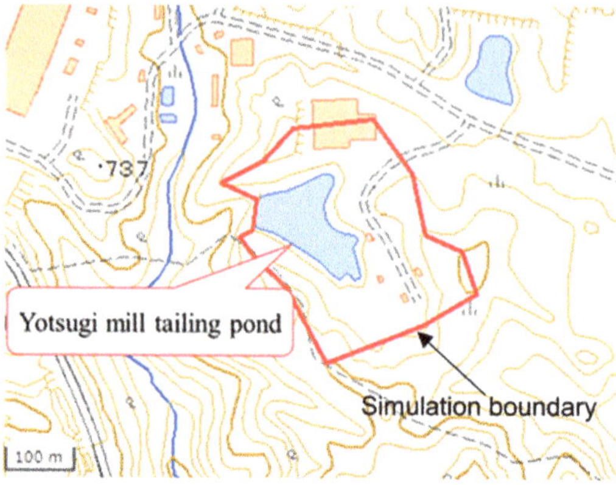

Figure 3. The boundary of the simulation area (the red line), which was set on the ridges surrounding the Yotsugi mill tailings pond.

2.3. Kinetics Model

In geochemical modeling, the kinetics model of both ferrous and arsenite ions oxidation was considered in this study. For the oxidation reaction of ferrous, the kinetics model used in the quantitative model is considered in Equation (4), as proposed by Singer and Stumm [29]:

$$\frac{d[\mathrm{Fe(II)}]}{dt} = -(k_1 + k_2[\mathrm{OH^-}]^2 P_{O2})[\mathrm{Fe(II)}] \quad (4)$$

where [Fe(II)] is the activity of ferrous ion; $[OH^-]$ is the activity of the hydroxide ion; P_{O2} is the oxygen partial pressure; and k_1 and k_2 are rate constants with reference values of 2.91×10^{-9} and 1.33×10^{12}, respectively. Equation (4) has been used by many researchers, including our group, to show that the kinetics of ferrous oxidation depend on pH and DO. Although the kinetic constants of k_1 and k_2 should be affected by several chemical and biogeochemical conditions, the usefulness of these values was confirmed under the atmosphere and without any biogeochemical influence.

For the oxidation reaction of arsenite, the kinetics model used in the quantitative model is reported in Equation (5):

$$\frac{d[As(III)]}{dt} = -k_3[As(III)] \quad (5)$$

where *[As(III)]* is the activity of arsenite; k_3 is a rate constant with a reference value of 3.0×10^{-3} [30].

The ORP in the Yotsugi mill tailings pond was almost stable at around 0.2 V. According to the Eh-pH diagram reported by Bednar et al. [31], the dominant species of arsenic was $HAsO_4^-$ as an arsenate ion. Bednar et al. [31] reported that arsenite was oxidized according to Equation (6) in the acidic pH area, when iron and arsenic ions co-existed.

$$H_2AsO_3^0 + H_2O + 2Fe^{3+} \rightarrow H_2AsO_4^- + 3H^+ + 2Fe^{2+} \quad (6)$$

However, it was considered that the oxidation reaction of arsenite did not proceed according to Equation (6) because our target pH area was a neutral pH area and iron ion was precipitated. Thus, we did not consider the oxidation reaction of arsenite, according to Equation (6) in this study.

Bednar et al. [31] also reported that the ratio of arsenate/total-arsenic in the acidic pH area was larger than that in the neutral pH area. The reason for this was the arsenite/arsenate adsorption reaction for iron (hydr)oxide (e.g., ferrihydrite and magnetite) in the neutral pH area [31–34]. Since iron (hydr)oxide adsorbed more arsenate than arsenite in the neutral pH area, the ratio of arsenate was small. It was considered that this arsenite/arsenate adsorption reaction occurred in our target system.

Based on the above discussion, we introduced Equation (5) to show the kinetics of the oxidation reaction of arsenite. Since the ORP was almost stable in the Yotsugi mill tailings pond, it was treated as a constant value and unified into the kinetics constant of k_3 in Equation (5).

2.4. Surface Complexation Model

A previous comparison between coprecipitation and adsorption experiments for arsenate adsorption on ferrihydrite with various As/Fe molar ratios confirmed that arsenate was predominantly adsorbed as surface precipitation when the As/Fe molar ratio was over 0.4, whereas it was predominantly adsorbed by surface complexation when the As/Fe ratio was below 0.4 [5,35]. In our previous study, we used the Dzombak and Morel model to model the adsorption of arsenate on ferrihydrite and found that the adsorption experimental results were successfully represented to the model when we combined the diffusion layer model with the Dzombak and Morel parameter. In this study, the As/Fe molar ratio in the Yotsugi mill tailings pond was less than 0.4 at all points, and surface complexation was considered to be the dominant process in the adsorption of arsenic onto ferrihydrite. Therefore, the Langmuir model with the coefficient from Dzombak and Morel model was used in this study. The Langmuir adsorption isotherm equation used in this study is shown in Equation (7):

$$W_s = \frac{W_{s\ max} aC}{1 + aC} \tag{7}$$

where W_s is the amount adsorbed (mol/mol-Fe), $W_{s\ max}$ is the saturation adsorption (mol/mol-Fe), a is the adsorption equilibrium constant, and C is the concentration (mg/L). The Langmuir constants are shown in Table 2.

Table 2. The Langmuir constants of arsenite and arsenate adsorption on ferrihydrite.

	Langmuir Constant	Amount of Saturated Adsorption [mol/mol-Fe]
Arsenate (As(V))	2.27×10^6	2.70×10^{-1}
Arsenite (As(III))	1.13×10^6	8.82×10^{-2}

3. Results and Discussion

3.1. Field Survey

The quality of each water sample obtained from the field surveys in summer and winter are shown in Tables 3 and 4. The average pH throughout the Yotsugi mill tailings pond and at each inlet was almost pH 7. The temperature of the water sample was almost stable at around 13 °C in winter, whereas it was 15 °C at point A, and gradually increased downstream up to 30 °C at point H in summer. Since point A was near the inlet and the surrounding area was an artificial wetland, it was directly affected by the raw water in the mine, which had a stable temperature of around 15 °C regardless of the season. On the other hand, since point H had a long residence time and the surrounding area was a pond, it was easily affected by the outside temperature. The EC was around 30 mS/cm in summer and around 0.2 mS/cm in winter.

The concentrations of arsenic and iron ions in summer and winter were mostly affected by inlets 2 and 3, which were located close to each other. These decreased towards the downstream area of the pond from point A to H.

Table 3. Properties of the water sample obtained during field survey in summer 2016.

Point	TEMP	pH (–)	DO	EC	ORP	Concentration (mg/L)									
						Si	Fe^{2+}	As	Na	K	Ca	Mg	Al	SO_4	Cl
I	15.6	6.5	8.7	17	0.1	24	4.1	6.1×10^{-3}	17	2.3	18	6.3	2.8×10^{-3}	14	35
Ii	11.9	6.5	8.8	31	0.03	1.1×10^2	16	1.4×10^{-2}	55	3.2	21	10	5.8×10^{-3}	1.3×10^2	32
Iii	12.9	6.6	4.9	7.7	0.2	40	0.6	1.3×10^{-3}	11	1.4	6.9	3.5	6.6×10^{-3}	8.8	31
Iv	18.3	6.5	2.5	0.1	0.3	13	0.4	9.9×10^{-3}	14	1.1	13	1.6	1.5×10^{-2}	25	29
A	14.8	7.0	4.2	28	0.2	1.1×10^2	17	1.4×10^{-3}	54	2.5	18	9.8	4.0×10^{-3}	1.4×10^2	18
B	17.1	6.8	2.0	23	0.2	79	1.6	1.4×10^{-2}	45	5.7	14	7.9	7.2×10^{-3}	1.1×10^2	21
C	18.6	6.6	3.6	17	0.3	37	0.2	7.9×10^{-3}	17	3.1	23	5.8	3.7×10^{-3}	46	16
D	19.1	7.1	0.5	21	0.03	32	3.5	5.8×10^{-3}	17	1.0	25	5.6	6.1×10^{-3}	47	16
E	22.0	6.5	0.4	42	0.02	32	6.0	7.6×10^{-3}	18	0.7	68	7.6	3.8×10^{-3}	2.3×10^2	20
F	26.7	7.0	3.1	0.5	0.1	31	0.1	2.7×10^{-3}	17	0.6	54	4.7	4.9×10^{-3}	1.8×10^2	20
G	32.1	7.2	6.2	0.3	0.2	29	2.2×10^{-3}	1.1×10^{-3}	16	0.3	53	4.4	2.3×10^{-3}	1.6×10^2	19
H	32.4	7.3	3.1	0.3	0.3	28	2.2×10^{-3}	1.2×10^{-3}	15	0.4	49	4.5	2.7×10^{-3}	1.5×10^2	39

TEMP: temperature (°C]; DO: dissolved oxygen (mg/L); EC: electric conductivity (mS/cm); and ORP: oxidation-reduction potential (V).

Table 4. Properties of the water sample obtained during field survey in winter 2016.

Point	TEMP	pH (–)	DO	EC	ORP	Concentration (mg/L)									
						Si	Fe^{2+}	As	Na	K	Ca	Mg	Al	SO_4	Cl
I	12.8	6.4	7.8	0.2	0.1	24	2.0	1.2×10^{-2}	9.9	1.3	13	3.5	0.2×10^{-4}	12	0.0
Ii	10.2	6.5	8.7	0.3	0.03	72	8.2×10^{-2}	8.2×10^{-3}	36	2.0	14	6.5	0.1×10^{-4}	98	0.0
Iii	10.8	6.6	4.7	0.1	0.3	35	0.4	5.6×10^{-3}	6.2	0.9	6.1	2.5	0.0	7.7	0.0
Iv	11.1	6.7	3.1	0.2	0.06	14	3.7	5.3×10^{-2}	11	1.5	16	1.6	0.3×10^{-4}	45	0.0
A	11.6	6.6	4.7	0.1	0.2	72	0.5	3.0×10^{-3}	13	1.0	6.9	3.0	0.0	48	0.0
B	12.0	7.0	6.7	0.2	0.2	35	0.2	2.0×10^{-3}	14	1.2	6.3	2.8	0.0	15	7.2
C	11.8	6.9	5.6	0.2	0.3	22	0.4	2.2×10^{-3}	10	1.4	14	3.1	0.0	14	11
D	12.3	6.6	1.0	0.2	0.2	18	0.8	3.5×10^{-3}	8.9	1.2	15	3.1	0.0	13	9.7
E	11.4	6.9	0.8	0.3	0.2	16	0.2	3.1×10^{-3}	8.9	0.9	24	2.8	0.0	32	8.8
F	12.2	6.7	0.8	0.7	0.01	19	0.3	2.5×10^{-3}	9.8	0.7	80	4.1	0.0	1.5×10^2	9.3
G	11.8	7.0	6.2	0.3	0.2	20	0.9	1.5×10^{-3}	9.3	1.2	21	3.4	0.0	0.02	7.7
H	11.9	7.0	3.3	0.3	0.3	22	0.1	1.5×10^{-3}	11	1.3	30	3.5	0.1×10^{-4}	94	0.0

TEMP: temperature (°C); DO: dissolved oxygen (mg/L); EC: electric conductivity (mS/cm); and ORP: oxidation-reduction potential (V).

From the concentrations silicon, iron, and arsenic ions are shown in Tables 3 and 4, we confirmed that ferrihydrite (pK= –3.191; $Fe(OH)_3 + 3H^+ = Fe^{3+} + 3H_2O$ for $Fe(OH)_3$) was saturated in the pond and ferric arsenate (pK= –0.4; $FeAsO_4{:}2H_2O + 3H^+ = Fe^{3+} + H_3AsO_4{}^0 + 2H_2O$ for $FeAsO_4{:}2H_2O$) and amorphous silica (pK= 2.71; $SiO_2 + 2H_2O = H_4SiO_4{}^0$ for SiO_2) were unsaturated [36]. Based on our previous study [4] and pre-investigation by chemical equilibrium calculation, including surface complexation model, it was confirmed that silicate precipitates were not formed and the effect of silicon for arsenic removal by ferrihydrite was very small and could be ignored at neutral pH, due to the low concentration of silicon ion in the pond. Thus, silicon could be used as a tracer to understand the mass transfer behavior in the Yotsugi mill tailings pond.

3.2. X-ray Diffraction Analysis

Figure 4 shows the XRD patterns of the sediment samples, ferrihydrite, and the original soil sample of the Yotsugi mill tailings pond in summer and winter 2016. From a previous paper [37], all XRD peaks of the original soil was that of gypsum ($CaSO_4{\cdot}2H_2O$). Comparing the peak of sediment samples and the original soil, the peak position of sediment samples was the same as that of the original sample. Additionally, the XRD patterns of sediment samples showed the broad peaks at 2θ 34.9° and 62.3°, which corresponded to the 2-line ferrihydrite ($Fe(OH)_3$) [38,39].

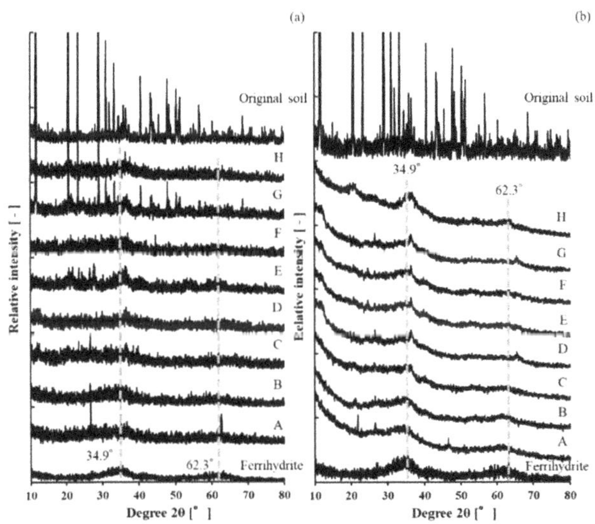

Figure 4. X-ray diffraction patterns of sediment samples (**a**) in summer and (**b**) winter.

The iron ion is known to be precipitated as several species (e.g., ferrihydrite, schwertmannite, and green rust). In our previous study, it was also confirmed that 2-line ferrihydrite is the most dominant as a fresh iron (hydr)oxide in AMD at neutral pH area, using XRD analysis [34]. The XRD pattern of sediment samples (Figure 4) was the same as that of the 2-line ferrihydrite reported in our previous paper [35]. Schwertmannite has been known as a good adsorbent of arsenic ion in AMD [40], but it was stable at acidic pH area and the chemical equilibrium calculation showed that ferrihydrite formation should be more dominant than schwertmannite due to the neutral pH area in the pond [41]. Schwertmannite has a specific peak around 2θ 36°, which was sharper than the 2-line ferrihydrite, but Figure 4 shows no clear evidence of the existence of schwertmannite [42]. Green rust is also known to remove many different kinds of anions from AMD; however, a more reducing atmosphere is necessary for green rust formation [43–45]. Therefore, it is suggested that all iron in Yotsugi mill tailings pond was precipitated as 2-line ferrihydrite. Several researchers have conducted a more detailed investigation of iron (hydr)oxide during ferrous-mediated autotrophic denitrification with some crystalline phases, such as goethite, hematite, akaganeite, and maghemite. In this study, denitrification was not dominant and Figure 4 provides no clear evidence of the existence of such crystalline phases; however, detailed detection of minor iron phases using Fourier transform Raman Spectroscopy infrared spectroscopy and a discussion of their effect on arsenic removal will be part of our future work [46,47].

The broad peak position of 2-line ferrihydrite did not change when the arsenic was removed by surface complexation for 2-line ferrihydrite in our previous study [35]. On the other hand, it changed from 34° to 28° when arsenic was removed by surface precipitation [5,35]. Thus, since there was no obvious peak shift around 28° in XRD patterns, it was confirmed that the arsenic removal mechanism in Yotsugi mill tailings pond was a simple surface complexation on the 2-line ferrihydrite.

3.3. Geofluid Analysis

The geofluid analysis simulation was performed using the precipitation given in the surface model based on elevation data. The inlet flow rates, precipitation, and evaporation measured during summer and winter are presented in Table 5. The precipitation and evaporation values were averaged for the whole simulation area (i.e., the pond and catchment area). The simulation results are presented in Figure 5. The simulation suggested that the influent from the catchment area flowed into the Yotsugi mill tailings pond in the lower part of the simulation area. The flow lines of the Yotsugi mill tailings pond on the northern and southern sides did not converge downstream, which indicated that these flows were separate.

Table 5. Inlet flow rates and average precipitation and evaporation in the Yotsugi mill tailings pond in summer and winter 2016.

Flow	Summer	Winter
Precipitation (mm/day)	6.6	5.8
Evaporation (mm/day)	1.6	0.7
Inlet flowrate point 1 (m^3/day)	169.8	254.1
Inlet flowrate point 2 (m^3/day)	34.9	29.4
Inlet flowrate point 3 (m^3/day)	11.6	16.2
Inlet flowrate point 4 (m^3/day)	7.52	5.45

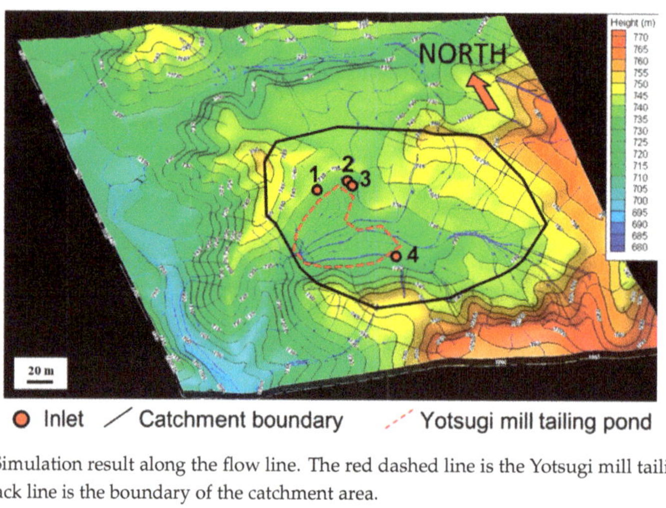

Figure 5. Simulation result along the flow line. The red dashed line is the Yotsugi mill tailings pond and the black line is the boundary of the catchment area.

3.4. Quantitative Modeling

The distribution map of the concentration of silicon ion in the summer and winter and the relationship between measured values and simulation results are shown in Figure 6. The concentration of silicon ion was greatly affected by the inflow from inlet 2, which was the discharge from the opencast mine (point ii). The simulation results matched well with the measured values (Figure 6c), indicating that the transport of silicon ion followed a mass transfer model without any reaction. These results also suggested that they were treated as flow field data and the dilution of silicon ion between points C and A occurred because of flow from the other inlets, which had lower concentrations and rainfall. The simulation results in summer showed that dilution contributed 74.6% to the concentration of silicon ion. Similarly, in winter, the contribution of dilution was 70.8%. The difference in percentages between summer and winter was assumed to be caused by differences in the inflow volumes of AMD to the Yotsugi mill tailings pond.

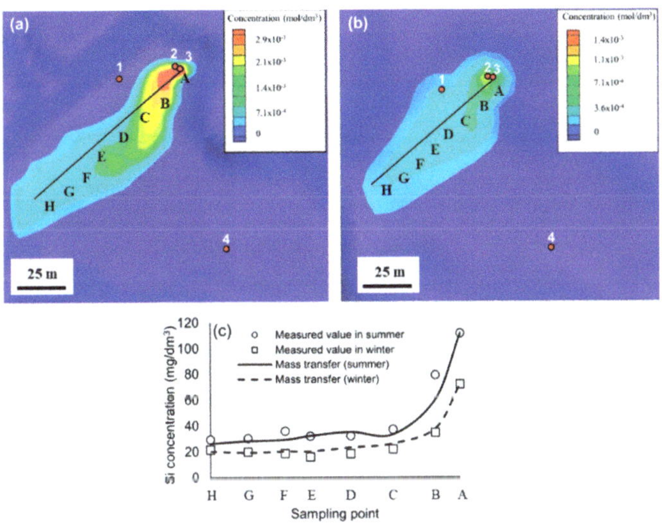

Figure 6. Simulation result: Distribution map of Si concentrations (**a**) in summer and (**b**) in winter, and (**c**) measured Si concentrations on lines A–H and the fitting results.

In the Yotsugi mill tailings pond, iron (hydr)oxides occurred because the pH was around 7. As the ferric (which was produced by the oxidation of ferrous) immediately precipitated as the 2-line ferrihydrite, it was considered that ferric ion was not included in the water. Thus, the ferric ion was not included in our model. The iron distribution was simulated by advection–diffusion using the oxidation reaction. Distribution maps of an iron in summer and winter were simulated using GETFLOWS (Figure 7). The concentration of an iron ion in the Yotsugi mill tailings pond in summer was mostly affected by AMD from inlet 2, which had the highest influent concentration among the inlets and a high flow rate. The concentration of the iron ion decreased in the flow direction and reached the lowest concentration at downstream point A. The concentration of iron ion was low in winter because the average concentration of iron ion from all of the inlets was low. The models for the mass transfers and reactions with measured values are depicted in Figure 7c,d. The concentration of iron ion dropped suddenly at point B, which indicated that the ferrous ion was oxidized to ferric ion and immediately precipitated. At several points, such as points D and E in summer and points C and D in winter, there were concentration increases. At these points, the concentration of iron ion was affected by the adjacent inflow from inlet 1. Overall, the measured values in summer and winter were successfully represented with the reaction model, which suggested that the kinetics model of oxidation of ferrous derived from the laboratory could be applied to the field. This result indicated that oxidation of ferrous was caused by the activity of the hydroxide ion and the oxygen partial pressure. In previous studies, it was reported that the oxidation of ferrous should be affected by not only chemistry but also biogeochemistry [32]. These factors should directly change ferrous speciation and concentration, DO, pH, and the kinetic constant, which should be affected by both the situation of dispersion of oxygen and the interface of precipitates in Equation (4). In our system, it was confirmed that the effect of biogeochemistry was small enough to be ignored, because the simple kinetics model of oxidation of ferrous with the kinetic constants but without any biogeochemistry influence (previously proposed by Singer and Stumn [29]) was quantitatively consistent with iron and arsenic ions distribution.

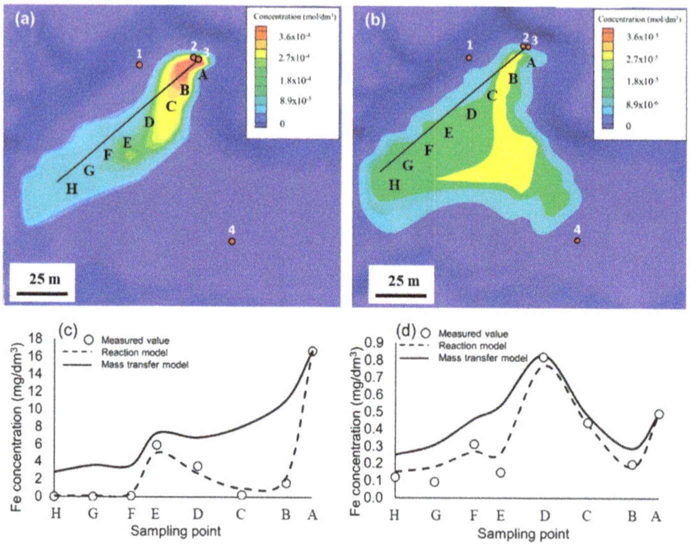

Figure 7. Simulation result: Distribution map of Fe concentrations (**a**) in summer, (**b**) in winter, and (**c**,**d**) measured Si concentrations on the lines A–H and the fitting results.

To model the flow field, the concentration distribution of the arsenic ion was simulated by advection–diffusion using the reactions. The simulation results for that in summer and winter are presented in Figure 8a,b. Generally, the concentration distribution of arsenic ion in summer and winter was consistent with that of the iron ion. As the water flowed further from the main inlet, the concentration gradually decreased. This pattern suggests that arsenite/arsenate adsorption occurred during coprecipitation with the 2-line ferrihydrite. Similar to iron ion, there was an increase in the concentration arsenic ion at certain points, which was caused by the inflow from inlet 1. This increase occurred at point E in summer and point D in winter. In Figure 8c,d, the simulation results using the reaction model successfully represented the concentration of arsenic, obtained from field survey, at each point, while that using the mass transfer model did not represent these. These results highlight that the decrease in concentration of the arsenic ion was caused by adsorption on the 2-line ferrihydrite. In addition, they showed that the Langmuir model using the constants, derived from the Dzombak and Morel model for arsenite/arsenate adsorption on the 2-line ferrihydrite and the oxidation reaction of arsenite using k_3 were applicable in the field. Thus, the main mechanism of arsenic removal was surface complexation for the 2-line ferrihydrite, as a chemical reaction in the Yotsugi mill tailings pond.

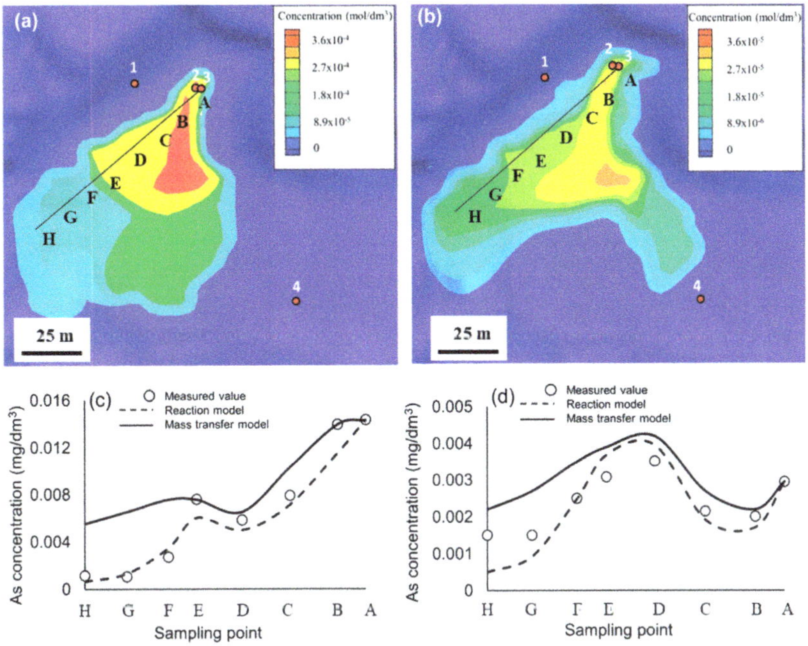

Figure 8. Simulation result: Distribution map of As concentrations (**a**) in summer, (**b**) in winter, and (**c**,**d**) measured Si concentrations on lines A–H and the fitting results.

From the above results, it could be seen that this model was effective to represent the trend of each element in the Yotsugi mill tailings pond. Thus, we could estimate the distribution of concentration of each element in the Yotsugi mill tailings pond using this model when several parameters, such as flow rate and evaporation, are modified due to climate change, in the future.

4. Conclusions

In the Ningyo-toge mine, AMD and several toxic element, such as iron and arsenic were observed. These flow into the Yotsugi mill tailings pond, which is an artificial wetland with a partial pond. In the artificial wetland and pond, ferrous ion is naturally oxidized and precipitates as a 2-line ferrihydrite. In this study, arsenic removal in the artificial wetland and pond through coprecipitation with the 2-line ferrihydrite was simulated and there was good agreement between the measured values and the simulation results. Field sediment sampling and analysis provided further confirmation of the 2-line ferrihydrite precipitation and arsenic retention. The XRD patterns for sediment samples from the artificial wetland and pond showed that 2-line ferrihydrite was present with broad peaks at 2θ 34.9° and 62.3°. The lack of a shift in the 2-line ferrihydrite peak at 34.9° to a lower 2θ degree suggested that the arsenic removal mechanism was surface complexation. This research showed that reaction models from laboratory experiments, such as oxidation of ferrous and arsenite ions and surface complexation, are applicable in the field. Our constructed model based on the removal mechanism successfully represented the concentration and distribution of each element in the Yotsugi mill tailings pond. These results suggest that the main removal mechanism of iron and arsenic ions in the Yotsugi mill tailings pond are inorganic oxidation, precipitation, and surface complexation, while the biogeochemical effect was negligible.

Author Contributions: Conceptualization, C.T., Y.T., and Y.O.; formal analysis, T.K., Y.K., K.S., and M.K.; investigation, T.K., Y.K., K.S., and M.K.; resources, C.T. and Y.O.; writing—original draft preparation, T.K. and Y.T.; writing—review and editing, C.T.; supervision, C.T.; project administration, C.T.; funding acquisition, C.T. All authors have read and agreed to the published version of the manuscript.

Funding: This research received no external funding.

Acknowledgments: We acknowledge the Japan Atomic Energy Agency (JAEA) for providing assistance during this research. Part of this work was performed as the activities of the Waseda Research Institute for Science and Engineering and Research Organization for Open Innovation Strategy, Waseda University.

Conflicts of Interest: The authors declare no conflict of interest.

References

1. Dold, B. Evolution of Acid Mine Drainage Formation in Sulphidic Mine Tailings. *Minerals* **2014**, *4*, 621–641. [CrossRef]
2. Johnson, D.B.; Hallberg, K.B. Acid mine drainage remediation options: A review. *Sci. Total. Environ.* **2005**, *338*, 3–14. [CrossRef]
3. Akcil, A.; Koldas, S. Acid Mine Drainage (AMD): Causes, treatment and case studies. *J. Clean. Prod.* **2006**, *14*, 1139–1145. [CrossRef]
4. Kato, T.; Yagisawa, M.; Matsuoka, M.; Tokoro, C.; Sakakibara, T.; Hayashi, K. Quantitative Modeling Incorporating Surface Complexation for the Treatment of Acid Mine Drainage. *Kagaku Kogaku Ronbunshu* **2017**, *43*, 207–212. [CrossRef]
5. Tokoro, C.; Kadokura, M.; Kato, T. Mechanism of arsenate coprecipitation at the solid/liquid interface of ferrihydrite: A perspective review. *Adv. Powder Technol.* **2020**, *31*, 859–866. [CrossRef]
6. Al Mamun, A.; Morita, M.; Matsuoka, M.; Tokoro, C. Sorption mechanisms of chromate with coprecipitated ferrihydrite in aqueous solution. *J. Hazard. Mater.* **2017**, *334*, 142–149. [CrossRef]
7. Tokoro, C.; Sakakibara, T.; Suzuki, S. Mechanism investigation and surface complexation modeling of zinc sorption on aluminum hydroxide in adsorption/coprecipitation processes. *Chem. Eng. J.* **2015**, *279*, 86–92. [CrossRef]
8. Ueda, H.; Masuda, N. An Analysis on Mine Drainage Treatment Cost and the Technical Development to Prevent Mine Pollution. *Shigen-to-Sozai* **2005**, *121*, 323–329. [CrossRef]
9. Otsuka, H.; Murakami, S.; Yamatomi, J.; Koide, R.; Tokoro, C. A predictive model for the future treatment of acid mine drainage with regression analysis and geochemical modeling. *J. MMIJ* **2014**, *130*, 488–493. [CrossRef]
10. Koide, R.; Tokoro, C.; Murakami, S.; Adachi, T.; Takahashi, A. A Model for Prediction of Neutralizer Usage and Sludge Generation in the Treatment of Acid Mine Drainage from Abandoned Mines: Case Studies in Japan. *Mine Water Environ.* **2012**, *31*, 287. [CrossRef]
11. Kato, T.; Fukushima, R.; Giuseppe, G.; Sato, K.; Yamagata, S.; Tokoro, C. Quantitative Modeling Incorporating Surface Complexation for Zinc Removal Using Leaf Mold. *J. Soc. Powder Technol. Jpn.* **2019**, *56*, 136–141. [CrossRef]
12. Gazea, B.; Adam, K.; Kontopoulos, A. A review of passive systems for the treatment of acid mine drainage. *Miner. Eng.* **1996**, *9*, 23–42. [CrossRef]
13. Tiruta-Barna, L.; Schiopu, N. Modelling inorganic biocide emission from treated wood in water. *J. Hazard. Mater.* **2011**, *192*, 1476–1483. [CrossRef] [PubMed]
14. Fagervold, S.K.; Chai, Y.; Davis, J.W.; Wilken, M.; Cornelissen, G.; Ghosh, U. Bioaccumulation of Polychlorinated Dibenzo-p-Dioxins/Dibenzofurans inE. fetidafrom Floodplain Soils and the Effect of Activated Carbon Amendment. *Environ. Sci. Technol.* **2010**, *44*, 5546–5552. [CrossRef]
15. Johnson, B.T.; Petty, J.D.; Huckins, J.N.; Lee, K.; Gauthier, J. Hazard assessment of a simulated oil spill on intertidal areas of the St. Lawrence River with SPMD-TOX. *Environ. Toxicol.* **2004**, *19*, 329–335. [CrossRef]
16. Valente, T.M.; Antunes, M.; Braga, A.S.; Prudêncio, M.I.; Marques, R.; Pamplona, J. Mineralogical attenuation for metallic remediation in a passive system for mine water treatment. *Environ. Earth Sci.* **2011**, *66*, 39–54. [CrossRef]
17. Furuta, S.; Ito, K.; Ishimori, Y. Measurements of radon around closed uranium mines. *J. Environ. Radioact.* **2002**, *62*, 97–114. [CrossRef]

18. Ishimori, Y. Time-integrated monitoring of radon progeny around a closed uranium mine in Japan. *J. Environ. Radioact.* **2007**, *93*, 51–61. [CrossRef]
19. Watanabe, K. Geochemical behavior of iron and manganese ions in the Ningyo-toge uranium deposit district, southwest Japan. *Chem. Geol.* **1987**, *60*, 299–307. [CrossRef]
20. Kitamura, A.; Kurikami, H.; Sakuma, K.; Malins, A.; Okumura, M.; Machida, M.; Mori, K.; Tada, K.; Tawara, Y.; Kobayashi, T.; et al. Redistribution and export of contaminated sediment within eastern Fukushima Prefecture due to typhoon flooding. *Earth Surf. Process. Landforms* **2016**, *41*, 1708–1726. [CrossRef]
21. Sakuma, K.; Kitamura, A.; Malins, A.; Kurikami, H.; Machida, M.; Mori, K.; Tada, K.; Kobayashi, T.; Tawara, Y.; Tosaka, H. Characteristics of radio-cesium transport and discharge between different basins near to the Fukushima Dai-ichi Nuclear Power Plant after heavy rainfall events. *J. Environ. Radioact.* **2017**, *169*, 137–150. [CrossRef] [PubMed]
22. Sakuma, K.; Malins, A.; Funaki, H.; Kurikami, H.; Niizato, T.; Nakanishi, T.; Mori, K.; Tada, K.; Kobayashi, T.; Kitamura, A.; et al. Evaluation of sediment and 137Cs redistribution in the Oginosawa River catchment near the Fukushima Dai-ichi Nuclear Power Plant using integrated watershed modeling. *J. Environ. Radioact.* **2017**, *182*, 44–51. [CrossRef] [PubMed]
23. Nagayasu, T.; Taki, T.; Fukushima, S. History and current situation of mine water treatment in Nigyo-toge uranium mine. *JAEA Technol.* **2017**, *31*, 1–74.
24. Furusho, Y.; Ono, M.; Yamada, M.; Ohashi, K.; Kitade, T.; Kuriyama, K.; Ohta, S.; Inoue, Y.; Motomizu, S. Advanced Solid Phase Extraction for Inorganic Analysis and Its Applications. *Bunseki Kagaku* **2008**, *57*, 969–989. [CrossRef]
25. Tosaka, H.; Ito, K.; Furuno, T. Fully coupled formulation of surface flow with 2-phase subsurface flow for hydrological simulation. *Hydrol. Process.* **2000**, *14*, 449–464. [CrossRef]
26. Mori, K.; Tada, K.; Tawara, Y.; Ohno, K.; Asami, M.; Kosaka, K.; Tosaka, H. Integrated watershed modeling for simulation of spatiotemporal redistribution of post-fallout radionuclides: Application in radiocesium fate and transport processes derived from the Fukushima accidents. *Environ. Model. Softw.* **2015**, *72*, 126–146. [CrossRef]
27. Tosaka, H.; Matsumoto, Y. An efficient reservoir simulation by the successive explicitization process. *J. Jpn. Assoc. Pet. Technol.* **1987**, *52*, 307–313. [CrossRef]
28. Tawara, Y.; Hazart, A.; Mori, K.; Tada, K.; Shimura, T.; Sato, S.; Yamamoto, S.; Asano, H.; Namiki, K. Extended two-phase flow model with mechanical capability to simulate gas migration in bentonite. *Geol. Soc. Lond. Spec. Publ.* **2014**, *400*, 545–562. [CrossRef]
29. Singer, P.C.; Stumm, W. Acidic Mine Drainage: The Rate-Determining Step. *Science* **1970**, *167*, 1121–1123. [CrossRef]
30. Oscarson, D.W.; Huang, P.M.; Hammer, U.T.; Liaw, W.K. Oxidation and sorption of arsenite by manganese dioxide as influenced by surface coatings of iron and aluminum oxides and calcium carbonate. *Water Air Soil Pollut.* **1983**, *20*, 233–244. [CrossRef]
31. Bednar, A.; Garbarino, J.; Ranville, J.; Wildeman, T. Effects of iron on arsenic speciation and redox chemistry in acid mine water. *J. Geochem. Explor.* **2005**, *85*, 55–62. [CrossRef]
32. Kiskira, K.; Papirio, S.; Van Hullebusch, E.D.; Esposito, G. Fe(II)-mediated autotrophic denitrification: A new bioprocess for iron bioprecipitation/biorecovery and simultaneous treatment of nitrate-containing wastewaters. *Int. Biodeterior. Biodegrad.* **2017**, *119*, 631–648. [CrossRef]
33. Parga, J.R.; Cocke, D.L.; Valenzuela-Garcia, J.L.; Gomes, J.A.; Kesmez, M.; Irwin, G.; Moreno, H.; Weir, M. Arsenic removal via electrocoagulation from heavy metal contaminated groundwater in La Comarca Lagunera México. *J. Hazard. Mater.* **2005**, *124*, 247–254. [CrossRef] [PubMed]
34. Ahoranta, S.H.; Kokko, M.; Papirio, S.; Ozkaya, B.; Puhakka, J.A. Arsenic removal from acidic solutions with biogenic ferric precipitates. *J. Hazard. Mater.* **2016**, *306*, 124–132. [CrossRef]
35. Tokoro, C.; Yatsugi, Y.; Koga, H.; Owada, S. Sorption Mechanisms of Arsenate during Coprecipitation with Ferrihydrite in Aqueous Solution. *Environ. Sci. Technol.* **2010**, *44*, 638–643. [CrossRef]
36. Allison, J.D.; Brown, D.S.; Novo-Gradac, K.J. Minteqa2/prodefa2—A geochemical assessment model for environmental systems: User manual supplement for version 4.0. *Environ. Res. Lab.* **1998**, *4*, 43–74.
37. Cole, W.F.; Lancucki, C.J. A refinement of the crystal structure of gypsum $CaSO_4 \cdot 2H_2O$. *Acta Cryst.* **1974**, *B30*, 921–929. [CrossRef]

38. Jia, Y.; Xu, L.; Fang, Z.; Demopoulos, G.P. Observation of Surface Precipitation of Arsenate on Ferrihydrite. *Environ. Sci. Technol.* **2006**, *40*, 3248–3253. [CrossRef]
39. Jia, Y.; Xu, L.; Wang, X.; Demopoulos, G.P. Infrared spectroscopic and X-ray diffraction characterization of the nature of adsorbed arsenate on ferrihydrite. *Geochim. Cosmochim. Acta* **2007**, *71*, 1643–1654. [CrossRef]
40. Fukushi, K.; Sasaki, M.; Sato, T.; Yanase, N.; Amano, H.; Ikeda, H. A natural attenuation of arsenic in drainage from an abandoned arsenic mine dump. *Appl. Geochem.* **2003**, *18*, 1267–1278. [CrossRef]
41. Bigham, J.; Schwertmann, U.; Traina, S.; Winland, R.; Wolf, M. Schwertmannite and the chemical modeling of iron in acid sulfate waters. *Geochim. Cosmochim. Acta* **1996**, *60*, 2111–2121. [CrossRef]
42. Cornell, R.M.; Schwertmann, H.C.U. *The Iron Oxides: Structure, Properties, Reactions, Occurences and Uses*, 2nd ed.; John Wiley & Sons: Hoboken, NJ, USA, 2003; pp. 345–364. ISBN 9783527302741.
43. Perez, J.P.H.; Freeman, H.; Schuessler, J.A.; Benning, L.G. The interfacial reactivity of arsenic species with green rust sulfate (GRSO4). *Sci. Total. Environ.* **2019**, *648*, 1161–1170. [CrossRef] [PubMed]
44. Al Mamun, A.; Onoguchi, A.; Granata, G.; Tokoro, C. Role of pH in green rust preparation and chromate removal from water. *Appl. Clay Sci.* **2018**, *165*, 205–213. [CrossRef]
45. Onoguchi, A.; Granata, G.; Haraguchi, D.; Hayashi, H.; Tokoro, C. Kinetics and mechanism of selenate and selenite removal in solution by green rust-sulfate. *R. Soc. Open Sci.* **2019**, *6*, 182147. [CrossRef] [PubMed]
46. Kiskira, K.; Papirio, S.; Mascolo, M.C.; Fourdrin, C.; Pechaud, Y.; Van Hullebusch, E.D.; Esposito, G. Mineral characterization of the biogenic Fe(III)(hydr)oxides produced during Fe(II)-driven denitrification with Cu, Ni and Zn. *Sci. Total. Environ.* **2019**, *687*, 401–412. [CrossRef]
47. Senko, J.M.; Dewers, T.A.; Krumholz, L.R. Effect of Oxidation Rate and Fe(II) State on Microbial Nitrate-Dependent Fe(III) Mineral Formation. *Appl. Environ. Microbiol.* **2005**, *71*, 7172–7177. [CrossRef]

minerals

MDPI

Review

Active Treatment of Contaminants of Emerging Concern in Cold Mine Water Using Advanced Oxidation and Membrane-Related Processes: A Review

Sébastien Ryskie [1], Carmen M. Neculita [1,2,*], Eric Rosa [1,3], Lucie Coudert [1,4] and Patrice Couture [5]

1 Research Institute on Mines and Environment (RIME), Université du Québec en Abitibi-Témiscamingue (UQAT), Rouyn-Noranda, QC J9X 5E4, Canada; Sebastien.Ryskie@UQAT.ca (S.R.); Eric.Rosa@uqat.ca (E.R.); Lucie.Coudert@uqat.ca (L.C.)
2 Canada Research Chair in Treatment of Contaminated Mine Water, RIME, UQAT, Rouyn-Noranda, QC J9X 5E4, Canada
3 Groupe de Recherche sur l'Eau Souterraine (GRES, Groundwater Research Group), RIME, UQAT, Amos, QC J9T 2L8, Canada
4 Canada Research Chair in Repurposing of Mine Waste, RIME, UQAT, Rouyn-Noranda, QC J9X 5E4, Canada
5 Centre Eau Terre Environnement, Institut National de la Recherche Scientifique (INRS), Québec, QC G1K 9A9, Canada; Patrice.Couture@ete.inrs.ca
* Correspondence: Carmen-Mihaela.Neculita@uqat.ca

Abstract: Responsible use and effective treatment of mine water are prerequisites of sustainable mining. The behavior of contaminants in mine water evolves in relation to the metastable characteristics of some species, changes related to the mine life cycle, and mixing processes at various scales. In cold climates, water treatment requires adaptation to site-specific conditions, including high flow rates, salinity, low temperatures, remoteness, and sensitivity of receiving waterbodies. Contaminants of emerging concern (CECs) represent a newer issue in mine water treatment. This paper reviews recent research on the challenges and opportunities related to CECs in mine water treatment, with a focus on advanced oxidation and membrane-based processes on mine sites operating in cold climates. Finally, the paper identifies research needs in mine water treatment.

Keywords: mine water treatment; contaminants of emerging concern (CECs); cold climates; advanced oxidation processes (AOPs); membranes

Citation: Ryskie, S.; Neculita, C.M.; Rosa, E.; Coudert, L.; Couture, P. Active Treatment of Contaminants of Emerging Concern in Cold Mine Water Using Advanced Oxidation and Membrane-Related Processes: A Review. *Minerals* **2021**, *11*, 259. https://doi.org/10.3390/min11030259

Academic Editor: Teresa Valente

Received: 23 December 2020
Accepted: 23 February 2021
Published: 2 March 2021

1. Introduction: Contaminants of Emerging Concern (CECs) in Mine Water

Sustainable mining implies a balance between economic profitability, safety, social acceptability, and environmental protection [1]. Given these factors, issues related to the mitigation of mining impacts on water resources inevitably arise. Optimizing low-cost water treatment approaches is critical to ensuring sustainable mining. This goal represents a major challenge as the global demand for mineral resources intensifies and diversifies [2], while the legislation regulating mining activities constantly evolves [3]. An integrated use of hydrological, geochemical, and isotopic tools in mining operations is also evolving [4]. Contaminants of emerging concern (CECs) are a central issue in this contemporary mining context.

In municipal and industrial wastewater, CECs include a wide range of highly soluble, persistent, and potentially toxic substances. Generally, there is limited knowledge about their concentrations (mostly because of poor knowledge on their quantification methods) and detrimental impacts [5]. These contaminants originate from natural or anthropic sources. Improved knowledge on the sources, transport, and spatiotemporal variability of CECs is an ongoing research need, especially for mixed-use watersheds, for better understanding of associated risks, and developing monitoring and mitigation strategies [6].

In mining, throughout history, new or emerging contaminants evolved with metallurgy challenges, scientific knowledge, and regulations [7,8]. As a result, new or CECs in mine water have specific features. They are not necessarily new chemicals and may have been present for a long time, but their presence or significance are only now being recognized, often because of increasing exploitation of low-grade ores, such as As from refractory gold mining [9]. The data about their characterization and toxicity are often scarce, the methods of detection are nonexistent or at an early stage, and there is no international definition of a new or emerging contaminant [10]. The CECs in mine water could even be contaminants that have already been treated but suddenly need to be mitigated to a new order of magnitude (especially in sensitive environments, such as cold climate (e.g., geographical isolation, extreme cold, strong winds, and erosion, salinity from de-icing agents or other, freeze-thaw cycles)). The definition of a CEC could also change geographically or for a different activity sector.

A review of Canadian legislation related to new contaminants or CECs in mine water and the development of treatment processes in the 1980s, 1990s, 2000s, and 2010s identified the following defining characteristics in each decade: (1) Before the 1980s, base metal mines were required to control metal concentrations in their effluents, but gold mines were not regulated [11]. Field-testing of cyanide destruction systems was reported (INCO or SO_2/air process discovery), followed by the testing, patenting, and implementing of all kinds of hydrogen peroxide-based processes. By the end of the 1980s, various types of water treatment systems were constructed and operated to control pH and treat dissolved metals, cyanides (and derivatives), total suspended solids, and so on. (2) In the 1990s, SO_2/air process was deployed at large scale. Acute aquatic toxicity was reportable but not regulated. There was a lot of research to lower copper and ammonia concentrations, but not much implementation of the findings. The 1990s also saw the first attempts to use metal precipitants. (3) In the 2000s, the International Cyanide Management Code (ICMC) and similar regulations meant that cyanide destruction now targeted the mill tailing pipe, not only final discharge, and the Metal Mining Effluent Regulations (MMER) meant that discharge water must not be acutely toxic. The decade also witnessed improved ammonia treatment as well as increased thiocyanates (SCN^-) biochemical treatment, and growing research on toxicity sources. (4) In the 2010s, the review of MMER saw the discussion and addition of new water targets, i.e., Metal and Diamond Mining Effluent Regulations (MDMER). Although Se was not mentioned, other parameters were tightened or introduced under MMER. Some mines in northern regions started to work on desalination projects and were required to meet the Canadian Council of Ministers of the Environment (CCME) criteria at mine site closure. In some countries, mining permits involved respecting sulfate parameters; and zero liquid discharge (ZLD) and zero liquid waste (ZLW) were introduced. The evolution of regulations in Canada's (MMER) and Quebec's (Directive 019) mining industry is presented in Figure 1.

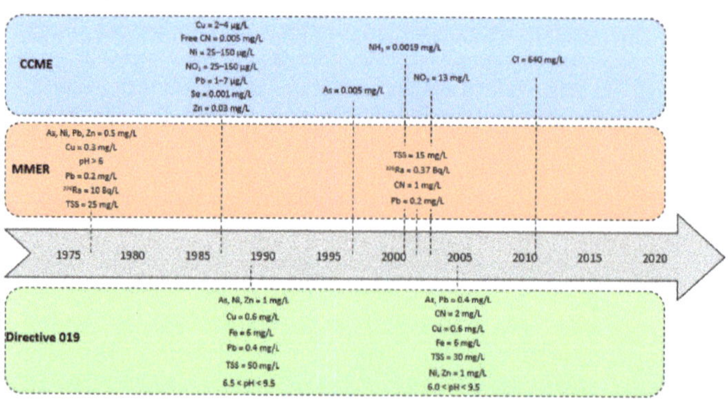

Figure 1. Regulation evolution for the mining industry in Canada [11–14]. CCME: Canadian Council of Ministers of the Environment; Metal Mining Effluent Regulations: MMER; TSS: Total Suspended Solids.

Therefore, the current basic classification of CECs in mine water includes new contaminants (e.g., rare earth elements (REE), radioactive elements, Se, Mn); contaminants of emerging interest (e.g., salinity, sulfate), for which environmental contamination issues were not fully comprehended earlier; and "well-documented" contaminants (e.g., As, thiosalts, N-based compounds, xanthates, SCN^-), for which new issues (e.g., persistent aquatic toxicity) recently emerged, particularly in cold climates (at low temperature and high salinity) [7,8,15]. By their very nature, CECs are of growing concern for the mining industry. The definitions of some CECs based on scientific knowledge of different substances in mine water are presented in Figure 2.

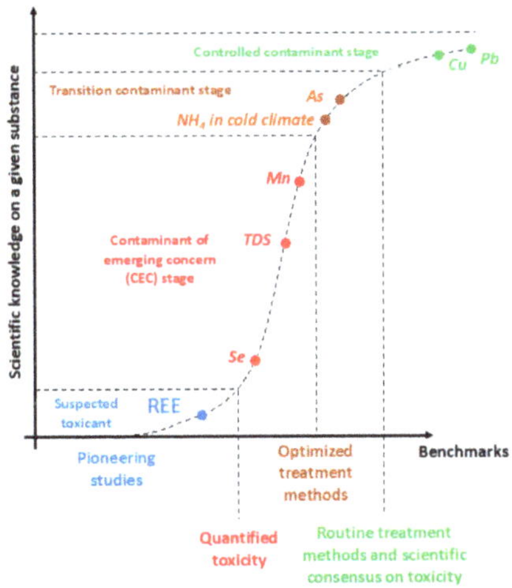

Figure 2. Contaminants of emerging concern (CECs) definitions based on scientific knowledge of different substances in mine water. REE: Rare Earth Elements; TDS: Total Dissolved Solids.

Based on the context outlined above, this paper provides an overview of the efficiency of selected processes and materials for the treatment of CECs from mine water. Background concentrations of selected CECs (As, Se, Mn, salinity, thiosalts, NH_3-N, and REE) in natural water are proposed as a basis for comparison with mine water. The treatment processes focus on As and Se, thiosalts, xanthates, N-based compounds (cyanides, NH_3-N, and SCN^-), and salinity. Regarding REE treatment, several comprehensive reviews are available, including two recent articles on the challenges and opportunities of their treatment and potential recovery from mine water [16–18]; thus, these elements will be not discussed in this review, even though they are of emerging concern. This review focuses on mine sites in cold climates to identify some priority issues related to the treatment of CECs. Questions relating to the treatment of CECs are of particular interest because of (1) the potential toxicity of these substances and (2) the gaps in the scientific knowledge required to define criteria for the discharge of mine water into the environment. The present study aims to contribute to the scientific efforts required to overcome this problem.

2. Characteristics of CECs

2.1. Background Concentrations of CECs

In this paper, As, Se, and Mn are considered as leading inorganic elements, given their importance as mine-related CECs and their known impacts on human health and the environment. In addition, xanthates, salinity, thiosalts, and nitrogen compounds were selected mainly because their ecotoxicological and cumulative effects remain uncertain even at low concentrations, especially with mixed contamination [7–9].

The geogenic sources of As include several minerals (e.g., arsenides, sulfides, oxides, arsenates, arsenites) that can be concentrated in mineralized areas [19]. In rocks, As typically ranges between 0.5 and 2.5 mg/kg [20], for an average crustal abundance of 1.5 mg/kg [21]. Common sources of As are As-bearing sulfides, which are often abundant in areas exploited for base metals, silver, and gold. In natural water, As mainly occurs as oxyanions of arsenite (As(III)) or arsenate (As(V)), while As-bearing organic species are generally less abundant [19]. In a review of worldwide data from large rivers, dissolved As concentrations of 0.11 to 2.71 μg/L, for a world average of 0.62 μg/L, were reported [22]. In groundwater, As concentrations can vary widely, with reported concentrations from <0.05 μg/L to 79.0 μg/L, based on 813 samples from European aquifers [23]. In acid mine drainage, As concentrations up to 12 g/L ($pH < 1.8$) and Fe up to 20 g/L, in Carnoulès abandoned mine site, France [24], or droplets of liquid on the arsenolite (As_2O_3) crust with extreme high As concentrations (80−130 g/L, pH close to 0), have been reported [25].

Se concentrations in rocks are generally low, and the average crustal abundance is 0.05 mg/kg [21]. Se can be enriched in phosphate rocks (up to 300 mg/kg), coal (1–20 mg/kg), and black shales (up to 600 mg/kg) [21,26]. In natural water, Se mainly occurs as selenite (Se(IV)) and selenate (Se(VI)) [21]. In European surface water, reported Se concentrations range from <0.01 μg/L to 15.0 μg/L, for a median value of 0.340 μg/L [27], while, in European groundwater, Se ranges from <0.015 to 247 μg/L, for a median value of 0.50 μg/L [23]. Se concentrations of 1.5 to 33 mg/L were reported in mine water [28].

Mn is the third most abundant transition metal on earth. The average concentrations found in natural water are 0.004 to 2 μg/L [7]. Mn is found in seven oxidation states in nature (0, 2+, 3+, 4+, 5+, 6+, and 7+), but the most common forms in water are Mn^{2+}, Mn^{3+}, and Mn^{4+}. Mine drainage, whether acidic or neutral, often contains a high concentration of Mn, which can have an undesirable impact on ecosystems. In mine water, Mn concentrations range from 0.02 to 352 mg/L, depending on the location and type of mineralogy [7]. Studies demonstrate that Mn can have effects on aquatic organisms, but they are little known, and the impact of Mn on aquatic toxicity remains to be clarified to better guide the selection of treatment methods [7].

Xanthates are the most used collectors in the flotation of sulfurous minerals, and they can be found in mine tailings in concentrations sufficient to have a toxic effect on aquatic fauna [29]. More than 11,000 metric tons of xanthates are consumed annually worldwide in the form of sodium isopropylxanthate (40%), sodium ethylxanthate (30%), sodium isobutylxanthate (15%), and potassium isopentylxanthate (10%) [30]. These molecules act as collectors by rendering the surfaces of the mineral particles hydrophobic and, thus, help them to cling to air bubbles during the flotation process [31]. Xanthates can be toxic to aquatic organisms such as *Daphnia magna* at concentrations of 0.1 to 1 mg/L [31,32]. The concentrations in mine water can vary depending on the type of ore processed. In general, 300 to 500 g of xanthates per tonne of ore is required to obtain satisfactory separation [33]. Residual concentrations ranging from 0.2 to 9 mg/L have been reported in process water [34]. Xanthates can also increase the bioaccumulation of metals by forming hydrophobic complexes with metals such as Zn, Cd, Pb, and Cu. These complexes facilitate the assimilation of metals by organisms through cell membranes [33]. The toxicity of xanthates is mainly linked to their degradation path, which causes the formation of carbon disulfide, which is volatile and slightly soluble in water. The xanthates have a half-life varying between 2 and 8 days, depending on the length of the alkyl chain, for a temperature of 15 °C. However, degradation in cold climates has not yet been studied [33].

Mine water salinity is an integrative parameter, which is characterized by the cumulative concentration of the most common ions (Na^+, K^+, Ca^{2+}, Mg^{2+}, Cl^-, $HCO_3{}^-$, $CO_3{}^{2-}$, and $SO_4{}^{2-}$) [35]. Salinity, electrical conductivity (EC), and total dissolved solids (TDS) are correlated. TDS may include an inorganic fraction (Na^+, K^+, Ca^{2+}, Mg^{2+}, Cl^-, $HCO_3{}^-$, $CO_3{}^{2-}$, and $SO_4{}^{2-}$) and an organic fraction (dissolved organic carbon—DOC) [36]. In mine water, TDS and salinity consist mainly of inorganic ions. The EC, which is defined by water's ability to allow the transport of electric charge, is an indication of the degree of water mineralization, providing information on dissolved elements (in ionic form) [37]. The high concentration of TDS results in an increase in the salinity of mine water. Salinity persists after the treatment of the mine water, which often even increases it. Moreover, salinity can have a detrimental impact on the aquatic toxicity depending on the species present in water and their concentrations. For example, the concentration of Ca^{2+} or water hardness can reduce the toxicity of Cl^- and SO_4^{2-} [38]. Once dissolved solids that form salinity are transported into the environment, they are likely to cause toxicity [39].

Thiosalts are metastable sulfur oxyanions, naturally present in concentrations ranging from very low to a few hundred mg/L, such as in highly acidic crater-lake water [40]. Thiosalts are the common intermediate species of sulfur oxidation in metal sulfides (e.g., pyrite, pyrrhotite), in the presence of sulfur-oxidizing bacteria [41,42]. The main thiosalts species are thiosulfate ($S_2O_3^{2-}$), trithionate ($S_3O_6^{2-}$), and tetrathionate ($S_4O_6^{2-}$), with $S_2O_3^{2-}$ being the dominant thiosalt species [43,44]. In mine water, thiosalts are mainly generated (up to 60%) during the milling and flotation of sulfide ores, and at high pH (9.4–10.7) and alkalinity [45]. Other factors that contribute to the acceleration of thiosalt formation include the following: sulfides content of ores (5 times more thiosalts generated in ores with 80 to 90% sulfides vs. 25 to 30% sulfides); residence time during flotation (proportional); temperature (double amount at 40 °C vs. 25 °C); agitation rate in flotation (accelerated oxidation kinetics); and grinding operation (significant reduction of thiosalts generation (86%) during grinding at pH 10.7 in a solution deaerated with nitrogen gas) [46]. A recent study evaluated the effect of freeze/thaw cycles on thiosalts concentrations using column tests with tailings mainly dominated by pyrrhotite and serpentine [47]. Results showed that $S_2O_3^{2-}$ could top up to 10 g/L for columns subjected to freeze/thaw cycles relative to 7.2 g/L for those at an ambient temperature. Additionally, $S_4O_6^{2-}$ reached concentrations of 2.6 g/L in freeze/thaw columns and 2.2 g/L in the ambient columns. Another study showed the increased reactivity of minerals during freeze/thaw cycles [48], thus, supporting previous observations [47].

Nitrogen compounds, including NH_3-N, NO_2^-, and NO_3^-, are naturally present in the environment, usually at low concentrations. Human activity tends to add more nitrogen compounds in natural streams. In the presence of phosphorus, the eutrophication of stagnant or low-flow water sources can often occur, even in cold climates [49]. In the mining industry, the two major sources of nitrogen compounds are explosives and cyanides. Nitrogen-based blasting agents (e.g., ANFO—ammonium nitrate fuel oil) are used in mining to extract the commodities from their ore bodies. The explosives that are not completely degraded in the explosion can be dissolved in the water and then pumped at the surface during the dewatering process [50]. In addition to explosives, which are common to all types of mines, in the gold and silver extraction process, cyanides are also needed for their efficient recovery (>90%, at 71 mg Au/t) [51].

Cyanidation, which was deemed a major advance at the end of the 19th century, remains the most efficient and least expensive separation process for gold and silver extraction, even though cyanides are toxic and nonselective, and generate effluents that are highly complex to treat [52,53]. The main issue related to these effluents is their mixed contamination with several groups of undesirable substances (e.g., cyanides and their derivatives, such as NH_3-N and SCN^-, metals), which requires multiple treatment steps for the final effluents to respect the physicochemical and toxicity discharge criteria. Two contaminants closely related to aquatic toxicity, even though not regulated by the Canadian law, are NH_3-N and SCN^- [52]. While the major source of NH_3-N (i.e. cyanides) is specific to gold mining, a nitrogen source associated with explosives (ANFO) is common to all mine effluents. The efficient treatment of NH_3-N and SCN^- requires their oxidation to bicarbonate and nitrate, in addition to sulfate (for SCN^-). The high flow rates of mine effluents, together with the very slow kinetics of NH_3-N oxidation, especially at low temperatures, supports the need for new, robust, adapted, and low-cost processes with low or no temperature and pH dependence [50], in addition to limited residual salinity creation.

Cyanides require treatment before wastewater is discharged into natural streams and can lead to the formation of NH_3-N [52]. If sulfur is present in the processed ores (as sulfide minerals and partially oxidized sulfur intermediates), toxic derivatives (SCN^-) can also be formed at high pH and alkalinity [51]. Even though SCN^- is less toxic than CN^-, its higher stability makes it more complex and difficult to treat [52].

The compilation of different contaminants, including their typical concentration in natural and mining environments, sources, and treatment methods is presented in Table 1. The schematic representation of the CECs evolution in the mining environment is presented in Figure 3.

Table 1. Compilation of different contaminants, including their typical concentration, source, and treatment methods.

Contaminant	Natural Environment Concentration (µg/L)	Mining Environment Concentration (mg/L)	Possible Source of Contamination	Applicable Processing Methods	References
Mn	0.004–2	0.02–352	Acid and neutral mine drainage	Co-precipitation Sorption Ion exchange Membrane filtration Oxidation and precipitation Biological	[7,54]
Se	<0.01–15	1.5–33	Mine drainage	Membrane filtration Ion exchange Evaporation Co-precipitation Electrocoagulation Photoreduction Adsorption Biological	[27,28,55,56]
Xanthates	N/A	0.2–9	Flotation process	Advanced oxidation processes Natural degradation	[34,57]
Thiosalts	N/A	<700	Acid mine drainage and mineral processing	Advanced oxidation processes Lime neutralization Biological Natural degradation Electrochemical oxidation Membrane filtration	[43,45,58]
Salinity (TDS)	<1000	<16,000	Mine drainage	Thermal processes Coagulation-flocculation Membrane filtration Aerobic treatment	[15,59,60]
SO_4^{2-}	3000–30,000	100–5000	Mine drainage	High density sludge treatment ChemSulphide process Biopaq – Bioteq Sulfatogenic bioreactor Passive treatment Sorption Dispersed alkaline substrates Membrane filtration	[15,61]
As	0.11–2.71	<130,000	Mine drainage and hydrometallurgical processes	Co-precipitation Membrane filtration	[9,25]

Table 1. *Cont.*

Contaminant	Natural Environment Concentration (µg/L)	Mining Environment Concentration (mg/L)	Possible Source of Contamination	Applicable Processing Methods	References
NH_3-N	<1000	46	Explosives and cyanide treatment	High pH stripping Membrane filtration Nitrification-denitrification Natural degradation Anammox Electro-oxidation Electrocoagulation Adsorption Advanced oxidation processes	[50,62–64]
NO_2^-	<1600	4.4	Explosives Biological treatment	Denitrification Membrane filtration Advanced oxidation processes	[50,64,65]
NO_3^-	<10,000	25–100	Explosives Cyanide treatment Biological treatment	Denitrification Membrane filtration	[50,65,66]
CNO^-	N/A	28	Cyanide treatment	Natural degradation Ozone Biological	[64,67]
SCN^-	N/A	168–680	Cyanide treatment	Ozone Electrochemical oxidation Biological treatment Ferrates Advanced oxidation processes	[52,64]

N/A: Not Applicable; TDS: Total Dissolved Solids.

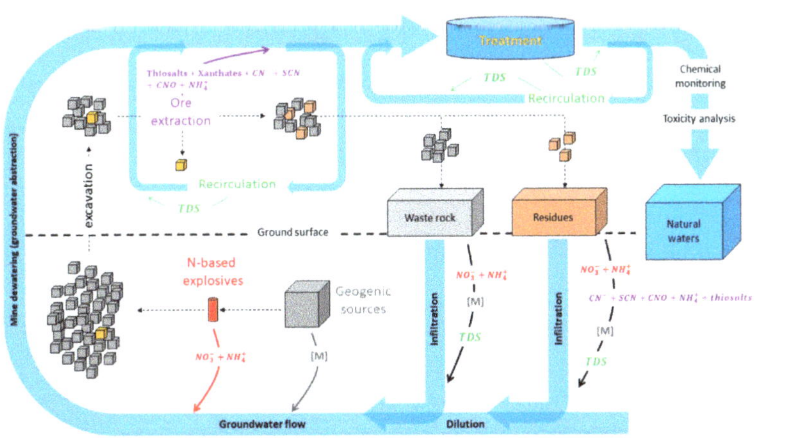

Figure 3. Schematic representation of the CECs evolution in the mining environment. [M]: metal or metalloid; TDS: Total Dissolved Solids.

2.2. Persistent Aquatic Toxicity

The aquatic toxicity of mine water is an important parameter to consider during mining operations. As an example of cold climate, the Canadian context is discussed. At the provincial level in Quebec, operating mining companies must comply with the guidelines set out in Directive 019 [68]. In Canada, at the federal level, they must comply with the Metal and Diamond Mining Effluent Regulations (MDMER) [14]. In addition to maximum authorized concentrations of several contaminants in mine water, the MDMER stipulates that the acute toxicity testing on rainbow trout (*Oncorhynchus mykiss*) and water fleas (*Daphnia magna*) must be performed once a month. Because *D. magna* is generally more sensitive to acutely lethal environments than rainbow trout, research often relies on *D. magna* as a representative indicator for acute toxicity. Even though *D. magna* is used widely in regulatory assessment, *D. pulex* is commonly found in Canadian lakes, streams, and rivers—that is, in almost all eutrophic and permanent watercourses [69]. Comparative evaluation of the acute toxicity for *D. magna* and *D. pulex* of sulfate-rich neutral mine water from active gold mines showed a greater sensitivity of *D. pulex* compared to *D. magna*, before and after the treatment by electrocoagulation, in toxicity tests using standard reconstituted hard or natural dilution water [70].

Water from mining activities tends to be highly charged with contaminants such as metal(loid)s, salinity, nitrogen compounds, and, in some cases, xanthates. Aquatic toxicity of some CECs (e.g., xanthates, thiosalts) in mine water is incompletely assessed and under-documented, as opposed to others (e.g., As, Se) for which the literature is abundant. A compilation of lethal (LC50) and effective (EC50) concentrations of the different contaminants for *D. magna* is presented in Table 2.

Several studies have demonstrated the impact of these various contaminants on aquatic fauna. Although mining companies are governed by standards for aquatic toxicity, even when the acute toxicity criteria for regulated species (*D. magna* and rainbow trout) are met, the entire wildlife food chain can be affected by chronic effects [71,72]. Depending on the contaminant speciation and water mixes, bioavailability and synergistic or competitive effects in their fixation on binding sites on a biotic ligand might occur [71].

Table 2. Lethal (LC50) and effective (EC50) concentrations of CECs reported for *Daphnia magna*.

Contaminant	LC50 (mg/L)	EC50 (mg/L)	Test duration (h)	Hardness (mg $CaCO_3$/L)	Temperature (°C)	pH	References
Mn	N/A	9.3 10.27	48	45 240	21	6.5–8.5	[73]
Se	N/A	0.71	48	72	21	6.5–8.5	[73]
Na-ethyl xanthate Na-isopropyl xanthate Na-isobutyl xanthate K-amyl xanthate K-pentyl xanthate	N/A	0.35 3.7 3.6 3.67 3.0	N/A	N/A	15	N/A	[33]
$S_2O_3^{2-}$ $S_4O_6^{2-}$	N/A	300 750	N/A	N/A	N/A	N/A	[45]
Ca^{2+}	N/A	52 560	48	45 240	21	6.5–8.5	[73]
Cl^-	2600	N/A	24	N/A	N/A	N/A	[74]
Na^+	N/A	1640 423.13	48	45 240	21	6.5–8.5	[73]
K^+	N/A	93 160.45	48	45 240	21	6.5–8.5	[73]
SO_4^{2-}	7000	N/A	48	N/A	N/A	N/A	[74]
As	N/A	2.4 7.4 74	48	72 45 240	21	6.5–8.5	[73]
Cu	0.004 0.012	N/A N/A	48 48	44 ± 4 150 ± 10	N/A N/A	N/A	[75]
Zn	0.82 0.1 0.655 0.3 1.29	N/A N/A N/A N/A N/A	48 24 24 48 48	N/A 45 196 44 ± 4 150 ± 10	N/A N/A N/A N/A N/A	N/A	[69] [76] [76] [75] [75]

Table 2. *Cont.*

Contaminant	LC50 (mg/L)	EC50 (mg/L)	Test duration (h)	Hardness (mg $CaCO_3$/L)	Temperature (°C)	pH	References
NH_3-N	1.980	N/A	24	N/A	20	7	[77]
NO_3^-	2047	N/A	48	156–172	N/A	N/A	[78]
N-NO_2^-	N/A	23	48	N/A	20	N/A	[79]
CNO^-	18	N/A	48	N/A	N/A	N/A	[80]
SCN^-	57.4 0.63–32	11.3 N/A	48 96	N/A 75	N/A 8–16	N/A	[52]

N/A: Not Applicable.

To reduce the dispersion of contaminants in the environment and prevent the deterioration of aquatic fauna, the federal government has introduced regulations. The MDMER requires that all metal mines produce an effluent that is nontoxic for rainbow trout and *D. magna*, in accordance with Environment and Climate Change Canada's methods of analysis. If there is mortality for more than 50% of the organisms in 100% of the whole effluent concentration during a period of 48 h for *D. magna* or 96 h for rainbow trout, the sample is considered as having failed the toxicity test [14]. If the lethality test is failed, the mining company must investigate the causes of toxicity.

The most common approach to determining the causes of toxicity is the Toxicity Reduction Evaluation (TRE) method. A step-by-step protocol was developed to help identify and confirm the causes and sources of toxicity and to eliminate them [58]. The first step of this protocol, known as Toxicity Identification Evaluation (TIE), is to identify the contaminant(s) responsible for the toxicity. Once the identification has been carried out, the next step is Source Identification (SI). The final step, Toxicity Treatability Evaluation (TTE), is to evaluate possible treatments or modify water management approaches to eliminate toxicity. During this process, it is important to have optimal communication and coordination between the various stakeholders, such as operators, toxicologists, chemists, and the engineering team, so that the TRE is a success [58].

Studies on the direct and indirect toxicity of thiosalts do not show a correlation between concentration and mortality of the species regulated by provincial and federal laws for acute and sublethal toxicity tests. In general, it is agreed that thiosalts are not directly acutely toxic at concentrations commonly found in mining effluents but rather indirectly due to the generation of latent acidity, which causes a decrease in pH [43]. Concentrations up to 4.1 g/L of $S_2O_3^{2-}$ did not entail acute toxicity to rainbow trout [45]. However, toxicity occurred when the pH of the water dropped to below 5. Direct toxicity test results consistently showed that sublethal toxicity, sublethal toxicity to *C. dubia* was high but total thiosalts concentrations were low, and vice versa [45]. The same report found that $S_2O_3^{2-}$ is more toxic to all organisms (96 h LC50 for *C. dubia* = 59 mg/L; 96 h LC50 for *L. minor* = 498 mg/L; 96 h LC50 for *P. promelas* = 665 mg/L) than $S_4O_6^{2-}$ (96 h LC50 for *C. dubia* = 562 mg/L; 96 h LC50 for *L. minor* > 901 mg/L; 96 h LC50 for *P. promelas* > 891 mg/L), which results in a non-additive toxicity of the polythionates, thus, explaining the absence of correlation between total thiosalts and toxicity [45]. The IC25 values for indirect toxicity are low (high toxicity) for all organisms when the pH drops below 5, except for *L. minor* (IC25 > 100% at pH 3.1; no toxicity) [45]. Confirmed toxicity at circumneutral pH indicates that other contaminants such as Cu and Se may add to thiosalts toxicity [45]. Further research on thiosalts toxicity is required considering the extreme concentrations recently reported [47] and the increased reactivity of minerals (and, therefore, leaching of other contaminants) during freeze/thaw cycles [48], especially for mines operating in cold climates. These studies should focus on the cooperative toxic effect with other metals and non-additive toxicity of polythionates.

Mixed contamination of mining effluents can entail complexities because of the interactions between the contaminants. The thermodynamic equilibrium between the mineral and chemical phases determines the speciation of the contaminants. These reactions are dependent on the temperature and the balance of the mineral, gas, and aqueous phases. Bacterial activities can also have a role as a biocatalyst for the various reactions that occur [81].

Several models of aquatic toxicity prediction that consider the mixture of different contaminants as well as the physiology of biota already exist. These models determine the speciation of inorganic elements according to the physicochemical parameters of water and then determine the LC50 values. For example, the Biotic Ligand Model (BLM) can be used for this type of prediction [82–84]. Mixed contamination can significantly contribute to aquatic toxicity. One of the most widely used bioassays in the world for acute aquatic toxicity is that of *D. magna*, one of the most sensitive crustaceans [85]. Examples of *D. magna* toxicity in the mining industry are presented in Table 3. Mixed contamination can sometimes have pronounced effects on *D. magna*. The combination of various contaminants according to different physicochemical properties modifies speciation, bioavailability, and equilibrium concentrations in the final effluent. For example, the hardness of water, a parameter that has a significant effect on the toxicity of metals, reduces the bioavailability of these metals and, therefore, increases the toxic concentration [83]. At high concentrations, the hardness can have an opposite effect by increasing osmotic stress in biota [86]. Calcium can also have a protective effect on *D. magna* by reducing the absorption of metals such as Ni, Zn, Se, and Cd [83,86].

Another study has shown that water with very high hardness and alkalinity and containing high concentrations of Ca^{2+}, Mg^{2+}, $CO_3{}^{2-}$, and $HCO_3{}^{-}$ can have a toxic effect on *D. magna*. A calcite shell ($CaCO_3$) forms over its entire surface and causes its disintegration following its death (LC50, 72 h). In this case, the head disintegrated first [87].

Table 3. Examples of *D. magna* toxicity in mine water.

Location	Exploitation Type	Parameters	*D. magna* Toxicity (Toxic Units—TU)	References
South Korea	Metal plating plant final effluent	pH = 7.58 DOC = 131.1 mg/L Hardness = 46 mg $CaCO_3$/L Cu dissolved = 0.36 mg/L Cl^- = 12,841 mg/L Br^- = 2307 mg/L	6.5	[88]
	After treatment by ion exchange	Cl^- = 2840 mg/L	<1	
South Korea	Final effluent of acid rock drainage treatment plant before pH correction	pH = 4.51 DO = 4.11 mg/L DOC = 3.87 mg/L Hardness = 2408 mg $CaCO_3$/L Al dissolved = 31.87 mg/L Cu dissolved $\leq$ DL Fe dissolved = 13.92 mg/L Zn dissolved = 1.39 mg/L	5.66	[88]
	After pH correction	pH = 7.0 Al dissolved = 0.093 mg/L Cu dissolved = 0.001 Fe dissolved = 0.011 mg/L Zn dissolved = 0.05 mg/L	<1	
South Korea	Mixed effluents of electronics plant	pH = 8.55 DOC = 139.5 mg/L Hardness = 178.8 mg $CaCO_3$/L Cu total = 1.5624 mg/L Cu dissolved = 0.4793 mg/L	35.46	[88]

Table 3. *Cont.*

Location	Exploitation Type	Parameters	*D. magna* Toxicity (Toxic Units—TU)	References
Canada (QC)	Final effluent, LaRonde mine, Agnico Eagle (04-03-1999 to 20-02-2001)	pH = 7.8–9.2 CN total = 0.005–0.36 mg/L CNO^- = 3.9–231 mg/L SCN^- = 73–293 mg/L NH_3-N = 20–88 mg/L SO_4^{2-} = 1350–2370 mg/L DOC = 35–93.6 mg/L Ca = 470–670 mg/L Cu = 0.02–0.14 mg/L Zn $\leq$ 0.01–0.26 mg/L	1–50.5	[89]
Canada (QC)	Final effluent, LaRonde mine, as reported by SGS Lakefield (06-05-2001)	pH = 8.05 TDS = 3320 mg/L CN total = 0.05 mg/L CNO^- = 21 mg/L SCN^- = 57 mg/L NH_3-N = 32.6 mg/L NO_2-N = 2.78 mg/L NO_3-N = 22 mg/L SO_4^{2-} = 2200 mg/L DOC = 12.4 mg/L Ca = 713 mg/L Cu = 0.016 mg/L Cu dissolved = 0.011 mg/L Na = 177 mg/L Zn = 0.02 mg/L Zn dissolved $\leq$ 0.01 mg/L Conductivity = 2710 µmhos/cm Hardness = 1820 mg $CaCO_3$/L DO = 9.4 mg/L	1.68	[89]
South Korea (Daeduck, Damyang county)	Mine drainage	pH = 6.1 Hardness = 16 mg $CaCO_3$/L Zn = 0.156 mg/L Pb = 0.0402 mg/L Cu $\leq$ DL Cd = 0.0013 mg/L	1.6	[75]
South Korea (Myungbong, Boseong county)	Mine drainage	pH = 7.8 Hardness = 50 mg $CaCO_3$/L Zn = 4.797 mg/L Pb = 0.0015 mg/L Cu = 0.0296 mg/L Cd = 0.0176 mg/L	22.9	[75]

DL: Detection Limit; DOC: Dissolved Organic Carbon; DO: Dissolved Oxygen; TDS: Total Dissolved Solids; TU: Toxicity units.

One way to predict the aquatic toxicity of mixed contamination is to sum the toxic units (TU). The TU is equal to the concentration of a contaminant in the water divided by the LC50. If the sum of the TUs of all the contaminants in the mixture is greater than 1, the water is considered toxic to the organism in reference. The validity of this principle was confirmed with real effluents, but the LC50 were corrected in relation to water hardness [75]. Considering the reduction in toxicity in response to an increase in hardness, formulas can be used to correct the LC50. Although this method predicts aquatic toxicity with a confidence rate of up to 89% in some cases, a high iron concentration can be an inhibiting factor in the estimation of toxicity because of amorphous iron precipitates [90]. A high DOC concentration can also have an important role in the complexation of metals and decreases the confidence rate of the prediction of aquatic toxicity by this method [88].

3. Treatment of CECs in Mine Water

Treatment strategies usually begin by identifying the problem, followed by setting objectives, and considering potential solutions based on constraints related to Best Available Technology Economically Achievable (BATEA), specific regulatory criteria, environmental impacts (residual contamination and aquatic toxicity), social acceptability, and license to operate [91,92]. The review of available literature showed that at least three main priorities can be identified in responsible management of mine water in cold climates: (1) development of sustainable treatment processes with limited energy consumption; (2) control of the residual salinity, after treatment; and (3) safe handling of produced sludge, after the recovery of potential economical elements or the immobilization of undesired ones. Contrary to temperate climates, the characteristics of cold climates contribute to additional challenges and opportunities in mine water treatment. The technologies presented in this paper can be used in cold climates with good treatment performance as the cold temperature has less impact on their treatability mechanisms compared to conventional treatments.

3.1. Advanced Oxidation Processes (AOPs)

Some AOPs fit into these requirements as they are efficient at low temperatures and allow for the complete oxidation of several contaminants in water using environmentally friendly oxidants, based on the hydroxyl radical, such as peroxone (H_2O_2+O_3) [93], Fenton-like (H_2O_2 and Fe^{3+} catalyst) [94,95], ferrates (Fe(VI)) [96], and O_3 microbubbles [63,64,97] or persulfate-based AOPs [98]. The peroxone was found more effective in the degradation of naphthenic acids and toxicity reduction than ozonation (at high doses), a costly treatment for oil sands process-affected water [99]. However, peroxone proved inefficient or inhibitory for NH_3-N treatment in a synthetic effluent [64].

In a recent study, the use of H_2O_2 to remove thiosalts from a synthetic and real mine effluent at 8 °C and 22 °C showed that a long reaction time was required for optimal efficiency [100]. The important role of temperature on the removal of thiosalts was confirmed. In fact, results showed that low temperatures enhance the stability of polythionates, but also seem to partially oxidize $S_2O_3^{2-}$ in the presence of H_2O_2 as oxidizing agent. At higher temperatures, 96% thiosalts removal from a real effluent was reached, but the oxidation of intermediate S species was incomplete and reaction time was 7 days.

The Fenton process consists of H_2O_2 activation with basically an iron catalyst to generate a hydroxyl radical, a powerful and nonselective, environmentally friendly oxidant. The Fenton-like process was found highly efficient (>99% of 1 g/L SCN^-) and a potential economical option for SCN^- partial oxidation (at pH 2.5–3) into cyanides to be returned to the leaching of gold and silver from their ores [94]. Consistently, at 21 °C, Fenton-like gave 84% and 22% of SCN^- and NH_3-N removal (at initial concentrations of 1 g/L and 40 mg/L, respectively), whereas at 4 °C the efficiency of SCN^- degradation decreased to 73% [95]. This AOP has major advantages, as the iron catalyst is easily available; the process is environmentally friendly and entails low residual toxicity and operating costs [101].

Electrochemical AOPs have been gaining popularity for limiting As toxicity, through oxidation of As(III) to As(V), and its subsequent removal. Electrochemical treatment methods have the advantage of using electrons as a reagent, thus, reducing the amount of added chemicals and residual salinity [102]. Bio-electrochemical AOPs have also received growing interest for organic CECs treatment in municipal wastewater because of the reduced cost of electricity. In fact, electro-Fenton (EF) and bio-electro-Fenton (bio-EF) are the eco-friendliest and most cost-efficient processes for the treatment of recalcitrant contaminants compared to other electrochemical methods [102,103]. In bio-EF, electrons are generated through microbial activity at the anode, and H_2O_2 is in situ generated at the cathode. H_2O_2 is then available to react with an iron catalyst to form hydroxyl radicals, as in the conventional Fenton process. In a study on EF in a dual chamber microbial fuel cell, results showed that As(III) removal efficiency was of 94% at optimal oxidation current efficiency. However, the removal process was incomplete for As(V) [103], and the final concentration did not meet the criteria specified by federal law [14].

Most publications focus on EF/bio-EF efficiency for the degradation of organic contaminants. Very little information is available on their performance for oxidation and removal of micropollutants, despite a promising application to the mining industry. Therefore, forthcoming research should focus on assessing EF and bio-EF performance on removal of oxidizable CECs in synthetic and especially real mine water, including As, NH_3-N, and thiosalts [104].

New research reported the satisfactory efficiency of ferrates for NH_3-N and SCN^- removal from highly contaminated mine effluents, but not simultaneously [105]. The wet Fe(VI) efficiency was evaluated on three synthetic and two real gold mine effluents contaminated by SCN^- and/or NH_3-N. Results showed that Fe(VI) oxidized more than 97% of SCN^-, while the NH_3-N increased up to 50%, after SCN^- oxidation in the presence of NH_3-N, within 1 h. A second step of NH_3-N treatment would, thus, be warranted. The main concerns, especially with wet Fe(VI) use, are the high pH and dissolved solids content of treated water (when high doses are required), that can adversely affect water quality downstream of discharge and its toxicity [96,106]. However, as the Fe(III) salt source for wet Fe(VI) production could be nitrates, chlorides, or sulfates, the residual contamination could vary: (1) from a four-time increase of residual nitrates with $Fe(NO_3)_3$, to (2) high residual chlorides with $FeCl_3$ and the oxidant used in the Fe(III) to Fe(VI) oxidation (sodium hypochlorite, NaClO), and (3) to no increase of residual sulfates with $Fe_2(SO_4)_3$ [106].

3.2. Ozone Microbubbles

The efficiency of several AOPs (O_3 microbubbles, UV, and H_2O_2) was also tested for NH_3-N removal from several synthetic (70 mg/L NH_3-N) and five actual mine effluents [64]. The tests started in batch mode, for optimizing the performance, followed by in continuous flow with one real effluent. Results showed that O_3 microbubbles gave the best efficiency. Indeed, more than 92.6% of the initial NH_3-N was treated, at pH 9, within 90 min. The use of Br^- as a catalyst increased treatment efficiency, whereas the combination of UV with O_3 was less efficient than O_3 alone under the conditions tested. Moreover, the presence of cyanides, cyanate, SCN^-, and metals adversely influenced NH_3-N removal efficiency, which ranged from 27.8 to 99.3%. Polishing final steps were required for the removal of the coloration that developed in some treated mine effluents. These results were consistent with previous findings on O_3 microbubbles that showed NH_3-N complete removal in a synthetic effluent (100 mg/L), at pH 9, in about 7.2 ks, whereas at pH 6 only 20 mg/L of NH_3-N were removed [63]. Moreover, they are consistent with reported findings on the oxidation of As(III) (50–200 µg/L) to As(V), at pH 7, within less than 25 min in a synthetic effluent [107]. The O_3 microbubbles process showed also satisfactory efficiency in thiosalts removal (99%) from a synthetic effluent [100]. The use of O_3 microbubbles for NH_3-N treatment in mine water has distinctive advantages, including the fast kinetics of the process, complete oxidation of several contaminants in mixing effluents, and no residual salinity creation in the treated water. Recently, the ultrasounds combined with ozonation and SrO-Al_2O_3 as a catalyst showed to be promising for NH_3-N treatment [108]. With a combination of ultrasounds at a frequency of 25 kHz vs. 270 W ultrasonic power and ozone, the NH_3-N conversion and N_2 gas yield were 83.2% and 51.8%, respectively.

The mode of diffusion of O_3 in water is an important process, allowing mass transfer to the gas-liquid interface. Indeed, due to the low solubility of O_3 in water, the injection method can increase the process efficiency, for example, by using microbubbles [109]. The microbubbles have a diameter of 10 to 50 µm, and a very high gas-liquid interfacial contact surface, then a very low ascent rate. The phenomenon that makes the use of microbubbles attractive for water treatment is the reduction in their size until their implosion in the liquid phase, unlike the coarse bubbles that always rise to the surface [109]. Two major advantages characterize microbubbles. First, there is higher ozone solubilization with higher internal pressure, forcing greater mass exchange at the gas-liquid interface. Ozone then forms hydroxyl radicals depending on the physicochemical properties of water. Second, microbubbles implosion phenomenon also generates hydroxyl radicals [109].

The use of AOPs such as ozone, microbubbles, and sonochemical cavitation in the mining sector is very rare. In recent years, these processes were also adapted for use with mine effluents, with satisfactory results. Comparative performance of ozone, microbubbles, peroxone (mixture of ozone and hydrogen peroxide), and sonochemical cavitation for mine water at different scales are presented in Table 4.

Table 4. Treatment performance of ozone, microbubbles, and ultrasound on mine water or compounds used in flotation processes.

Treatment Type	Industry	Influent	Effluent	Scale	References
Ozone microbubbles followed by coagulation-flocculation	Gold mine underground water	pH = 6.7 Eh = 445 mV T = 22 °C NH_3-N = 22 mg/L NO_2^- = 5.5 mg/L NO_3^- = 185 mg/L Cu = 0.154 mg/L Fe = 1.06 mg/L Mn = 3.41 mg/L Zn = 9.6 mg/L	pH = 8.7 Eh = 410 mV T = 22 °C NH_3-N = 0.78 mg/L $NO_2^- \leq$ DL NO_3^- = 180 mg/L Cu = 0.01 mg/L Fe = 0.16 mg/L Mn = 0.58 mg/L Zn = 0.2 mg/L	Pilot in continuous flow	[64]
Ozone microbubbles	Gold mine process water after cyanide destruction	pH = 9 Eh = 295 mV T = 22 °C NH_3-N = 34.6 mg/L NO_2^- = 6 mg/L NO_3^- = 48 mg/L CNO^- = 15.8 mg/L SCN^- = 135 mg/L Total CN = 0.04 mg/L Cu = 1.52 mg/L Fe = 0.5 mg/L Mn = 0.3 mg/L Zn = 0.29 mg/L	pH = 9.3 Eh = 382 mV T = 36 °C NH_3-N = 5.4 mg/L $NO_2^- \leq$ DL NO_3^- = 151 mg/L CNO^- = 12.5 mg/L $SCN^- \leq$ DL Total CN = 0.05 mg/L Cu $\leq$ DL Fe $\leq$ DL Mn = 0.06 mg/L Zn $\leq$ DL	Pilot in batch mode	[64]
$O_3 + H_2O_2$	Gold mine process water	CN = 172.5 mg/L pH = 11 Turbidity = 56 NTU TOC = 137.8 mg/L DOC = 496 mg/L	CN = 0.08 mg/L pH = 7 Turbidity = 68 NTU TOC = 24.5 mg/L DOC = 118.5 mg/L	Laboratory	[110]
Air micro-nanobubbles	Mine effluent	T = 21 °C pH = 11.99 Turbidity = 170 NTU Conductivity = 4.06 mS/cm DO $\leq$ 0.1 mg/L TS = 4740 mg/L TSS = 251 mg/L Pb = 51.3 mg/L Zn = 17.601 mg/L	T = 22 °C pH = 11.72 Turbidity = 15.93 NTU Conductivity = 2.39 mS/cm DO = 8.53 mg/L TS = 4120 mg/L TSS = 31 mg/L Pb = 0.98 mg/L Zn = 0.128 mg/L	Laboratory	[111]
O_3 + ultrasound	Synthetic water with surfactant	pH = 8.3 Sodium lauryl sulfate = 100 mg/L	pH = 8.3 Sodium lauryl sulfate = 16.7 mg/L	Laboratory	[112]
Ozone microbubbles	Synthetic water with As(III)	pH = 7 As(III) = 200 µg/L As(V) $\leq$ DL	pH = 7 As(III) $\leq$ DL As(V) = 200 µg/L	Laboratory	[107]

DL: Detection Limit; TOC: Total Organic Carbon; DOC: Dissolved Organic Carbon; DO: Dissolved Oxygen; TS: Total Solids; TSS: Total Suspended Solids.

More research is required to systematically evaluate the performance of O_3 microbubbles process with mine effluents at low temperatures and scaling-up conditions. Further techno-economic studies are also necessary prior to full-scale mine site applications.

3.3. Membrane Filtration

Membrane-based processes include microfiltration (MF), ultrafiltration (UF), nanofiltration (NF), reverse osmosis (RO), vacuum membrane distillation (VMD), and forward osmosis (FO). A combination of these filtration techniques is often required to ensure good performance and minimize clogging membranes. Membrane filtration processes treat a variety of contaminants, from suspended solids and colloidal material to more persistent and soluble ones as salinity [59,113].

The nature of contamination determines the type of membrane to be used for effective treatment. Direct membrane filtration, or FO, achieves good contaminant removal rates by reducing release rates and treatment costs [114]. In addition, VMD increases the recovery rate of permeate when used alone or in combination with conventional filtration types [115–117].

The operating principle of all these processes lies in the pressure gradient exerted on the membranes, commonly called transmembrane pressure (TMP) [113,118]. The feed can pass directly through the membrane but is usually recirculated at high speed on the face of the membrane (crossflow configuration) [118]. This type of operation is recommended because the turbulence generated causes erosion and makes it possible to reduce the accumulation of solids on the membrane [118]. Membrane processes, their separation mechanisms, the materials used to manufacture the membranes, and their typical treatment objectives are presented in Table 5.

Table 5. Technical characteristics of the different types of membrane process [113,118,119].

Process	Separation Mechanism	Material/Type	Typical Transmembrane Pressure	Process	Separation Mechanism	Material/Type
Microfiltration (MF)	Separation by sieving through macropores (>50 nm)	Polymer and inorganic/Porous	10–100	90–99+	Removal of suspended matter and coarse colloidal particles including microorganisms	Pore size Exclusion at the membrane interface
Ultrafiltration (UF)	Separation by sieving through mesopores (2–50 nm)	Polymer and inorganic/Porous	50–300	85–95+	Removal of coarse molecules in solution and colloidal particles in suspension including bacteria and macromolecules such as proteins	
Nanofiltration (NF)	Separation by a combination of charge rejection, diffusion of solubility, and sieving through micropores (<2 nm)	Polymer and inorganic/Dense	200–1500	75–90+	Removal of multivalent ions and specific charged or polar molecules	Solution Diffusion through the membrane
Reverse osmosis (RO)	Separation based on the difference in solubility and diffusion rate of water and solutes	Polymer/Dense	500–8000	60–90	Removal of low molecular weight compounds such as inorganic ions	

The membranes can be made of polymers based on materials such as polyacrylonitrile (PAN) and polyvinylidenediflouride (PVDF). The cost of these membranes is relatively low, and their life cycle is around seven years [113]. Inorganic membranes are more resistant to chemical and thermal stress than polymer-based membranes. They are divided into four categories according to their constituents: ceramic membranes, glass membranes, metal membranes, and zeolite membranes. These are mainly used for MF and UF [113].

Two types of membrane configuration dominate the saline water treatment market with NF and RO: spiral wound and fine hollow fiber commonly known as spiral wound (SW) and hollow fine fiber (HFF). SW elements are constructed with flat membrane sheets and materials that provide mechanical strength. The materials of these membranes can be cellulosic or non-cellulosic. Cellulose acetate membranes are composed of two layers of different shapes of the same polymer; they are called asymmetric [120]. Composite membranes are two layers of different polymers separated from a porous substrate, which is often polysulphone [120]. The materials used in the manufacture of HFF can be polyaramide or a mixture of cellulose acetate. The membranes usually have an external diameter between 100 and 300 μm and an internal diameter between 50 and 150 μm. Saline water is inserted into the media from the outside of the HFF; by applying pressure, the permeate passes through and is then collected [120].

Clogging of membranes is the main problem related to their use. Indeed, the presence of elements that clog the pores of the membranes reduces their effectiveness and even their lifespan. For example, the presence of organic species and suspended matter can have a high clogging effect [121]. Calcium sulphate and calcium carbonate can also clog the membranes [122]. In the treatment of mine water, the main clogging elements were identified as Sb, Al, Si, and Na [119]. Pretreatment steps can help overcome this problem [119]. Higher water temperatures can reduce the clogging effect as the membrane pores expand. By expanding, contaminants can more easily pass through the membrane and, thus, reduce the quality of the permeate [123].

The membrane cleaning methods depend on whether the clogging is reversible or not. If the clogging is reversible, the use of physical methods such as a pulsed reverse flow with purge, vibrations with ultrasound, air, or CO_2 jets, or reverse permeation at regular intervals can clear the pores. This type of cleaning tends to have less impact on the degradation of the membranes and their lifespan than a chemical cleaning [118,122]. Chemical cleaning uses a chemical to react with the sealing layer to facilitate physical cleaning. The most used membrane cleaners are alkaline or acidic cleaners, surfactants, and saline solutions; their choice is related to the type of sealant [122]. The use of ozone also reduces clogging by organic matter [121,124]. The parameters to consider during chemical cleaning are product concentration, pH, temperature, pressure, flow rate, and cleaning time [122].

Reported efficiency of membranes for mine water at different scales is presented in Table 6.

For example, the MF and UF can be used to remove total suspended solids (TSS) and colloidal chemical oxygen demand as pretreatment prior to RO, which can remove salinity [59]. The management of RO reject is an important consideration because of its high TDS concentrations. A treatment of this brine can reduce the reject and can be used as byproduct with resale value [125]. The VMD technology that uses hydrophobic membrane and heat can be a good way to concentrate the RO brine and recover metals. The removal of 99.9% of TDS with VMD technology was reported. The mine water tested contained 2332 mg/L of TDS, 14.4 mg/L of Ca, 2.72 mg/L of Mg, 1.92 mg/L of Fe, and 3.38 mg/L of Al. Removal efficiency was higher than 95% for all these elements after 90 min [126]. The EC of the treated water could be lower than 50 μS/cm [116]. The VMD can be used as a treatment for low concentration of heavy metals and to achieve better performance than conventional treatment [127].

As newer technology, FO has promising potential for the desalination of high salinity mine water streams (60–240 g/L TDS). FO utilizes a thin film composite membrane and draws solutions to recover the metals. The principle of FO is based on the osmotic difference between a dilute feed solution (FS, i.e., contaminated water to be treated) and a more concentrated draw solution (DS, with higher osmotic potential than FS), which is diluted during the salinity treatment. The next step of the process—the separation of clean water from the diluted draw solution—is energy consuming and, as a result, the limiting and decisive step in FO overall feasibility. A crystallization process can then be used to recover the metals and salts after the separation [128]. The major advantages of FO for water recovery are the non-selectivity (high rejection of a wide variety of contaminants), less fouling of membrane, the simpler osmotic cells, and the overall lower cost (no external hydraulic pressure is applied for the water to cross the membrane relative to pressure-driven processes such as in RO) [128]. FO was found efficient in metal removal from acid mine drainage close or over 98% using NaCl-DS [129]. However, with mine water, the reverse flux of ions from the DS into the FS, especially when they can react with feed solutes, could prove one of the major downsides of this process, in addition to the precipitation of some secondary minerals on the separating membrane when inorganic salts are used in DS (e.g., NH_4HCO_3) [129]. More recently, the FO membranes showed to be promising in power production by the pressure-retarded osmosis (PRO) process [130]. Notably, osmotic energy generation (using PRO) was proposed more than 70 years ago, but it was limited by the lack of effective membranes. As recently as 2009, a prototype plant was constructed in Norway, but the project was terminated in 2014 due to technology immaturity [131].

Table 6. Treatment efficiency using membrane filtration with mine water at different scales.

Treatment	Industry	Influent	Effluent (Permeate)	Scale	References
MF + NF + RO	Gold mine pressure oxidation process water effluent	pH = 1.46 EC = 28.07 mS/cm TSS = 571 mg/L TDS = 23 973 mg/L Cu = 156.8 mg/L Co = 40.01 mg/L Ni = 256.8 mg/L Ca = 487.6 mg/L Mg = 2561 mg/L Fe = 436.5 mg/L Mn = 105.5 mg/L Al = 348 mg/L As = 34.6 mg/L Total Acidity = 10.28 g $CaCO_3$/L Free Acidity = 6.89 g $CaCO_3$/L SO_4^{2-} = 21 480 mg/L	pH = 2.56 EC = 0.79 mS/cm TSS = 0 mg/L TDS = 192 mg/L Cu = 0.22 mg/L Co = 0.09 mg/L Ni = 1.24 mg/L Ca = 1.61 mg/L Mg = 11.96 mg/L Fe = 0.56 mg/L Mn = 0.13 mg/L Al = 0.63 mg/L As = 2 mg/L - - SO_4^{2-} = 270 mg/L	Pilot	[132]
NF	Acid mine drainage from abandoned mercury mine	Fe = 515 mg/L Al = 23 mg/L As = 6 mg/L Hg = 2.3 µg/L SO_4^{2-} = 2300 mg/L pH = 2.47 ORP = 592 mV DO = 3.3 mg/L	Fe = 7.5 mg/L Al = 1.7 mg/L As = 0.08 mg/L - SO_4^{2-} = 245 mg/L - - -	Pilot	[133]
MF + NF	Acidic bioleaching mining waste process	Co = 4.04 mg/L Ge = 1.33 mg/L Mo = 14.30 mg/L Re = 3.26 mg/L Cu = 54 mg/L Fe = 1980 mg/L Zn = 720 mg/L	Co = 0.04 mg/L Ge = 1.18 mg/L Mo = 0.39 mg/L Re = 2.96 mg/L Cu = 0.74 mg/L Fe = 15 mg/L Zn = 7.8 mg/L	Laboratory	[134]
RO	Final effluent of antimony mine in operation	Turbidity = 39.4 NTU TDS = 7910 mg/L Sb = 49.8 mg/L As = 0.038 mg/L Ni = 0.052 mg/L Zn = 0.052 mg/L Fe = 0.05 mg/L Cd = 0.0001 mg/L Cr = 0.001 mg/L Cu = 0.002 mg/L Pb = 0.001 mg/L	Turbidity = 1 NTU TDS = 218 mg/L Sb = 0.1 mg/L As = 0.001 mg/L Ni = 0.001 mg/L Zn = 0.001 mg/L Fe = 0.05 mg/L Cd = 0.0001 mg/L Cr = 0.001 mg/L Cu = 0.001 mg/L Pb = 0.001 mg/L	Industrial	[119]

ORP: Oxydo-reduction potential.

The use of membrane filtration processes is common in the field of desalination for drinking water but less in the mining field. In recent years, research has made it possible to apply this technology to mine water and obtain satisfactory results [135].

4. Challenges and Opportunities in Mine Water Treatment in Cold Climates

Additional challenges in cold climates mean that system designs for effective performance of mine water treatment must also properly integrate specific characteristics, including the following: (1) high sensitivity and limited resilience of ecosystems, (2) unusual faster dissolution of carbonates at low temperatures, (3) the increased salinity in response to the very narrow window of water flow (2–3 months/year), (4) high costs of freshwater, (5) enhanced detrimental impacts of salinity on treatment efficiency and solubility of minerals from the gangue, (6) accelerated clogging of membranes, (7) limited knowledge about the suitable management of residual materials in a continuously evolving context (e.g., the increasing thickness of the active layer of permafrost), and (8) slower kinetics of most of the chemical processes. The impact from the mixing of different effluents on aquatic toxicity is also a big challenge. All these challenges must be targeted for future research to ensure sustainable mining in the North.

Nevertheless, few opportunities are arising in relation to mine water treatment in cold climates. Energy production from salinity, such as osmotic power generation by PRO using seawater brine as the DS and wastewater retentate as the feed, showed potential as a renewable energy source [130,136]. The use of promising membrane filtration technologies can help to recover the metals and reduce the salinity. The membrane fouling with high calcium and sulfate concentrations (i.e., gypsum scaling) is still an issue with some membrane materials. Further studies are necessary to solve these issues [137]. Finally, some AOPs could limit the creation of residual salinity and allow the simultaneous treatment of mixed contaminants.

5. Conclusions

This paper provided an overview of the issues related to the removal of CECs from mine water, with focus on As, Mn, Se, salinity, thiosalts, xanthates, and N-based compounds (cyanides, NH_3-N, and SCN^-) in cold climates. The background concentrations of As, Mn, Se, salinity, thiosalts, xanthates, and N-based compounds in natural water were first discussed. The performance of AOPs and membrane filtration processes in CECs treatment in mine water were then emphasized. Further studies are required for optimal treatment performance under cold climate conditions and high flow rates, including field-scale tests. In the contemporary context, the mining industry's ability to strengthen its capacity to adapt to increasingly stringent criteria related to CECs is central to sustainable development. Such adaptation will necessarily depend on close collaboration between mining operators and scientists conducting applied research, and the challenges will most likely increase in complexity in the very near future, which highlights the critical need for intensifying research on the treatment of CECs.

Author Contributions: Conceptualization, original draft preparation, writing—review and editing, S.R., C.M.N., E.R., L.C., and P.C.; supervision, C.M.N., E.R., and P.C. All authors have read and agreed to the published version of the manuscript.

Funding: Natural Sciences and Engineering Research Council of Canada: RDCPJ530091; Canada Research Chairs: 950-231412; Canada Research Chairs: 950-232194.

Institutional Review Board Statement: Not applicable.

Informed Consent Statement: Not applicable.

Data Availability Statement: Not applicable.

Acknowledgments: This research was supported by the Natural Sciences and Engineering Research Council of Canada (NSERC), Canada Research Chairs Program, and RIME UQAT-Polytechnique industrial partners, including Agnico Eagle, Canadian Malartic Mine, Iamgold, Raglan Mine–Glencore, Newmont, and Rio Tinto. The authors would like thank M.D., senior metallurgist at Agnico Eagle Mines Ltd., for his valuable contribution on the review of historical and new CECs in mine water. The authors also wish to thank A.R.L. and J.D. for their contributions during the preparation of the manuscript.

Conflicts of Interest: The authors declare no conflict of interest. The funders had no role in the design of the study; in the collection, analyses, or interpretation of data; in the writing of the manuscript, or in the decision to publish the results.

References

1. Laurence, D. Establishing a sustainable mining operation: An overview. *J. Clean. Prod.* **2011**, *19*, 278–284. [CrossRef]
2. Krausmann, F.; Gingrich, S.; Eisenmenger, N.; Erb, K.H.; Haberl, H.; Fischer-Kowalski, M. Growth in global materials use, GDP and population during the 20th century. *Ecol. Econ.* **2009**, *68*, 2696–2705. [CrossRef]
3. Roach, B.; Walker, T.N. Aquatic monitoring programs conducted during environmental impact assessments in Canada: Preliminary assessment before and after weakened environmental regulation. *Environ. Monit. Assess.* **2017**, *189*, 109. [CrossRef]
4. Wolkersdorfer, C.; Nordstrom, D.K.; Beckie, R.D.; Cicerone, D.S.; Elliot, T.; Edraki, M.; Valente, T.; Alves França, S.C.; Kumar, P.; Oyarzún Lucero, R.A.; et al. Guidance for the integrated use of hydrological, geochemical, and isotopic tools in mining operations. *Mine Water Environ.* **2020**, *39*, 204–228. [CrossRef]
5. Sauvé, S.; Desrosiers, M. A review of what is an emerging contaminant. *Chem. Cent. J.* **2014**, *8*, 15. [CrossRef]
6. Fairbairn, D.J.; Arnold, W.A.; Barber, B.L.; Kaufenberg, E.F.; Koskinen, W.C.; Novak, P.J.; Rice, P.J.; Swackhamer, D.L. Contaminants of emerging concern: Mass balance and comparison of wastewater effluent and upstream sources in a mixed-use watershed. *Environ. Sci. Technol.* **2016**, *50*, 36–45. [CrossRef]
7. Neculita, C.M.; Rosa, E. A review of the implications and challenges of manganese removal from mine drainage. *Chemosphere* **2019**, *214*, 491–510. [CrossRef]
8. Neculita, C.M.; Coudert, L.; Rosa, E.; Mulligan, C. Future prospects for treating contaminants of emerging concern in water and soils/sediments. In *Advanced Nano-Bio Technologies for Water and Soil Treatment*; Filip, J., Cajthaml, T., Najmanová, P., Černík, M., Zbořil, R., Eds.; Springer Nature Switzerland AG: Cham, Switzerland, 2020; pp. 589–605.
9. Coudert, L., Bondu, R.; Rakotonimaro, T.V.; Rosa, E.; Guittonny, M.; Neculita, C.M. Treatment of As-rich mine effluents and produced residues stability: Current knowledge and research priorities for gold mining. *J. Hazard. Mater.* **2020**, *386*, 121920. [CrossRef]
10. OECD. *New and Emerging Water Pollutants arising from Agriculture*; OECD: Paris, France, 2012; p. 49.
11. Fisheries and Environment Canada. *Metal Mining Liquid Effluent Regulations and Guidelines*; Report EPS 1-WP-77-1; Fisheries and Environment Canada: Ottawa, ON, Canada, 1977.
12. D019. *Directive 019 Sur l'Industrie Minière*; Government of Quebec: Quebec, QC, Canada, 2005. Available online: https://www.environnement.gouv.qc.ca/milieu_ind/directive019/directive019-2005.pdf (accessed on 18 October 2020).
13. CCME Summary Table. Available online: http://st-ts.ccme.ca/en/index.html (accessed on 18 October 2020).
14. Minister of Justice. *MMER (Metal and Diamond Mining Effluent Regulations)*; SOR/2002–222; Government of Canada: Ottawa, ON, Canada, 2019.
15. Ben Ali, H.E.; Neculita, C.M.; Molson, J.W.; Maqsoud, A.; Zagury, G.J. Performance of passive systems for mine drainage treatment at low temperature and high salinity: A review. *Miner. Eng.* **2019**, *134*, 325–344. [CrossRef]
16. Royer-Lavallée, A.; Neculita, C.M.; Coudert, L. Removal and potential recovery of rare earth elements from mine water. *J. Ind. Eng. Chem.* **2020**, *89*, 47–57. [CrossRef]
17. Costis, S.; Mueller, K.; Coudert, L.; Neculita, C.M.; Reynier, N.; Blais, J.F. Recovery potential of rare earth elements from mining and industrial residues: A review and case studies. *J. Geochem. Explor.* **2021**, *221*, 106699. [CrossRef]
18. Olías, M.; Cánovas, C.; Basallote, M.D.; Lozano, A. Geochemical behaviour of rare earth elements (REE) along a river reach receiving inputs of acid mine drainage. *Chem. Geol.* **2018**, *493*, 468–477. [CrossRef]
19. Smedley, P.L.; Kinniburgh, D.G. A review of the source, behaviour and distribution of arsenic in natural waters. *Appl. Geochem.* **2002**, *17*, 517–568. [CrossRef]
20. Mandal, B.K.; Suzuki, T. Arsenic round the world: A review. *Talanta* **2002**, *58*, 201–235. [CrossRef]
21. Plant, J.A.; Kinniburgh, D.G.; Smedley, P.L.; Fordyce, F.M.; Klinck, B.A. Arsenic and selenium. In *Treatise on Geochemistry*; Holland, H.D., Turekian, K.K., Eds.; Pergamon: Oxford, UK, 2003; pp. 17–66.
22. Gaillardet, J.; Viers, J.; Duprée, B. Trace elements in river waters. In *Treatise on Geochemistry*; Drever, J.I., Ed.; Pergamon: Oxford, UK, 2003; pp. 225–272.
23. Shand, P.; Edmunds, W.M. The baseline inorganic chemistry of European groundwaters. In *Natural Groundwater Quality*; Edmunds, M.W., Shand, P., Eds.; Blackwell Publishing: Hoboken, NJ, USA, 2008; pp. 22–58.

24. Giloteaux, L.; Duran, R.; Casiot, C.; Bruneel, O.; Elbaz-Poulichet, F.; Goni-Urriza, M. Three-year survey of sulfate-reducing bacteria community structure in Carnoulès acid mine drainage (France), highly contaminated by arsenic. *FEMS Microbiol. Ecol.* **2013**, *83*, 724–737. [CrossRef]
25. Majzlan, J.; Plášil, J.; Škoda, R.; Gescher, J.; Kögler, F.; Rusznyak, A.; Küsel, K.; Neu, T.R.; Mangold, S.; Rothe, J. Arsenic-rich acid mine water with extreme arsenic concentration: Mineralogy, geochemistry, microbiology and environmental implications. *Environ. Sci. Technol.* **2014**, *48*, 13685–13693. [CrossRef] [PubMed]
26. Fordyce, F. Selenium geochemistry and health. *Ambio* **2007**, *36*, 94–97. [CrossRef]
27. Flem, B.; Reimann, C.; Fabian, K.; Birke, M.; Filzmoser, P.; Banks, D. Graphical statistics to explore the natural and anthropogenic processes influencing the inorganic quality of drinking water, ground water and surface water. *Appl. Geochem.* **2018**, *88*, 133–148. [CrossRef]
28. Ji, Y.; Li, L.; Wang, Y. Selenium removal by activated alumina in batch and continuous-flow reactors. *Water Environ. Res.* **2020**, *92*, 51–59. [CrossRef]
29. Sun, Z.; Forsling, W. The degradation kinetics of ethyl-xanthate as a function of pH in aqueous solution. *Miner. Eng.* **1997**, *10*, 389–400. [CrossRef]
30. Xu, Y.; Lay, J.; Korte, F. Fate and effects of xanthates in laboratory freshwater systems. *Bull. Environ. Contam. Toxicol.* **1988**, *41*, 683–689. [CrossRef] [PubMed]
31. Alto, K.; Broderius, S.; Smith, L. *Toxicity of Xanthates to Freshwater Fish and Invertebrates*; Minnesota Environmental Quality Council: Minneapolis, MN, USA, 1978.
32. Rostad, C.E.; Schmitt, C.J.; Schumacher, J.G.; Leiker, T.J. An exploratory investigation of polar organic compounds in waters from a lead–zinc mine and mill complex. *Water Air Soil Pollut.* **2011**, *217*, 431–443. [CrossRef]
33. Bach, L.; Nørregaard, R.D.; Hansen, V.; Gustavson, K. *Review on Environmental Risk Assessment of Mining Chemicals Used for Mineral*; Scientific Report; DCE–Danish Centre for Environment and Energy: Roskilde, Denmark, 2016.
34. Muzinda, I.; Schreithofer, N. Water quality effects on flotation: Impacts and control of residual xanthates. *Miner. Eng.* **2018**, *125*, 34–41. [CrossRef]
35. Van Dam, R.A.; Harford, A.J.; Lunn, S.A.; Gagnon, M.M. Identifying the cause of toxicity of a saline mine water. *PLoS ONE* **2014**, *9*, e106857. [CrossRef]
36. Pinto, P.X.; Al-Abed, S.R.; Balz, D.A.; Butler, B.A.; Landy, R.B.; Smith, S.J. Bench-scale and pilot-scale treatment technologies for the removal of total dissolved solids from coal mine water: A review. *Mine Water Environ.* **2016**, *35*, 94–112. [CrossRef]
37. Cohen, B.; Lazarovitch, N.; Gilron, J. Upgrading groundwater for irrigation using monovalent selective electrodialysis. *Desalination* **2018**, *431*, 126–139. [CrossRef]
38. Simmons, J.A. Toxicity of major cations and anions (Na^+, K^+, Ca^{2+}, Cl^-, and $SO_4{}^{2-}$) to a macrophyte and an alga. *Environ. Toxicol. Chem.* **2012**, *31*, 1370–1374. [CrossRef] [PubMed]
39. Bowell, R.J. *A Review of Sulfate Removal Options for Mine Waters*; SRK Consulting: Cardiff, Wales, UK, 2004; 24p.
40. Takano, B.; Ohsawa, S.; Glover, R.B. Surveillance of Ruapehu Crater Lake, New Zealand by aqueous polythionates. *J. Volcanol. Geoth. Res.* **1994**, *60*, 29–57. [CrossRef]
41. Johnson, D.B. Chemical and microbiological characteristics of mineral spoils and drainage waters at abandoned coal and metal mines. *Water Air Soil Pollut.* **2003**, *3*, 47–66. [CrossRef]
42. Range, B.M.K.; Hawboldt, K.A. Adsorption of thiosulphate, trithionate, tetrathionate using biomass ash/char. *J. Environ. Chem. Eng.* **2018**, *6*, 5401–5408. [CrossRef]
43. Miranda-Trevino, J.C.; Pappoe, M.; Hawboldt, K.; Bottaro, C. The importance of thiosalts speciation: Review of analytical methods, kinetics, and treatment. *Crit. Rev. Environ. Sci. Technol.* **2013**, *43*, 2013–2070. [CrossRef]
44. Range, B.M.K.; Hawboldt, K.A. Removal of thiosalt/sulfate from mining effluents by adsorption and ion exchange. *Min. Proc. Ext. Met. Rev.* **2019**, *40*, 79–86. [CrossRef]
45. Schwartz, M.; Vigneault, B.; McGeer, J. *Evaluating the Potential for Thiosalts to Contribute to Toxicity in Mine Effluents*; Report presented for Thiosalts Consortium, Project: 602591, Report CANMET-MMSL 06-053 (CR); CANMET Mining and Mineral Sciences Laboratories (CANMET-MMSL): Ottawa, ON, Canada, 2006; p. 101.
46. Wasserlauf, M.; Dutrizac, J.E. *The Chemistry, Generation and Treatment of Thiosalts in Milling Effluents: A Non-Critical Summary of CANMET Investigations 1976–1982*; Report: 82-4E; CANMET: Ottawa, ON, Canada, 1982; p. 104.
47. Schudel, G.; Plante, B.; Bussière, B.; McBeth, J.; Dufour, G. The effect of arctic conditions on the geochemical behaviour of sulphidic tailings. In Proceedings of the Tailings and Mine Waste, Vancouver, BC, Canada,, 17–20 November 2019.
48. Jouini, M.; Neculita, C.M.; Genty, T.; Benzaazoua, M. Freezing/thawing effects on geochemical behaviour of residues from acid mine drainage passive treatment systems. *J. Water Process. Eng.* **2020**, *33*, 101807. [CrossRef]
49. Chlot, S. Nitrogen and Phosphorus Interactions and Transformations in Cold-Climate Mine Water Recipients. Ph.D. Thesis, Lulea University of Technology, Lulea, Sweden, 2013.
50. Jermakka, J.; Wendling, L.; Sohlberg, E.; Heinonen, H.; Vikman, M. Potential technologies for the removal and recovery of nitrogen compounds from mine and quarry waters in subarctic conditions. *Crit. Rev. Environ. Sci. Technol.* **2015**, *45*, 703–748. [CrossRef]
51. Mudder, T.I.; Botz, M.M.; Smith, A. *Chemistry and Treatment of Cyanidation Wastes*, 2nd ed.; Mining Journal Books Limited: London, UK, 2001.

52. Gould, D.W.; King, M.; Mohapatra, B.R.; Cameron, R.A.; Kapoor, A.; Koren, D.W. A critical review on destruction of thiocyanate in mining effluents. *Miner. Eng.* **2012**, *34*, 38–47. [CrossRef]
53. Johnson, C.A. The fate of cyanide in leach wastes at gold mines: An environmental perspective. *Appl. Geochem.* **2015**, *57*, 194–205. [CrossRef]
54. Marsidi, N.; Hasan, H.A.; Abdullah, S.R.S. A review of biological aerated filters for iron and manganese ions removal in water treatment. *J. Water Process. Eng.* **2018**, *23*, 1–12. [CrossRef]
55. Staicu, L.C.; Morin-Crini, N.; Crini, G. Desulfurization: Critical step towards enhanced selenium removal from industrial effluents. *Chemosphere* **2017**, *172*, 111–119. [CrossRef]
56. Golder Associates Inc. *Literature Review of Treatment Technologies to Remove Selenium from Mining Influenced Water*; Teck Coal Limited: Calgary, AB, Canada, 2009.
57. Molina, G.C.; Cayo, C.H.; Rodrigues, M.A.S.; Bernardes, A.M. Sodium isopropyl xanthate degradation by advanced oxidation processes. *Miner. Eng.* **2013**, *45*, 88–93. [CrossRef]
58. Novak, L.; Holtze, K.; Wagner, R.; Feasby, G.; Liu, L. Guidance Document for Conducting Toxicity Reduction Evaluation (TRE) Investigations of Canadian Metal Mining Effluents. Prepared ESG International Inc. and SGC Lakefield for TIME Network: Guelph, ON, Canada; Lakefield, ON, Canada, 2002; p. 85.
59. Lefebvre, O.; Moletta, R. Treatment of organic pollution in industrial saline wastewater: A literature review. *Water Res.* **2006**, *40*, 3671–3682. [CrossRef] [PubMed]
60. Nordstrom, D.K.; Blowes, D.W.; Ptacek, C.J. Hydrogeochemistry and microbiology of mine drainage: An update. *Appl. Geochem.* **2015**, *57*, 3–16. [CrossRef]
61. Katz, M. *The Canadian sulphur problem. Sulphur and Its Inorganic Derivatives in the Canadian Environment*; National Research Council of Canada, NRC Associate Committee on Scientific Criteria for Environmental Quality: Ottawa, ON, Canada, 1977; pp. 21–67.
62. CCME (Canadian Water Quality Guidelines for the Protection of Aquatic Life). *Ammonia*; Canadian Council of Ministers of the Environment: Winnipeg, MB, Canada, 2010; pp. 1–8.
63. Khuntia, S.; Majumder, S.K.; Ghosh, P. Removal of ammonia from water by ozone microbubbles. *Ind. Eng. Chem. Res.* **2013**, *52*, 318–326. [CrossRef]
64. Ryskie, S.; Gonzalez-Merchan, C.; Genty, T.; Neculita, C.M. Efficiency of ozone microbubbles for ammonia removal from mine effluents. *Miner. Eng.* **2019**, *145*, 106071. [CrossRef]
65. Health Canada. Guidelines for Canadian drinking water quality: Guideline technical document-Nitrate and nitrite. In *Water and Air Quality Bureau*; Healthy Environments and Consumer Safety: Ottawa, ON, Canada, 2013; p. 128.
66. Baker, J.A.; Gilron, G.; Chalmers, B.A., Elphick, J.R. Evaluation of the effect of water type on the toxicity of nitrate to aquatic organisms. *Chemosphere* **2017**, *168*, 435–440. [CrossRef]
67. di Biase, A.; Wei, V.; Kowalski, M.S.; Bratty, M.; Hildebrand, M.; Jabari, P.; Oleszkiewicz, J.A. Ammonia, thiocyanate, and cyanate removal in an aerobic up-flow submerged attached growth reactor treating gold mine wastewater. *Chemosphere* **2020**, *243*, 125395. [CrossRef] [PubMed]
68. D019. *Directive 019 Sur l'Industrie Minière*; Government of Quebec: Quebec, QC, Canada, 2012. Available online: https://www.environnement.gouv.qc.ca/milieu_ind/directive019/directive019.pdf (accessed on 18 October 2020).
69. Shaw, J.R.; Dempsey, T.D.; Chen, C.Y.; Hamilton, J.W.; Folt, C.L. Comparative toxicity of cadmium, zinc, and mixtures of cadmium and zinc to daphnids. *Environ. Toxicol. Chem.* **2006**, *25*, 182–189. [CrossRef]
70. Foudhaili, T.; Jaidi, R.; Neculita, C.M.; Rosa, E.; Triffault-Bouchet, G.; Veilleux, E.; Coudert, L.; Lefebvre, O. Effect of the electrocoagulation process on the toxicity of gold mine effluents: A comparative assessment of *Daphnia magna* and *Daphnia pulex*. *Sci. Total Environ.* **2020**, *708*, 134739. [CrossRef]
71. Couture, P.; Pyle, G. Field studies on metal accumulation and effects in fish. In *Homeostasis and Toxicology of Essential Metals*; Wood, C.M., Farrell, A.P., Brauner, C.J., Eds.; Fish Physiology: Volume 31, Part 1; Academic Press: Cambridge, MA, USA, 2012; pp. 417–473.
72. Simate, G.S.; Ndlovu, S. Acid mine drainage: Challenges and opportunities. *J. Environ. Chem. Eng.* **2014**, *2*, 1785–1803. [CrossRef]
73. Okamoto, A.; Yamamuro, M.; Tatarazako, N. Acute toxicity of 50 metals to *Daphnia magna*. *J. Appl. Toxicol.* **2015**, *35*, 824–830. [CrossRef]
74. Vinot, H.; Larpent, J. Water pollution by uranium ore treatment works. *Hydrobiologia* **1984**, *112*, 125–129. [CrossRef]
75. Yim, J.H.; Kim, K.W.; Kim, S.D. Effect of hardness on acute toxicity of metal mixtures using *Daphnia magna*: Prediction of acid mine drainage toxicity. *J. Hazard. Mater.* **2006**, *138*, 16–21. [CrossRef] [PubMed]
76. USEPA (US Environmental Protection Agency). *Ambient Water Quality Criteria for Zinc*; Office of Research and Development: Washington, DC, USA, 1980.
77. USEPA. *Aquatic Life Ambient Water Quality Criteria for Ammonia–Freshwater*; Office of Water: Washington, DC, USA, 2013.
78. CCME. *Nitrate Ion*; Canadian Council of Ministers of the Environment: Winnipeg, MB, Canada, 2012; pp. 1–17.
79. Eytcheson, S.A.; LeBlanc, G.A. Hemoglobin levels modulate nitrite toxicity to *Daphnia magna*. *Sci. Rep.* **2018**, *8*, 1–8. [CrossRef] [PubMed]
80. Dauchy, J.W.; Waller, T.W.; Piwoni, M.D. Acute toxicity of cyanate to *Daphnia magna*. *Bull. Environ. Contam. Toxicol.* **1980**, *25*, 194–196. [CrossRef] [PubMed]
81. Banks, D.; Younger, P.L.; Arnesen, R.T.; Iversen, E.R.; Banks, S.B. Mine-water chemistry: The good, the bad and the ugly. *Environ. Geol.* **1997**, *32*, 157–174. [CrossRef]

82. Paquin, P.R.; Gorsuch, J.W.; Apte, S.; Batley, G.E.; Bowles, K.C.; Campbell, P.G.; Galvez, F. The biotic ligand model: A historical overview. *Comp. Biochem. Phys. C.* **2002**, *133*, 3–35. [CrossRef]
83. Deleebeeck, N.M.; De Schamphelaere, K.A.; Heijerick, D.G.; Bossuyt, B.T.; Janssen, C.R. The acute toxicity of nickel to *Daphnia magna*: Predictive capacity of bioavailability models in artificial and natural waters. *Ecotoxicol. Environ. Safe.* **2008**, *70*, 67–78. [CrossRef]
84. Kozlova, T.; Wood, C.M.; McGeer, J.C. The effect of water chemistry on the acute toxicity of nickel to the cladoceran *Daphnia pulex* and the development of a biotic ligand model. *Aquat. Toxicol.* **2009**, *91*, 221–228. [CrossRef]
85. Persoone, G.; Baudo, R.; Cotman, M.; Blaise, C.; Thompson, K.C.; Moreira-Santos, M.; Han, T. Review on the acute *Daphnia magna* toxicity test—Evaluation of the sensitivity and the precision of assays performed with organisms from laboratory cultures or hatched from dormant eggs. *Knowl. Manag. Aquat. Ec.* **2009**, *393*, 1–29. [CrossRef]
86. Sivula, L.; Vehniainen, E.R.; Karjalainen, A.K.; Kukkonen, J.V.K. Toxicity of biomining effluents to *Daphnia magna*: Acute toxicity and transcriptomic biomarkers. *Chemosphere* **2018**, *210*, 304–311. [CrossRef]
87. Bogart, S.J.; Woodman, S.; Steinkey, D.; Meays, C.; Pyle, G.G. Rapid changes in water hardness and alkalinity: Calcite formation is lethal to *Daphnia magna*. *Sci. Total Environ.* **2016**, *559*, 182–191. [CrossRef]
88. Kang, S.W.; Seo, J.; Han, J.; Lee, J.S.; Jung, J. A comparative study of toxicity identification using *Daphnia magna* and *Tigriopus japonicus*: Implications of establishing effluent discharge limits in Korea. *Mar. Pollut. Bull.* **2011**, *63*, 370–375. [CrossRef] [PubMed]
89. Wagner, R.; Liu, L.; Grondin, L. *Toxicity Treatment Evaluation of Mine Final Effluent—Using Chemical and Physical Treatment Methods*; Technical Paper; SGS Minerals Services: Lakefield, ON, Canada, 2002.
90. Lee, S.H.; Kim, I.; Kim, K.W.; Lee, B.T. Ecological assessment of coal mine and metal mine drainage in South Korea using *Daphnia magna* bioassay. *Springer Plus* **2015**, *4*, 518. [CrossRef]
91. Kratochvil, D. Sustainable water treatment technologies for the treatment of acid mine drainage. In Proceedings of the International Conference on Acid Rock Drainage (ICARD), Ottawa, ON, Canada, 20–26 May 2012.
92. MEND (Mine Environment Neutral Drainage). *Study to Identify BATEA for the Management and Control of Effluent Quality from Mines*; MEND Report 3.50.1 prepared by HATCH for MEND Program; MEND: Etobicoke, ON, Canada, 2014; p. 614.
93. Miklos, D.B.; Remy, C.; Jekel, M.; Linden, K.G.; Drewes, J.E.; Hubner, U. Evaluation of advanced oxidation processes for water and wastewater treatment: A critical review. *Water Res.* **2018**, *139*, 118–131. [CrossRef] [PubMed]
94. Budaev, S.L.; Batoeva, A.A.; Tsybikova, B.A. Effect of Fenton-like reactions on the degradation of thiocyanate in water treatment. *J. Environ. Chem. Eng.* **2014**, *2*, 1907–1911. [CrossRef]
95. Gonzalez-Merchan, C.; Genty, T.; Bussière, B.; Potvin, R.; Paquin, M.; Benhammadi, M.; Neculita, C.M. Influence of contaminant to hydrogen peroxide to catalyzer molar ratio in the advanced oxidation of thiocyanates and ammonia using Fenton-based processes. *J. Environ. Chem. Eng.* **2016**, *4*, 4129–4136. [CrossRef]
96. Yates, B.J.; Zboril, R.; Sharma, V.K. Engineering aspects of ferrate in water and wastewater treatment: A review. *J. Environ. Sci. Heal.* **2014**, *49*, 1603–1614. [CrossRef]
97. Khuntia, S.; Majumder, S.M.; Ghosh, P. Microbubbles-aided water and wastewater purification: A review. *Rev. Chem. Eng.* **2012**, *28*, 191–221. [CrossRef]
98. Lee, J.; von Gunten, U.; Kim, J.H. Persulfate-based advanced oxidation: Critical assessment of opportunities and roadblocks. *Environ. Sci. Technol.* **2020**, *54*, 3064–3081. [CrossRef]
99. Meshref, M.N.A.; Klamerth, N.; Islam, M.S.; McPhedran, K.N.; Gamal El-Din, M. Understanding the similarities and differences between ozone and peroxone in the degradation of naftenic acids: Comparative performance for potential treatment. *Chemosphere* **2017**, *180*, 149–159. [CrossRef]
100. Gervais, M.; Dubuc, J.; Paquin, M.; Gonzalez-Merchan, C.; Genty, T.; Neculita, C.M. Comparative efficiency of three advanced oxidation processes for thiosalts treatment in mine impacted water. *Miner. Eng.* **2020**, *152*, 106349. [CrossRef]
101. Bokare, A.D.; Choi, W. Review of iron-free Fenton-like systems for activating H_2O_2 in advanced oxidation processes. *J. Hazard. Mater.* **2014**, *275*, 121–135. [CrossRef]
102. Babu, D.S.; Nidheesh, P.V. A review on electrochemical treatment of arsenic from aqueous medium. *Chem. Eng. Commun.* **2021**, *208*, 389–410. [CrossRef]
103. Wang, X.Q.; Liu, C.P.; Yuan, Y.; Li, F.B. Arsenite oxidation and removal driven by a bio-electro-Fenton process under neutral pH conditions. *J. Hazard. Mater.* **2014**, *275*, 200–209. [CrossRef] [PubMed]
104. Olvera-Vargas, H.; Dubuc, J.; Wang, Z.; Coudert, L.; Neculita, C.M.; Lefebvre, O. Electro-Fenton beyond the degradation of organics: Treatment of thiosalts in contaminated mine water. *Environ. Sci. Technol.* **2021**, *55*, 2564–2574. [CrossRef] [PubMed]
105. Gonzalez-Merchan, C.; Genty, T.; Bussière, B.; Potvin, R.; Paquin, M.; Benhammadi, M.; Neculita, C.M. Ferrates performance in thiocyanates and ammonia degradation in gold mine effluents. *Miner. Eng.* **2016**, *95*, 124–130. [CrossRef]
106. Gonzalez-Merchan, C.; Genty, T.; Paquin, M.; Gervais, M.; Bussière, B.; Potvin, R.; Neculita, C.M. Influence of ferric iron source on ferrate's performance and residual contamination during the treatment of gold mine effluents. *Miner. Eng.* **2018**, *127*, 61–66. [CrossRef]
107. Khuntia, S.; Majumder, S.K.; Ghosh, P. Oxidation of As(III) to As(V) using ozone microbubbles. *Chemosphere* **2014**, *97*, 120–124. [CrossRef] [PubMed]
108. Liu, C.; Chen, Y.; He, C.; Yin, R.; Liu, J.; Qiu, T. Ultrasound-enhanced catalytic ozonation oxidation of ammonia in aqueous solution. *Int. J. Env. Res. Public Health* **2019**, *16*, 2139. [CrossRef]

109. Xiong, X.; Wang, B.; Zhu, W.; Tian, K.; Zhang, H. A review on ultrasonic catalytic microbubbles ozonation processes: Properties, hydroxyl radicals generation pathway and potential in application. *Catalysts* **2019**, *9*, 10. [CrossRef]
110. Esparza, J.M.; Cueva, N.C.; Pauker, C.S.; Jentzsch, P.V.; Bisesti, F.M. Combined treatment using ozone for cyanide removal from wastewater: A comparison. *Revista Internacional de Contaminacion Ambiental* **2019**, *35*, 459–467. [CrossRef]
111. Vicente, C.; Valverde Flores, J. Removal of lead and zinc from mining effluents by applying air micro-nanobubbles. *J. Nanotechnol.* **2017**, *1*, 73–78. [CrossRef]
112. Ghanem, H.; Kravchenko, V.; Makedon, V.; Shulha, O.; Oleksandr, S. Preliminary water purification from surfactants and organic compounds through ozone oxidation, intensified by electrical impulses. In Proceedings of the 2019 IEEE 6th International Conference on Energy Smart Systems (ESS), Kyiv, Ukraine, 17–19 April 2019.
113. Elshorbagy, W.; Chowdhury, R. *Water Treatment*; InTech: London, UK, 2013; p. 392.
114. Hube, S.; Eskafi, M.; Hrafnkelsdóttir, K.F.; Bjarnadóttir, B.; Bjarnadóttir, M.Á.; Axelsdóttir, S.; Wu, B. Direct membrane filtration for wastewater treatment and resource recovery: A review. *Sci. Total Environ.* **2020**, *710*, 136375. [CrossRef]
115. Drioli, E.; Ali, A.; Macedonio, F. Membrane distillation: Recent developments and perspectives. *Desalination* **2015**, *356*, 56–84. [CrossRef]
116. Andrés-Mañas, J.A.; Ruiz-Aguirre, A.; Acién, F.G.; Zaragoza, G. Assessment of a pilot system for seawater desalination based on vacuum multi-effect membrane distillation with enhanced heat recovery. *Desalination* **2018**, *443*, 110–121. [CrossRef]
117. Karanasiou, A.; Kostoglou, M.; Karabelas, A. An experimental and theoretical study on separations by vacuum membrane distillation employing hollow-fiber modules. *Water* **2018**, *10*, 947. [CrossRef]
118. Speed, D. Environmental aspects of planarization processes. In *Advances in Chemical Mechanical Planarization (CMP)*; Elsevier: Amsterdam, The Netherlands, 2016; pp. 229–269.
119. Samaei, S.M.; Gato-Trinidad, S.; Altaee, A. Performance evaluation of reverse osmosis process in the post-treatment of mining wastewaters: Case study of Costerfield mining operations, Victoria, Australia. *J. Water Process. Eng.* **2020**, *34*, 101116. [CrossRef]
120. Watson, I.C.; Morin, O.; Henthorne, L. *Desalting Handbook for Planners*; Desalination Research and Development Program Report; Bureau of Reclamation: Washington, DC, USA, 2003; p. 72.
121. Van Geluwe, S.; Braeken, L.; Van der Bruggen, B. Ozone oxidation for the alleviation of membrane fouling by natural organic matter: A review. *Water Res.* **2011**, *45*, 3551–3570. [CrossRef] [PubMed]
122. Aguiar, A.; Andrade, L.; Grossi, L.; Pires, W.; Amaral, M. Acid mine drainage treatment by nanofiltration: A study of membrane fouling, chemical cleaning, and membrane ageing. *Sep. Purif. Technol.* **2018**, *192*, 185–195. [CrossRef]
123. Andrade, L.; Aguiar, A.; Pires, W.; Miranda, G.; Amaral, M. Integrated ultrafiltration-nanofiltration membrane processes applied to the treatment of gold mining effluent: Influence of feed pH and temperature. *Sep. Sci. Technol.* **2017**, *52*, 756–766. [CrossRef]
124. Fujioka, T.; Hoang, A.T.; Okuda, T.; Takeuchi, H.; Tanaka, H.; Nghiem, L.D. Water reclamation using a ceramic nanofiltration membrane and surface flushing with ozonated water. *Int. J. Env. Res. Public Health* **2018**, *15*, 799. [CrossRef]
125. Loganathan, K.; Chelme-Ayala, P.; Gamal El-Din, M. Pilot-scale study on the treatment of basal aquifer water using ultrafiltration, reverse osmosis and evaporation/crystallization to achieve zero-liquid discharge. *J. Environ. Manag.* **2016**, *165*, 213–223. [CrossRef]
126. Sivakumar, M.; Ramezanianpour, M.; O'Halloran, G. Mine water treatment using a vacuum membrane distillation system. *APCBEE Proc.* **2013**, *5*, 157–162. [CrossRef]
127. Ji, Z. Treatment of heavy-metal wastewater by vacuum membrane distillation: Effect of wastewater properties. *IOP Conf. Ser. Earth Environ. Sci.* **2018**, *108*, 042019. [CrossRef]
128. Kolliopoulos, G.; Shum, E.; Papangelakis, V.G. Forward osmosis and freeze crystallization as low energy water recovery processes for a water-sustainable industry. *Environ. Process.* **2018**, *5*, 59–75. [CrossRef]
129. Vital, B.; Bartacek, J.; Ortega-Bravo, J.; Jeison, D. Treatment of acid mine drainage by forward osmosis: Heavy metal rejection and reverse flux of draw solution constituents. *Chem. Eng. J.* **2018**, *332*, 85–91. [CrossRef]
130. Altaee, A.; AlZainati, N. Novel thermal desalination brine reject-sewage effluent salinity gradient for power generation and dilution of brine reject. *Energies* **2020**, *13*, 1756. [CrossRef]
131. Chung, T.S.; Luo, L.; Wan, C.F.; Cui, Y.; Amy, G. What is next for forward osmosis (FO) and pressure retarded osmosis (PRO)? *Sep. Purif. Technol.* **2015**, *156*, 856–860. [CrossRef]
132. Ricci, B.C.; Ferreira, C.D.; Aguiar, A.O.; Amaral, M.C. Integration of nanofiltration and reverse osmosis for metal separation and sulfuric acid recovery from gold mining effluent. *Sep. Purif. Technol.* **2015**, *154*, 11–21. [CrossRef]
133. Sierra, C.; Saiz, J.R.Á.; Gallego, J.L.R. Nanofiltration of acid mine drainage in an abandoned mercury mining area. *Water Air Soil Pollut.* **2013**, *224*, 1734. [CrossRef]
134. Meschke, K.; Hofmann, R.; Haseneder, R.; Repke, J.-U. Membrane treatment of leached mining waste–A potential process chain for the separation of the strategic elements germanium and rhenium. *Chem. Eng. J.* **2020**, *380*, 122476. [CrossRef]
135. Naidu, G.; Ryu, S.; Thiruvenkatachari, R.; Choi, Y.; Jeong, S.; Vigneswaran, S. A critical review on remediation, reuse, and resource recovery from acid mine drainage. *Environ. Pollut.* **2019**, *247*, 1110–1124. [CrossRef] [PubMed]
136. Skilhagen, S.E.; Dugstad, J.E.; Aaberg, R.J. Osmotic power—Power production based on the osmotic pressure difference between waters with varying salt gradients. *Desalination* **2008**, *220*, 476–482. [CrossRef]
137. Shaffer, D.L.; Werber, J.R.; Jaramillo, H.; Lin, S.; Elimelech, M. Forward osmosis: Where are we now? *Desalination* **2015**, *356*, 271–284. [CrossRef]

minerals

MDPI

Article

Removal of Pollutants from an AMD from a Coal Mine by Neutralization/Precipitation Followed by "In Vivo" Biosorption Step with the Microalgae *Scenedesmus* sp.

Karine Batista dos Santos, Vítor Otacílio de Almeida, Jéssica Weiler and Ivo André Homrich Schneider *

Laboratório de Tecnologia Mineral e Ambiental, Programa de Pós-Graduaçao em Engenharia de Minas, Metalúrgica e de Materiais, Escola de Engenharia, Universidade Federal do Rio Grande do Sul, Porto Alegre 90040-060, Brasil; karineebatista@gmail.com (K.B.d.S.); vitor01almeida@hotmail.com (V.O.d.A.); jessica.weiler18@gmail.com (J.W.)

* Correspondence: ivo.andre@ufrgs.br; Tel.: +55–51–3308–7882

Received: 24 June 2020; Accepted: 4 August 2020; Published: 12 August 2020

Abstract: This work evaluates the benefits of a complementary treatment step of acid mine drainage (AMD) using the algae *Scenedesmus sp.* in terms of algae biomass production, residual metal removal, and the toxicity of the discharged water. Conventional treatment by neutralization/precipitation of an AMD from a coal mine in Brazil was conducted with $Ca(OH)_2$ at pH 8.7. Algal growth studies were performed in the treated AMD, with and without a nutrient supply. The raw effluent and treatments were compared in terms of residual concentration of metals and sulfate, conductivity, and toxicity with the *Allium cepa* and *Daphnia magna* test organisms. The results show that the conventional treatment allowed a major metal removal, reduction in the conductivity, and good indices in the toxicological parameters evaluated. The biosorption with in vivo microalgae improved the quality of the effluent for residual metals. No significant toxicity was observed to *Allium cepa* in all treatments performed, while the *Daphnia magna* test indicated a reduction in toxicity after the biosorption step. It was concluded that algae growth can be carried out in treated mine waters, providing algae biomass and helping to achieve the standards for water discharge.

Keywords: acid mine drainage; conventional treatment; biosorption; algae; metal removal; toxicity

---

1. Introduction

Active chemical treatment by neutralization has been widely applied by the coal mining industry to prevent the damages caused by acid mine drainage. The process involves the addition of alkaline reagents to increase the pH and precipitate the dissolved metals as hydroxides [1–3]. Recently, Skousen et al. [4] pointed out six chemicals commonly used to treat AMD: limestone—$CaCO_3$, hydrated lime—$Ca(OH)_2$, pebble quicklime—CaO, soda ash—Na_2CO_3, caustic soda—NaOH, and ammonia—NH_3. Considering the acidic mine water treatment plants in Brazil, the most commonly used neutralizing reagent is hydrated lime, raising the pH to 8.7. Calcium chemicals are preferred since they are cheaper and allow the partial removal of the amount of sulfate ions present in the wastewater [5].

This practice was validated by recent studies applying the reagents $Ca(OH)_2$ and NaOH at pH 7.0 and 8.7 to treat acid mine drainage from a coal mine [6,7]. All neutralizing treatments resulted in a satisfactory reduction in the metal concentration, with the best results achieved using $Ca(OH)_2$ at pH 8.7, although Mn and As remained above or near the maximum limits for discharge according to the Brazilian legislation. Additional treatment steps can also be applied for AMD control when greater treatment efficiency is needed [3]. Studies suggested the use of algae to remove residual metals from AMD [7–9], since they can act as metal accumulators through bioaccumulation (active mechanism) and/or biosorption (passive mechanism). Algae species of genera such as *Spirulina*, *Chlorella*, *Scenedesmus*, *Cladophora*, *Oscillatoria*, *Anabaena*, and *Phaeodactylum tricornutum* showed capacity to remove a considerable content of metals from AMD [9]. Algae ponds could also generate a possible biomass use [10] and, in particular, the *Scenedesmus* genera are greatly used for the production of biodiesel, biogas, and proteins [11,12].

Indeed, there is a growing trend in the different uses of treated AMD. Many of the studies aim to integrate AMD treatment systems with the recovery of dissolved metals and the synthesis of products [13–22]. However, there are considerably fewer works concerning the use of discharged water. The main initiatives are for water reuse in the industrial systems [5,23,24], as a resource for drinking water [22], to produce electricity [25], and for algae growth [7,26–28]. We should point out that the study of Pereira et al. [7] was carried out with the *Scenedesmus sp.* strain in treated coal mine water, but not exploring the total algae growth capacity, given that there was a lack of nutrients in the system, mainly in terms of nitrogen and phosphorus.

The objective of this work was to evaluate the benefits of an additional step in AMD treatment by growing the algae *Scenedesmus* sp. in vivo. The water substrate was attained from a coal mine in south Brazil after conventional treatment by neutralization/precipitation with $Ca(OH)_2$ at pH 8.7. The treated effluent was subjected to algal growth both with and without the addition of nutrients. The quality of the treated water was analyzed in terms of conductivity, metals, and sulfate concentration, as well as in terms of toxicological effects for *Allium cepa* and *Daphnia magna*.

2. Materials and Methods

2.1. AMD Sample

The sample of AMD used in this work was provided by a coal mining company located in the municipality of Figueira, northeast of the State of Paraná, Brazil (23°49′15″ S; 50°25′50″ W). The coal seams are explored in the Paraná Basin and are part of the Permian Rio Bonito Formation [29]. Coal waste is highly associated with iron sulfides, with a content of pyrite ranging up to 15%. Reject surface dumps are also associated with toxic pollutants such as Zn, Mn, Pb, and As, which are commonly dissolved in water when the pH decreases [30].

2.2. AMD Treatments

The conventional treatment of AMD was conducted based on the treatment carried out by Brazilian mining companies. First, the pH was raised to 8.7 by adding the alkaline reagent (calcium hydroxide—$Ca(OH)_2$) under stirring and monitoring the pH with a pH meter, followed by precipitation of the metals as hydroxides. The supernatant was filtered and then separated for the additional treatment with algae.

The algae species selected was *Scenedesmus* sp. due to its known metal absorption capacity [9,31], and its use as a biofuel resource [11,12]. Microalgae cultures for inoculation were maintained in 1000 mL Erlenmeyer flasks with the Guillard modified culture medium at 25 °C and 3910 lux and were used when the growth reached the exponential phase. Two scenarios for *Scenedesmus* sp. growth in the treated AMD were observed: (a) without addition of nutrients; and (b) with the addition of 0.1% (volume) of the macronutrients supply of Guillard medium composed of: 36.76 g L^{-1} of $CaCl_2.2H_2O$, 36.97 g L^{-1}, $MgSO_4.7H_2O$, 12.6 g L^{-1} of $NaHCO_3$, 8.71 g L^{-1} of K_2HPO_4, 85.01 g L^{-1} of $NaNO_3$ and 28.42 g L^{-1} of $Na_2SiO_3.7H_2O$ [32]. Erlenmeyer flasks (250 mL) containing AMD and 1% algae inoculum were incubated on an orbital shaker (model CT-712RN Cientec) at 125 rpm, 27.5 °C, with a photoperiod of 16 h/day under illumination with a light intensity of 6200 lux. Dry weight and optical density (OD) were measured daily to determine the microalgal growth and relationship between the dry weight and optical density. OD was measured at 570 nm wavelength (UV1100 spectrophotometer Pro-tools), and the dry matter was measured by filtering the samples through 0.7 μm pre-weighted membranes, which were dried at 60 °C for 24 h.

Thus, the samples (each one in triplicate) were submitted to physical-chemical and toxicological analyses, as illustrated in Figure 1.

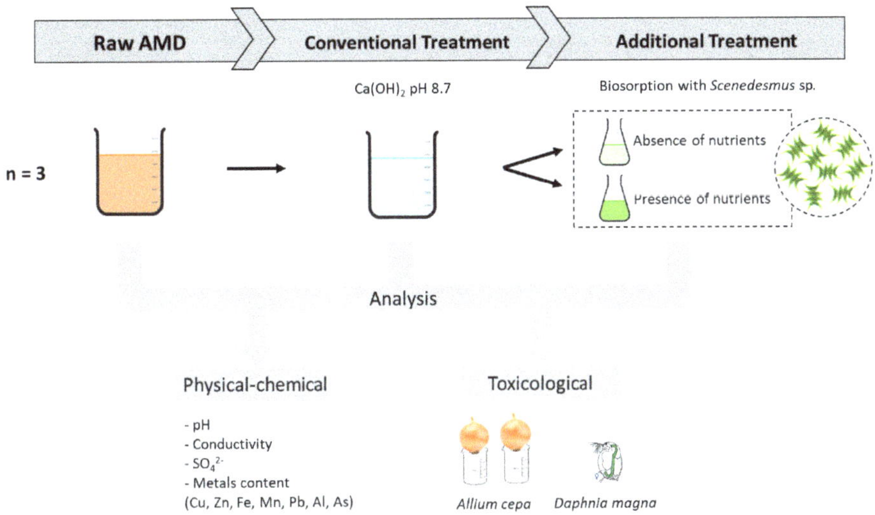

Figure 1. Illustrative scheme of this work, including the treatments and analysis performed.

2.3. *Physical-Chemical and Toxicological Analysis*

Raw AMD and treated waters were analyzed for pH, conductivity and metals content (Cu, Zn, Fe, Mn, Pb, Al, As), sulfate content, and toxicological tests with *Allium cepa* and *Daphnia magna*.

The pH of the suspension was measured using a bench pH meter (AKSO model 86505, São Leopoldo, Brazil). Electrical conductivity was measured using a Zetasizer (Malvern Panalytical, Malvern, UK). Metal analyses were carried out by sample preservation with HNO_3 and by an issuing optical spectrophotometer with inductively coupled plasma (ICP-OES). The sulfate content was measured using the turbidimetric method. All analyses followed the procedures described in the "Standard Methods for Examination of Water and Wastewater" [33].

The *Allium cepa* root growth test was carried out with onions of the same origin and approximate sizes. The outer scales of the onions were removed and the bulb gently scraped. Bulbs were kept in touch with deionized water in beaker glasses for 24 h at room temperature to stimulate root growth. After this period, the bulbs were placed at each test condition and maintained in contact with the same for 72 h under the light and following the procedure of Fiskesjö [34]. The test was performed in triplicate and a control group was adopted using deionized water. The number, as well as the mass and the length of the three largest roots in each bulb, were measured. The results of the different tests conditions were compared with the control in terms of relative growth index (RGI) (Equation (1)), the inhibition index (II) (Equation (2)) and the germination index (GI) (Equation (3)). The phytotoxicity was considered when there was a significant decrease between the test and control groups:

$$RGI = \frac{Lm}{L_0} \tag{1}$$

$$II = \left(1 - \frac{Lm}{L_0}\right) \times 100 \tag{2}$$

$$GI = \frac{G}{G_0} \times RGI \times 100 \quad (3)$$

where Lm is the average length of the roots in the samples, L_0 is the average length of the roots in the control, G is the germination value in the samples, and G_0 is the germination rate in the control.

Acute toxicity tests for *Daphnia magna* microcrustaceans were performed in triplicate and in accordance with the Guideline for Testing of Chemicals using *Daphnia sp.* OECD 202 [35], using the statistical technique EPA 821-R2-012 [36]. This test assesses the immediate toxic effects of a sample on the mobility of this microcrustacean species and is based on the exposure of organisms to a series of dilutions of the sample for 48 h. The effect on organisms is expressed by a loss of mobility. The result is calculated by a statistical procedure and represented by the concentration effect to 50% of the population (EC-50-48 h).

2.4. Statistical Analysis

The results were presented in terms of mean and standard deviation. Results were assessed using analysis of variance (ANOVA) with significance level p-value < 0.05 and the Tukey test to compare the differences between averages.

3. Results and Discussion

Chemical parameters of the AMD used in this study, and the results of the AMD treated for each treatment in terms of pH, conductivity, sulfates, and metals (iron, aluminum, manganese, zinc, lead, arsenic, and copper), are compiled in Table 1. These results were compared with the current Brazilian environmental legislation, CONAMA 430/2011 [37].

Table 1. Characterization of acid mine drainage (AMD) after conventional treatment—neutralization/precipitation with $Ca(OH)_2$ at pH 8.7—and additional treatment with growth of *Scenedesmus* sp. algae.

	Raw AMD		Conventional Treatment $Ca(OH)_2$ pH 8.7		Additional Treatment				CONAMA 430/2011
					Algae		Algae + Macronutrients		
	Mean	sd	Mean	sd	Mean	sd	Mean	sd	
pH	2.33	0.15	8.7	0.0	5.55	0.21	5.75	0.7	
Conductivity (mS cm^{-1})	7.79	0.02	5.28	0.07	4.84	0.08	4.34	0.10	–
SO_4^{2-} (mg L^{-1})	7410.40	36.94	2727.60	19.49	1325.75	5.59	1250.40	3.82	–
Fe (mg L^{-1})	611.38	5.11	0.90	0.02	0.08	0.11	0.05	0.07	15
Al (mg L^{-1})	269.37	11.67	0.08	0.08	0.24	0.33	0.30	0.43	–
Zn (mg L^{-1})	62.65	0.55	0.10	0.10	0.06	0.06	0.08	0.04	5
Mn (mg L^{-1})	37.98	0.36	1.54	0.40	0.46	0.27	0.45	0.11	1
Pb (mg L^{-1})	0.41	0.02	0.15	0.05	0.06	0.08	0.02	0.03	0.5
As (mg L^{-1})	0.85	0.05	0.30	0.31	0.14	0.18	0.12	0.14	0.1
Cu (mg L^{-1})	nd	–	nd	–	nd	–	nd	–	1
Metal content * (mg L^{-1})	982.64	15.16	3.07	0.75	1.03	0.28	1.02	0.73	–

* Sum of the metals Fe, Al, Zn, Mn, Pb, As, and Cu. sd: standard deviation. nd: not detected. Database: Santos [38] and results partially published in Pereira et al. [7].

The conventional treatment was satisfactory considering the pH correction, 60% reduction in the sulfate content, and almost total removal of Fe, Al and Zn. However, for Mn and As, the treatment efficiency was not sufficient to achieve the limits stablished by the legislation. Thereafter, additional treatment with algae was investigated for polishing and removing the remaining metals. The AMD after conventional treatment was submitted to contact with *Scenedesmus* sp. and their growth was monitored over 10 days for each scenario in terms of biomass (Figure 2).

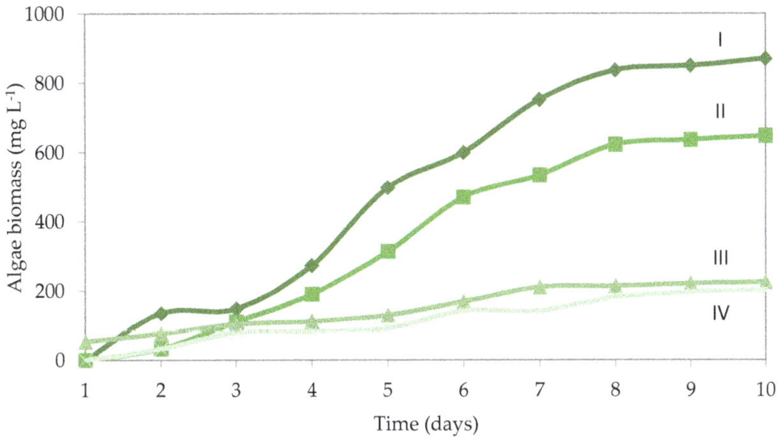

Figure 2. Growth of algae biomass as a function of time for Guillard medium—I; AMD after conventional treatment with nutrients—II; AMD after conventional treatment without nutrients—III; and for the control (water purified by reverse osmosis)—IV.

At the end of ten days of growth, 880 mg L^{-1} of algae biomass was achieved in the Guillard nutrients solution, 650 mg L^{-1} in the treated AMD with nutrient supply, 230 mg L^{-1} in the treated AMD without addition of nutrients, and 200 mg L^{-1} in distilled water. Algal growth in AMD was successful when nutrients were added, reaching values (75%) close to those obtained in an ideal culture medium, in this case the Guillard medium. Without the addition of nutrients, the growth was similar to the control (water only).

After algal growth, the water quality parameters were even more satisfactory for conductivity, sulfates, and the remaining metals. The removal of Mn was effective, reaching values below that established for discharge in water resources in Brazil [37]. There was also a decrease in the concentration of arsenic, but this was still above the limits applied by the legislation.

The differences between the adopted treatment procedures are depicted in Figure 3 in terms of final sulfate concentration and metal content (sum of Fe, Al, Zn, Mn, Pb, As, and Cu). It is possible to observe that the conventional treatment provides a partial removal of sulfates and an almost complete removal of these metals. Regarding sulfates, each treatment step allowed a statistically significant improvement, reaching the greatest efficiency with algae growth with the addition of nutrients. For metals, the same was observed, with the difference being that the addition of nutrients, with an increase in algae biomass production, did not provide a refinement. This may be because the metal removal is already quite high from the conventional treatment. The metal uptake was 8.9 mg g^{-1} of algae (in dry basis) in the case where no nutrients were added, similar to results obtained in other studies [39]. When nutrients were added, the mass of algae increased, but the removal of metals was almost the same, resulting in a metal uptake of 3.15 mg g^{-1} of algae, which is probably below the biosorption accumulation capacity of this *Scenedesmus sp.* The optimal conditions for the metal uptake by microalgae are due to many factors, including pH, metal speciation, the initial concentration of metal ions, and the mass of algae present, with an adsorption limit [40].

In the present case, the approximate values in the residual concentration of metals with and without nutrients may be explained by the low concentration of metal ions in the effluent after the conventional treatment. After this treatment step, the sum of transition metals plus arsenic and aluminum is approximately 3 mg L^{-1}. Considering the amount of algae biomass of 0.65 g L^{-1} in the treated AMD with nutrient supply, 0.23 g L^{-1} in the treated AMD without addition of nutrients, and literature data [40], the adsorption would fits in the beginning of the equilibrium curve plotted in the Cartesian plane, where there is a little variation of the residual pollutant in water solution with the increase in pollutant accumulation onto the biomass-assuming Langmuir adsorption model, which suits for biosorption with *Scenedesmus* and other microalgae. At these low concentrations, the metal ions listed in Table 1 are at a disadvantage to compete with other ions present in the solutions for the metal binding sites. Still, we cannot rule out that the residual elements may be chemically bound to soluble organic molecules excreted by the algae, making them not available for passive adsorption. This set of results indicates that both passive and active mechanism of metal removal may occur by in vivo *Scenedesmus* growth on treated AMD, even though quantitative values were not determined and this would be the subject of investigation on the continuity of this study. Although there was no significant difference in metal removal, the addition of nutrients enhances the biomass growth, which can be interesting from another perspective. The use of nutrients should be considered in scenarios where algae biomass can be used as raw material for the production of biofuel and biofertilizers [28,41,42].

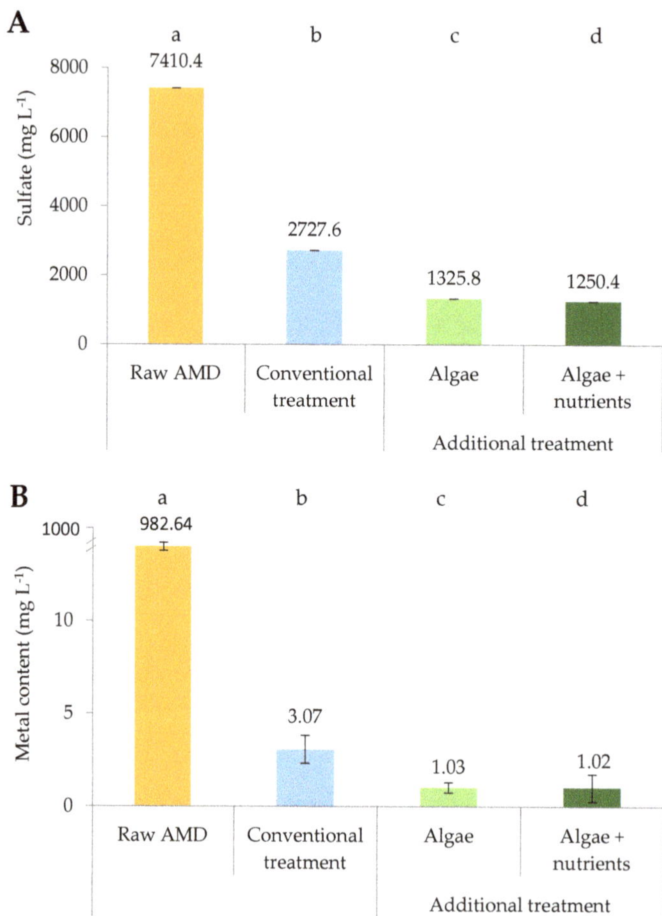

Figure 3. Comparison between raw AMD, conventional and additional treatments in terms of sulfate (**A**) and metal content (**B**). Means that differ significantly (Tukey's test, $p < 0.05$) are indicated by different lowercase letters.

Nevertheless, although physical-chemical analysis is important to attend the legal framework for effluent discharge, they do not predict toxicity to living organisms. The environmental effects and behavior of substances should also be evaluated by biological assays [43,44]. In this specific work, it was carried out with the toxicological bio-indicators *Allium cepa* (onion) and *Daphnia magna*.

The toxicity results with *Allium cepa* were based on relative growth, inhibition, and germination indices of the roots (Table 2). In raw AMD, it was possible to observe the total inhibition in onion root growth, probably due to the high acidity of the medium and high metal concentrations. These values are in accordance with the study of Geremias et al. [45], which also suggested that the phytotoxic effect could be associated with low pH values (pH = 2.5) and the expressive concentration of metals such as iron, aluminum, manganese, zinc, and lead, present in the effluent, that could be incorporated by the plant. After conventional chemical treatment with $Ca(OH)_2$ at pH 8.7, the effluent was non-toxic. There was an adequate root growth, both in length and mass, being in fact superior to the control (negative II values). Similar results were attained for the effluent after the additional treatment step with algae. In other words, the polishing step with algae did not modify the behavior of the onion root growth, either negatively or positively. Regarding the germination index, according to the classification adopted by Zucconi et al. [46], it was also not possible to observe a difference between treatments, which were classified as absent or low toxicity.

Table 2. Relative growth index (RGI), inhibition index (II), germination (G) and germination indices (GI) of *Allium cepa* roots for control (H_2O); raw AMD; AMD after conventional treatment; and AMD after additional treatment with algae without and with nutrients addition.

	RGI	II (%)	Toxicity	G (%)	GI (%)	Toxicity
Raw AMD	0	100	Inhibition	0	0	High
Conventional treatment	1.37	−37	Stimulation	71	97	Absent or low
			Additional treatment			
Algae	1.16	−17	No significant effect	74	86	Absent or low
Algae + nutrients	1.32	−32	Stimulation	90	119	Absent or low

Toxicity criterion (II): inhibition—RGI between 0 and 0.8, no significant effect—RGI between 0.8 and 1.2, stimulation—RGI greater than 1.2. Toxicity criterion (GI)—absent or low toxicity—GI greater than 80%, moderate toxicity—GI between 50 and 80%; high toxicity—GI less than 50%.

Table 3 shows the results obtained for the acute toxicity tests for the *Daphnia magna* microcrustaceans exposed to raw AMD, and the respective treatments by conventional and additional treatment with biosorption of algae *Scenedesmus* sp. The toxicity results show that raw acid drainage presented a toxicity factor (TF) of 400 and concentration of effect at 50% of the population (CE-50) of 6.25%, which classified it as an extremely toxic effluent. Similar results were presented by Kang et al. [47], who showed that acid mine drainage is highly toxic to microcrustaceans. This high toxicity may be due to the acidic pH, considerably below the no effect pH range for *Daphnia magna* (pH 4–11.5), and the high concentration of metals [48,49]. After conventional treatment, the TF values improved to 4, and the CE-50 was greater than 100%, classifying the samples as slightly toxic. The remaining toxic effect may be related to the Mn and As concentrations still present after treatment, since the studied species has sensitivity to these metals [50,51]. It may also be related to the conductivity, since the organism is sensitive to conductivity values greater than 2 mS cm^{-1} [52]. After the additional treatment with algae, there was another noticeable decrease in FT, reaching the value of 2 for both situations considered—with and without nutrients addition. The CE-50 was greater than 100%, framing it as slightly toxic.

The results show that microalgae may be suitable candidates for simultaneous remediation and sustainable biomass production in aqueous environments fed by treated acid mine drainage—at least from those of the coal mining industry. These efforts may help to attain the environmental standards that regulate the mining sector [53,54] and to contribute to the existing tendency of the mineral sector to look for sustainable practices, such as zero waste emission or further use of residual materials according to the economic model of the circular economy [55,56].

Table 3. Toxicity assays for *Daphnia Magna* microcrustaceans exposed to raw AMD; AMD after conventional treatment; and AMD after additional treatment with algae, with and without nutrients addition.

	TF	CE-50 48 h (%)	Toxicity
Raw AMD	400	6.25	Extremely toxic
Conventional treatment	4	>100	Slightly toxic
	Additional treatment		
Algae	2	>100	Slightly toxic
Algae + nutrients	2	>100	Slightly toxic

Toxicity criterion: extremely toxic—CE-50 < 25%, highly toxic—CE-50 between 25 and 50%, moderately toxic—CE-50 between 50 and 65%, and slightly toxic—CE-50 > 65% [46].

4. Conclusions

The biosorption process with the in vivo microalgae *Scenedesmus* sp proved to be efficient as a polishing step in AMD treatment systems considering both physical-chemical and toxicological analyses. After 10 days of algal growth, the *Scenedesmus* sp biomass (on a dry basis) concentration can reach 230 mg L^{-1} without the use of nutrients, and 650 mg L^{-1} with the addition of nutrients. Regardless of the use or not of nutrients, the biosorption process contributed significantly to reducing the total metal and sulfate content, as well as the conductivity.

Concerning the toxicological studies, the organisms tested in this work proved to be suitable as bioindicators to evaluate the performance of AMD treatment procedures in coal mining, each one presenting certain particularity. The growth of *Allium cepa* roots presented the advantage of a simple test and allowed a differentiation between treated and untreated AMD, but showed a low sensitivity among the various treatment scenarios. The acute toxicity test with *Daphnia magna* showed a higher sensitivity, allowing the differentiation of results between the raw and treated AMD, as well as between the different types of treatment applied.

Author Contributions: K.B.d.S.: Conceptualization, Investigation, Formal analysis, Writing—original draft. V.O.d.A.: Formal analysis, Investigation. J.W.: Investigation, Writing—original draft, Writing—review & editing. I.A.H.S.: Conceptualization, Formal analysis, Resources, Writing—review & editing. All authors have read and agreed to the published version of the manuscript.

Funding: This research was funded by Coordenação de Aperfeiçoamento de Pessoal de Nível Superior (CAPES) grant number [88882.345791/2019-01] and Conselho Nacional de Desenvolvimento Científico e Tecnológico (CNPQ) grant number [310369/2016-9] and [160570/2019-0].

Acknowledgments: This work was supported by the following Brazilian funding agencies: Coordenação de Aperfeiçoamento de Pessoal de Nível Superior (CAPES) and Conselho Nacional de Desenvolvimento Científico e Tecnológico (CNPQ). The authors are also grateful to Jean Carlo Salomé dos Santos Menezes for providing AMD samples.

Conflicts of Interest: The authors declare no conflict of interest.

References

1. Kontopoulos, A. Acid mine drainage control. In *Effluent Treatment in the Mining Industry*; Castro, S.H., Vergara, F.S.M.A., Eds.; University of Concepciòn: Concepción, Chile, 1998; pp. 57–118. ISBN 956-227-156-0.
2. Skousen, J.; Rose, A.; Geidel, G.; Foreman, J.; Evans, R.; Hellier, W. Handbook of Technologies for Avoidance and Remediation of Acid Mine Drainage. Available online: https://wvwri.wvu.edu/files/d/c2e42b2b-e40d-4ada-8bad-3c264d867e76/99-handbook-avoidance-remediation.pdf (accessed on 24 June 2020).
3. Kefeni, K.K.; Msagati, T.A.M.; Mamba, B.B. Acid mine drainage: Prevention, treatment options, and resource recovery: A review. *J. Clean. Prod.* **2017**, *151*, 475–493. [CrossRef]
4. Skousen, J.G.; Ziemkiewicz, P.F.; McDonald, L.M. Acid mine drainage formation, control and treatment: Approaches and strategies. *Extr. Ind. Soc.* **2019**, *6*, 241–249. [CrossRef]
5. Silveira, A.N.; Silva, R.; Rubio, J. Treatment of acid mine drainage (AMD) in South Brazil: Comparative active processes and water reuse. *Int. J. Miner. Process.* **2009**, *93*, 103–109. [CrossRef]

6. Santos, K.B.; Almeida, V.O.; Weiler, J.; Menezes, J.C.S.S.; Schneider, I.A.H. Ecotoxicity with *Allium cepa* to determine the efficiency of conventional ARD treatment by neutralization/precipitation from a brazilian coal mine. *Proc. IMWA* **2019**, 269–274.
7. Pereira, T.C.B.; dos Santos, K.B.; Lautert-Dutra, W.; de Souza Teodoro, L.; de Almeida, V.O.; Weiler, J.; Schneider, I.A.H.; Bogo, M.R. Acid mine drainage (AMD) treatment by neutralization: Evaluation of physical-chemical performance and ecotoxicological effects on zebrafish (*Danio rerio*) development. *Chemosphere* **2020**, *253*, 126665. [CrossRef]
8. Freitas, A.P.P.; Schneider, I.A.H.; Schwartzbold, A. Biosorption of heavy metals by algal communities in water streams affected by the acid mine drainage in the coal-mining region of Santa Catarina state, Brazil. *Miner. Eng.* **2011**, *24*, 1215–1218. [CrossRef]
9. Bwapwa, J.K.; Jaiyeola, A.T.; Chetty, R. Bioremediation of acid mine drainage using algae strains: A review. *South Afr. J. Chem. Eng.* **2017**, *24*, 62–70. [CrossRef]
10. Edmundson, S.J.; Wilkie, A.C. Landfill leachate—A water and nutrient resource for algae-based biofuels. *Environ. Technol.* **2013**, *34*, 1849–1857. [CrossRef] [PubMed]
11. Leite, G.B.; Abdelaziz, A.E.M.; Hallenbeck, P.C. Algal biofuels: Challenges and opportunities. *Bioresour. Technol.* **2013**, *145*, 134–141. [CrossRef] [PubMed]
12. Ghorbani, A.; Rahimpour, M.R.; Ghasemi, Y.; Raeissi, S. The biodiesel of microalgae as a solution for diesel. *Energies* **2018**, *11*, 950. [CrossRef]
13. Wei, X.; Viadero, R.C.; Buzby, K.M. Recovery of iron and aluminum from acid mine drainage by selective precipitation. *Environ. Eng. Sci.* **2005**, *22*, 745–755. [CrossRef]
14. Chen, T.; Yan, B.; Lei, C.; Xiao, X. Pollution control and metal resource recovery for acid mine drainage. *Hydrometallurgy* **2014**, *12*, 147–148. [CrossRef]
15. Hedrich, S.; Johnson, D.B. Remediation and selective recovery of metals from acidic mine waters using novel modular bioreactors. *Environ. Sci. Technol.* **2014**, *48*, 12206–12212. [CrossRef]
16. Menezes, J.C.S.S.; Silva, R.A.; Arce, I.S.; Schneider, I.A.H. Production of a poly-ferric sulphate chemical coagulant by selective precipitation of iron from acidic coal mine drainage. *Mine Water Environ.* **2009**, *28*, 311–314. [CrossRef]
17. Menezes, J.C.S.S.; Silva, R.A.; Arce, I.S.; Schneider, I.A.H. Production of a poly-alumino-iron sulphate coagulant by chemical precipitation of a coal mining acid drainage. *Miner. Eng.* **2010**, *23*, 249–251. [CrossRef]
18. Silva, R.D.A.; Castro, C.D.; Vigânico, E.M.; Petter, C.O.; Schneider, I.A.H. Selective precipitation/UV production of magnetite particles obtained from the iron recovered from acid mine drainage. *Miner. Eng.* **2012**, *29*, 22–27. [CrossRef]
19. Silva, R.D.A.; dos Santos Menezes, J.C.S.; Lopes, F.A.; Kirchheim, A.P.; Schneider, I.A.H. Synthesis of a goethite pigment by selective precipitation of iron from acidic coal mine drainage. *Mine Water Environ.* **2017**, *36*, 386–392. [CrossRef]
20. Macías, F.; Pérez-López, R.; Caraballo, M.A.; Cánovas, C.R.; Nieto, J.M. Management strategies and valorization for waste sludge from active treatment of extremely metal-polluted acid mine drainage: A contribution for sustainable mining. *J. Clean. Prod.* **2017**. [CrossRef]
21. De AlmeidaSilva, R.; Secco, M.P.; Lermen, R.T.; Schneider, I.A.H.; Hidalgo, G.E.N.; Sampaio, C.H. Optimizing the selective precipitation of iron to produce yellow pigment from acid mine drainage. *Miner. Eng.* **2019**, *141*, 1057–1066. [CrossRef]
22. Akinwekomi, V.; Maree, J.P.; Masindi, V.; Zvinowanda, C.; Osman, M.S.; Foteinis, S.; Mpenyana-Monyatsi, L.; Chatzisymeon, E. Beneficiation of acid mine drainage (AMD): A viable option for the synthesis of goethite, hematite, magnetite, and gypsum—Gearing towards a circular economy concept. *Miner. Eng.* **2020**, *148*, 106204. [CrossRef]
23. Bonnélye, V. Acid Mine Drainage: Case Study of One of the Largest Copper Mine Sites in the World, Ozwater15. Available online: https://pdfs.semanticscholar.org/4909/b6b7814b58c391ea9c54306bb9b7c4c834c4.pdf (accessed on 6 August 2020).
24. Chesters, S.P.; Morton, P.; Fazel, M. Membranes and minewater-waste or revenue stream. In Proceedings of the IMWA 2016 - Mining Meets Water - Conflicts and Solutions, Leipzig, Germany; 11–15 2016; pp. 1310–1322.
25. Cheng, S.; Dempsey, B.A.; Logan, B.E. Electricity generation from synthetic acid-mine drainage (AMD) water using fuel cell technologies. *Environ. Sci. Technol.* **2007**, *41*, 8149–8153. [CrossRef] [PubMed]

26. Palma, H.; Killoran, E.; Sheehan, M.; Berner, F.; Heimann, K. Assessment of microalga biofilms for simultaneous remediation and biofuel generation in mine tailings water. *Bioresour. Technol.* **2017**, *234*, 327–335. [CrossRef]
27. Urrutia, C.; Yañez-Mansilla, E.; Jeison, D. Bioremoval of heavy metals from metal mine tailings water using microalgae biomass. *Algal Res.* **2019**, *43*, 101659. [CrossRef]
28. Abinandan, S.; Subashchandrabose, S.R.; Venkateswarlu, K.; Megharaj, M. Sustainable iron recovery and biodiesel yield by acid-adapted microalgae, *Desmodesmus sp.* MAS1 and *Heterochlorella sp.* MAS3, grown in synthetic acid mine drainage. *ACS Omega* **2020**, *5*, 6888–6894. [CrossRef]
29. Levandowski, J.; Kalkreuth, W. Chemical and petrographical characterization of feed coal, fly ash and bottom ash from the Figueira Power Plant, Paraná, Brazil. *Int. J. Coal Geol.* **2009**, *77*, 269–281. [CrossRef]
30. Campaner, V.P.; Luiz-Silva, W.; Machado, W. Geochemistry of acid mine drainage from a coal mining area and processes controlling metal attenuation in stream waters, southern Brazil. *An. Acad. Bras. Cienc.* **2014**, *86*, 539–544. [CrossRef]
31. Mehta, S.K.; Gaur, J.P. Use of algae for removing heavy metal ions from wastewater: Progress and prospects. *Crit. Rev. Biotechnol.* **2005**, *25*, 113–152. [CrossRef]
32. Guillard, R.R.L. Culture of phytoplankton for feeding marine invertebrates. In *Culture of Marine Invertebrate Animals*; Smith, W.L., Chanley, H.H., Eds.; Plenum Press: New York, NY, USA, 1975; pp. 29–60.
33. Eaton, A.D.; Clesceri, L.S.; Rice, E.; Greenbaerg, A.E.; Franson, M.A.H. *Standard Methods for Examination of Water and Wastewater*, 21th ed.; American Public Health Association: Washington DC, USA, 15 October 2005.
34. Fiskesjö, G. The Allium test—An alternative in environmental studies: The relative toxicity of metal ions. *Mutat. Res. Fundam. Mol. Mech. Mutagen.* **1988**, *197*, 243–260. [CrossRef]
35. OECD. *Test No. 202: Daphnia sp. Acute Immobilisation Test, OECD Guidelines for the Testing of Chemicals, Section 2*; OECD Publishing: Paris, France, 2004. [CrossRef]
36. Edition, F. *Methods for Measuring the Acute Toxicity of Effluents and Receiving Waters to Freshwater and Marine Organisms*, 5th ed.; US Environmental Protection Agency US EPA: Washington, DC, USA, 2002; Volume 232, p. 266.
37. Ministério do Meio Ambiente, Conselho Nacional do Meio Ambiente, Brasil. *Resolução CONAMA n° 430/2011*; Conselho Nacional do Meio Ambiente, Ministério do Meio Ambiente, Brasil: PBrasília, Brazil, 2011.
38. Santos, K. Tratamento de drenagem ácida de minas por neutralização/precipitação-efeito do pH, agente neutralizante e tratamento complementar por biossorção com microalgas. Master's Thesis, Universidade Federal do Rio Grande do Sul, Porto Alegre, Brazil, 2020.
39. Kumar, K.S.; Dahms, H.U.; Won, E.J.; Lee, J.S.; Shin, K.H. Microalgae—A promising tool for heavy metal remediation. *Ecotoxicol. Environ. Saf.* **2015**, *113*, 329–352. [CrossRef] [PubMed]
40. Dönmez, G.Ç.; Aksu, Z.; Öztürk, A.; Kutsal, T. A comparative study on heavy metal biosorption characteristics of some algae. *Process Biochem.* **1999**, *34*, 885–892. [CrossRef]
41. Priyadarshani, I.; Rath, B. Commercial and industrial applications of micro algae—A review. *J. Algal Biomass Utln.* **2012**, *3*, 89–100.
42. Mehrabadi, A.; Craggs, R.; Farid, M.M. Wastewater treatment high rate algal ponds (WWT HRAP) for low-cost biofuel production. *Bioresour. Technol.* **2015**, *184*, 202–214. [CrossRef]
43. Chamorro, S.; Barata, C.; Piña, B.; Casado, M.; Schwarz, A.; Sáez, K.; Vidal, G. Toxicological analysis of acid mine drainage by water quality and land use bioassays. *Mine Water Environ.* **2018**, *37*, 88–97. [CrossRef]
44. Yim, J.H.; Kim, K.W.; Kim, S.D. Effect of hardness on acute toxicity of metal mixtures using *Daphnia magna*: Prediction of acid mine drainage toxicity. *J. Hazard. Mater.* **2006**, *138*, 16–21. [CrossRef]
45. Geremias, R.; Bortolotto, T.; Wilhelm-Filho, D.; Pedrosa, R.C.; de Fávere, V.T. Efficacy assessment of acid mine drainage treatment with coal mining waste using *Allium cepa* L. as a bioindicator. *Ecotoxicol. Environ. Saf.* **2012**, *79*, 116–121. [CrossRef]
46. Zucconi, F.; Forte, M.; Monaco, A.; De Bertoldi, M. Biological evaluation of compost maturity. *Biocycle* **1981**, *22*, 27–29.
47. Kang, S.W.; Seo, J.; Han, J.; Lee, J.S.; Jung, J. A comparative study of toxicity identification using *Daphnia magna* and *Tigriopus japonicus*: Implications of establishing effluent discharge limits in Korea. *Mar. Pollut. Bull.* **2011**, *63*, 370–375. [CrossRef]

48. Arambašić, M.B.; Bjelić, S.; Subakov, G. Acute toxicity of heavy metals (copper, lead, zinc), phenol and sodium on *Allium cepa* L., *Lepidium sativum* L. and *Daphnia magna* St.: Comparative investigations and the practical applications. *Water Res.* **1995**, *29*, 497–503. [CrossRef]
49. Seco, J.I.; Fernández-Pereira, C.; Vale, J. A study of the leachate toxicity of metal-containing solid wastes using *Daphnia magna*. *Ecotoxicol. Environ. Saf.* **2003**, *56*, 339–350. [CrossRef]
50. Soucek, D.J.; Cherry, D.S.; Trent, G.C. Relative acute toxicity of acid mine drainage water column and sediments to *Daphnia magna* in the Puckett's Creek watershed, Virginia, USA. *Arch. Environ. Contam. Toxicol.* **2000**, *38*, 305–310. [CrossRef] [PubMed]
51. De Bisthoven, L.J.; Gerhardt, A.; Soares, A.M.V.M. Chironomidae larvae as bioindicators of an acid mine drainage in Portugal. *Hydrobiologia* **2005**, *532*, 181–191. [CrossRef]
52. Benzie, J.A.H. Cladocera: The Genus DAPHNIA (including DAPHNIOPSIS) (*Anomopoda: Daphniidae*). Guides to the Identification of the Microinvertebrates of the Continental Waters of the World. *Q. Rev. Biol.* **2005**, *21*. [CrossRef]
53. Thomashausen, S.; Maennling, N.; Mebratu-Tsegaye, T. A comparative overview of legal frameworks governing water use and waste water discharge in the mining sector. *Resour. Policy* **2018**, *55*, 143–151. [CrossRef]
54. Rezaie, B.; Anderson, A. Sustainable resolutions for environmental threat of the acid mine drainage. *Sci. Total Environ.* **2020**, *717*, 137211. [CrossRef]
55. Ellen MacArthur Foundation. *Towards the Circular Economy—Economic and Business Rationale for an Accelerated Transition*; Available online: https://www.ellenmacarthurfoundation.org/publications/towards-the-circular-economy-vol-1-an-economic-and-business-rationale-for-an-accelerated-transition (accessed on 6 August 2020).
56. Singh, S.; Sukla, L.B.; Goyal, S.K. Mine waste & circular economy. *Mater. Today Proc.* **2020**, 1–8. [CrossRef]

minerals

Article

Algae in Acid Mine Drainage and Relationships with Pollutants in a Degraded Mining Ecosystem

Patrícia Gomes [1], Teresa Valente [1,*], Teresa Albuquerque [2], Renato Henriques [1], Núria Flor-Arnau [3], Jorge Pamplona [1] and Felipe Macías [4]

[1] Campus de Gualtar, Institute of Earth Sciences-Pole of University of Minho, University of Minho, 4710-057 Braga, Portugal; patricia_s_gomes@hotmail.com (P.G.); rhenriques@dct.uminho.pt (R.H.); jopamp@dct.uminho.pt (J.P.)
[2] CERNAS/QRural, Polytechnic Institute of Castelo Branco and Institute of Earth Sciences-Pole of University of Évora, University of Évora, 7000-345 Évora, Portugal; teresal@ipcb.pt
[3] Department of Vegetal Biology (Botany Unit), Faculty of Biology, University of Barcelona, 08014 Barcelona, Spain; n.flor@ub.edu
[4] Departament of Edaphology and Agricultural Chemistry, Faculty of Biology, University of Santiago de Compostela, 15782 Santiago de Compostela, Spain; felipe.macias.vazquez@usc.es
* Correspondence: teresav@dct.uminho.pt

Abstract: Acid mine drainage represents an extreme environment with high concentrations of potentially toxic elements and low pH values. These aquatic habitats are characterised by harsh conditions for biota, being dominated by acidophilic organisms. The study site, São Domingos mine, located in one of the largest metallogenetic provinces in the world, the Iberian Pyrite Belt, was closed without preventive measures. To identify the algae species and understand the relationships with abiotic parameters of the ecosystem, water and biological material were collected and analysed. Digital terrain models were obtained with an unmanned aerial vehicle for geomorphological and hydrologic characterisation of the mine degraded landscape. The results show two types of algal colours that seem to represent different degrees of photosynthetic activity. Optical and scanning electron microscopy revealed 14 taxa at the genus level, divided into eight classes. The genus *Mougeotia* is the most abundant multicellular algae. With respect to unicellular algae, diatoms are ubiquitous and abundant. Abiotic analyses expose typical features of acid mine drainage and support an inverse relationship between chemical contamination and biological diversity. Factorial correspondence analysis indicates three groups of attributes and samples by their relationship with specific toxic elements. This analysis also suggests a close association between *Spirogyra* and Pb, together composing a structurally simple ecosystem. The highest contamination in the river system is related to the hydrologic patterns obtained from photogrammetric products, such as the digital surface model and flow map accumulation, indicating the input of leachates from the section having the finest sulfide-rich wastes. Information about the algae community and their association with flow patterns of toxic elements is a relevant tool from a biomonitoring perspective.

Keywords: acid mine drainage; *Mougeotia*; acidophilic algae; photogrammetric products; factorial correspondence analysis; ecological monitoring; Iberian Pyrite Belt

Citation: Gomes, P.; Valente, T.; Albuquerque, T.; Henriques, R.; Flor-Arnau, N.; Pamplona, J.; Macías, F. Algae in Acid Mine Drainage and Relationships with Pollutants in a Degraded Mining Ecosystem. *Minerals* **2021**, *11*, 110. https://doi.org/10.3390/min11020110

Received: 7 December 2020
Accepted: 19 January 2021
Published: 22 January 2021

1. Introduction

Acid mine drainage (AMD), typical of sulfide environments, is promoted by mining activities and often presents features of extreme degree of pollution [1,2]. Usually, the affected systems show high concentrations of potentially toxic elements (PTE), low pH and presence of iron colloids, causing water turbidity and nutrient deficiency. These characteristics contribute to stress conditions that affect biota by inducing a reduction in biodiversity [3]. Therefore, these environments are dominated mostly by prokaryotes and eukaryotes that are acidophilic and acid-tolerant, composing a structurally simple ecosystem [4]. These organisms play important roles in safeguarding primary production and can be divided into two groups: chemoautotrophic (mainly bacteria) and photoautotrophic (such as acidophilic algae). Both interfere with the mobility of dissolved chemical species in the aquatic medium. It is known from scientific literature that algae have mechanisms to avoid toxic effects, exhibiting tolerance to metals [2,5]. Furthermore, some algae, namely the diatoms, which are known for their ability to survive in different types of metal-polluted waters [6,7], can be used in monitoring processes. Some species have optimal growth limits under acidic conditions and therefore they are good markers of contamination generated by AMD [8]. In line with that, the European Water Framework Directive (WFD, European Commission 2000) introduced ecological criteria for determination of water quality, reinforcing the importance of ecological indicators for the study of human impact in water bodies. Thus, studies of biotic systems in acidified waters have been undertaken in various regions of the world [9], and also in the Iberian Pyrite Belt (IPB) [4,7,10,11], including the São Domingos mining area [12].

Since contamination and rehabilitation processes based on ecological procedures are paramount, it is crucial to have a thorough knowledge of these extreme survival communities in AMD. So, the objectives of this research study are to investigate and identify the community of algal flora in the aquatic ecosystem that receives the drainage from São Domingos mining area and correlate it to abiotic features. For this, hydrochemical properties as well as hydrological and geomorphological characteristics are fundamental and will be considered. Photogrammetric products and statistical tools, such as factorial correspondence analysis, are used to determine possible relationships between algae distribution and flow patterns of potentially toxic elements. In this way, this study intends to contribute to the development of biomonitoring plans that could be applied in abandoned mining sites with AMD contamination.

2. Materials and Methods

2.1. Study Area

São Domingos mine, characterized by sulfide deposits, is located in the south of Portugal, and is part of the IPB Metallogenic Province (Figure 1). It represents a typical scenario of mining contamination by old exploitations in the IPB. Mining started in pre-Roman times and ended in 1966 when it was abandoned without rehabilitation measures. This area tends to be characterized by its low pH and extremely high concentrations of heavy metals, mainly Cu, Fe, and Zn but also rich in other less abundant elements that are burdensome to the environment such as Sb, As, Hg, and Pb [13]. Infrastructure including ore-processing plants and machinery are abandoned through the study area, leading to different types of wastes disseminated through the mining complex [14]. Specifically, the mining area is covered by two streams: São Domingos and Mosteirão streams. The Chança River, primary effluent of Guadiana River, especially its water dam is the final receptor of the acidic discharges (Figure 1). The intense red-orange colour associated with high concentrations of iron and deposition of ochre-precipitates is characteristic of São Domingos waters [15].

With a typical Mediterranean climate, the region has two distinct seasons: a wet period, from November to March, and a dry period, from May to September. The annual average air temperature is 17.6 °C, and annual precipitation is 559 mm, with hot summers (30 °C) and cold winters (12 °C) [16].

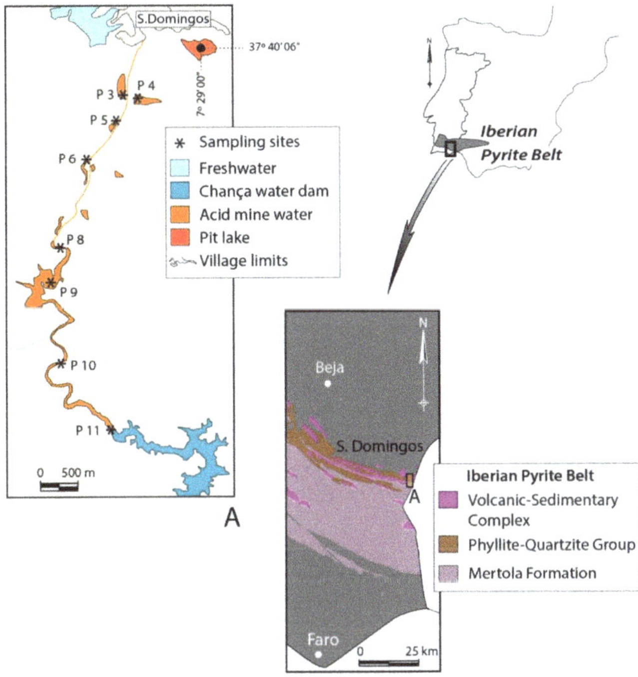

Figure 1. Map of the study area with representation of the geological context and the river system, with location of the sampling points. Adapted from [3].

2.2. Geomorphological Characterization

In order to make the geomorphological characterization of the area, a survey campaign was conducted using an unmanned aerial vehicle (UAV). In total, an area of 3,813,157.52 m^2 was surveyed, with a maximum width of 1.58 km and a maximum length of 4.12 km (Figure 2).

Figure 2. Aerial view of the São Domingos mining area. The yellow line represents the boundaries of the section surveyed by the UAV (Imagery from Google Earth). Location approximately at coordinates 37.65965 lat., −7.50506 long. (WGS 84).

The unmanned aerial vehicle used was a Phantom 4 Pro®, DJi™, Shenzhen, Guangdong, China. Due to the limited battery life for each flight, a total of five flights were performed to ensure photogrammetric coverage of the entire area (Figure 2). Each flight was performed at 150 m high, with flight paths in parallel rows, using a front overlap of 75% and a side overlap of 70%. This was done using the MapPilot® flight planning application from Maps made Easy™, which has the functionality of keeping the UAV flight at a constant altitude relative to the topographic surface, using NASA's SRTM altitude models. The final photogrammetric products obtained were a digital surface model (DSM) and an orthophotomap, with a ground sampling distance (GSD) of 4.1 cm/pixel. The photogrammetric processing was performed using the Metashape® application from Agisoft™. In this work no ground control points were collected using GNSS equipment due to the difficulty of safe access to some areas included in the survey, which would allow a centimetric accuracy of the photogrammetric products [17]. However, the purpose of this work does not require such high precision. Thus, to ensure sub-metric accuracy of the photogrammetric products, GCPs were collected from survey grade products, with the latitude and longitude extracted from the 2015 orthophotomaps produced by the Portuguese General Directorate of the Territory, with a ground sampling distance of 10 cm/pixel, and the altitude extracted from topographic surveys carried out by the Portuguese army to produce the official Portuguese topographic cartography. A total of 9 GCPs were used, evenly distributed throughout the area, allowing the production of photogrammetric products (DSM and orthophotomap) with average errors of 0.36 m in x axis, 0.17 m in y axis and 0.10 m in z axis, with the RMS error being 0.41 m.

The morphological analysis of the terrain was done using the open-source GIS application Quantum GIS. The hydrological analysis, namely the runoff patterns and flow accumulation, was done using the "r.terraflow" routine [18,19], included in the GRASS processing module of Quantum GIS. A detailed description of the flow accumulation calculation can be found at https://grass.osgeo.org/grass78/manuals/r.terraflow.html.

2.3. Water Sampling and Analytical Methods

Samples were collected in the hydrological year of 2016–2017 (monthly from October to September), at eight different locations according to the scheme shown in Figure 1. Determination of pH, temperature (°C), and electrical conductivity (EC µS/cm, at 25 °C) were recorded on-site, operating a portable meter Thermo Scientific Orion. Two sterilized polyethylene containers were used at each point to store superficial water: one was filled with 0.45 µm filtered water and acidified with concentrated acid nitric to obtain a pH < 2, in order to prevent metals precipitation and bacterial growth during transportation to the laboratory; the other was collected to be used for the remaining analyses.

The sulfate was determined by the turbidimetric method [20] and total acidity and alkalinity were analysed by volumetric determination [21]. Concentrations of selected elements (Al, Mn, Fe, As, Cu, Zn, Cr, Ni, Co, Cd, Pb, Mg, Ca, K) were obtained by inductively coupled plasma optical or mass spectrometry (ICP-OES/MS). These analyses were performed at Activation Laboratory, Lda (Actlabs, Ancaster, ON, Canada), including duplicate samples and blanks to check precision, whereas accuracy was obtained by using certified standards. All the reagents used were of analytical grade or of Suprapur quality (Merck, Darmstadt, Germany). The standard solutions were Merck AA Certificate. Milli-Q water was used in all the experiments.

2.4. Algae Community—Sampling and Identification

With the purpose of studying and identifying the algae community, sampling, treatment, and analyses of phytoplankton were performed according with European Committee for Standardization norms (2003). Algae were collected at the same time and place that the water samples. Specifically, samples were taken based on the macroscopic presence of algae and to study their diversity. They were obtained by scraping the substrates and by dredging the water surface through a polyethylene container. At each site, two samples were taken: one represents in vivo and the other was preserved with Kew solution immediately after collection. Subsequently, they were accommodated in a portable refrigerator at 4 °C to avoid degradation, being analysed when arrived to the laboratory. Taxonomic identification was achieved by optical microscopy (Leica Microscope DMRB with UEYE coupled camera) and scanning electronic microscopy (SEM) (Evo LS15 Zeiss) based on morphological features.

2.5. Statistical Analysis

Factorial correspondence analysis (FCA) was applied to dataset in order to find possible relationships between the variables and the samples. FCA belongs to a group of factorial extraction methods whose primary goal is to uncover the underlying pattern of relationships within a dataset. The aggregation of standardized parameters was developed by Benzécri at the beginning of the 1970s [22,23]. This is done by rearranging the data into a smaller number of uncorrelated "components" or "factors" that are extracted from the data by statistical transformations. Such transformations involve diagonalizing the same type of variable similarity matrix, such as a correlation or variance-covariance matrix. Each factor describes a certain amount of the statistical variance of the analysed data and is interpreted according to the intercorrelated variables. The main advantage of FCA is that symmetry is given to the data matrix [23], enabling the simultaneous study of correlations within and between variables and samples. A detailed discussion of the theory underlying the FCA is beyond the scope of this article, but its application in this matter is rather straightforward. For computation purposes, ANDAD version 7.12, CVRM/IST: Lisboa, Portugal [24] was used.

3. Results and Discussion

3.1. Photogrammetric Products

Through a survey using the UAV, high-resolution images of the area that borders the São Domingos stream were obtained. Figure 3 shows the digital surface model and an example of the typical pattern of surface flow accumulation in one of the most affected sections of the surveyed area.

The area is morphologically dominated by a V-shaped valley, more or less open, which talweg constitutes the bottom of the São Domingos stream. The mining waste deposits with contaminants, resulting from mining activity, are predominantly on the east side of the stream. These deposits are exposed to precipitation and can be easily mobilized by surface run-off in the direction of the talweg, which means that the flow is mainly in the direction of the riverbed of São Domingos stream. Topographic blockages to surface runoff are usually small elevations made up of the waste deposits or acidic lagoons. The section illustrated in Figure 3 has one of these lagoons, which is completely covered by sulfate efflorescences during dry season. This is a section of special interest, due to the presence of large amounts of fine sulfide-rich wastes and due to this lagoon, cyclically subject to the dissolution of the sulfates blooms.

The preferential flow accumulation in acidic lagoons and in São Domingos stream is notorious. Considering that all the topography along the valley is convergent to the São Domingos stream, the surface run-off and the underground run-off will invariably lead to inputs of contaminants in the stream. As surveyed, this stream has a maximum elevation of 133.378 m and a minimum of 85.819 m, with an extension of 3148.6 m. These values correspond to an average slope of the talweg of 2.73%. This slope leads to a general drainage of the São Domingos stream basin in the direction of the Chança dam. As an example, the P5 and P8 sampling sites receive the leachates from the areas most affected by the presence of very fine wastes.

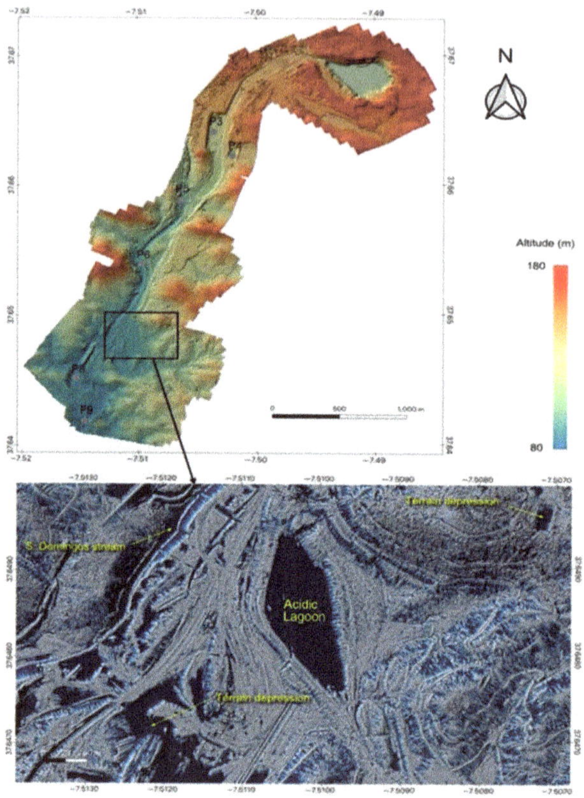

Figure 3. Digital Surface Model of the surveyed area obtained by UAV (**upper**); Section of the flow accumulation map (**lower image**) in the area with the finest sulfide wastes. In the lower part of the figure, the blue lines are flow lines of surface runoff and the blue continuous zones are areas of flow accumulation. Coordinates in WGS 84.

3.2. Water Physical-Chemical Characterization

The statistical summary of the hydrochemical parameters are presented in Table 1. The results show that the lowest pH average is 2.5, found at P5. On the other hand, P11 represents the sampling point with the highest average pH value (4.0). The same pattern can be observed consistently for electrical conductivity (EC) and sulfate, which have higher values for P5 (4035 µS/cm and 3155 mg/L, respectively) and lower values for P11 (741 µS/cm and 301 mg/L, respectively). Between these two sites, there is a decrease in sulfate concentration, reflecting a reduction of contamination, as a function of attenuation processes, such as dilution and the development of mineralogical neoformations. Specifically, efflorescences of Fe and Al-sulfates and precipitation of jarosite, the most abundant iron hydroxysulfate in this system, are the main mineralogical control factors [15].

Table 1. Statistical summary of parameters analysed monthly across the hydrological year. Min = Minimum; Max = Maximum; EC = electrical conductivity (µS/cm); all concentrations are in mg/L.

Samples		pH	EC	T °C	SO_4^{-2}	Al	As	Ca	Cr	Cu	Fe	K	Mg	Mn	Ni	Co	Zn	Cd	Pb
P3	Mean	2.7	2877.2	23.3	1608.3	148.858	0.075	93.409	0.071	12.721	71.260	3.684	71.819	7.564	0.197	0.669	7.109	0.028	0.030
	Min	2.3	228.0	16.2	661.5	67.872	0.026	44.817	0.030	5.750	19.677	1.281	32.784	3.039	0.100	0.319	3.265	0.014	0.005
	Max	3.5	4378.0	29.6	2760.5	251.000	0.160	163.000	0.100	22.200	170.000	8.380	169.000	14.700	0.483	1.300	10.855	0.041	0.051
P4	Mean	2.7	2654.8	23.3	1728.2	135.585	0.081	59.266	0.063	1.660	95.830	4.071	117.758	22.614	0.340	1.025	63.778	0.085	0.017
	Min	2.3	310.0	14.5	552.3	42.306	0.015	19.784	0.020	0.589	29.084	2.333	39.060	6.996	0.056	0.342	20.550	0.050	0.011
	Max	3	4230.0	31.3	2600.4	199.073	0.240	105.000	0.100	3.210	160.007	6.190	189.000	34.265	0.738	1.555	96.076	0.118	0.030
P5	Mean	2.5	4034.7	23.2	3155.3	288.885	1.506	101.635	0.121	32.429	253.137	1.393	105.460	15.584	0.309	1.277	62.080	0.373	0.102
	Min	2.1	251.0	13.7	865.7	4.830	0.061	14.842	0.040	0.381	3.020	0.600	8.440	0.594	0.050	0.334	0.977	0.033	0.020
	Max	2.8	6943.0	30.0	6564.9	675.000	6.260	205.000	0.285	83.900	801.000	2.262	252.000	37.200	1.040	3.020	199.000	1.300	0.286
P6	Mean	3	1734.8	23.2	1001.9	83.493	0.849	50.865	0.062	9.427	70.664	4.869	47.338	5.065	0.129	0.389	19.837	0.099	0.289
	Min	2.6	117.3	15.6	68.2	1.757	0.003	9.639	0.001	0.378	0.995	1.966	8.374	0.165	0.033	0.077	0.595	0.002	0.031
	Max	4	4320.0	31.9	2955.1	354.446	9.498	86.900	0.119	48.057	459.577	8.400	108.104	11.346	0.360	1.479	155.883	0.819	0.626
P8	Mean	2.8	3392.7	22.2	2498.0	211.161	0.119	117.307	0.103	16.886	220.407	4.785	81.232	10.737	0.207	0.895	31.697	0.080	0.307
	Min	2.2	181.9	14.7	58.5	0.075	0.060	10.480	0.050	0.166	1.201	0.421	8.554	0.162	0.100	0.100	0.687	0.004	0.011
	Max	4.7	7320.0	35.1	5731.9	497.000	0.299	256.000	0.171	38.000	499.000	29.142	170.389	24.152	0.499	1.913	82.000	0.194	1.181
P9	Mean	2.9	1728.7	22.4	836.4	67.196	0.108	47.504	0.062	7.090	66.351	5.825	33.874	3.699	0.099	0.336	10.848	0.044	0.407
	Min	2.4	204.0	16.8	110.4	4.809	0.021	13.724	0.020	0.590	4.676	2.483	9.781	0.666	0.048	0.069	1.544	0.011	0.152
	Max	3.5	3186.0	30.2	1641.5	131.366	0.279	86.456	0.100	13.343	166.920	29.143	63.881	7.355	0.137	0.652	23.491	0.080	0.765
P10	Mean	2.9	3199.5	20.3	2055.2	103.202	0.113	195.022	0.060	6.803	67.240	4.123	181.316	12.684	0.176	0.563	15.995	0.072	0.135
	Min	2.4	185.7	8.0	151.3	9.741	0.030	16.088	0.020	1.144	1.779	0.100	11.363	0.613	0.062	0.061	1.944	0.019	0.005
	Max	3.7	8598.0	26.2	6753.2	310.774	0.320	658.000	0.100	15.791	303.574	11.000	651.026	42.800	0.638	1.770	45.852	0.204	0.477
P11	Mean	4	740.8	20.5	300.6	26.502	0.067	52.233	0.053	2.321	14.231	5.790	33.537	2.799	0.071	0.147	4.103	0.019	0.129
	Min	2.9	168.7	12.6	58.4	0.100	0.007	15.833	0.003	0.019	0.330	1.330	10.500	0.140	0.005	0.003	0.042	0.002	0.001
	Max	6.3	1527.0	28.4	790.0	98.570	0.149	130.137	0.100	11.023	78.501	29.145	73.218	9.803	0.109	0.434	12.616	0.054	0.393

Of the metals, Fe, Al, and Zn presented the highest concentrations, with Fe achieving maximum of 801 mg/L in P5. The dominance of these metals has been reported in other AMD affected sites around the world [25–27]. According to Soyol-erdene et al. [28], the high concentrations obtained for Al and Mg can be explained by the abundant dissolution of host felsic rocks, allowed by the strong acidity of the medium. As observed for in situ parameters (pH and EC), the selected PTE present higher concentrations in P5, followed by P8 (Table 1) in dry season. A previous study [3] indicated the highest amount of chlorophyll in these two sampling points, leading us to believe that algae present mechanisms that confer them tolerance and even preference to toxic factors. As mentioned before, the contact of water with the different types of mining wastes disposed around the acidic lagoons [29], explains the highest contamination, as observed by the flow accumulation map (Figure 3).

3.3. *Algae Community Study*

3.3.1. Field Evidence

AMD systems represent harsh conditions where only a small number of specialized alga can develop [6]. In this way, as reported by [10] in previous studies conducted at IPB, two distinct types of algal colours were observed in São Domingos mine: green and brown-purplish (Figure 4). The first was the most abundant type occupying the majority of sampling points, where the water flow was low (Figure 4a,b). They were attached to fixed materials, such as twigs and rocks, presenting long filaments that trailed along the water. According to [3], abiotic characteristics such a lotic or lentic environment, are factors with important influence on algae fixation and colonization. The same observations were realized by [30]. The filamentous brown-purplish algae blooms (Figure 4c) are more evident at P3, where the water flow is low. In Figure 4d it is possible to observe a detail of gaseous effervesce formation, which are oxygen bubbles from photosynthesis.

It is possible that these distinctive colours indicate different species and/or different degrees of algae activity. According to Gomes et al. [3], in this case, brown-purplish algae seems to exhibit clear senescent properties and algae with fresh biomass presents green colour. With excess nutrients and favourable temperature, phytoplankton can multiply rapidly, forming a bloom (Figure 4f). In this case, the water turns greenish briefly, from one to two days, depending on the temperature, and becomes brownish when the plankton depletes the nutrients and begins to die. However, in the middle of these filamentous algae, diatoms can also occur, that present a golden or brownish colour, and are also designated by gold algae, contributing to this appearance [10]. Often, algal blooms hold suspended particles, which give them ochre shades (Figure 4e). According to Valente and Gomes [6], this phenomenon can be explained by the dense community of algae that often acts as a barrier for iron oxyhydroxide precipitates.

No important differences in the macroscopic appearance of algae were observed during the monthly sampling, leading us to observe that the same type of algae inhabits in certain locations at São Domingos mine, regardless of the season. However, it was possible to observe that the occurrence of filamentous species is higher during the dry summer months, when most physicochemical parameters are extreme, especially in P3 and P5. These results are in accordance with [3,11], that revealed the same pattern. When the conditions are more extreme, higher photosynthetic production rates were analysed, following the exuberance of the blooms.

Figure 4. (**a**) Macroscopic observations of intense red-orange colour, reddish-yellow iron-precipitates and the presence of typical green algae observed in waters channels; (**b**) detail of small green algae in the sediment; (**c**) presence of layer brown-purplish intercalated with green algae filaments; (**d**) purplish brown colour of senescent algae with oxygen bubbles; (**e**) algae filaments with brown shades in low water course (P5); (**f**) bloom of algae patches, coexisting with efflorescent salts, and covering the surface bed at P3; (**g**) green algae located on rocky substrate in a small seep.

3.3.2. Identification by Optical Microscopy

Figure 5 shows photomicrographs of several genera of multi and unicellular algae found at different sampling sites. Figure 5a shows large crystals in interface with algal filaments. The same effect was previously reported by Valente et al. [31]. Figure 5d further exposes the interaction of algae and sediments, which supply soluble elements and compounds, fundamental for photosynthesis processes. Moreover, it is possible that elements such as metals may be retained by salt efflorescences [15].

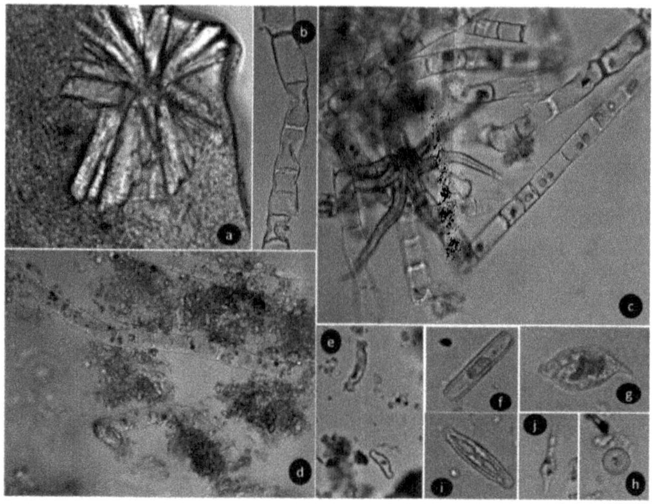

Figure 5. Photomicrographs including potentially identified algae at Sao Domingos mine. (**a**) unidentified radiating crystals formed around biological material, showing mineral-algae interaction; (**b**) *Mougeotia sp.* image without membrane organelles; (**c**) radial structure (possible plant hair) overlapping on algae filaments; (**d**) *Ulothrix sp.* with loss of membrane organelles and surrounded by sediment; (**e**) *Eunotia exigua*; (**f**) *Pinnularia sp.*; (**g**) *Euglena mutabilis*; (**h**) *Cyclotella sp.*; (**i**) *Pinnularia sp.*; (**j**) *Euglena mutabilis (?)*. All images are from P3 sampling point, with the exception of image (**b**) from P11. Magnification 40×, except (**b**,**g**,**i**) which are 100×.

The multicellular algae, mostly belonging to the genus *Mougeotia* in P3, were quite degraded (Figure 5b). This assumption was made by the absence of cellular organelles within the cells, which provides senescence properties, and gives the brown-purplish colour due to the lack of chlorophyll.

In this previous Figure 5d, it is also visible algae from the *Ulothrix* genus. Despite presenting organelles, it seems to show already some signs of oxidative stress. The remaining images show different unicellular algae, representing mainly diatom species. Figure 5g exhibits a single cell of *Euglena mutabilis*. This alga is commonly reported in AMD. However, it was rarely observed in this study.

3.3.3. Mineral-Algae Interactions (SEM Study)

Observations in SEM (Figure 6) show communities of pioneer species forming dense biofilm networks, holding sediment particles able to retain metals in the clay size fraction [32]. The presence of Fe, S, and Al is common (Figure 6a,b), revealing the relationship between acidophilic algae and iron oxyhydroxides and hydroxysulfates. Tubular mineral formations (Figure 6a) were often observed in association with the algae (Figure 6a). These hollow iron-rich tubes, also reported by Valente et al. [33], could be indicative of aquatic insects or biogenic iron-precipitates [34,35], confirming the contribution of acidophilic organisms in biomineralisation processes.

In accordance, several studies reported the absorption of metals by algae, and in some cases, this occurs external to the cells using precipitation as a protective mechanism [8]. Also, this suggests that these species can even use the referred inorganic components as propellants of photosynthesis production. According to Prygiel and Coste [36], some elements such as silica have importance related to the increase of diatom biomass, as it supports the formation of the silica carapace that surrounds the whole cell, called frustule (Figure 6c).

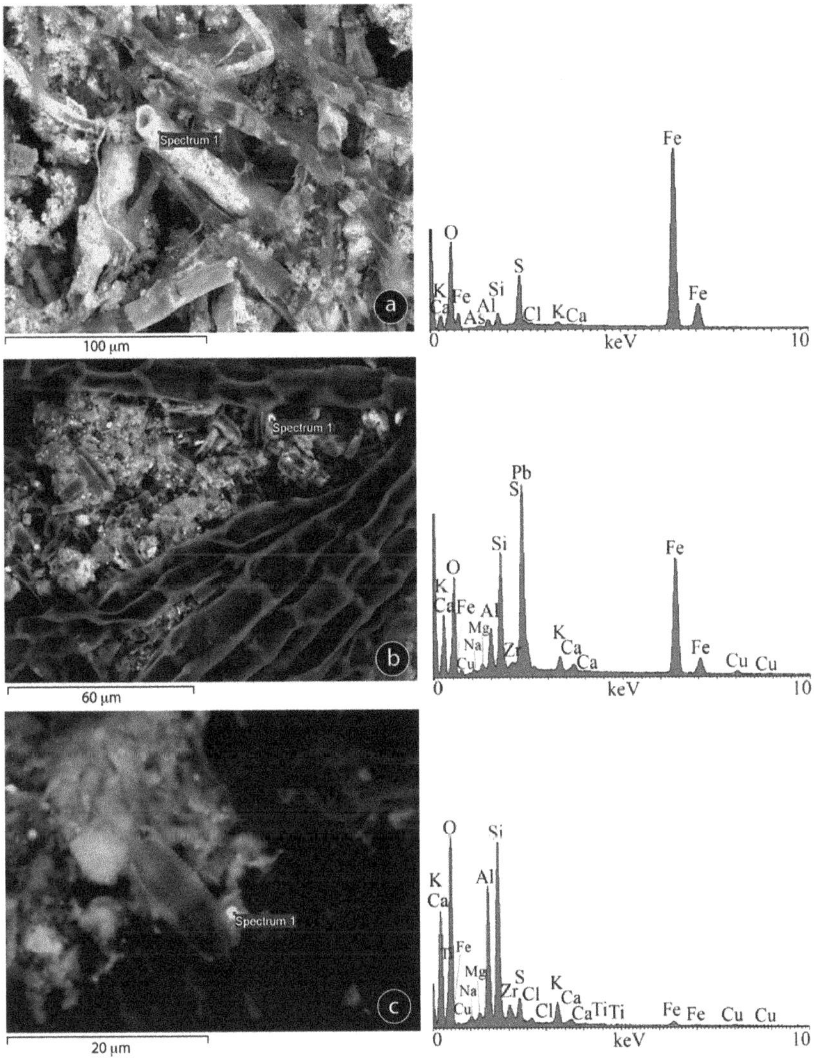

Figure 6. SEM (ES) micrographs and EDS analysis showing the relation between acidophilic algae and iron oxyhydroxides and hydroxysulfates. (**a**) Interconnection between algal filaments and iron cylindrical tubes; (**b**) multicellular and unicellular algae in the sediment; (**c**) Intergrowth morphology of *Pinnularia sp.* in sediment bearing Si and Al.

The autotrophic organisms promote the increase of organic matter, which further leads to colonization by other species. So, the established cycle formed in the dependence of these communities can influence the characteristics of the water ecosystem. In this way, they can represent the first stage of biological stabilization and reverse the processes through natural attenuation of the AMD effects [37,38]. Therefore, this acidophilic colonization could contribute to remediation processes through the stabilization/absorption/adsorption of some PTE.

3.3.4. Spatial Distribution of Algae Taxa

Table 2 presents an inventory with a relative abundance of the observed taxa. The communities of algae were represented by 14 taxa at the genus level, and were divided into eight classes: Chlorophyceae (*Mougeotia, Spirogyra,* and *Chlamydomonas*); Ulvophyceae (*Klebsormidium*); Euglenophyceae (*Euglena*); Bacillariophyceae (*Pinnularia, Eunotia, Navicula* and *Cyclotella*); Chlorophyceae (*Characium*); Chrysophyceae (*Dinobryon*); Ulothricophyceae (*Oedogonium* and *Ulothrix*) and Trebouxiophyceae (*Chlorella*). These results are in agreement with those found by [12] in São Domingos, that were based on the genera *Eunotia, Pinnularia, Euglena, Chlamydomonas,* and *Ulothrix*.

Table 2. Taxonomic classification of the identified material in the effluent collected through the year. +++ very abundant ++ abundant + isolated specimens.

Genus	P3	P4	P5	P6	P8	P9	P10	P11
Mougeotia sp.	+++	+	++	+	++		+	++
Ulothrix sp.	++						+	
Spirogyra sp.	+		+	+	+			+
Klebsormidium sp.	+							+
Euglena mutabilis	+						+	
Pinnularia sp.	+++	+++	+++	+++	+++	+++	+++	+++
Eunotia exigua	+++	+++	+++	+++	+++	+++	+++	+++
Navicula sp.	+++	+++	+++	+++	+++	+++	+++	+++
Oedogonium	++							
Chlamydomonas sp.								+++
Characium sp.						+		
Dinobryon sp.								+++
Chlorella sp.	+							
Cyclotella sp.	+							
Bacteria	+++					+++		+++

The sites with the greatest diversity of acidophilic species are P3 and P11. Therefore, the more acidic/contaminated the site, the less biological diversity it presents. A decrease in biodiversity with the increase of acidity was also observed by other authors [12]. It should also be noted that these points represent lentic environments, and as mentioned above, they have great importance in algae retention, while P6 and P9 are characterised by being lotic environments.

The diatoms are the group with major representation, being detected in all points, highlighting the best adaptation to the range of pH values, and to the variety of concentration of metals and sulfate. Diatoms are than followed by multicellular algae *Mougeotia* and *Spirogyra*.

Bacteria (without identification) were seen in large number in P3, P9, and P11. They were spherical and stick-shaped with high mobility. These organisms are known to stimulate contamination [12].

The presence of typically indicative *Euglena mutabilis* was only rarely detected in P3 and P10. According to Valente [8], they are usually found floating in the water, or colonizing ochre products in streambeds and acidic pools. Amils et al. [1] suggest that *Euglena* may not survive in extremely acidic environments, such as São Domingos mine. These results are in accordance with the present observations, as *Euglena* was only detected at least contaminated sites. Nonetheless, [39] reveals that *Euglena mutabilis* is common in small acidic water bodies and is frequently abundant in waters contaminated by heavy metals, with low pH. They can also be found in unpolluted to moderately polluted water, being acidobiont.

The scarcity of *Euglena mutabilis* in São Domingos may also be related to the nature of the mineralogical substrate. Indeed, [6] suggests that this alga is preferentially associated with the hydroxysulfate schwertmannite, which in turn is rarely identified in this ecosystem.

Despite the extreme hydrochemical conditions observed in São Domingos, when compared with other geomicrobiological works carried out in AMD regions [37], the obtained results indicate a considerable variability in terms of algae taxa.

3.3.5. Relationships between Hydrochemistry, Sample Location and Algae

Factorial correspondence analysis (FCT) was used as multivariate statistical tool to find potential relationships between abiotic and biotic features of the ecosystem. The mean concentration in each sampled point of the analysed hydrochemical attributes (columns) together with the eight samples (rows) was used as active individuals (Table 1) and so, the ones used for the new factorial components' construction. The biological variables (Algae) were recodified, from zero to three, regarding the correspondent abundance (Table 2) in each sampled point. Figure 7 shows the results of the application of FCT.

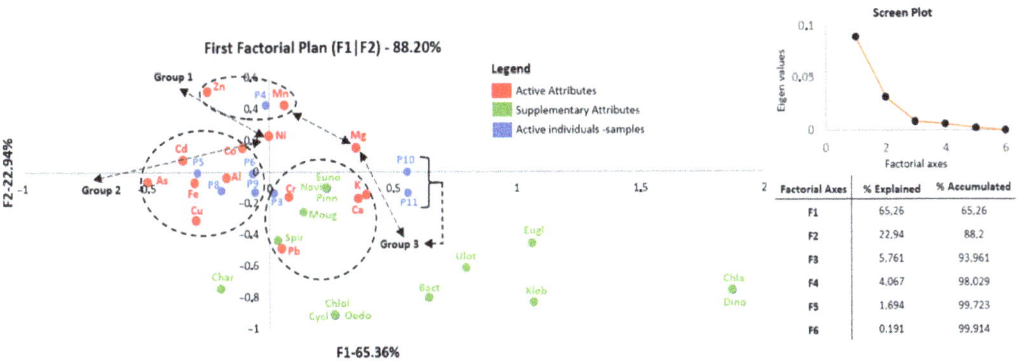

Factorial Axes	% Explained	% Accumulated
F1	65.26	65.26
F2	22.94	88.2
F3	5.761	93.961
F4	4.067	98.029
F5	1.694	99.723
F6	0.191	99.914

Figure 7. Factorial correspondence analysis output between abiotic and biotic features of the ecosystem. Moug—*Mougeotia*; Ulot—*Ulothrix*; Spir—*Spirogyra*; Kleb—*Klebsormidium*; Eugl—*Euglena*; Pinn—*Pinnularia*; Euno—*Eunotia*; Navi—*Navicula*; Oedo—*Oedogonium*; Chla—*Chlamydomonas*; Char—*Characium*; Dino—*Dinobryon*; Chlol—*Chlorella*; Cycl—*Cyclotella*; Bact—Bacteria.

The screen plot and the table in Figure 7 indicate that the first two axes are controlled by 88% of the total variance. The projection of the attributes and samples in the factorial plan highlights an association in three main groups. The first one is explained by the second Factor (F2) while the second and third groups are explained by the first Factor (F1), respectively negative and positive coordinates.

Group 1 comprises the sample P4, which distances itself from all others and shows affinity with the concentrations in Zn and Mn. Group 2 includes the samples P5, P6, P8, and P9, which show affinities with Fe, As, Cd, Cu, Co, and Al. Group 3 contains the sample P3, displaying affinity with Cr, Pb, K, and Ca. The supplementary projection of the algae indicates a clear overlap with group 3, mainly concerning to the diatoms *Pinnularia*, *Navicula*, and *Eunotia*, as well as with the filamentous algae *Spirogyra* and *Mougeotia*. These two algae, and especially *Spirogyra*, when projected, are very close to Pb that is more significant in P3. No relationships are observable between the other algae and the associated chemical composition or with any of the other samples.

These associations reflect the spatial distribution of the samples and the algae communities. For example, P4 (Figures 1 and 3) is the only point located on the east bank of the stream, away from the remaining sampling sites. Due to its location, this lagoon receives drainage from waste-dumps very enriched in jasper, with a particular geochemical signature, marked here by the elements Zn and Mn. Group 2 proposes evidence of the effect of drainage from the largest accumulations of very fine sulfide-rich wastes, located near points P5 to P9, as shown by the patterns of surface flow observed in Figure 3. Thus, the affinity with the elements from the oxidative dissolution of sulfides (e.g., Fe, As, Cu) is justified. The extremely acidic conditions in this section of the mining complex facilitate the dissolution of the host rocks, explaining the joining of Al. The projection of Ni indicates its similar influence over groups 1 and 2. The same is observable for Mg (abundant from dissolution of ubiquitous host rocks), which affects groups 1 and 3, as well as the samples P10 and P11.

P10 and P11 are the furthest samples from the primary sources of sulfides. The projection of these samples in the factorial plan highlights the effect of this downstream location (Figure 1) and then stressing the geochemical dynamics of the system concerning the fate the elements. Therefore, regarding the chemical composition, these samples are distant (Figure 7) relatively to the elements of group 2, mobilized from sulfides, upstream. These elements are retained along the river stream, so that in P10 and P11 only some proximity to elements such as K, Ca and Mg is verified. P3 is also a particular site, since it is a water reservoir that receives drainage from several emergences and waste-dumps with a complex-mixed composition (including, phylites, slag, gossan, and sulfide mineralized blocks). This seems to result in a particular geochemical signature in the AMD that highlights a close relationship between Pb and the algae *Spirogyra*. This result points to the possibility of using *Spirogyra* as a bioindicator for the presence of Pb.

4. Conclusions

Aquatic habitats in the São Domingos mining area show ecological evidence of AMD contamination, namely blooms of acidophilic algae. Field observations show a variety of colours, which are related not only to the abiotic properties of the system, but also to the state of the algae communities. Algae senescence was detected by the brown-purplish colour, indicative of photosynthetic pigments loss, as well as of all the cellular organelles. The study revealed 14 taxa at the genus level, divided into eight classes. The diatoms represent the most common group of unicellular algae, whereas the genus *Mougeotia* prevails in multicellular algae.

SEM observations show mineral-algae interactions, namely through the presence of tubular iron-rich precipitates that could result from biogenic contribution. The present study also points out the scarcity of schwertmannite as being responsible for the rarity of *Euglena mutabilis*, which is very common in other AMD systems.

Thus, the ecological manifestations are in accordance with the hydrochemical, mineralogical, and hydrologic features of the system. The sites P5 and P8 receive the drainage from proximal highly reactive fine wastes and, consequently, they present the highest concentration of sulfate and PTE. The highest algae variability was observed at intermediate points concerning chemical contamination, coinciding also with lentic environments (P3 and P11). Nevertheless, the more acid/contaminated the site, the lesser is the algal diversity. Therefore, drainage flowing from the finest accumulation of sulfide wastes appears to control the development, maintenance, and proliferation of acidophilic algae.

The results obtained through factorial correspondence analysis showed an association between the location of the samples, their chemistry and the distribution and diversity of algal communities. Moreover, they suggest a close relationship between *Spirogyra* and Pb. These results should help to establish long-term monitoring protocols, applicable during and after environmental rehabilitation of the mining complex and respective river system.

Author Contributions: Conceptualization, P.G. and T.V.; data curation, P.G., T.V., R.H., N.F.-A., J.P., and F.M.; formal analysis, T.A., R.H., N.F.-A., and F.M.; funding acquisition, T.V.; investigation, P.G., T.V., T.A., R.H., N.F.-A., J.P., and F.M.; methodology, P.G., T.V., T.A., and R.H.; supervision, T.V.; visualization, J.P.; writing—original draft, P.G., T.V., T.A., R.H., N.F.-A., J.P., and F.M. All authors have read and agreed to the published version of the manuscript.

Funding: This work was funded by FCT (Fundação para a Ciência e Tecnologia) by the research fellowship under the POCH (Programa Operacional Capital Humano) supported by the European Social Fund and National Funds of MCTES (Ministério da Ciência, Tecnologia e Ensino Superior) with reference SFRH/BD/108887/2015. This work was co-funded by FCT through projects UIDB/04683/2020, UIDP/04683/2020 and Nano-MINENV 029259 (PTDC/CTA-AMB/29259/2017).

Institutional Review Board Statement: Not applicable.

Informed Consent Statement: Not applicable.

Data Availability Statement: Not applicable.

Acknowledgments: Patrícia Gomes acknowledges FCT (Science and Technology Foundation, Portugal) by the research fellowship under the POCH (Programa Operacional Capital Humano) supported by the European Social Fund and National Funds of MCTES (Ministério da Ciência, Tecnologia e Ensino Superior) with reference SFRH/BD/108887/2015. This work was co-funded by FCT through projects UIDB/04683/2020, UIDP/04683/2020 and Nano-MINENV 029259 (PTDC/CTA-AMB/29259/2017). The authors are grateful to the three anonymous reviewers and to the academic editor for their valuable contributions to improving the manuscript.

Conflicts of Interest: The authors declare no conflict of interest.

References

1. Gomes, P.; Valente, T. Physical and chemical conditions for colonization by Euglena mutabilis: Case studies in two acid mine drainage sites. In Proceedings of the IMWA: Mine Water: Technological and Ecological Challenges, Perm, Russia, 15–19 July 2019; Wolkersdorfer, C., Khayrulina, E., Polyakova, S., Bogush, A., Eds.; International Mine Water Association: Wendelstein, Germany, 2019; pp. 419–424.
2. Gross, W. Ecophysiology of algae living in highly acidic environments. *Hydrobiology* **2000**, *433*, 31–37. [CrossRef]
3. Gomes, P.; Valente, T.; Geraldo, D.; Ribeiro, C. Photosynthetic pigments in acid mine drainage: Seasonal patterns and associations with stressful abiotic characteristics. *Chemosphere* **2020**, *239*, 124774. [CrossRef]
4. Valente, T.; Rivera, M.J.; Almeida, S.F.P.; Delgado, C.; Gomes, P.; Grande, A.; de la Torre, M.L.; Santisteban, M. Characterization of water reservoirs affected by acid mine drainage: Geochemical, mineralogical, and biological (diatoms) properties of the water. *Environ. Sci. Pollut. Res.* **2015**, *23*, 6002–6011. [CrossRef]
5. Lessmann, D.; Fyson, A.; Nixdorf, B. Phytoplankton of the extremely acidic mining lakes of Lusatia (Germany) with pH $\leq$ 3. *Hydrobiology* **2000**, *433*, 123–128. [CrossRef]
6. Valente, T.; Gomes, C.L. The role of two acidophilic algae as ecological indicators of acid mine drainage sites. *J. Iber Geol.* **2007**, *33*, 283–294.
7. Luís, A.T.; Durães, N.; de Almeida, S.F.P.; da Silva, E.F. Integrating geochemical (surface waters, stream sediments) and biological (diatoms) approaches to assess AMD environmental impact in a pyritic mining area: Aljustrel (Alentejo, Portugal). *J. Environ. Sci.* **2016**, *42*, 215–226. [CrossRef]
8. Valente, T. Modelos de Caracterização de Impacte Ambiental para Escombreiras Reactivas-Equilíbrio e Evolução de Resíduos de Actividade Extractiva. Ph.D. Thesis, University of Minho, Braga, Portugal, 2004; 301p.
9. Levings, C.D.; Varela, D.E.; Mehlenbacher, N.M.; Barry, K.L.; Piercey, G.E.; Guo, M.; Harrison, P.J. Effect of an acid mine drainage effluent on phytoplankton biomass and primary production at Britannia Beach, Howe Sound, British Columbia. *Mar. Pollut. Bull.* **2005**, *50*, 1585–1594. [CrossRef]
10. Sabater, S.; Buchaca, T.; Cambra, J.; Catalan, J.; Guasch, H.; Ivorra, N.; Munõz, I.; Navarro, E.; Real, M.; Romaní, A. Structure and function of benthic algal communities in an extremely acid river. *J. Phycol.* **2003**, *39*, 481–489. [CrossRef]
11. Amils, R.; Gonzalez-Toril, E.; Aguilera, A.; Rodríguez, N.; Fernandez-Remolar, D.; Gomez, F.; García-Moyano, A.; Malki, M.; Oggerin, M.; Sanchez-Andrea, I.; et al. From Río Tinto to mars. The terrestrial and extraterrestrial ecology of acidophiles. *Adv. Appl Microbiol.* **2011**, *77*, 41–70. [CrossRef]
12. Wolowski, K.; Turnau, K.; Henriques, F.S. The algal fora of an extremely acidic, metal-rich drainage pond of São Domingos pyrite mine (Portugal). *Cryptogam. Algol.* **2008**, *29*, 313–324.
13. Tavares, M.T.; Abreu, M.M.; Vairinho, M.M.; Sousa, A.J.; Quental, L. Comportamento geoquímico de alguns elementos vestigiais na envolvente das Minas de S.; Domingos, Alentejo: Áreas da Tapada e do Telheiro. *Rev. Ciências Agrárias* **2009**, *32*, 182–194.
14. Quental, L.; Sousa, A.J.; Marsh, S.; Abreu, M.M. Identification of materials related to acid mine drainage using multi-source spectra at S. Domingos Mine, southeast Portugal. *Int J. Remote Sens.* **2013**, *34*, 1928–1948. [CrossRef]
15. Gomes, P.; Valente, T.; Grande, J.A.; Cordeiro, M. Occurrence of sulphate efflorescences in São Domingos mine. *Comun. Geol.* **2017**, *104*, 83–89.
16. Abreu, M.; Tavares, M.T.; Batista, M.J. Potential Use of *Erica andevalensis* and *Erica australis* in Phytoremediation of Sulphide Mine Environments: São Domingos, Portugal. *J. Geochem. Explor.* **2008**, *96*, 210–222. [CrossRef]
17. Gonçalves, J.A.; Henriques, R. UAV photogrammetry for topographic monitoring of coastal areas. *ISPRS J. Photogramm. Remote Sens.* **2015**, *104*, 101–111. [CrossRef]
18. Arge, L.; Chase, J.S.; Halpin, P.N.; Toma, L.; Vitter, J.S.; Urban, D.; Wickremesinghe, R. Flow computation on massive grids. *Proc. ACM Work Adv. Geogr. Inf. Syst.* **2001**, 82–87. [CrossRef]
19. Arge, L.; Chase, J.S.; Halpin, P.N.; Toma, L.; Vitter, J.S.; Urban, D.; Wickremesinghe, R. Flow computation on massive grid terrains. *GeoInformatica Int. J. Adv. Comput. Sci. Geogr. Inf. Syst.* **2003**, *7*, 283–313.
20. AWWA. *Standard Methods for the Examination of Water and Wastewater*, 18th ed.; American Public Health Association: Washington, DC, USA, 1992.
21. ASTM Committee on Standards. *Standard Practice for Cleaning Laboratory Glassware, Plasticware and Equipment Used in Microbiological Analysis*; American Society for Testing and Materials: West Conshohocken, PN, USA, 1992; Volume 5.
22. Benzécri, J.P. L'Analyse des Correspondences. *Cah. Anal. Données* **1977**, *2*, 125–142.

23. Pereira, H.G.; Brito, M.G.; Albuquerque, T.; Ribeiro, J. *Geostatistical Estimation of a Summary Recovery Index for Marble Quarries, Geostatistics Troia'92*; Kluwer Academic Publishers: Dordrecht, The Netherlands, 1993; Volume 2.
24. Sousa, P.; Sousa, J. *ANDAD, Version 7.12 Copyright*; CVRM/IST: Lisboa, Portugal, 2000.
25. Robbins, E.I.; Rodgers, T.M.; Alpers, C.N.; Nordstrom, D.K. Ecogeochemistry of the subsurface food web at pH 0–2.5 in Iron Mountain, California, U.S.A. *Hydrobiology* **2000**, *433*, 15–23. [CrossRef]
26. Robbins, E.I.; Anderson, J.E.; Podwysocki, M.H.; Nord, G.L. Seasonal variations in spectral reflectance of microbial flocculates, precipitates, and oil-like films associated with neutral and acidic mine drainage. In *Environmental Monitoring and Biodiagnostics of Hazardous Contaminants*; Healy, M., Wise, D.L., Moo-Young, M., Eds.; Springer: Dordrecht, The Netherlands, 2001. [CrossRef]
27. Anderson, J.E.; Robbin, E.I. Spectral reflectance and detection of iron-oxide precipitates associated with acidic mine drainage. *Photogramm. Eng. Remote Sens.* **1998**, *64*, 1201–1208.
28. Soyol-Erdene, T.O.; Valente, T.; Grande, J.A.; de la Torre, M.L. Mineralogical controls on mobility of rare earth elements in acid mine drainage environments. *Chemosphere* **2018**, *205*, 317–327. [CrossRef] [PubMed]
29. Cordeiro, M. Caracterização ambiental do complexo mineiro de São Domingos–Cartografia de Infra-Estruturas e Impacte Sobre o Meio Hídrico. Master's Thesis, University of Minho, Braga, Portugal, 2017; 84p.
30. Smucker, N.J.; Vis, M.L. Acid mine drainage affects the development and function of epilithic biofilms in streams. *J. N. Am. Benthol. Soc.* **2011**, *30*, 728–738. [CrossRef]
31. Valente, T.; Gomes, P.; Pamplona, J.; de la Torre, M.L. Natural stabilization of mine waste-dumps—Evolution of the vegetation cover in distinctive geochemical and mineralogical environments. *J. Geochem. Explor.* **2012**, *123*, 152–161. [CrossRef]
32. Gomes, P.; Valente, T.; Sequeira Braga, M.A.; Grande, J.A.; de la Torre, M.L. Enrichment of trace elements in the clay size fraction of mining soils. *Environ. Sci. Pollut. Res.* **2016**, *23*, 6039–6045. [CrossRef] [PubMed]
33. Valente, T.; Antunes, I.M.; Sequeira Braga, M.A.; Neiva, A.; Santos, A.; Moreno, F. Mobility control of uranium and other potentially toxic elements in mine waters by ochre-precipitates. In Proceedings of the IMWA: Mine Water: Technological and Ecological Challenges, Perm, Russia, 15–19 July 2019; Wolkersdorfer, C., Khayrulina, E., Polyakova, S., Bogush, A., Eds.; International Mine Water Association: Wendelstein, Germany, 2019; pp. 458–462.
34. Peng, X.; Chen, S.; Xu, H. Formation of biogenic sheath-like Fe oxyhydroxides in a near-neutral pH hot spring: Implications for the origin of microfossils in high-temperature, Fe-rich environments. *J. Geophys. Res. Biogeosci.* **2013**, *118*, 1397–1413. [CrossRef]
35. Ishihara, H.; Hashimoto, H.; Taketa, E.; Suzuki, T.; Mandai, K.; Kunoh, H.; Takada, J. Silicon-Rich, Iron Oxide Microtubular Sheath Produced by an Iron-Oxidizing Bacterium, Leptothrix sp. Strain OUMS1, in Culture. *Minerals* **2014**, *4*, 565–577. [CrossRef]
36. Prygiel, J.; Coste, M. *Guide Méthodologique pour la Mise en Oeuvre de l'Indice Biologique Diatomées NF T 90–354*; Agence de l'eau Artois Picardie: Douai, France, 2000; p. 340.
37. Freitas, A.P.P.; Schneider, I.A.H.; Schwartzbold, A. Biosorption of heavy metals by algal communities in water streams affected by the acid mine drainage in the coal-mining region of Santa Catarina state, Brazil. *Miner. Eng.* **2011**, *24*, 1215–1218. [CrossRef]
38. Kumar, R.N.; McCullough, C.D.; Lund, M.A.; Larranaga, S.A. Assessment of factors limiting algal growth in acidic pit lakes-a case study from Western Australia, Australia. *Environ. Sci. Pollut. Res.* **2015**, *23*, 5915–5924. [CrossRef]
39. Wolowski, K. Phylum Euglenophyta. In *Freshwater Algal Flora of the British Isles*; John, D.M., Whitton, B.A., Brook, A.J., Eds.; Cambridge University Press: Cambridge, UK, 2002; pp. 144–179.

minerals

MDPI

Article

The Evolution of Pollutant Concentrations in a River Severely Affected by Acid Mine Drainage: Río Tinto (SW Spain)

Manuel Olías *, Carlos R. Cánovas, Francisco Macías, María Dolores Basallote and José Miguel Nieto

Department of Earth Sciences/Research Center on Natural Resources, Health and the Environment (RENSMA), Faculty of Experimental Sciences, University of Huelva, Campus 'El Carmen' s/n, 21071 Huelva, Spain; carlos.ruiz@dgeo.uhu.es (C.R.C.); francisco.macias@dgeo.uhu.es (F.M.); maria.basallote@dct.uhu.es (M.D.B.); jmnieto@uhu.es (J.M.N.)

* Correspondence: manuel.olias@dgyp.uhu.es

Received: 13 May 2020; Accepted: 29 June 2020; Published: 30 June 2020

Abstract: The Río Tinto, located in the Iberian Pyrite Belt (SW Spain), constitutes an extreme case of pollution by acid mine drainage. Mining in the area dates back to the Copper Age, although large-scale mining of massive sulfide deposits did not start until the second half of the 19th century. Due to acidic mining discharges, the Río Tinto usually maintains a pH close to 2.5 and high concentrations of pollutants along its course. From a detailed sampling during the hydrological year 2017/18, it was observed that most pollutants followed a similar seasonal pattern, with maximum concentrations during autumn due to the washout of secondary soluble sulfate salts and minimum values during large flood events. Nevertheless, As and Pb showed different behavior, with delayed concentration peaks. The dissolved pollutant load throughout the monitored year reached 5000 tons of Fe, 2600 tons of Al, 680 tons of Zn, and so on. While most elements were transported almost exclusively in the dissolved phase, Fe, Pb, Cr, and, above all, As showed high values associated with particulate matter. River water quality data from 1969 to 2019 showed a sharp worsening in 2000, immediately after the mine closure. From 2001 on, an improvement was observed.

Keywords: sulfide mining; Iberian Pyrite Belt; rivers; pollutant load

1. Introduction

The Río Tinto is a relatively small river (101 km long) known worldwide for its extreme conditions (in Spanish "Tinto" means dark red). From its source to its mouth at the Ria of Huelva estuary (Figure 1), its waters usually maintain a pH value of around 2.5 and very high concentrations of Fe and other toxic elements. No plants, fish, or other higher aquatic organisms can survive in these conditions, with the exception of some invertebrates [1]. Nevertheless, the river water is rich in extremophile organisms (bacteria, algae, fungi, and protists), with a remarkably high eukaryotic diversity [1–4].

Mining wastes containing sulfide minerals can generate acid mine drainage (AMD) [5,6]. Under natural conditions, most sulfides remain underground within an anoxic environment, where they are stable and highly insoluble. The mining activities cause a sharp increase in the oxidation of sulfides due to their contact with oxygen and water, releasing acidity and large amounts of sulfates and toxic metals (e.g., Fe, Cu, Zn, Cd, Pb) into the aquatic medium. The oxidation of sulfide minerals is accelerated by microbial catalysis of the oxidation of aqueous ferrous iron and sulfide [7,8]. This process leads to the generation of significant discharges of acid leachates with very high concentrations of toxic elements [9]. The generation of AMD depends on the types and quantities of sulfide and carbonate minerals present, as well as the dissolution rates of these minerals [10]. Pollution of water resources by AMD is a major worldwide environmental problem associated with the mining of sulfide and coal deposits, affecting many sites [11,12].

The Río Tinto's headwaters are located in the Iberian Pyrite Belt (IPB), a metallogenic region that runs from the Seville and Huelva provinces (SW Spain) to the Portuguese coast and has one of the largest concentrations of giant and supergiant polymetallic massive sulfide deposits on Earth [13]. Metal exploitation in the IPB, including the Río Tinto mines, began in the third millennium BC (the Copper Age). Mining consisted of extracting copper from carbonates, oxides, and some sulfides (chalcocite and covellite) located close to the land surface [14]. During the Roman period, mining increased. After a long time of little activity, large-scale mining began in the second half of the 19th century and continued until 2000, when exploitation ceased due to falling mineral prices [15]. In 2015 the Río Tinto mine reopened, beginning a new period of activity. Currently, the headwaters of the Río Tinto are considered a protected zone, becoming a tourist attraction due to the special characteristics and colors of its waters, the rich industrial heritage, and the spectacular landscape left by the mining.

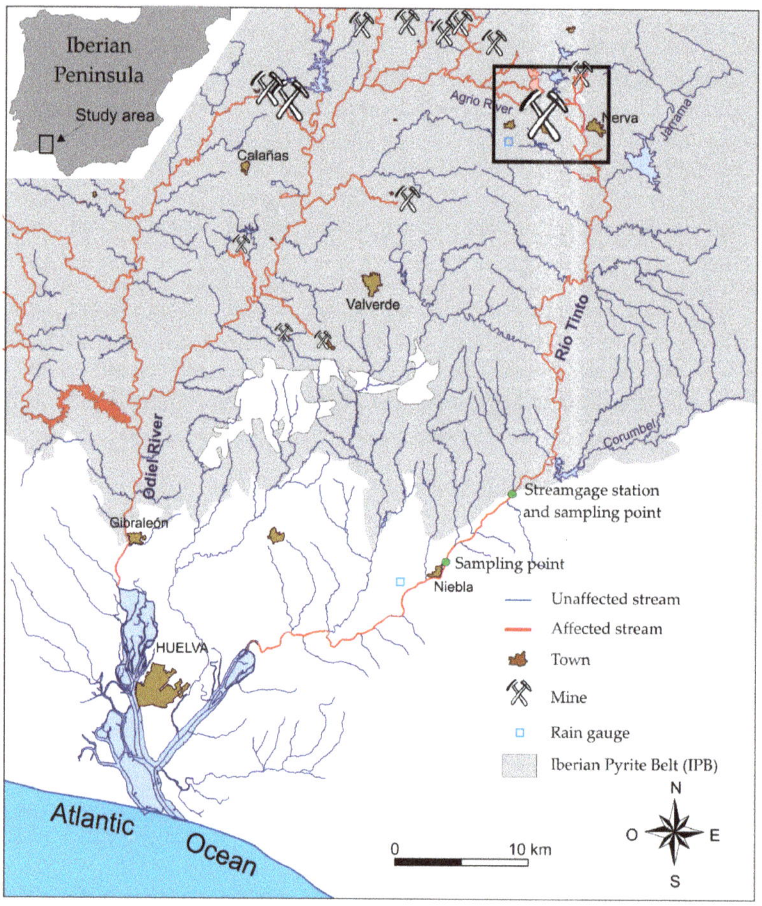

Figure 1. Map of Río Tinto drainage network. The inner rectangle indicates the area shown in Figure 2. The size of the mine symbol (crossed hammers) is proportional to the amount of ore extracted.

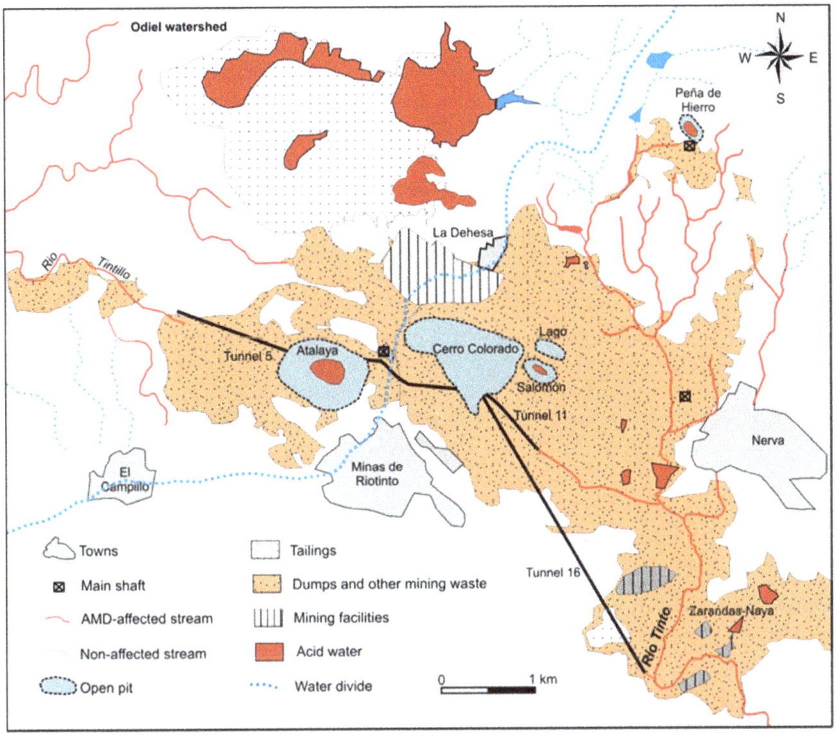

Figure 2. Detailed map of Río Tinto mining district.

The Odiel River (Figure 1) is also deeply affected by AMD, although the pollutant concentrations are lower than those of the Río Tinto [16,17]. The high concentrations of pollutants transported by both rivers have a severe environmental impact on the estuary, where most of the pollutants precipitate when the acid waters are neutralized by mixing with alkaline seawater. High concentrations of metal(loid)s such as As, Cd, and Pb have been measured in the estuarine waters, sediments, and biota [18–21]. Nevertheless, some of the most conservative metals under circumneutral pH conditions, such as Zn, Cd, and Mn, reach the Atlantic Ocean or even the Mediterranean Sea [22,23].

Although there are numerous cases worldwide of streams contaminated by AMD, to our knowledge there are no other rivers that maintain such extreme conditions as the Río Tinto along its almost 100 km. Another similar case of an extremely affected stream could be the King River (Tasmania, Australia), which receives acidic mine discharges (around 2 tons/day of Cu) from Mount Lyell [24,25]. Nevertheless, the length affected by AMD and dissolved concentrations are lower than those of Río Tinto. In this context, the main goal of this work is to study the seasonal variations of the pollutant concentrations and loads to investigate the main processes controlling toxic element behavior in rivers with extreme acidic conditions.

2. Site Description

The Río Tinto watershed runs mainly over the IPB (Figure 1), except at its lower course, where more recent Neogene materials outcrop. The IPB consists of three main geological units: the Phyllite-Quartzite Group (PQ), the Volcano-Sedimentary Complex (VSC), and the Culm Group (CG). The PQ group consists of a thick sequence of phyllites and sandstones of upper Devonian age [13,26]. The materials of the VSC (upper Devonian-lower Carboniferous age) are composed of a volcanic sequence with alternating episodes of felsic (dacites and, in a lower proportion, rhyolites) and mafic rocks (basalts) intercalated in a sedimentary sequence (mainly phyllites). The VSC hosts numerous sulfide deposits, mainly composed of pyrite (FeS_2) with sphalerite (ZnS), galena (PbS), chalcopyrite ($CuFeS_2$), arsenopyrite (FeAsS), and other sulfides in much lower proportions. Finally, the CG is a detrital unit of synorogenic turbidites of shales and conglomerates of Carboniferous age. There are no carbonate rocks in the IPB, so the acidity generated by AMD processes is not neutralized; this, together with the intensive mining developed, causes the extreme pollution of most streams in the area.

The climate is of Mediterranean type, with cold winters and warm summers. The annual mean temperature is close to 17 °C and rainfall ranges from 800 mm at the north of the basin to 550 mm at the south. The rainiest period of the year is commonly from November to January, while during the summer rainfall is practically nonexistent. Nevertheless, rainfall exhibits a high inter- and intra-annual variability, with long droughts and intense wet periods. The Río Tinto's main tributaries are the Jarrama and Corumbel Rivers (Figure 1), both regulated by reservoirs. According to the water authorities, the mean river discharge is 7.2 m^3/s, although, due to the absence of permeable rocks in the watershed, the river flow presents high variability, depending on the rainfall regime [16].

Nowadays, the Río Tinto's source is considered to be located at the foot of a sulfide dump of Peña de Hierro mine, northeast of the Río Tinto mines (Figure 2). Nevertheless, the main acidic discharges come from the much larger Rio Tinto mine, which has a surface area of 21 km^2 covered by waste dumps, tailings, flooded open pits, slags, and other mining facilities (Figure 2). In addition, there is a dense and complex network of underground mine galleries. There is geological, archaeological, and historical evidence indicating that, although natural sulfide oxidation in the zone dates back 24 million years [27], the current river conditions are mainly due to large-scale mining since the second half of 19th century, while natural processes of acid rock drainage can be considered negligible [15]. Despite all this evidence, there are some works claiming that the Río Tinto's conditions are mainly caused by natural processes [28].

Río Tinto mines also constitute the main source of pollution in the Odiel River watershed [29,30], through several streams running to the west (Figure 2). The current exploitation is focused on the Cerro Colorado open pit (Figure 2). According to the environmental requirements for the reopening, the mining company is obligated to progressively reduce the acidic discharges to the Odiel River, where a water reservoir is being constructed [31]. Thus, a reduction of 30% must be achieved before the third year from the start of exploitation, 50% before the sixth year, and 100% before the tenth year. However, there is a large surface covered by mining wastes producing acidic leachates at the Zarandas-Naya zone (Figure 2), which is outside the current exploitation area, and therefore, no remediation measures are considered.

In the Río Tinto, downstream of the mining area, a decrease in most dissolved element concentrations is observed, mainly by dilution with clean waters from tributaries [32]. Moreover, the Fe concentration diminishes in greater proportion because of the precipitation of Fe oxyhydroxy sulfates along the river, mainly schwertmannite and jarosite [32,33]. Other types of abundant secondary minerals precipitating on the riverbanks and the mining area are soluble efflorescent salts formed by water evaporation [34], being especially abundant at the end of the summer.

3. Methods

3.1. Data Acquisition

In order to study the seasonal variation in pollutant concentrations and loads of the Río Tinto, high-resolution sampling was carried out during one hydrological year (from September 2017 to August 2018) at the streamgage station, located 8.9 km upstream at Niebla (Figure 1). This location was determined by the availability of streamflow data and the possibility of having a place to protect the sampling equipment. An autosampler (Teledyne ISCO) was programmed with a frequency ranging from 2 to 24 h, depending on the weather forecast. Only those samples reflecting variations in electrical conductivity (EC) were selected for analysis (n = 143). The distribution of collected samples within the study period is highly variable, mainly linked to rainfall-induced hydrochemical variations (from 65 samples collected in March 2018 to only two samples in the summer). The autosampler was equipped with a 24-bottle container. Cross-contamination between pumping cycles was avoided by scheduled purge sequences. Bottles were previously washed in 10% (*v*/*v*) HNO_3 and Milli-Q water before sampling. Manual samples coinciding with the visits for the collection of the automatic samples were taken. Physicochemical parameters such as pH, EC, oxidation-reduction potential (ORP), and temperature were measured in each sample using a previously calibrated Crison MM40+ multimeter (HACH LANGE GmbH, Düsseldorf, Germany). ORP readings in the field were converted to redox potential values with reference to the standard hydrogen electrode (Eh), according to Nordstrom and Wilde [35]. Selected samples (according to changes in EC) were filtered through 0.45-µm Millipore Teflon filters and subsequently acidified with suprapure HNO_3 Merk to pH < 1 and stored in the dark at 4 °C until analysis. Unfiltered aliquots (n = 53) were taken from selected samples and acidified to study metal particulate transport by the river. Thus, the difference between the concentrations in the filtered and unfiltered samples is considered to be associated with the particulate matter.

Rainfall data belonging to the weather station network of the Andalusian Government were obtained from two stations located north and south of the Río Tinto basin (Figure 1). Flow data were obtained every 15 min from a streamgage station (Figure 1). Historical data from the Water Authority for the Río Tinto at Niebla (Figure 1) were collected in 1969-2019 for pH and electrical conductivity (EC) and from 1980 (when concentration data were available on a regular basis) to 2019 for As, Cd, Cu, Fe, Mn, Pb, SO_4, and Zn (n = 669; an average of 17 samples per year).

3.2. Analysis

Samples were analyzed at the R&D laboratories at the University of Huelva (Huelva, Spain) by Inductively Coupled Plasma-Atomic Emission Spectroscopy (ICP-AES) for major elements, and by Inductively Coupled Plasma-Mass Spectroscopy (ICP-MS) for trace elements. Detection limits were 200 µg/L for Al, Ca, Fe, K, Mn, Mg, Na, and S; 50 µg/L for Zn; 5 µg/L for Cu; 1 µg/L for Ba and Sr; and 0.1 µg/L for As, Cd, Co, Cr, Ni, and Pb. The accuracy of the analyses was checked with NIST-1640 certified reference materials. Homemade standards from certified materials were also used during each analysis sequence to check the accuracy. The analytical precision was assessed by performing analyses in triplicate, being better than 5% in each case. The average balance error obtained by the PHREEQC code was 2.4% (interquartile range between −0.4% and 4.8%).

3.3. Pollutant Load

The calculation of river loads introduces inevitable uncertainties: whereas continuous records of streamflow are commonly available from flow gauging stations, concentration data frequently only represent individual samples [36]. Several procedures have been employed to reliably estimate loads based on limited concentration data [16,37]. Owing to the existence of high number of collected samples coinciding with dissolved concentration variations, the pollutant loads delivered by the Río Tinto were calculated, interpolating concentrations between two consecutives samples according to the following equation:

$$C_i = \frac{C_n - C_{n-1}}{t_n - t_{n-1}}(t_i - t_{n-1}) + C_{n-1}$$

where C_i is the concentration at time i, C_{n-1} and C_n are the concentrations of the available samples before and after time i and t_n and t_{n-1} are the dates of samples C_n and C_{n-1}.

Thus, the concentrations were obtained every 15 min and multiplied by the corresponding river flow for obtaining the pollutant load. These 15-min data were added for each month to obtain the monthly loads.

3.4. Saturation Indices and Multivariate Analysis

Saturation indices (SI) of water were obtained using the PHREEQC code v3.4 (US Geological Survey, Denver, CO, USA) [38]. Thermodynamic constants were taken from the geochemical MINTEQA2 database (USEPA Environmental Research Laboratory, Athens, GA, USA) [39]. The equilibrium constants (K_e) of schwertmannite were obtained from [40,41]. On the other hand, a principal component analysis (PCA) was performed on the data to evaluate the statistical relationships between multivariate datasets [42,43]. Previously, a normality test was performed on samples, which evidenced the non-normality of all variables. Then, Spearman's correlation coefficient was used to determine the significant relationships between data.

4. Results and Discussion

4.1. Evolution during the Hydrological Year 2017/18

The annual rainfall in the study period was close to the mean value of the area (517 and 814 mm in the south and north rain gauges, respectively). Nevertheless, the rainfall distribution through the year was irregular (Figure S1). The autumn and winter were anomalously dry (mean of 175 mm between September 1 and February 27), while an intense rainy period was recorded between February 28 and March 20 (mean of 333 mm). The wet period continued in April (100 mm).

River flow was below 0.025 m^3/s in September 2017 (Figure S1). Scarce rainfall recorded from October to February provoked some discharge peaks, with a maximum close to 6 m^3/s. On February 27, the river flow was only 0.13 m^3/s, but intense rains recorded afterwards provoked a sharp flow increase, reaching a value of 51.5 m^3/s on March 1. Discharges were very high until March 22 and at the beginning of April. After that, the streamflow decreased progressively down to 0.001 m^3/s at the end of August. In addition, frequent and sharp daily increases due to water releases from the Jarrama and Corumbel reservoirs were observed during the spring (Figure S1).

Values of pH during the autumn and winter were close to 2.5 and increased up to around 3 with the floods in March, although with some fluctuations (Table 1 and Figure 3). Nevertheless, the highest values were reached in April, with a maximum pH of 4.22. Electrical conductivity values underwent a fast and progressive increase with the first October rainfalls, from 4.7 to 11.4 mS/cm just four days after the beginning of rainfalls, due to the dissolution of efflorescent salts precipitated during the summer. These values were higher than those previously observed during the autumn in the Río Tinto, ranging between 8 and 9 mS/cm, with pH values between 2.3 and 3.6 [34,44]. Throughout autumn and winter, there was a progressive decrease in EC to approximately 3.5 mS/cm in February, with some increases also linked to the dissolution of remaining evaporitic salts by rains. With the intense rains of March and April, EC values notably diminished to a minimum of 0.38 mS/cm, although with sharp variations in the response to changes in streamflow. From April onwards, a progressive increase in EC values up to a value of around 4 mS/cm in the summer was observed (Figure 3).

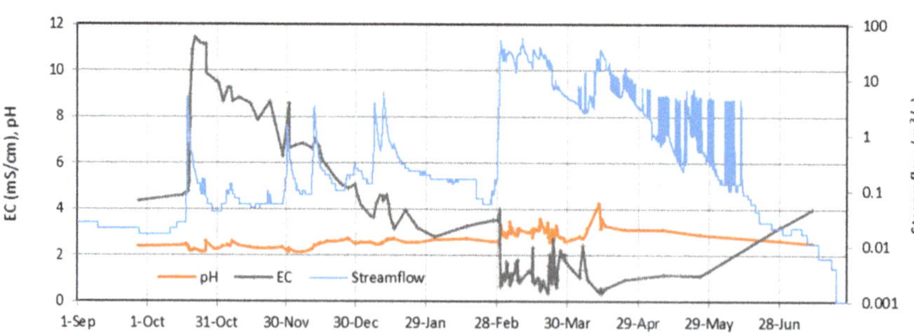

Figure 3. Evolution of pH, electrical conductivity (EC), and streamflow during 2017/18.

The basic statistics of dissolved concentrations can be seen in Table 1. The most abundant metal was Fe (median value of 145 mg/L), followed by Mg (67 mg/L), Al (65 mg/L), and lower concentrations of Ca, Na, Zn, Cu, and Mn. Among trace elements, the high dissolved concentrations of As, Cd, Co, Ni, and Pb stand out (maximum values of 3.31, 0.74, 5.33, 1.06, and 1.07 mg/L, respectively). Mean concentrations were much higher than the median ones (Table 1), showing that dissolved elements did not exhibit a normal distribution (especially As, with a mean value 11 times higher than the median), as indicated by Shapiro-Wilk and Kolmogorov-Smirnov tests.

The first rainfalls in October caused a notable increase in the dissolved concentrations because of the redissolution of the riverbank secondary soluble sulfate salts, reaching their maximum values except for Ba, Na, and Sr (Figure 4 and Figure S2). The increase was especially significant for sulfate (from 2.4 to 12.2 g/L), Fe (from 162 to 1529 mg/L), and As (from 21.6 to 411 μg/L). These secondary minerals are mainly sulfates (such as melanterite, copiapite, and coquimbite), precipitated during dry periods, and play an important role in metal fluxed from areas affected by AMD processes, since they can be either a sink or a source for acidity and toxic metals [34,45,46]. After this peak, the EC and concentrations of most elements progressively decreased, but with some small increases due to several moderate rainy events recorded in the autumn and winter (Figure 4 and Figure S2).

Table 1. Basic statistic of physicochemical parameters and dissolved concentrations for the hydrological year 2017/18.

Statistic	pH	EC	Eh	Al	Ca	Cu	Fe	K	Mg	Mn	Na
		mS/cm	mV	mg/L	mg/L	mg/L	mg/L	mg/L	mg/L	mg/L	mg/L
Mean	2.74	3.54	740	166	63	29	377	1.6	178	16	34
Median	2.70	2.21	730	65	25	15	145	1.3	67	5.3	18
25th percentile	2.47	1.02	665	18	13	4.1	26	1.0	20	1.5	11
75th percentile	3.02	4.98	794	209	100	40	674	1.9	250	22	50
Min.	2.12	0.38	536	4.6	8.8	1.3	0.3	<0.5	9.2	0.6	7.4
Max.	4.22	11.42	892	740	219	115	1529	9.1	731	66	310
Statistic	**SO_4**	**Si**	**Zn**	**As**	**Ba**	**Cd**	**Co**	**Cr**	**Ni**	**Pb**	**Sr**
	mg/L	mg/L	mg/L	μg/L	μg/L	μg/L	μg/L	μg/L	μg/L	μg/L	μg/L
Mean	2888	19	45	374	15	144	890	42	218	166	159
Median	1123	7.2	16	33	13	51	412	9.3	65	110	63
25th percentile	347	5.0	4.4	5.1	8.4	14	101	2.3	19	71	37
75th percentile	4221	24	62	344	17	187	1294	72	340	205	223
Min.	124	2.9	1.2	0.3	<1	5.1	34	<0.2	6.8	19	25
Max.	12,156	96	205	3309	54	739	5332	225	1065	1068	708

However, As showed different behavior, reaching its maximum dissolved concentrations in mid-December (Figure 4), coinciding with the maximum values of the Fe/SO_4 ratio (Figure 5A). Arsenic is strongly adsorbed/coprecipitated with Fe oxyhydroxysulfates [47,48], which is an important attenuation process in waters far from the mine sites. This peak must be related to the arrival of less evolved water from the mining area, with less Fe precipitation and higher concentrations of both Fe/SO_4 and As (Figure 5B). From previous tracer tests in the adjacent Odiel River, values of water velocity of 10 cm/s and 0.5 cm/s for flows of 1.1 m^3/s and 0.035 m^3/s, respectively, were obtained [49]. Considering an approximate water velocity of 1 cm/s and taking into account the distance between the sampling point and the Río Tinto mines (46 km), the water's travel time from the mining area to the sampling point would be 53 days, which approximately coincides with the delay observed between the first autumn rains and the peak concentration of As.

Due to the scarcity of rainfall during the winter, at the end of February dissolved concentrations for most elements were similar to those commonly recorded at the end of the summer (Figure 4 and Figure S2). With the intense rainfalls of March, there was an initial increase in concentrations due to the washout of soluble sulfate salts, followed by a general decrease, showing later sudden fluctuations in response to successive flood events. Although the highest discharges occurred at the beginning of March, the lowest concentrations for most elements were reached in mid-April due to the progressive washout of sulfide oxidation products. From mid-April onwards, concentration values began to recover, coinciding with decreasing river flows (Figure 4 and Figure S2).

On the other hand, Pb and Ba showed a different evolution (Figure 4 and Figure S2) and their dissolved concentrations increased notably during the peak floods of March and April. This behavior may be related to the solubility control exerted by barite and anglesite or other Pb minerals [17]. In this sense, in the Río Tinto headwaters it was deduced using geochemical modeling that suggested that the dissolved Pb concentrations were limited by the equilibrium with anglesite below pH 1.5 and coprecipitation with a jarosite phase above this pH [50]. Thus, the low content of sulfate during floods would enhance the dissolved concentrations of Ba and Pb. Saturation indices (SI) for anglesite showed values close to −1 throughout the year, although with some fluctuations, while barite was oversaturated (SI between 0 and 1, Figure S3). Previous studies in Wales also reported high fluxes of Pb during stormflows after extended periods of dry weather, due to the encouraged oxidation of Pb sulfide in the mine spoils [51]. However, high concentrations of Pb were also observed following wet conditions [51]. This could be explained by the hypothesis of solubility control, which has significant implications for the load of Pb transported by the rivers, as will be explained further.

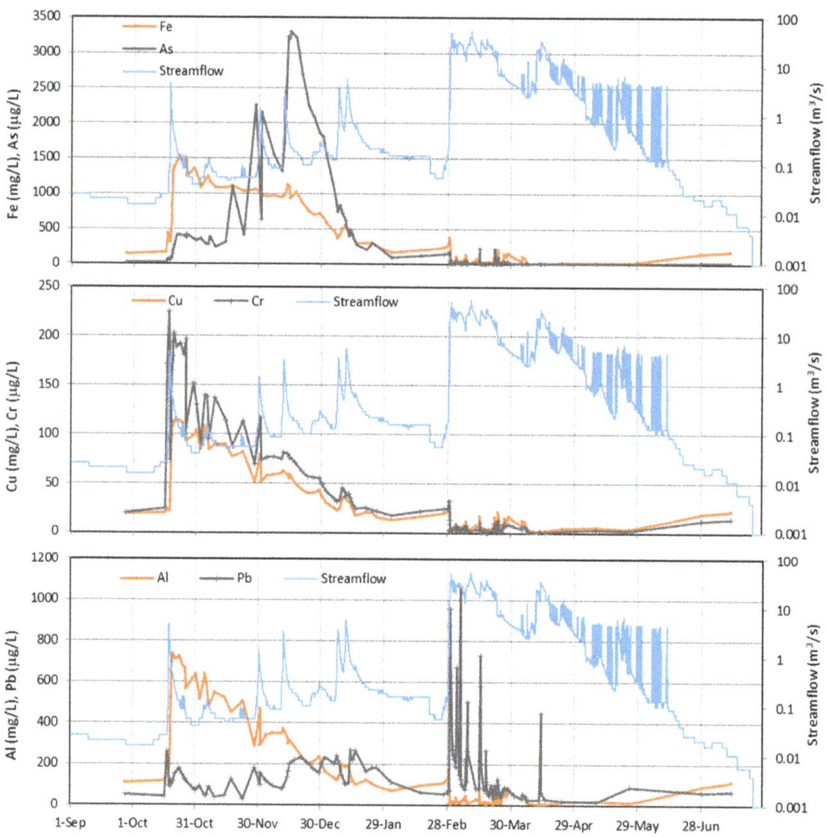

Figure 4. Evolution of dissolved Fe, As, Cr, Cu, Al, and Pb in 2017/18.

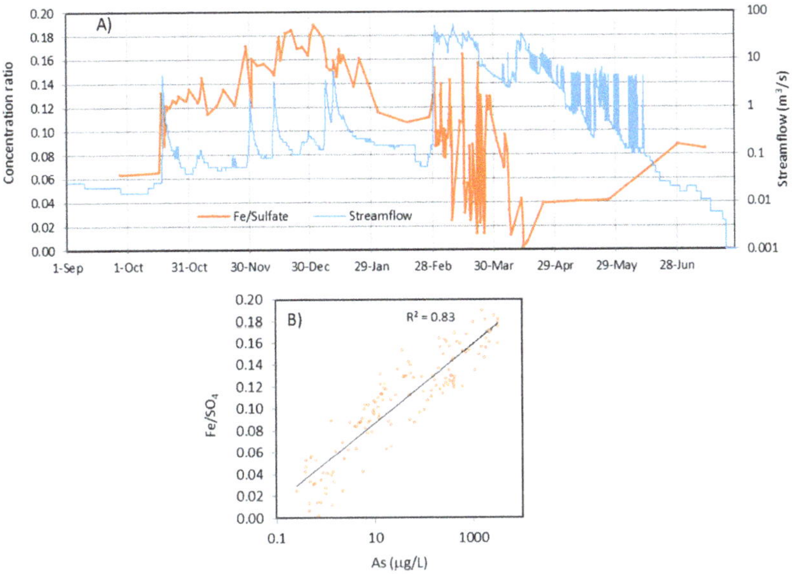

Figure 5. (**A**) Evolution of Fe/SO_4 ratio in 2017/18 and (**B**) relationship between As concentration and Fe/SO_4 ratio.

Most elements (Al, Cu, Fe, Cd, Co, Cr, Mg, Mn, Ni, SO_4, and Zn) exhibited a high correlation with EC (Spearman correlation coefficient > 0.98) due to their similar behavior over the course of the year (Figure S2). However, some slight differences can be observed. For instance, lower values of Cu/Zn ratio (<0.6) were observed from September to November (Figure S4), while maximum values (>1) occurred during the March floods. This pattern may be related to the preferential removal of Cu over Zn by melanterite, and possibly by other soluble sulfate salts' precipitation [17,52].

Dissolved concentrations showed highly negative correlations with streamflow for most elements (Figure 6), with a Spearman coefficient between 0.80 and 0.87 for Al, Ca, Cd, Co, Cu, Mg Mn, Na, Ni, SO_4, Sr, and Zn, and lower values for Fe (0.77), Cr (0.78), and As (0.69). Lead showed a low positive correlation (0.23, significant at $p < 0.05$), while there was no significant correlation for Ba and K and streamflow. When the streamflow in a river increases, a decrease in the concentration of dissolved substances is generally observed due to the dilution of the less-concentrated surface runoff, although the transported load can increase [11,51,53]. Elements with the highest correlations are those that are more conservative in the river water, while the lower correlation of Fe, Cr, and As must be associated with precipitation (Fe) and coprecipitation/sorption processes, especially affecting As, but also Cr. Saturation indexes for schwertmannite showed oversaturation or subsaturation, depending on the equilibrium constant used, while for jarosite, oversaturated values were observed during the autumn and winter and subsaturated ones during the spring (Figure S3). The affinity of Cr to be removed during Fe precipitation has been previously observed in the Río Tinto [17] and other AMD-affected systems [54–56]. On the other hand, K concentration in AMD water may be controlled by jarosite precipitation [54], not showing any relationship with streamflow, as well as Ba.

All these results are verified in the performed PCA (Figure 7). The first component explains most of the variance (79%) and seems to be controlled by the element concentration (except for Ba, K, and Pb), while the negative part covers the streamflow. The second component (only 8.4% of the variance) is mainly controlled by Pb and Ba concentrations and, with a lower weight, K. Arsenic and, to a lesser extent, Fe and Cr show slightly different behavior than most elements due to the precipitation/coprecipitation/sorption processes explained above.

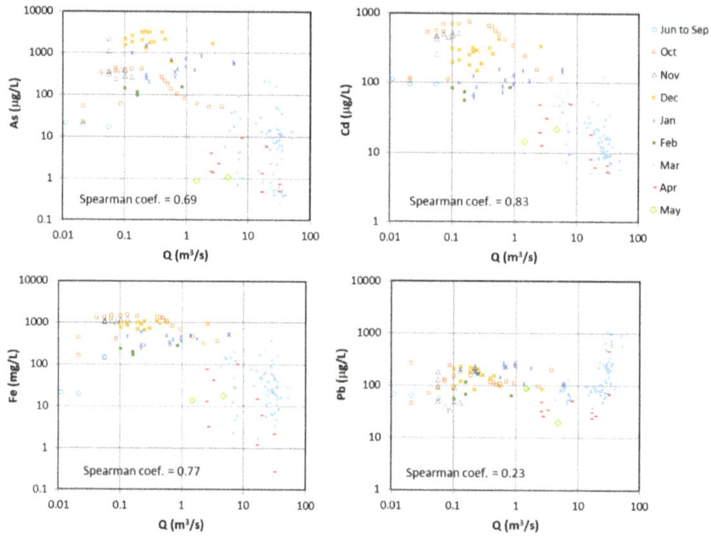

Figure 6. Relationships between streamflow (Q) and dissolved concentrations of As, Cd, Fe, and Pb.

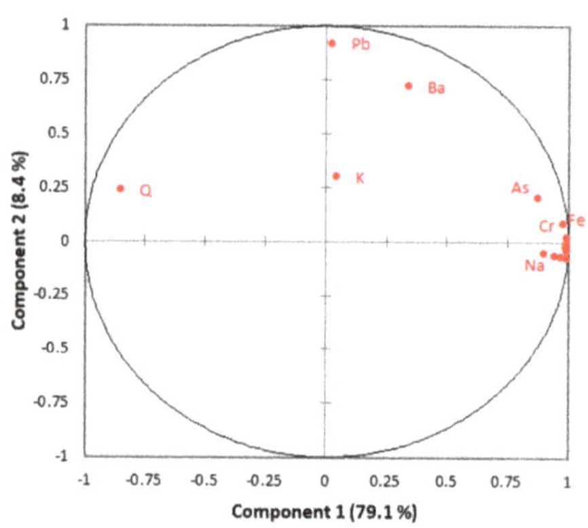

Figure 7. PCA of the analytical data and streamflow (Q). Only variables with a distinguishable position are identified.

However, for those elements exhibiting a high correlation with river flow, dissolved concentrations can vary by approximately one order of magnitude for Cd and Pb and three orders of magnitude for As and Fe with the same streamflow value (Figure 6). In addition, seasonal differences can be seen, with the lowest concentrations in April while the highest were in October for most elements (including Cd and Fe), in December for As and in March for Pb (Figure 6).

The percentages of dissolved concentrations over the total were obtained from the difference in concentrations between unfiltered and filtered aliquots (Table 2). Most elements (mainly those with the highest correlation with the EC) displayed percentages of dissolved concentration higher than 98%, which means that they were almost exclusively transported by the dissolved phase. On the contrary, As was mainly transported by the particulate phase linked to Fe precipitates, as mentioned above (Figure 5). Values for Fe and Ba were close to 50%, while Cr and Pb were mainly transported by the dissolved phase, although with a significant role of particulate matter (10–30%).

Table 2. Median value of the percentage of dissolved over total concentration.

Median	Element/Compound
>98%	Al, Ca, Cd, Co, Cu, Mg, Mn, Na, Ni, SO_4, Sr, Zn
90–98%	K, Si
70–90%	Cr, Pb
50–70%	Fe
30–50%	Ba
<10%	As

Table 3 shows the dissolved load obtained for the hydrological year 2017/18. The load has only been calculated for elements whose concentrations were above the detection limit in all samples. The Río Tinto transported almost 5000 tons of Fe, 2600 tons of Al, 680 tons of Zn, 560 tons of Cu, and 230 tons of Mn. Among trace elements, the transport of Pb and Co stands out (15 tons), although the median concentration of Pb was much lower than that of Co (110 and 412 μg/L, respectively). Significant loads of As, Cd, and Ni (between 2.1 and 3.6 tons) and other toxic elements were also recorded. These values are similar to those obtained from correlations between flow and element concentrations for the period 1995–2003 [16]. The most striking difference is that obtained for As (12 versus 3.6 tons in this work), which could be explained by the anomalous rainfall distribution observed in 2017/18, together with the extreme As values observed in 2000, coinciding with the cessation of mining (see next section). A high difference was also observed for Al (1224 versus 2593 tons), but for the period 1995–2003 it is acknowledged that Al data had high uncertainty [16].

Table 3. Dissolved pollutant load obtained for 2017/18.

Element	Tons	Element	Tons
Al	2593	Fe	4963
As	3.6	Mn	234
Cd	2.1	Ni	3.0
Co	15	Pb	15
Cr	0.44	SO_4	49,420
Cu	556	Zn	683

The seasonal distribution of dissolved pollutant loads in the Rio Tinto in 2017/18 is shown in Figure 8. Approximately 30% of the annual amount of Al, Cd, Co, Cu, Ni, and Zn was transported from October to February, even though the water contribution during this period was only 4%. However, some small differences among these elements can be seen: for example, the percentage of Cu (26%) was lower than that of Zn (33%) in this period because of the differences in the preferential coprecipitation of Cu in melanterite compared to Zn (Figure S4). It can also be observed that from October to February, 78% of As, 49% of Cr, and 43% of Fe was delivered to the Río Tinto estuary. On the contrary, only 4% of the total Pb delivered by the river in the whole year was transported during this period. Nevertheless, the particulate transport of Cr, Pb, Fe, and, especially, As may be high compared with the dissolved transport, according to the percentages shown in Table 3. A recent work found significant transport of particulate As, Cr, and Pb in the middle course of the Río Tinto [57].

During the March floods, the Río Tinto carried the highest dissolved loads for all elements studied except for As (Figure 8). The first significant floods after summer, normally occurring at the end of autumn or beginning of winter, represent a high percentage of the annual pollutant load of the river [34,58]. The high Pb load (86% of the total of the year) stands out due to its increase in concentration during floods. March floods caused the total washout of sulfide oxidation weathering products from the mining zone. Consequently, although the water contribution in April was high (25% of the total), the dissolved loads carried by the river were much lower, especially for As, Fe, Cr, and Pb (<10%; Figure 8). This indicates that the pyrite oxidation products from the mining zone were not completely washed out until the March floods. Thus, two stages could be differentiated: (1) A washout with the first rainfall after the summer of salts precipitated on the surface of the mining zone and riverbanks, which produces the maximum pollutant concentrations; and (2) total removal of sulfate salts from the mining area after the winter floods, which is less apparent. In the latter case, this could be related to the dissolution during intense rainy periods of: (a) less soluble sulfate minerals, (b) salts precipitated within waste-rock dumps, which can be found as cement filling the interclast porosity and/or as a coating on the particles, and (c) salts from mine voids due to water table elevation [46,51].

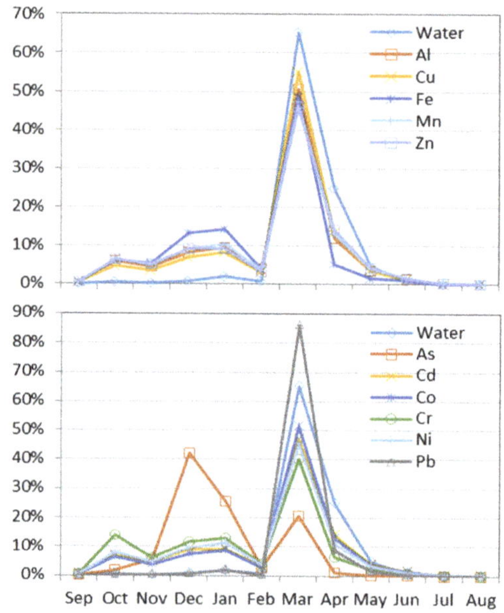

Figure 8. Percentages of the monthly distribution of the pollutant load in 2017/18. The water contribution percentage is also shown for comparison.

4.2. *Long-Term Evolution*

The basic statistics of the chemical evolution of the Río Tinto from the official water quality monitoring network are shown in Table 4. The conditions are more extreme (higher EC and concentrations and lower pH values) than obtained for 2017/18 (Table 1). These differences are explained by: (1) The sampling point of the official water quality monitoring network (Niebla; Figure 1) being located 9 km downstream of the sampling point for 2017/18 (streamgage station; Figure 1) and, therefore, being affected to a greater extent by dilution with tributaries; (2) different sampling strategies: while samples in the historical dataset were collected periodically, sampling during 2017/18 aimed at determining the concentration variations during the floods and, consequently, the highest number of samples was taken during the rainiest months; and (3) lower levels of pollutants since the mine closure in 2000, as discussed below.

Regarding the temporal evolution of pH and EC since 1968, some samples with pH above 5, coinciding with the lowest EC values during very rainy periods, stand out (Figure 9). On the other hand, pH values tend to increase, and EC values tend to decrease over time. This change is especially evident from 2000 (Figure 9), coinciding with the mine closure (after this year pH values often exceed 3, while before this date this occurs only very sporadically). Likewise, the EC values were at a maximum in 2000. Since then values have rarely exceeded 6 mS/cm, whereas before 2000 this value was frequently surpassed.

Table 4. Basic statistics on pH and EC (1968–2019) and dissolved concentrations for some elements (1980–2019).

Statistic	pH	EC	Cu	Fe	Mn	SO_4	Zn	As	Cd	Pb
		mS/cm	mg/L	mg/L	mg/L	mg/L	mg/L	μg/L	μg/L	μg/L
Mean	2.6	3.5	28	413	13	2801	63	1153	278	205
Median	2.5	3.0	20	219	9	1946	39	90	160	120
25th percentile	2.4	2.0	12	107	5	1076	18	16	82	74
75th percentile	2.7	4.6	31	491	15	3511	70	550	290	207
Min.	1.7	0.2	0.02	0.02	0.09	58	0.02	<dl	<dl	<dl
Max.	6.8	15	365	5080	113	25,256	730	50,000	5340	1630

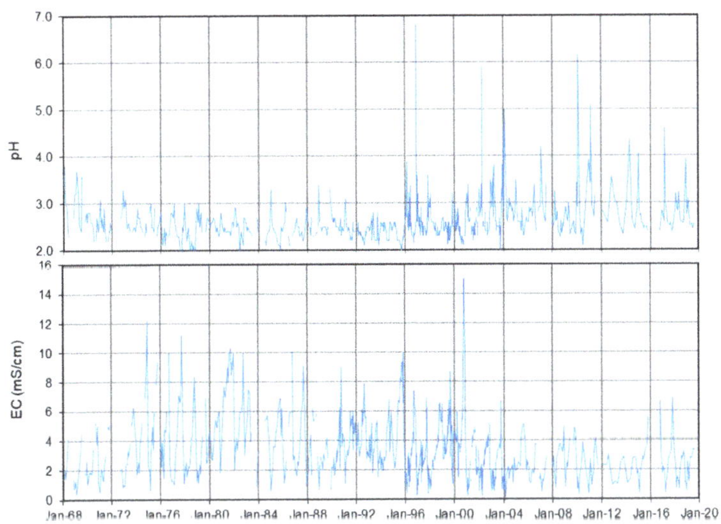

Figure 9. Evolution of pH and EC from 1968 to 2019.

Concerning metal concentrations, there is also a worsening of water quality in 2000 (Figure S5), reaching concentrations of up to 5 g/L of Fe, 590 mg/L of Zn, and 50 mg/L of As. This pollutant peak was due to the closure of Rio Tinto mines and the cessation of environmental control works [16]. These peaks are observed in other mines worldwide due to the rebound effect [40], although in the case of Río Tinto this is very sharp. These episodes of high pollution are followed by periods of decreasing concentrations [59]. Consequently, pollutant concentrations in the Río Tinto suffered a significant decrease that seems to be maintained until now (Figure S5). Concerning the reopening of the Río Tinto mine in 2015, from the available data we have not been able to observe a clear influence on the river quality (Figure 9 and Figure S5).

5. Conclusions

The evolution of pollutant concentrations and loads in 2017/18 was controlled by the rainfall distribution. Although the amount of rainfall during this year was close to average values, its distribution throughout the year was irregular: very scarce rainfall was recorded from October to February, while there was an intense rainy period in March. Due to the washing out of efflorescent salts that precipitated during the summer at the surface of the mining zone and along the riverbanks, the maximum concentrations of EC and most elements (Al, Ca, Fe, Cd, Co, Cu, Mg, Mn, Na, Ni, SO_4, Si, and Zn) occurred just after the first autumn rainfalls. On the other hand, the maximum concentration of As was recorded in December, which seems to be linked to the arrival of acidic waters with lower Fe precipitation at the sampling point. This behavior is due to the strong tendency of As to coprecipitate/be adsorbed by Fe oxyhydroxysulfate minerals. Chromium may also be affected by this process, although to a lesser degree than As. The maximum concentrations of Ba and Pb occurred during the March and April floods, linked to the low sulfate concentrations, which seems to control the solubility of these elements by the precipitation of barite/anglesite or other minerals. On the contrary, the high floods produced a second washout process (masked by strong runoff dilution processes) of less soluble sulfates and/or soluble salts precipitated within the waste dumps or the mining voids. Consequently, the minimum concentrations for most elements were reached in late March or April.

The detailed sampling carried out in 2017/18 allowed us to obtain a reliable estimate of the load of dissolved pollutants transported by the Río Tinto: 5000 tons of Fe, 2600 of Al, 683 of Zn, and so on. Approximately 30% of the annual amount of Al, Cd, Co, Cu, Ni, and Zn was transported between October and February due to the washout of evaporitic soluble sulfate salts, although the water contribution was only 4%. These values were even higher for Fe (43%), Cr (49%), and, above all, As (78%). Except for this latter element, the highest loads occurred during the intense floods of March. This is especially relevant for Pb (86% of the annual load) due to its concentration increasing strongly during floods, unlike most elements. In April, although the streamflow was high, the pollutant loads were lower due to the total washout of secondary soluble salts. On the other hand, the particulate phase may play an important role in the river metal transport of Cr, Fe, Pb, and, above all, As.

Regarding the long-term evolution, the cessation of mining activities in 2000 produced a worsening of conditions and extreme concentrations were reached (up to 5 g/ L of Fe, 50 mg/L of As, and so on). After that, there was a slight improvement in water quality in terms of pollutant loads. However, the AMD generation in the mining area is expected to continue for many hundreds of years [59], and the Río Tinto will continue to transport very high amounts of toxic metals to the Huelva estuary.

Supplementary Materials: The following are available online at http://www.mdpi.com/2075-163X/10/7/598/s1, Figure S1: (A) Evolution of rainfall and streamflow during the hydrological year 2017/18. (B) Detailed evolution of daily discharge increase from 4–8 June (corresponding to the inner square in (A)), Figure S2: Evolution of dissolved concentrations in 2017/18 for all studied elements, Figure S3: Evolution of saturation indices for some secondary minerals precipitating from AMD in 2017/18, Figure S4: Evolution of the Cu/Zn ratio in 2017/18, Figure S5: Evolution of dissolved concentrations of As, Cd, Cu, Fe, Mn, Pb, SO_4, and Zn from 1980 to 2019 (a potential trend line for each element is shown).

Author Contributions: Conceptualization, M.O., J.M.N., and C.R.C.; methodology, C.R.C., F.M., and M.D.B.; writing—original draft preparation, M.O. and C.R.C.; writing—review and editing, J.M.N., F.M., and M.D.B. All authors have read and agreed to the published version of the manuscript.

Funding: This work was supported by the Spanish Ministry of Economy and Competitiveness through the research projects CAPOTE (CGL2017-86050-R) and SCYRE (CGL2016-78783-C2-1-R). M.D. Basallote thanks the Spanish Ministry of Science and Innovation for the postdoctoral fellowship granted under application reference IJC2018-035056-I. F. Macías was funded by the R&D FEDER Andalucía 2014-2020 call through the project RENOVAME (FEDER; UHU-1255729).

Acknowledgments: The authors would like to thank to Eduardo Navarrete (Junta de Andalucía) for providing access to the sampling point and flow data information and the four anonymous reviewers, whose comments and suggestions allowed us to significantly improve the manuscript.

Conflicts of Interest: The authors declare no conflict of interest.

References

1. Sabater, S.; Buchaca, T.; Cambra, J.; Catalan, J.; Guasch, H.; Ivorra, N.; Muñoz, I.; Navarro, E.; Real, M.; Romaní, A. Structure and function of benthic algal communities in an extremely acid river 1. *J. Phycol.* **2003**, *39*, 481–489. [CrossRef]
2. López-Archilla, A.I.; Marín, I.; Amils, R. Microbial community composition and ecology of an acidic aquatic environment: The Tinto River, Spain. *Microb. Ecol.* **2001**, *41*, 20–35. [CrossRef] [PubMed]
3. Zettler, L.A.A.; Gómez, F.; Zettler, E.; Keenan, B.G.; Amils, R.; Sogin, M.L. Eukaryotic diversity in Spain's River of Fire. *Nature* **2002**, *417*, 137. [CrossRef] [PubMed]

4. Gomes, P.; Valente, T.; Geraldo, D.; Ribeiro, C. Photosynthetic pigments in acid mine drainage: Seasonal patterns and associations with stressful abiotic characteristics. *Chemosphere* **2020**, *239*, 124774. [CrossRef] [PubMed]
5. Dold, B. Evolution of acid mine drainage formation in sulphidic mine tailings. *Minerals* **2014**, *4*, 621–641. [CrossRef]
6. Chopard, A.; Marion, P.; Mermillod-Blondin, R.; Plante, B.; Benzaazoua, M. Environmental impact of mine exploitation: An early predictive methodology based on ore mineralogy and contaminant speciation. *Minerals* **2019**, *9*, 397. [CrossRef]
7. Evangelou, V.P.; Zhang, Y. A review: Pyrite oxidation mechanisms and acid mine drainage prevention. *Crit. Rev. Environ. Sci. Technol.* **1995**, *25*, 141–199. [CrossRef]
8. Blowes, D.W.; Jambor, J.L.; Hanton-Fong, C.J.; Lortie, L.; Gould, W.D. Geochemical, mineralogical and microbiological characterization of a sulphide-bearing carbonate-rich gold-mine tailings impoundment, Joutel, Québec. *Appl. Geochem.* **1998**, *13*, 687–705. [CrossRef]
9. Nordstrom, D.K.; Blowes, D.W.; Ptacek, C.J. Hydrogeochemistry and microbiology of mine drainage: An update. *Appl. Geochem.* **2015**, *57*, 3–16. [CrossRef]
10. Elghali, A.; Benzaazoua, M.; Bouzahzah, H.; Bussière, B.; Villarraga-Gómez, H. Determination of the available acid-generating potential of waste rock, part I: Mineralogical approach. *Appl. Geochem.* **2018**, *99*, 31–41. [CrossRef]
11. Jarvis, A.P.; Davis, J.E.; Orme, P.H.; Potter, H.A.; Gandy, C.J. Predicting the benefits of mine water treatment under varying hydrological conditions using a synoptic mass balance approach. *Environ. Sci. Technol.* **2018**, *53*, 702–709. [CrossRef] [PubMed]
12. Dore, E.; Fancello, D.; Rigonat, N.; Medas, D.; Cidu, R.; Da Pelo, S.; Frau, F.; Lattanzi, P.; Marras, P.A.; Meneghini, C. Natural attenuation can lead to environmental resilience in mine environment. *Appl. Geochem.* **2020**, 104597. [CrossRef]
13. Leistel, J.; Marcoux, E.; Thiéblemont, D.; Quesada, C.; Sánchez, A.; Almodóvar, G.; Pascual, E.; Sáez, R. The volcanic-hosted massive sulphide deposits of the Iberian Pyrite Belt Review and preface to the Thematic Issue. *Miner. Depos.* **1997**, *33*, 2–30. [CrossRef]
14. Nocete, F.; Álex, E.; Nieto, J.M.; Sáez, R.; Bayona, M.R. An archaeological approach to regional environmental pollution in the south-western Iberian Peninsula related to Third millennium BC mining and metallurgy. *J. Archaeol. Sci.* **2005**, *32*, 1566–1576. [CrossRef]
15. Olías, M.; Nieto, J.M. Background conditions and mining pollution throughout history in the Río Tinto (SW Spain). *Environments* **2015**, *2*, 295–316. [CrossRef]
16. Olías, M.; Cánovas, C.R.; Nieto, J.; Sarmiento, A.M. Evaluation of the dissolved contaminant load transported by the Tinto and Odiel rivers (South West Spain). *Appl. Geochem.* **2006**, *21*, 1733–1749. [CrossRef]
17. Cánovas, C.; Olías, M.; Nieto, J.; Sarmiento, A.; Cerón, J. Hydrogeochemical characteristics of the Tinto and Odiel Rivers (SW Spain). Factors controlling metal contents. *Sci. Total Environ.* **2007**, *373*, 363–382. [CrossRef]
18. Cánovas, C.R.; Basallote, M.D.; Borrego, P.; Millán-Becerro, R.; Pérez-López, R. Metal partitioning and speciation in a mining-impacted estuary by traditional and passive sampling methods. *Sci. Total Environ.* **2020**, 137905. [CrossRef]
19. López-González, N.; Borrego, J.; Morales, J.; Carro, B.; Lozano-Soria, O. Metal fractionation in oxic sediments of an estuary affected by acid mine drainage (south-western Spain). *Estuar. Coast. Shelf Sci.* **2006**, *68*, 297–304. [CrossRef]
20. Usero, J.; Izquierdo, C.; Morillo, J.; Gracia, I. Heavy metals in fish (Solea vulgaris, Anguilla anguilla and Liza aurata) from salt marshes on the southern Atlantic coast of Spain. *Environ. Int.* **2004**, *29*, 949–956. [CrossRef]
21. Vicente-Martorell, J.J.; Galindo-Riaño, M.D.; García-Vargas, M.; Granado-Castro, M.D. Bioavailability of heavy metals monitoring water, sediments and fish species from a polluted estuary. *J. Hazard. Mater.* **2009**, *162*, 823–836. [CrossRef] [PubMed]
22. Van Geen, A.; Adkins, J.; Boyle, E.; Nelson, C.; Palanques, A. A 120-yr record of widespread contamination from mining of the Iberian pyrite belt. *Geology* **1997**, *25*, 291–294. [CrossRef]
23. Baconnais, I.; Rouxel, O.; Dulaquais, G.; Boye, M. Determination of the copper isotope composition of seawater revisited: A case study from the Mediterranean Sea. *Chem. Geol.* **2019**, *511*, 465–480. [CrossRef]
24. Koehnken, L. Mount Lyell remediation research and demonstration program-final report. *Superv. Sci. Rep.* **1997**, *126*, 104.

25. Fergusson, L. A 12-month field trial to remediate an exposed "Tailings Beach" in Tasmania. *Resour. Environ.* **2014**, *4*, 238–245. [CrossRef]
26. Tornos, F. Environment of formation and styles of volcanogenic massive sulfides: The Iberian pyrite belt. *Ore Geol. Rev.* **2006**, *28*, 259–307. [CrossRef]
27. Essalhi, M.; Sizaret, S.; Barbanson, L.; Chen, Y.; Lagroix, F.; Demory, F.; Nieto, J.M.; Saez, R.; Capitan, M.A. A case study of the internal structures of gossans and weathering processes in the Iberian Pyrite Belt using magnetic fabrics and paleomagnetic dating. *Miner. Depos.* **2011**, *46*, 981–999. [CrossRef]
28. Amils, R. Lessons learned from thirty years of geomicrobiological studies of Río Tinto. *Res. Microbiol.* **2016**, *167*, 539–545. [CrossRef]
29. Sánchez España, J.; Pamo, E.L.; Santofimia, E.; Aduvire, O.; Reyes, J.; Barettino, D. Acid mine drainage in the Iberian Pyrite Belt (Odiel river watershed, Huelva, SW Spain): Geochemistry, mineralogy and environmental implications. *Appl. Geochem.* **2005**, *20*, 1320–1356. [CrossRef]
30. Sarmiento, A.M.; Nieto, J.M.; Olías, M.; Cánovas, C.R. Hydrochemical characteristics and seasonal influence on the pollution by acid mine drainage in the Odiel river Basin (SW Spain). *Appl. Geochem.* **2009**, *24*, 697–714. [CrossRef]
31. Olías, M.; Nieto, J.M.; Sarmiento, A.M.; Cánovas, C.R.; Galván, L. Water quality in the future Alcolea reservoir (Odiel River, SW Spain): A clear example of the inappropriate management of water resources in Spain. *Water Resour. Manag.* **2011**, *25*, 201–215. [CrossRef]
32. Cánovas, C.R.; Olías, M.; Nieto, J.M. Metal (loid) attenuation processes in an extremely acidic river: The Rio Tinto (SW Spain). *Water Air Soil Pollut.* **2014**, *225*, 1795. [CrossRef]
33. Ferris, F.; Hallbeck, L.; Kennedy, C.; Pedersen, K. Geochemistry of acidic Rio Tinto headwaters and role of bacteria in solid phase metal partitioning. *Chem. Geol.* **2004**, *212*, 291–300. [CrossRef]
34. Cánovas, C.; Hubbard, C.; Olías, M.; Nieto, J.; Black, S.; Coleman, M.L. Hydrochemical variations and contaminant load in the Río Tinto (Spain) during flood events. *J. Hydrol.* **2008**, *350*, 25–40. [CrossRef]
35. Nordstrom, D.; Wilde, F. Reduction–oxidation potential (electrode method). In *National Field Manual for the Collection of Water Quality Data*; Book 9, Chapter 6.5; US Geological Survey Techniques of Water-Resources Investigations; US Geological Survey: Reston, VA, USA, 1998; 20p.
36. Walling, D.; Webb, B. Estimating the discharge of contaminants to coastal waters by rivers: Some cautionary comments. *Mar. Pollut. Bull.* **1985**, *16*, 488–492. [CrossRef]
37. Quilbé, R.; Rousseau, A.N.; Duchemin, M.; Poulin, A.; Gangbazo, G.; Villeneuve, J.-P. Selecting a calculation method to estimate sediment and nutrient loads in streams: Application to the Beaurivage River (Québec, Canada). *J. Hydrol.* **2006**, *326*, 295–310. [CrossRef]
38. Parkhurst, D.L.; Appelo, C. *Description of Input and Examples for PHREEQC Version 3: A Computer Program for Speciation, Batch-Reaction, One-Dimensional Transport, and Inverse Geochemical Calculations*; US Geological Survey: Reston, VA, USA, 2013; pp. 2328–7055.
39. Allison, J.D.; Brown, D.S.; Novo-Gradac, K.J. *MINTEQA2/PRODEFA2, A Geochemical Assessment Model for Environmental Systems: Version 3.0 User's Manual*; Environmental Research Laboratory Office of Research and Development; U.S. Environmental Protection Agency: Athens, GA, USA, 1991.
40. Bigham, J.; Schwertmann, U.; Traina, S.; Winland, R.; Wolf, M. Schwertmannite and the chemical modeling of iron in acid sulfate waters. *Geochim. Cosmochim. Acta* **1996**, *60*, 2111–2121. [CrossRef]
41. Yu, J.-Y.; Heo, B.; Choi, I.-K.; Cho, J.-P.; Chang, H.-W. Apparent solubilities of schwertmannite and ferrihydrite in natural stream waters polluted by mine drainage. *Geochim. Cosmochim. Acta* **1999**, *63*, 3407–3416. [CrossRef]
42. Elghali, A.; Benzaazoua, M.; Bussière, B.; Genty, T. Spatial mapping of acidity and geochemical properties of oxidized tailings within the former Eagle/Telbel mine site. *Minerals* **2019**, *9*, 180. [CrossRef]
43. Dmitrijeva, M.; Cook, N.J.; Ehrig, K.; Ciobanu, C.L.; Metcalfe, A.V.; Kamenetsky, M.; Kamenetsky, V.S.; Gilbert, S. Multivariate statistical analysis of trace elements in pyrite: Prediction, bias and artefacts in defining mineral signatures. *Minerals* **2020**, *10*, 61. [CrossRef]
44. Cánovas, C.; Olías, M.; Nieto, J.; Galván, L. Wash-out processes of evaporitic sulfate salts in the Tinto river: Hydrogeochemical evolution and environmental impact. *Appl. Geochem.* **2010**, *25*, 288–301. [CrossRef]
45. Valente, T.M.; Gomes, C.L. Occurrence, properties and pollution potential of environmental minerals in acid mine drainage. *Sci. Total Environ.* **2009**, *407*, 1135–1152. [CrossRef]

46. Carbone, C.; Dinelli, E.; Marescotti, P.; Gasparotto, G.; Lucchetti, G. The role of AMD secondary minerals in controlling environmental pollution: Indications from bulk leaching tests. *J. Geochem. Explor.* **2013**, *132*, 188–200. [CrossRef]
47. Casiot, C.; Morin, G.; Juillot, F.; Bruneel, O.; Personné, J.C.; Leblanc, M.; Duquesne, K.; Bonnefoy, V.; Elbaz-Poulichet, F. Bacterial immobilization and oxidation of arsenic in acid mine drainage (Carnoulès creek, France). *Water Res.* **2003**, *37*, 2929–2936. [CrossRef]
48. Fukushi, K.; Sasaki, M.; Sato, T.; Yanase, N.; Amano, H.; Ikeda, H. A natural attenuation of arsenic in drainage from an abandoned mine dump. *Appl. Geochem.* **2003**, *18*, 1267–1278. [CrossRef]
49. Olías, M.; Riera, J.; Galván, L.; Cánovas, C.R.; Pérez-López, R.; Macías, F.; Sarmiento, A.M.; Cruz, P.; Carrero, S.; Bonnail, E.; et al. Obtención de los parámetros hidrodinámicos en un tramo del río Odiel (Huelva) mediante un ensayo de trazadores. In Proceedings of the IX Simposio sobre el Agua en Andalucía, Málaga, Spain, 4–6 November 2015; pp. 1123–1133.
50. Hubbard, C.G.; Black, S.; Coleman, M.L. Aqueous geochemistry and oxygen isotope compositions of acid mine drainage from the Río Tinto, SW Spain, highlight inconsistencies in current models. *Chem. Geol.* **2009**, *265*, 321–334. [CrossRef]
51. Byrne, P.; Reid, I.; Wood, P.J. Stormflow hydrochemistry of a river draining an abandoned metal mine: The Afon Twymyn, central Wales. *Environ. Monit. Assess.* **2013**, *185*, 2817–2832. [CrossRef]
52. Alpers, C.N.; Nordstrom, D.K.; Thompson, J.M. Seasonal variations of Zn/Cu ratios in acid mine water from Iron Mountain, California. In *Environmental Geochemistry of Sulfide Oxidation*; Alpers, C.N., Blowes, D.W., Eds.; American Chemical Society: New York, NY, USA, 1994; pp. 324–344.
53. Runkel, R.L.; Kimball, B.A.; Nimick, D.A.; Walton-Day, K. Effects of flow regime on metal concentrations and the attainment of water quality standards in a remediated stream reach, Butte, Montana. *Environ. Sci. Technol.* **2016**, *50*, 12641–12649. [CrossRef]
54. Accornero, M.; Marini, L.; Ottonello, G.; Zuccolini, M.V. The fate of major constituents and chromium and other trace elements when acid waters from the derelict Libiola mine (Italy) are mixed with stream waters. *Appl. Geochem.* **2005**, *20*, 1368–1390. [CrossRef]
55. Regenspurg, S.; Peiffer, S. Arsenate and chromate incorporation in schwertmannite. *Appl. Geochem.* **2005**, *20*, 1226–1239. [CrossRef]
56. Choppala, G.; Burton, E.D. Chromium (III) substitution inhibits the Fe (II)-accelerated transformation of schwertmannite. *PLoS ONE* **2018**, *13*, e0208355. [CrossRef] [PubMed]
57. Abramov, S.M.; Tejada, J.; Grimm, L.; Schädler, F.; Bulaev, A.; Tomaszewski, E.J.; Byrne, J.M.; Straub, D.; Thorwarth, H.; Amils, R. Role of biogenic Fe (III) minerals as a sink and carrier of heavy metals in the Rio Tinto, Spain. *Sci. Total Environ.* **2020**, *718*, 137294. [CrossRef] [PubMed]
58. Cánovas, C.R.; Olias, M.; Vazquez-Suñé, E.; Ayora, C.; Nieto, J.M. Influence of releases from a fresh water reservoir on the hydrochemistry of the Tinto River (SW Spain). *Sci. Total Environ.* **2012**, *416*, 418–428. [CrossRef] [PubMed]
59. Younger, P.L. The longevity of minewater pollution: A basis for decision-making. *Sci. Total Environ.* **1997**, *194*, 457–466. [CrossRef]

minerals

MDPI

Article

UAS-Based Hyperspectral Environmental Monitoring of Acid Mine Drainage Affected Waters

Hernan Flores [1,*], Sandra Lorenz [1], Robert Jackisch [1], Laura Tusa [1,2], I. Cecilia Contreras [1,2], Robert Zimmermann [1,2] and Richard Gloaguen [1]

[1] Helmholtz-Zentrum Dresden-Rossendorf, Helmholtz Institute Freiberg for Resource Technology, Chemnitzer Str. 40, 09599 Freiberg, Germany; s.lorenz@hzdr.de (S.L.); r.jackisch@hzdr.de (R.J.); l.tusa@hzdr.de (L.T.); i.contreras@hzdr.de (I.C.C.); r.zimmermann@hzdr.de (R.Z.); r.gloaguen@hzdr.de (R.G.)

[2] TheiaX, HZDR Innovation GmbH, Bautzner Landstraße 400, 01328 Dresden, Germany

* Correspondence: h.flores@hzdr.de; Tel.: +49-351-260-4473

Abstract: The exposure of metal sulfides to air or water, either produced naturally or due to mining activities, can result in environmentally damaging acid mine drainage (AMD). This needs to be accurately monitored and remediated. In this study, we apply high-resolution unmanned aerial system (UAS)-based hyperspectral mapping tools to provide a useful, fast, and non-invasive method for the monitoring aspect. Specifically, we propose a machine learning framework to integrate visible to near-infrared (VNIR) hyperspectral data with physicochemical field data from water and sediments, together with laboratory analyses to precisely map the extent of acid mine drainage in the Tintillo River (Spain). This river collects the drainage from the western part of the Rio Tinto massive sulfide deposit and discharges large quantities of acidic water with significant amounts of dissolved metals (Fe, Al, Cu, Zn, amongst others) into the Odiel River. At the confluence of these rivers, different geochemical and mineralogical processes occur due to the interaction of very acidic water (pH 2.5–3.0) with neutral water (pH 7.0–8.0). This complexity makes the area an ideal test site for the application of hyperspectral mapping to characterize both rivers and better evaluate contaminated water bodies with remote sensing imagery. Our approach makes use of a supervised random forest (RF) regression for the extended mapping of water properties, using the samples collected in the field as ground-truth and training data. The resulting maps successfully estimate the concentration of dissolved metals and related physicochemical properties in water, and trace associated iron species (e.g., jarosite, goethite) within sediments. These results highlight the capabilities of UAS-based hyperspectral data to monitor water bodies in mining environments, by mapping their hydrogeochemical properties, using few field samples. Hence, we have demonstrated that our workflow allows the rapid discrimination and mapping of AMD contamination in water, providing an essential basis for monitoring and subsequent remediation.

Keywords: hyperspectral; remote sensing; machine learning; unmanned aerial system; acid mine drainage; random forest regression; post-mining

Citation: Flores, H.; Lorenz, S.; Jackisch, R.; Tusa, L.; Contreras, I.C.; Zimmermann, R.; Gloaguen, R. UAS-Based Hyperspectral Environmental Monitoring of Acid Mine Drainage Affected Waters. *Minerals* **2021**, *11*, 182. https://doi.org/10.3390/min11020182

Academic Editor: Teresa Valente

Received: 27 November 2020

Accepted: 4 February 2021

Published: 9 February 2021

1. Introduction

Acid mine drainage (AMD) is an environmental phenomenon that can occur either by the natural exposure of sulfide minerals to weathering conditions or as a consequence of certain mining activities. Lottermoser [1] defines AMD as a process whereby low pH mine water is formed from the oxidation of sulfide minerals. These acidic and metal-enriched waters can negatively affect the natural ecosystem's quality and aquatic life. Mainly impacted areas are rivers, lakes, estuaries, and coastal waters. AMD's advancement can take years or decades and can spatially continue for centuries [1]. Therefore, such an environmental problem needs to be carefully monitored and ideally, remediated. Several efforts have been applied in order to monitor the spatial distribution and intensity of contamination by AMD, commonly involving systematic sampling and laboratory analysis of stream sediment followed by interpolation of the results in assembled distribution maps [2,3], however, such approaches can be time-consuming, costly, and with limited spatial coverage.

Previous studies have shown the benefits of remote sensing for many environmental monitoring purposes. In relation to AMD detection and monitoring, multi and hyperspectral sensors has been widely used, due to the distinctive spectral absorption features of iron-minerals present in the visible to shortwave infrared region of the electromagnetic spectrum [4]. These studies have covered a wide range of spatial dimensions (scales) depending on the platform used for data acquisition: including satellite studies [4] for water reservoirs protection [5] and indirect pH estimations by mapping iron-bearing minerals precipitated on the stream bed [6] to airborne surveys over mine tailings [7–10]. Laboratory-scale spectral studies over AMD minerals have been performed for reference spectral libraries [11] as well as the study of spectral signatures of surface waters [12]. The emergent use of unmanned aerial systems (UAS), like multicopters coupled with lightweight hyperspectral sensors has become a tool to collect data at a higher spatial resolution than most of aircraft and satellite counterparts, resulting in greater precision (higher spatial resolution of a scene enabling the investigation of down to a few centimeters pixel size) [13]. Most recently [14] implemented is the use of UAS-hyperspectral imaging for high-resolution, multi-temporal mapping of proxy minerals for AMD in the Sokolov lignite region, Czech Republic.

Our study focuses on mapping hydrogeochemical properties to assess the extent of AMD in waters. In this paper, we integrate UAS-borne visible to near-infrared (VNIR) hyperspectral data with physicochemical parameters of water (field and laboratory measurements) to map acidity, redox potential, and metal concentrations in surface water. We propose a novel machine learning approach to map gradual changes occurring in the water properties based on their spectral signature. A multivariate regression problem is solved by a random forest (RF) algorithm which uses constrained training data to predict the properties for the entire surface water area. We use water quality samples and field spectroscopy from ground-control points to fully assess the accuracy of the method, by deriving from analytical studies and in-situ measurements the training data for the regression model. The obtained hydrogeochemical maps can be used to monitor water bodies surrounding mining ecosystems, by targeting sources and/or acidic contamination in water, promoting its continuous supervision or assisting in the selection of the most adequate remediation treatment.

1.1. Test Site

To test the approach, the confluence between the Odiel River and the Tintillo River in the Huelva Province (Iberian, Pyrite Belt, Southern Spain) (Figure 1a–c) was selected with the aim of monitoring different water behaviors in the analysis. The Tintillo River emerges in the mining area of the Rio Tinto mines and flows downstream for 10 km before being discharged into the Odiel River. After this union, the Odiel river is affected with large amounts of acidity and dissolved metals transferred by the Tintillo River. At the confluence of these rivers (Figure 1e), different geochemical and mineralogical processes occur due to the interaction of very acidic water (pH 2.5–3.0) with neutral water (pH 7.0–8.0). The hyperspectral survey covers the three branches of the confluence as detailed in Figure 1d.

1.2. Geological Framework

The Iberian Pyrite Belt (IPB) is located in a north-vergent fold and thrust belt of late Variscan age. Its extension includes Setubal (Portugal) to the north of Seville (Spain) [15]. The typical geological facies includes phyllites/quartzites, followed by slates, basalt sills, felsic volcanic (rhyolites and dacites), and Culm series (greywackes and slates) [16] with an absolute lack of carbonate or alkaline materials [17]. The IBP belongs to the South Portuguese Zone of the Hercynian Iberian Massif and according to [18] is formed by upper Palaeozoic materials that can be divided into three lithological groups: (i) the phyllite–quartzite group (PQ), formed by a thick sequence of shales and sandstones, (ii) the volcano-sedimentary complex (VSC), including a mafic–felsic volcanic sequence interstratified with shales, and (iii) the culm group where shales, sandstones, and conglomerates prevail.

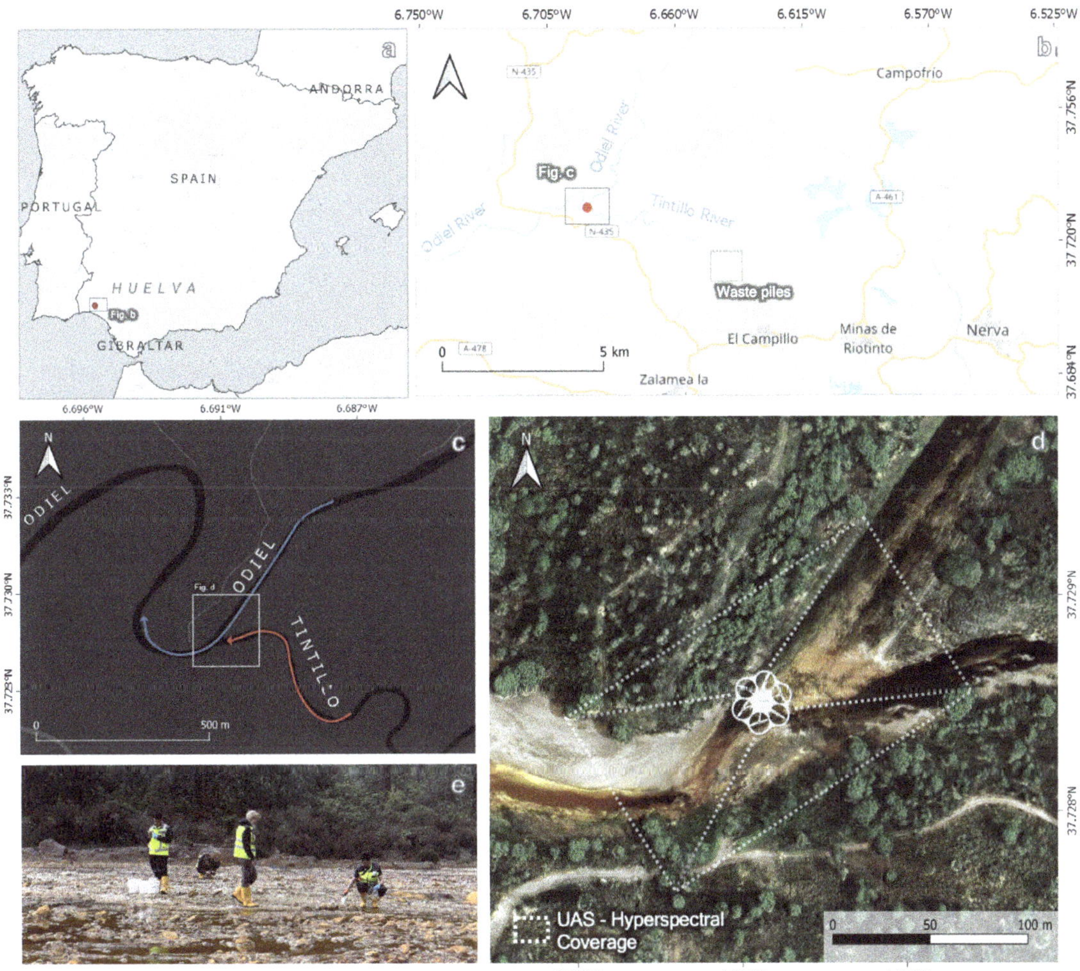

Figure 1. (**a**–**c**) Location of the investigated area in the Odiel River (Huelva, Spain). (**d**) Orthophoto of the Odiel-Tintillo confluence showing the unmanned aerial system (UAS)-borne hyperspectral scanned scene. (**e**) Panorama of the field sampling campaign and members of the HZDR-HIF—Exploration Department in the test site.

The southern part of the Odiel river flows across Neogene marly sediments of the Guadalquivir depression. In the northern part of the basin, plutonic and metamorphic rocks from the Ossa–Morena Zone also outcrop [19]. The strata bound, volcanogenic massive sulfide lenses are hosted in felsic volcanic of Upper Devonian to Lower Carboniferous ages [15]. Zones of chloritic and argillitic alteration are correlated with those lenses. Stockwork zones occur underneath the lenses in the vicinity of faults. A gossan usually forms in the cap-rock above [18].

The IPB region has been mined for copper and smaller amounts of manganese, iron, and gold since the Bronze Age [20]. Related to VSC are many massive polymetallic sulfide deposits (>80). Pyrite (FeS_2) is the most abundant mineral in these deposits, with sphalerite (ZnS), galena (PbS), chalcopyrite ($CuFeS_2$), arsenopyrite ($FeAsS$) in minor quantities and other sulfides combined with Cd, Sn, Ag, Au, Co, and Hg [19]. Therefore, pyrite is abundant both on mine waste and rock surfaces, and AMD is present in most streams throughout the region [19]. The natural geological conditions, the long mining history of the IPB added to the lack of natural alkaline materials makes the capacity to neutralize the acidity of the acidic streams very limited [21].

1.3. Hydrology and Climatology

The hydrological characteristics of the Huelva region are typical of a semi-arid climate [17]. The precipitation is characterized by intense, short rain events in autumn, not so intense dry winters, and dry springs and summers. The largest drainage basin from Huelva is the Odiel River Basin, which covers an area of about 2300 km^2. The Odiel River begins in the Sierra de Aracena and, as well as the Tinto river, flows down to the Ría of Huelva estuary [19].

The length of the Odiel river is about 140 km (from the headwaters to the mouth) with a 600 m drop along its way. Most of the annual precipitation (812 mm) occurs in the winter months (between October and January) [19]. These climate variations affect the hydrological behavior of the Odiel River throughout the year. The geochemistry is influenced by inter-annual water flows dependent on fluctuations in rainfall and evapotranspiration [17].

The Tintillo River is located in the north of the Huelva province, and drains an area of about 57 km^2. The length of this river is about 10 km and is mainly formed by acid sulfate waters coming from the base of a large, sulfide-bearing waste-rock pile ($\sim$1 km long, $\sim$40 m high) situated in the surrounding area. This river feeds the Odiel River with acidity and dissolved metals. In this confluence, several geochemical and mineralogical processes typical of acid water mixed with neutral water occur [21]. These acid waters affect the chemical balance of the Odiel and Tinto fluvial systems, with the transfer of large amounts of acids and dissolved metals (Fe, Al, Mn, Cu, Zn, Cd, Pb), As, and SO_4^{2-} [20]. Even though the Odiel river receives small discharges of acid mine water emanating from several abandoned mines of the IPB (i.e., Concepcion, San Platon, Esperanza, and La Poderosa-El Soldado) [22], the vast amount of acidity and metals of Tintillo consumes all the alkalinity and makes Odiel waters remain hyperacidic downstream until it reaches the Atlantic Ocean (70 km downstream) on the coast of Huelva [21,22].

Sánchez [17] report a main annual flood in autumn, which carries AMD from solubilized iron sulfide into the river. The sediments decrease in water content and are continuously oxidized (Figure 2a) during winter, spring, and summer. The metals produced by the mine waste stay as particulate matter along the river watershed. The dissolved metals increase during dry seasons and decrease during more humid-wet seasons [19]. There is a considerable difference in the pH ranges of the water along the Odiel River that can be used to monitor its quality [23].

During summer conditions, the low precipitation rate causes a decrease in the pH values and an increase of AMD contamination in areas closer to the mine waste heaps. The intense dark red color of acidic waters is related to their geochemistry, depending on the the precipitation of dissolved iron and aluminum in the mine water, mainly controlled by pH value [17]. The colloidal Fe^{3+} is then responsible for the red color of very acid waters (see Figure 2b), whereas Fe^{2+} gives a more greenish color, and white color when aluminum precipitates [23]. Transitions between the three hydro-geochemical facies (see Figure 2c) can happen within very small distances and are mapped in this study.

Several efforts have been made to reduce the environmental impact of acid mine drainage in the region, including geotechnical stabilization and revegetation of waste piles, construction of rainwater drainage systems, and sealing of mine adits, as well as passive treatments such as Anoxic Limestone Drainage and anaerobic compost wetlands. Nevertheless, these tests have not been enough due to chemical and climatic effects such as high acidity and metal contents with seasonal variability of water discharge [17].

Figure 2. Three pictures of the study area, (**a**) detail of the sediments and the characteristic color of forming layers; (**b**) detail of the distinctive water color during the sampling campaign (17 April 2019); (**c**) detail of the different water color bands, formed by the mixing of waters at the confluence.

2. Data Acquisition

2.1. UAS-Borne Hyperspectral Data

The hyperspectral data used in this study were acquired with the frame-based Senop Oy 'Rikola' Hyperspectral (HS) Camera (Figure 3a). The size and weight of this HS camera have allowed it to be gimbal-mounted on a customized multi-copter (Tholeg THO-R-PX8) (Figure 3b). This setup provides a stable platform at the time of image acquisition and allows for a sufficient integration time for the spectral imaging sensor.

Figure 3. Overview of the unmanned aerial system (UAS) used for image acquisition: (**a**) The Rikola hyperspectral (HS) camera [24] mounted on the (**b**) Tholeg UAS.

The Rikola sensor provides frame-based images in the VNIR spectral range between 504 and 900 nm. The specific parameters used for this survey are detailed in Table 1. Raw data are acquired as a function of wavelength and digital number which is subsequently converted to radiance using the Rikola hyperspectral imager software [24].

Table 1. Sensor specifications of the hyperspectral camera for the UAS survey.

Parameter	Value
Image Resolution	1010 × 648 Pixel
Spectral Bands	50
Field of View (FOV)	36.5°
Focal length	∼9 mm
Spectral resolution	∼10 nm, FWHM
Spectral sampling	∼8 nm
Weight	720 g

FWHM (full width at half maximum).

2.1.1. Flight Set-Up

All the hyperspectral images were acquired within 2 flights, the first one in a parallel line to Odiel River (before and after the confluence) and a second one for scanning the Tintillo River. All acquired images are assumed nadir position.

The flight parameters were defined in the field (see Table 2) and a pre-defined flight-path was uploaded to the UAS. Before the flight, the calibration panels (black, gray, and white; PVC) with known spectral characteristics were placed underneath the flight line for the subsequent conversion from radiance to reflectance during pre-processing stage.

Table 2. Acquisition parameters for flight operations on the scanned area.

Parameter	Value
Flight altitude (above take-off)	50 m
Ground resolution	3.25 cm/px
Number of Flights	2
Flight duration (total)	12 min
Number of scenes	140
Area covered	14,000 m^2

2.1.2. Pre-Processing

The hyperspectral data were pre-processed using the python-based MEPHySTo toolbox [25]. The toolbox performs a set of steps over each single image in order to co-register, correct the geometric effects (lens- and topographic correction), and fit each HS snapshot using the high-precision geo-referenced orthophoto generated with a structure-from-motion multi-view-stereo (SfM-MVS) workflow. Quality control of the pre-processed single-images was performed for removing defective scenes (e.g., distorted or out-of-size images), when keypoints for matching were not found. After this verification, the mosaic is constructed.

As the final step in the pre-processing chain, the hyperspectral radiance mosaic was converted to reflectance. Here, a simple empirical line method was applied to the dataset [26] using the known spectra from ground-reference calibration PVC panels. A gray panel proved successful for the radiometric correction since all the HSI snapshots showed low reflectance. The Rikola camera has two detectors for covering the wavelength range. This provokes a spectral-shift (a flat section) in the spectral data, thus bands in the spectral range 624–671 nm have been removed, and in consequence not considered in the processing chain of this research. The resultant reflectrance data were smoothed using the Savitzky–Golay Filter, with a window size of five and a second degree polynomial in order to remove noise [27].

2.2. Ground-Truth Data

A total of 15 control points were established within the scanned area for reference. The distribution of these points was designed to be equidistant (Figure 4a), as long as the conditions of personal security allowed it. Field measurements (Figure 4b–e) and analytical techniques were the basis for the validation and accomplishment of the delivered maps of this research.

Figure 4. Field sampling strategy: (**a**) geographical location of the sampled spots, (**b**) water sampling, (**c**) sediment sampling and in-situ measurement of (**d**) pH, temperature, redox, and (**e**) spectroscopy on the river water.

2.2.1. Field Measurements and Sampling

Two sets of water samples were taken for chemical analyses. The first one for measuring major ions (Ca^{2+}, Mg^{+}, K^{+}, SO_4^{2-}, Cl^{-}, Fe, amongst others) were taken with 60 mL-syringes and immediately filtered with 0.45 µm cellulose acetate membrane filters, stored in 50 mL-polyethylene bottles and acidified down to pH < 2 with concentrated HNO_3. The second batch of samples for ionic measurements, were taken directly in 50 mL-polyethylene bottles (flushed three times before sampling) and fully filled. All samples were stored in covered containers in the dark and were refrigerated during transport until analysis in the laboratory. Sediment and samples of 200 to 500 g were obtained using a PVC shovel and kept in a transparent polyethylene bag. The PVC shovel was washed with distilled water between each sampling point. Samples and measurements were collected at a spacing of approximately 15 m, in the center of the channel, as long as the characteristics of the site allowed it.

The on-site measurements, pH, and temperature (T) were measured with a VWR pH110 meter which also incorporates a temperature sensor and the redox potential with a Redox-

One-Bar Measuring Chain (Ahlborn FY96RXEK) with Pt–Ag/AgCl electrode system. The mV measurement was corrected for temperature and adjusted to a potential relative to that of the standard hydrogen electrode (see Method [28]). The pH meter was calibrated using Hanna standard solutions (pH 4.01 and pH 7.01) and the redox meter was checked using Hanna standard solutions (240 mV and 470 mV). The probes were placed directly in the river water (approximately 10 cm deep) with gentle agitation, and the measurement was made once the reading was stabilized (Method 2550: [29]). Back in the laboratory, the Electrical Conductivity (EC) was measured for each sample with a WTW 3320 Multimeter prior to elemental concentration analyses.

For the spectral point-measurements on the water surface, a Spectral Evolution PSR-3500 high-resolution full range hand-held spectroradiometer was used. It provides spectral information in the wavelength range between 350 nm and 2500 nm (covering the visible (VIS), the near-infrared (NIR), and the short wave infrared (SWIR) part of the electromagnetic spectrum) with a spectral resolution of 3 nm in VNIR and 8 nm in the SWIR. A total of three flexible PVC panels in a white shade with known spectra (reflectance >97.0 $\pm$ 0.3%) were submerged in the river water, spatially distributed to cover the three parts of the confluence. These are monochrome synthetic 50 $\times$ 50 cm panels which are spectrally homogeneous and have nearly Lambertian reflectance. Water depth measurements at the 15 sampled points and at the position of the PCV panels presented a similar depth (10–12 cm). Water depth along the scanned river area was approximately constant.

Calibration of the device was done using a near-Lambertian reflector (PTFE, >99% in the VNIR, >95% in the SWIR). This calibration was done under sun/ambient illumination, meaning that the signal is calibrated on the actual irradiance (thus, atmospheric influences are ignored). The sensor tip was placed as close as possible to the surface of the water, where each panel was submerged, using sun-light illumination (Figure 4d). For each spectral measurement, the integration time was set to 60 ms and 10 individual spectra were consecutively averaged.

2.2.2. Analytical Techniques

Spectral and mineralogical characterization of solid and water samples was performed at the Helmholtz Institute Freiberg in the Spectroscopy and X-ray laboratories, respectively. Water sample analyses were carried out by the Water Laboratory of the Department of Mining and Special Civil Engineering in the Technical University Bergakademie Freiberg.

For SO_4^{2-} and Cl^- water content a spectrophotometer (Hach DR3900) was used [30]. The method determines elemental concentration via absorbance of the light when interacting with the compounds present in the sample solution. Test-stripes were used regularly to correctly dilute the sample for the measurement inside the detection limits of the equipment. Microwave Plasma Atomic Emission Spectroscopy (4200 MPAES from Agilent) was used to quantify the concentration of the dissolved metals (Al, Cd, Co, Cu, Fe, Mg, Mn, Ni, Pb, Zn). After emitting a microwave-plasma that atomizes the created aerosol sample, electrons are excited and the device is able to (1) detect the emitted light at specific wavelengths (spectrum-emission lines) characteristic for each element and (2) measure the intensity of the emitted light which is compared to known emission-concentrations lines of elements in calibration curves [31].

The mineralogical characterization of the solid samples was achieved through powder X-ray diffraction (XRD) using a PANalytical Empyrean diffractometer equipped with a Co-tube, Fe-Filter, and a PIXcel 3Dmedipix area detector. Samples were first ground with an Agate mortar and then split into (3–4 g) representative samples for the next grinding step in a McCrone mill (Retsch) where the required sample grain size (4 µm) is achieved for analysis. Using the backloading technique, the sample was filled in the sample holder for measurement. Qualitative mineral phase identification was based on the ICDD PDF-4+2019 database [32] and the PANalytical HighScore 3.0.4 software. Raw data were analyzed using the package BGMN/Profex v. 3.10.2 [33], using the Rietveld method for obtaining quantitative mineralogical data. X-ray fluorescence (XRF) was used for elemental composition analysis. Additional preparation steps were required for preparing fused bets. After the sample was (4 μm) sized, the sample was further dried overnight at 105 °C before preparing for calcination at 950 °C in ceramic crucibles. The samples in the crucibles were weighted before and after calcination for the loss of ignition calculation. After this step, a fraction of 1 g of the calcinated sample was mixed with 8 g of Lithium Tetraborate (LiT) into a platinum crucible for the preparation of fused bets. Prior to this, the platinum crucible went into the Electrical Fusion Instrument at (1050 °C) for 40 min for the lithium and sample fusion. Furthermore, a batch of spectral measurements on the dried sediments samples was carried out with a Spectral Evolution PSR-3500 spectroradiometer in order to characterize the spectral mineralogy of the samples and to compile an endmember library for validation of the hyperspectral data acquired with the Rikola camera. The spectral measurements were done using the contact probe (8 mm spot size) with artificial illumination, directly on the surface rock sample. Each spectra consists of 10 individual measurements, consecutively taken and averaged. Calibration and conversion to reflectance was carried out using a pre-calibrated PTFE panel (>99% in VNIR, >95% in SWIR). Three spots were measured per sample over a 5 × 5 cm area to account for lithological heterogeneities. Calibration and conversion to radiance was carried out every 20 measurements with the Spectralon panel.

3. Methodological Framework

Adapted processing steps were implemented for the integration of hyperspectral images with the hydrogeochemical and mineralogical data available. The workflow in Figure 5 shows the proposed machine learning-based system in which the first step is splitting the hyperspectral image into the two main surface components to map: the river flow path (water pixels) and the exposed sediments in the river borders (soil pixels). Subsequently, training data are derived out of the chemical analysis results from water samples and mineralogical results from sediment samples.

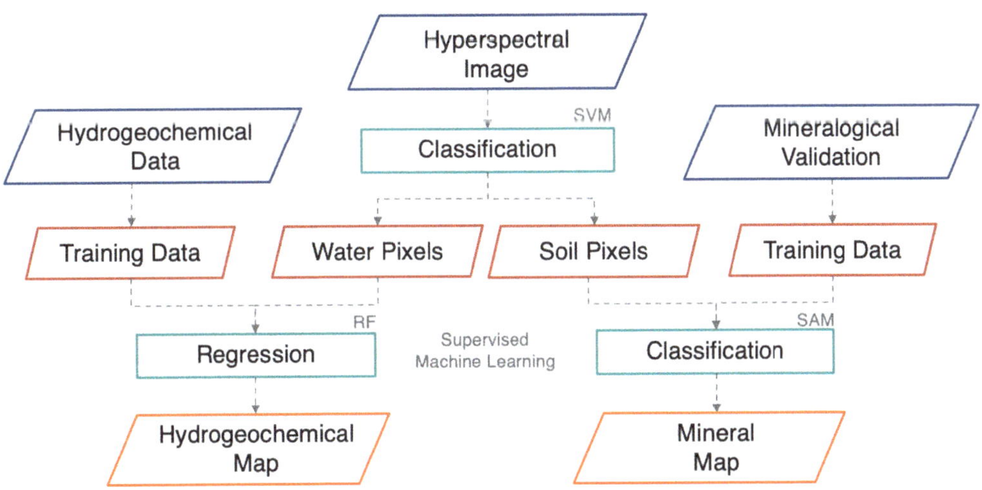

Figure 5. Implemented processing chain over the hyperspectral images for mapping.

After masking the HSI data for each surface component and preparing the training data, a different supervised machine learning method is applied over each mask: A regression model for the pixels belonging to the river flow path to get a continuous output where gradual changes in water properties can be mapped, and a classification model for the pixels belonging to the exposed soil to get a discrete mineral distribution map on the river borders.

3.1. *Surface Classification*

The first step in the proposed approach consists of identifying the main surface components in the hyperspectral image. This is achieved by using a Support Vector Machine (SVM) classification based on features extracted by the novel Orthogonal Total Variation Component Analysis (OTVCA; [34]). OTVCA is an unsupervised feature extraction method which uses a total variation optimization of a non-convex cost function with an orthogonal restriction. The dimensions of the hyperspectral data are reduced and the spatial information is extracted while edges and structures are preserved [35]. This allows the differences in the main surface materials of the study area to be highlighted, discriminating the river flow path from the exposed soil, and also vegetation pixels, which are later removed. The accuracy of the Support Vector Machine algorithm is increased when classifying spectral data after feature extraction [35].

3.1.1. Training Data

The training dataset was collected by manually labeling a small number of pixels within the image based on both spectral image inspections and field observations using the main distinctive classes of land cover (water, vegetation and soil). In order to increase the number of training pixels per class, 75 to 100 extra pixels surrounding each of the georeferenced training pixels have been considered in each of the created region of interest per class.

3.1.2. Support Vector Machine (SVM)

SVM fits a separating hyperplane (defined as the class boundary) that segregates the feature space in two classes with the largest margin for each class [36]. An optimization problem is solved by structural risk minimization for identifying the aforementioned hyperplane. Only the samples that are closest to the class boundaries are required to train the classifier, the so-called support vectors. The classification results of SVM can have high accuracy even when only a small number of training samples are available [35–37]. The first conception of an SVM classifier was a binary linear method for the separation of two classes [38].

For the application of the algorithm over complex classification scenarios, where the boundaries are non-linear, kernel methods have been widely implemented for an extended SVM nonlinear separation method. In this sense, the separation is done after input data are transformed by a kernel function in a high-dimensional feature space, where the samples could be linearly separated. The Kernel function employed in this work was the Gaussian radial basis function (RBF) kernel [36]. For obtaining the optimal hyperplane, the only needed parameters are the gamma γ (spread of the RBF kernel) and the regularization parameter C (value for quantifying the amount of penalty along with the SVM optimization). In terms of simplifying a multiclass classification problem into several binary subproblems, different methods such as the one-against-one (OAO) or the one-against-all (OAA) strategies are commonly used for solving the simplification problem [38]. In this workflow, SVM is implemented using LibSVM by [39] with an RBF kernel fed with the features extracted by the OTVCA for isolating river flow path pixels from sediments and vegetation.

3.2. *Hydrogeochemical Maps*

A regression has been applied to the water masked pixels of the HS image. A classification for parameters that change gradually within an area, could yield maps that only show properties by clusters and lead to over/underestimate values for the rest of pixels, following the supervised-training dataset. In this sense, it is more convenient to perform a regression over the hyperspectral data since the output values will be continuous for each pixel where the property is unknown. A random forest (RF) algorithm is used to solve the multivariate regression problem and predict the values for five different water parameters (pH, redox potential, Fe, SO_4^{2-}, and Al). The approach aims to use data from the ground-validation points to predict the spatial distribution of the water properties over the entire river flow path.

3.2.1. Training Data

The training data for the regression model were selected manually, using the coordinates where the field measurements were performed (Figure 6a). A total of 15 ground-truth points were available for the data integration. Around 400 pixels surrounding each of these points and an additional supporting set of pixels (black colored) were selected as initial training dataset for the prediction model. For this selection, characteristic features apparent in the image spectra and the analysis of validation spectra in regions uniform within the extracted OTVCA features were considered. Subsequently, the physicochemical data (field and laboratory based) associated to each station were assigned to the training dataset, creating a multi-variable matrix for feeding the regression algorithm.

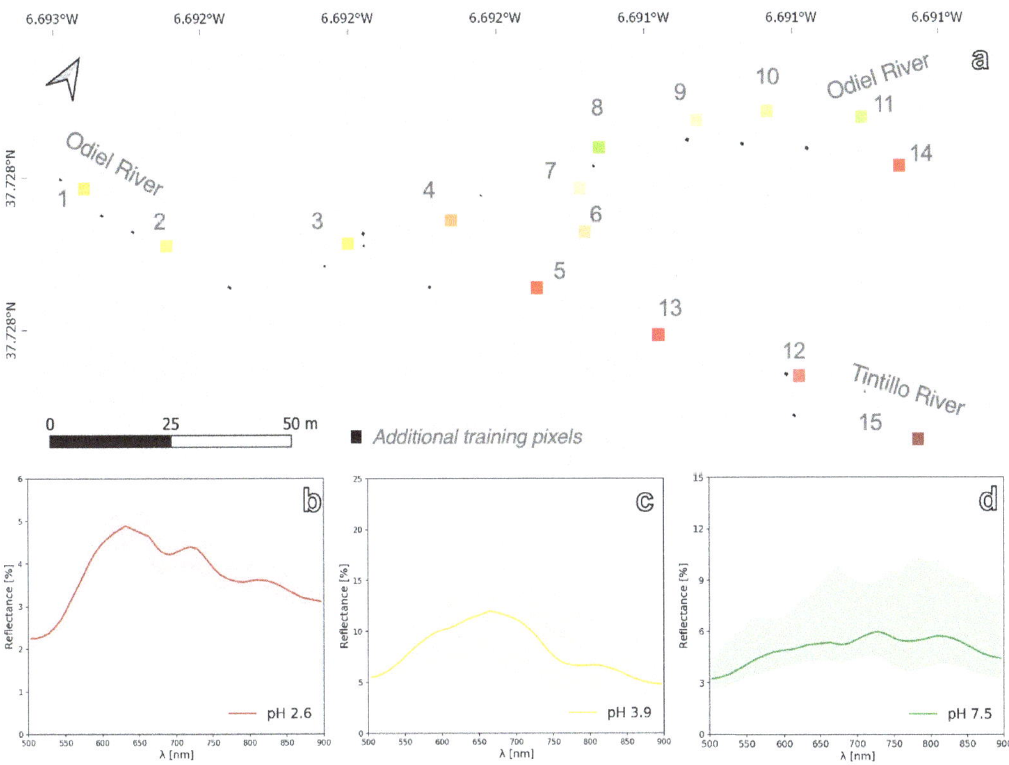

Figure 6. (**a**) Created training dataset based on 15 ground-truth georeferenced stations. Mean spectrum of the clustered pixels used in the training dataset and standard deviation plotted as shaded envelope. (**b**) pH 2.6 ($\sigma = 0.16$), (**c**) pH 3.9 ($\sigma = 2.01$), and (**d**) pH 7.5 ($\sigma = 0.98$).

Examples of spectra from the training spots of the UAS hyperspectral data are shown in Figure 6. Only the spectra of three group of pixels are shown for simplicity, one from each part of the river confluence. Groups of pixels corresponding to more acidic pH (Figure 6b) show a higher absorption feature in the 500–650 nm range, as do the slightly less acidic (Figure 6c). Unlike the acidic water pixels, the spectra related to neutral pH values (Figure 6d) lack this feature and are relatively flat. Overall, reflectance values are below 20%, as expected from the dark-red color of waters and the river bed sediments.

For assessing the performance of the regression algorithm, two validation phases have been implemented. The first one or internal validation consisted of using the same initial training dataset where each class is randomly re-sampled into a test and training dataset (75% training and 25% for test/validation). Additionally, a final validation (referred in this work as the external validation) is performed using a less populated test dataset. This second test dataset is created considering only around 100 pixels neighboring the original-georeferenced pixels where the field measurements were taken. R-squared (R^2) is calculated on both test datasets to assess the performance of the regression map.

3.2.2. Random Forest (RF)

RF is a learning method often used in classification and regression problems, in which a set of decision trees are trained and their individual results are then combined through a voting process [35]. The idea of a decision tree predictive model is to break up a complex decision into a union of several simpler decisions, hoping the final solution obtained this way would resemble the intended desired solution [40]. While the predictions of a single tree are highly sensitive to noise in its training set, the average of many trees is not that sensitive as long as the trees are not correlated [35]. Random forests for regression are created by growing trees depending on a random vector, so the tree predictor takes numerical values rather than class labels [41]. The general technique of bootstrap aggregating is used for training the model. Bootstrap aggregating is used for training data creation by resampling the original dataset in a random fashion with replacement, leading to a more efficient performance of the model [35]. Different from most of the machine learning algorithms, random forest process needs two parameters to be set for generating a prediction model: the number of regression trees and the number of evidential features which are used in each node to make the regression trees grow [42].

3.3. *Mineral Maps*

A supervised classification using Spectral Angle Mapper (SAM) has been applied for mineralogical mapping of the pixels extracted as exposed soil after the surface classification (dry sediments and border crust of the confluence area). The aim of this exercise is to identify the mineralogical changes in the soil related to the varying composition of the proximal water bodies.

3.3.1. Endmember Spectral Library

Reference spectra to be used by the SAM classifier are obtained from laboratory spectral point-measurements over field sediment samples (Figure 7a). The endmember spectra collection are then subjected to wavelength analyses in the available range (Figure 7b) in order to identify the main occurring VNIR active iron-bearing minerals such as hematite, goethite, jarosite, and schwertmannite. Spectral interpretation of the mineralogy is validated with the XRD results and compared to reference mineral spectral libraries (USGS [43] and AMD-minerals [11]). The resultant endmember spectral library is then resampled to the available range of the Rikola HS sensor range (504–900 nm) (Figure 7c).

Two regions are gray-shaded in Figure 7b to analyze the shapes and wavelength positions of each mineral in the charge transfer (ligand to metal charge transfer) transition and those triggered by the crystal field effects (transitions of electrons from lower to higher energy states) [44]. Hematite characteristically has a narrower absorption at wavelengths surrounding 880 nm, while goethite has a broader feature with wavelengths around 920 nm or greater [44]. This feature associated with crystal field absorption around 900 nm is also found in the jarosite and schwertmannite spectral curves. However, the spectra shoulders around 650 nm, associated with the electronic charge transfer of Fe^{3+} and Fe^{2+} [45], allow further distinction for schwertmannite which has no known inflection point at 650 nm and spectral peak located at 738 nm [11]. The peak location at 720 nm and a small distinctive absorption feature at 2264 nm confirms spectral identification for jarosite [11].

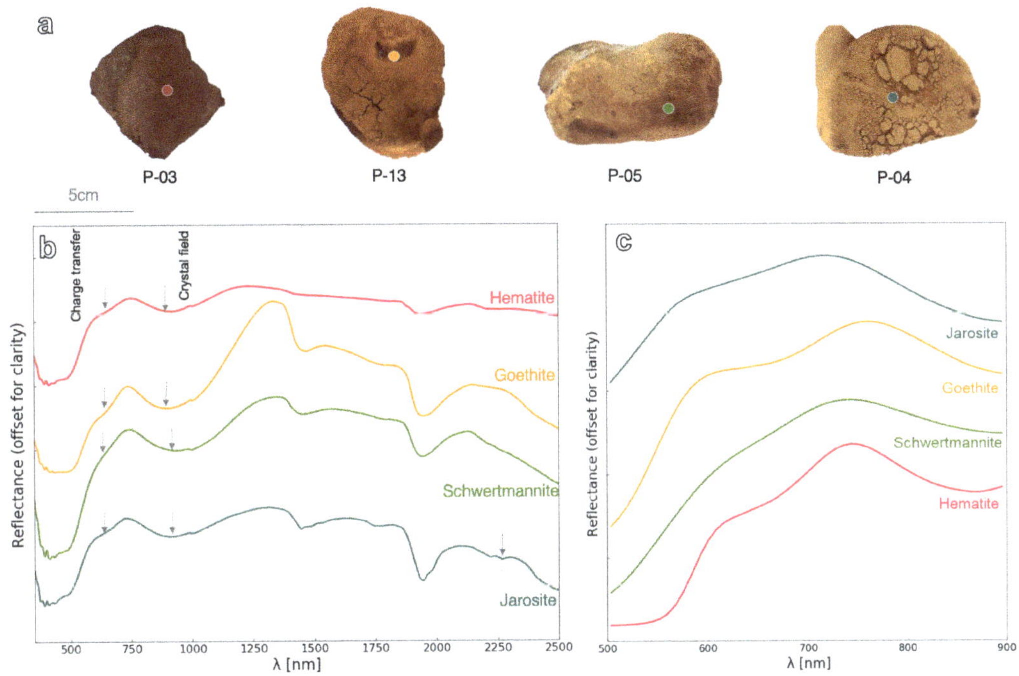

Figure 7. (**a**) Field samples, indicating the point-spectral measurements. (**b**) Spectral curves for the main minerals found in the region (hematite, goethite, jarosite, and schwertmannite), indicating prominent absorption features of each. (**c**) Reference spectra resampled to Rikola spectral range (504–900 nm).

3.3.2. Spectral Angle Mapper (SAM)

The spectral angle mapper is a supervised classification algorithm, measuring the similarity between a known reference spectrum and the unknown image spectra by calculating the angle (in radians) between the two spectra as they are expressed as vectors in a n-dimensional coordinate system, being n equal to the number of available bands [46]. By choosing smaller angles, the closer will be the matching to the reference spectra. After the comparison, each pixel is assigned to the class that exhibits the smallest SAM angle. Therefore, the threshold range for the similarity analysis in this work, relied on very small SAM values (0.05–0.08 rad) for performing the classification.

4. Results

4.1. Hydrogeochemical Maps

The analytical results from water samples can be split according to each branch of the river confluence: (1) Tintillo, (2) Odiel before, and (3) Odiel after the confluence with Tintillo River. The Tintillo River is characterized by very acidic pH values (2.5–2.6) and high electrical conductivity (EC) values (9000–9500 µS/cm). The redox conditions are mainly governed by the Fe^{2+} to Fe^{3+} oxidation rate, values around 575 mV were found and are characteristic of oxidized aqueous environments. The chemical behavior of the Odiel River samples changed drastically from neutral pH values (6.8–7.9) (before confluence) passing through a transition phase (4.5–4.7) and finally decreasing to pH values of 2.8–3.9 after the confluence. The detailed water parameters for each of the samples can be found in the Table A1 in the Appendix A.

The redox conditions vary strongly in the samples from Odiel River, around 200 mV in the initial branch to around 520 to 570 mV after the mixing. In general, the EC values are lower in comparison to Tintillo waters, before the confluence (373–645 µS/cm), and increasing up to 9460 µS/cm after. Samples from the Tintillo River were brownish to light-brown in color, samples from Odiel River before the confluence were colorless, and samples from Odiel River after the confluence were yellow-brown in color. The most significant contents are those of SO_4, Fe, Zn, Pb, Mn, Mg, Al, Cl, Cd, Cu, Ni, Co (Figure 8a).

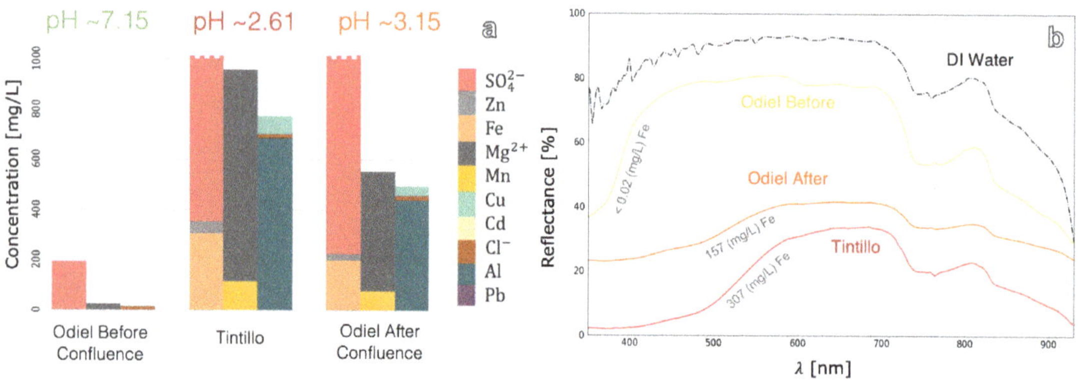

Figure 8. (**a**) Clustered mean values of the main elements in the water samples from the three river sections. The SO_4^{2-} values for Tintillo and Odiel affected by acid mine drainage (AMD) exceed 9000 mg/L (marked bars), for better visualization, the concentration axis shows only up to 1000 mg/L. (**b**) Field spectra on surface water obtained with the hand-held spectrometer (Spectral Evolution PSR-3500). Gray lines are located at prominent absorption feature wavelengths: 436 nm, 494 nm, and 763 nm.

The chemical composition of affected waters include extreme concentrations of dissolved SO_4^{2-} (up to 11,000 mg/L) and metals (up to 325.99 mg/L Fe, 729 mg/L Al, 884 mg/L Mg, 143 mg/L Zn, 76.2 mg/L Cu, and 122 mg/L Mn). Other trace elements such as Pb, Cd, Ni, and Co were always close to or below the detection limits. The detailed elemental composition for each sample can be found in the appendices (Table A2).

Although iron speciation could not be quantified, the highly oxidizing and acidic conditions (pH 2.5–2.8) of Tintillo and Odiel waters indicate the prevalence of Fe^{3+}. Dissolved Fe^{2+} can be rapidly oxidized to Fe^{3+}, characteristic of bacterially catalyzed oxidation, and leading to the appearance of both dissolved and particulate Fe^{3+} which turns the acidic waters a deep red color [17]. A pH value around 3 favors the hydrolysis and precipitation of Fe^{3+} ions, which takes place usually in the form of very fine-grained schwertmannite (with traces of jarosite) [17]. These minerals are mineralogically meta-stable and usually are transformed into goethite or hematite [47].

The spectral curves for the in-situ water point measurements are shown in Figure 8b. Each measurement corresponds to a different river branch of the studied confluence. The DI water spectral line was acquired under laboratory conditions and was used as a reference to describe the spectral features of pure water. The DI spectra is featureless until the appearance of two broad absorption features at 763 nm and 975 nm which are characteristic of the water molecule in the liquid phase [48,49]. These absorption features are produced by the combination of three fundamental vibrational modes of the water molecule [49]. After these features, the spectral curve decreases to virtually 100 % absorption. Thus, water spectral analysis in this study was focused in the VNIR range (350–1000 nm).

The spectral behavior at the 3 different river locations was characterized by an increasing absorption feature range between 350 and 650 nm and a peak reflectance around 640 nm. The absorption features of the three-river water curves from 700 nm onwards were matching to those in the pure water line (DI). The main differences found in the field water spectral measurements are in the 400–600 nm range, probably associated with iron species [11]. The Odiel River before confluence showed the smallest absorption with a very slight inflection point at 436 nm. In the case of the Tintillo River spectral curve, the absorption feature is broader and the inflection point at 436 nm is slightly less pronounced followed for a second absorption feature at 494 nm. The curve for the third point in the Odiel River after the influence of Tintillo showed similar absorption features than the Tintillo curve at 436 nm and 494 nm, but the absorption between 450–520 nm is higher.

The computed SVM classification (Figure 9b) efficiently identified the water path, exposed muddy and dry soil based on features extracted by the OTVCA Figure 9a. The parameters for finding the optimal hyperplane were determined using five-fold cross validation, resulting in C = 6.4980 and $\gamma = 0.0442$. The HSI mosaic was cropped to the water mask (Figure 9c) and OTVCA computed again.

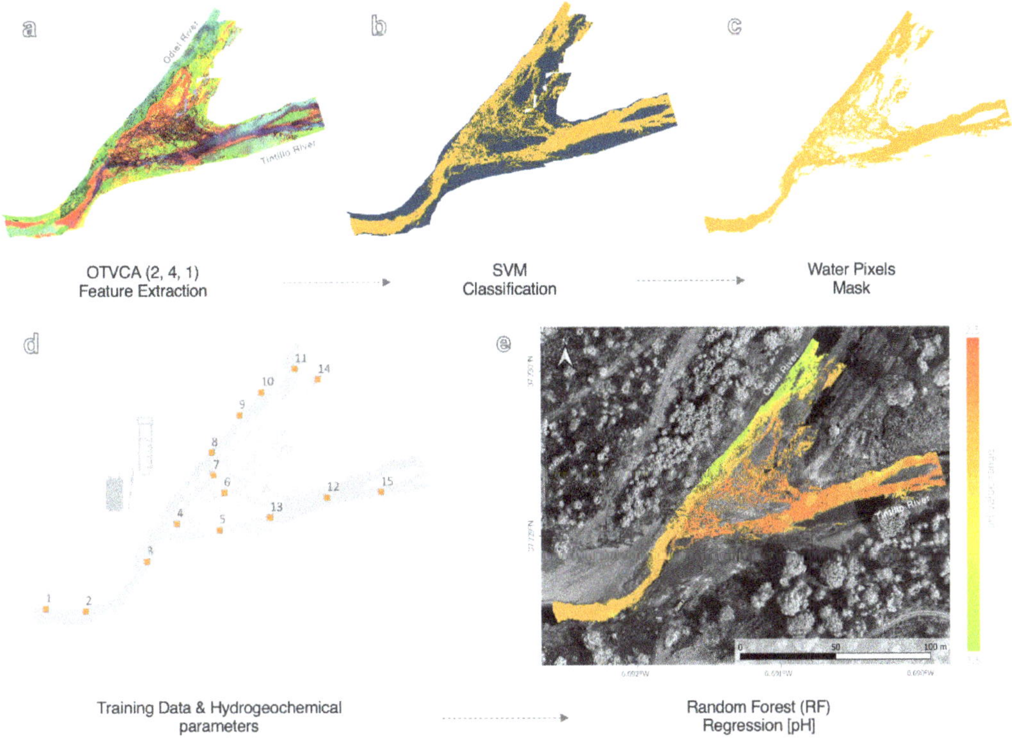

Figure 9. Results of the UAS-hyperspectral data analysis for the water masked pixels. (**a**) Orthogonal Total Variation Component Analysis (OTVCA) feature extraction displaying bands (2,4,1). (**b**) Support Vector Machine (SVM) classification for extracting water pixels belonging to the river flow path and sediment borders. (**c**) Water mask (HSI mosaic was cropped and OTVCA was reapplied). (**d**) Training data preparation (georeferenced ROIs over ground-samples and water parameter allocation). (**e**) Random forest (RF) regression resultant map displaying the pH value range.

For the training dataset, five variables (pH, redox, Fe, Al, and SO_4^{2-} concentration) were considered in the training matrix Figure 9d. Figure 9e shows the obtained pH regression map, which clearly show the gradient in the pH value of water, and reveals the strong influence of the Tintillo that reach Odiel's cleaner part, probably caused by transport of the AMD contamination via groundwater flow or flooding during the heavy raining events of the region. The derived collection of maps for assessing the river water in Figure 10 makes it possible to see the fluctuations present in its quality by combining different criterion in the environmental assessment.

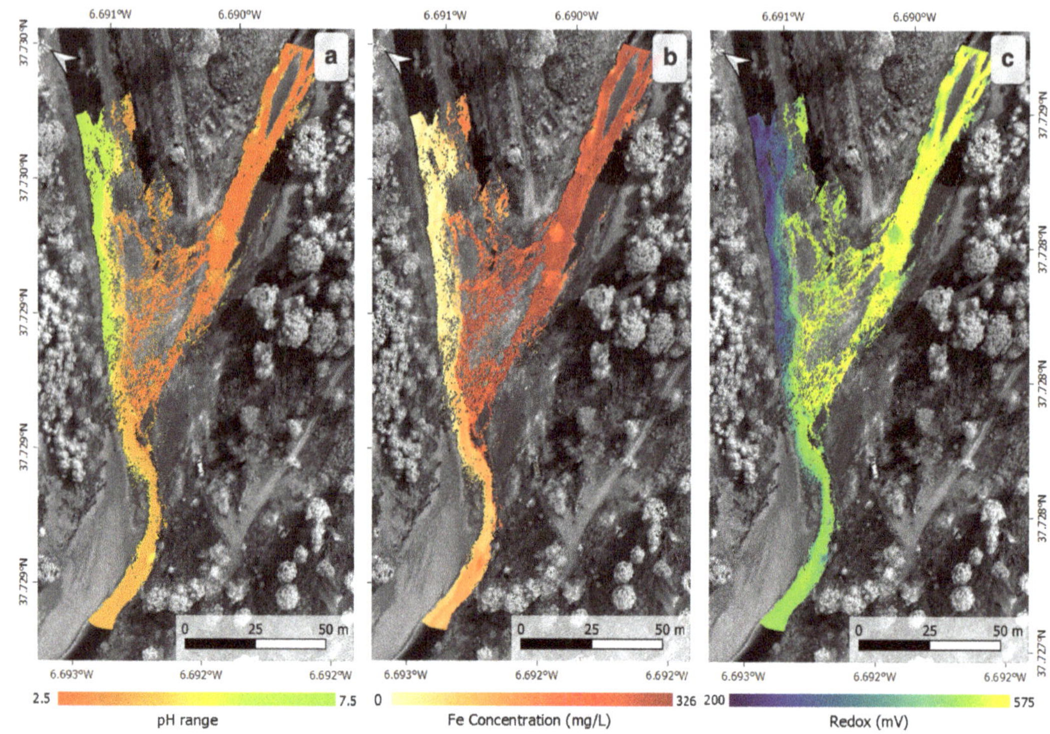

Figure 10. Multivariate RF regression maps. (**a**) pH conditions, (**b**) Fe concentration in water, and (**c**) redox conditions. Map were rotated 55° for better visualization.

Table 3 shows the resultant R^2 values for the five variables that fed the RF algorithm. The so-called internal validation show for all the variables a close to 1 R^2, showing a good correlation among the predicted values in comparison with the pixels only used as the validation set. Although during the external validation of the regression the R^2 values decreased, especially for variables with a wider range of values (Al, Fe, and SO_4^{2-}, the performance of the regression remained good.

Table 3. R^2 values after regression are given for each variable predicted in the obtained surface water maps.

Variables	Internal Validation R^2	External Validation R^2
pH	0.99	0.73
Redox	0.99	0.82
SO_4^{2-}	0.99	0.61
Al	0.99	0.68
Fe	0.99	0.66

4.2. Mineral Maps

The major visual feature of the collected sediment samples was the characteristic yellow to reddish-brown color, a product of the AMD influence in the region. The mineralogy of this precipitate consists mainly of Fe-phases precipitated from the Fe dissolved in AMD originating from pyrite oxidation in the mine site. Figure 11 shows the mineral composition of four of the main representative samples from the study area. The matrix is dominated by tectosilicates ranging from 36 to 58 wt%, the mass fractions are associated with quartz and feldspars (microcline and albite), and some sheet silicates suchas muscovite or chlorite, while the iron-bearing minerals (hematite, goethite, jarosite, and schwertmannite) are lower in mass fraction since they occur mainly on top of the layer in the form of fine-grained coatings.

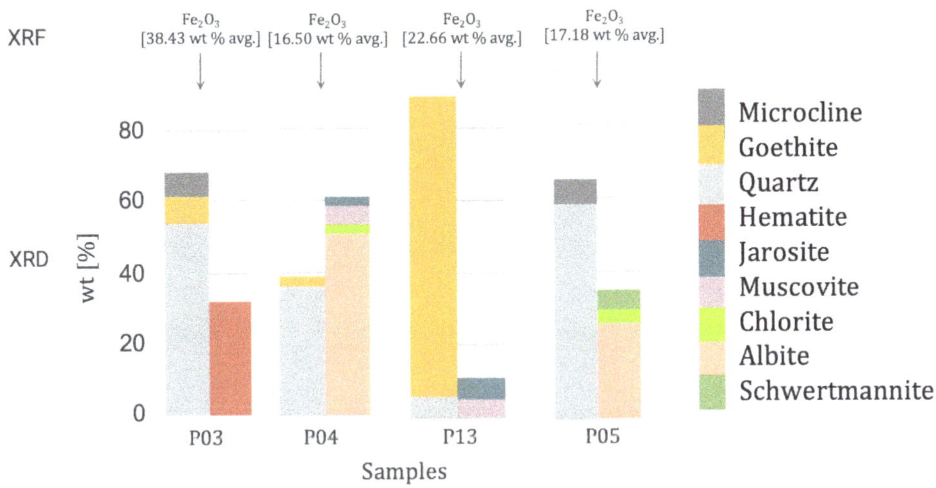

Figure 11. Chart of the main mineral fractions of four selected samples determined by the XRD mineralogical analyses. Results of the iron-fraction after XRF analysis are given.

Highly oxidized minerals controlled by acidic pH (2.5–3), such as goethite (P13; 83 wt%) or hematite (P03, 32 wt%), cover the sediments on top of the river bank, while less oxidized minerals, such as jarosite (P04; 3 wt%) cover sediments in association with goethite, on the transition stage from metastable mineral to a stable form. More hydrated minerals, such as schwertmannite (P05; 6 wt%), grow over fine-grained sediments closer to less acidic waters at the beginning of the confluence, expected in pH ranges from 3.0 to 4.5 [5]. The detection of schwertmannite could be challenging since it is poorly crystalline in comparison to jarosite and goethite [17]. Schwertmannite fraction in P05 is identified by looking at their specific absorption features in the spectral measurement of the sample. The specific conditions for the precipitation of iron-bearing secondary products can vary widely according to the levels of pH, SO_4 concentration, and paragenetic relationships [5].

XRF results of elements and compounds match with the minerals revealed by the XRD analysis. Results are given in their oxide forms. Aluminum (Al_2O_3), iron (Fe_2O_3), silicon (SiO_2), and titanium (TiO_2) are the prevalent constituents of the bulk geochemistry of the samples. The iron fraction for the selected samples (Figure 11) are P013 (38.43 wt% avg. Fe_2O_3); P021 (22.66 wt% avg. Fe_2O_3); P025 (17.18 wt% avg. Fe_2O_3); P016 (16.50 wt% avg. Fe_2O_3). XRF iron values for these samples are range between 16–38 %, showing that AMD minerals occur mainly on the top layer of the sediments. The greatest Fe content and metal concentrations are found in the most acidic leachates from the Tintillo River, which favors the presence of iron in aqueous form. From the collection of minerals detectable by the mineralogical diffraction analysis (Figure 11), only a small selection of them can be detected with VNIR spectroscopy. This is because quartz, feldspar, or pyrite show no characteristic absorption features in this range. However, spectral curves of the hand specimens (Figure 7b) revealed distinctive absorption feature depths of the iron minerals (goethite, jarosite, hematite, and schwertmannite) present in the samples.

Figure 12 presents the classification result at an angle of 0.08 radians for the occurring endmembers. The resultant map shows that goethite and hematite coatings cover the coarser clasts on top of the Tintillo River bars. Jarosite tends to develop over the pebbles and sands of the sides river bars on the border crust close to the Odiel river after the confluence and schwertmannite is present along the sides of the river bars over fine-grained sands closer to water in the left border of the upper part of the Odiel River.

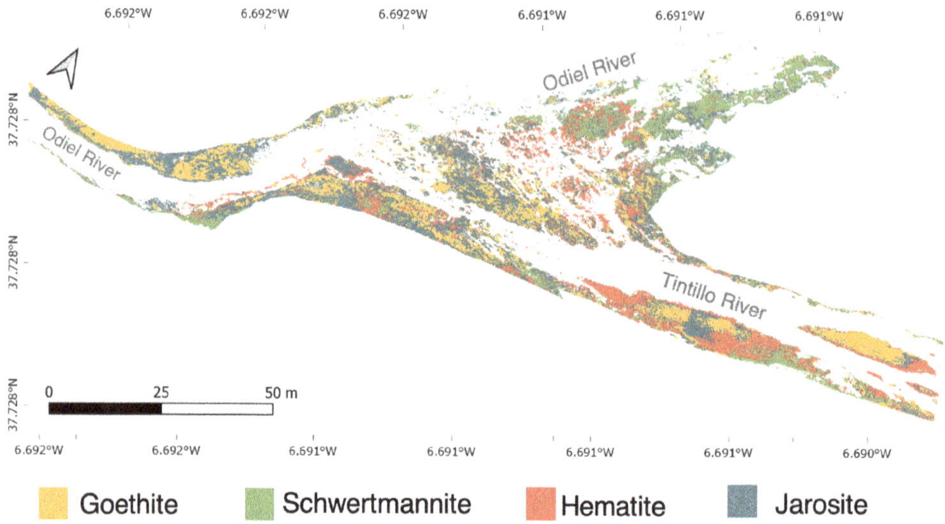

Figure 12. Spectral Angle Mapper (SAM) supervised classification for the four endmembers minerals in the Tintillo–Odiel region.

5. Discussion

In this paper, we propose a framework to monitor water pollution produced by AMD integrating UAS-HSI data, field measurements, and geochemical analyses. Compiled geochemical information from water samples is used as reference data to train a regression model to up-scale the detailed information coming from the sampled spots to the entire area.

5.1. Results Assessment

As the Tintillo–Odiel area is a rapidly changing environment, imaging spectroscopy of natural water body conditions could be affected by water depth differences constantly changing due to climatic parameters, water dynamics, and seasonal conditions. This study was performed under spring conditions (April) and the measured water depth was approximately constant (10–12 cm) over the sampled spots. Hence, the strong absorption features found in the water spectra between 400 and 700 nm might be mostly influenced by the presence of Fe^{3+} ions which turns the acidic waters a deep red color. This absorption was deeper, where the iron content was greater. However, it is possible that predicted values towards the borders of the river could be influenced by the slight decrease on the water depth. The catalog of maps created in this research allows evaluation of the degree of contamination in the area covering water and sediments. While the hydrogeochemical maps enable visualization of the gradual variations of the water properties, the mineralogical maps show the spatial distribution of the iron minerals that can be used as a substitute to detect the contamination and the degree of acidity. The maps resulting from this study should be understood as simplified images of hydrogeochemical and mineralogical trends in a very spatially heterogeneous environment at all scales changing rapidly over time, not only annually or seasonally but even weekly.

This study involves steps and processes that may be inevitably affected by biases, as in most analytical studies that require human interaction. The quality of the original hyperspectral data may be affected by parameters set on site. (e.g., set-up geometry, frame rate, integration time, spectral sampling interval, band range, and bandwidth). Then, during data processing, various parameters must be entered manually, such as (1) the number of features to be extracted for OTVCA; (2) the quantity (radius) of the sampled pixels to be included as training data for classification or regression algorithms; (3) precision errors during laboratory analysis. Regardless of the constraints, the information provided by the data integration portion of this study represents a significant benefit for the better evaluation of AMD affected ecosystems. Mine waters may vary strongly in composition and be dominated by diverse dissolved constituents and thermodynamic conditions according to the geological environment in which they originate [1]. Similar to optical studies of natural water bodies, the intrinsic and apparent optical properties of AMD are influenced by a combination of physical, chemical and biological processes, which may affect their spectral response. The application of this method in other study areas, with less degree of AMD and deeper water conditions is encouraged in upcoming studies.

5.2. Relevance

Overall, the field acquisition was fast, taking about one day. Hyperspectral image acquisition took only 12 min approximately and the rest of the time was used for taking the validation measurements. Ground-truth sampling was demonstrated to be vital to provide as accurate and precise maps as possible. The integration of spectral information combined with further analytical procedures contributed to better interpretation of the remote-spectral information, increasing the reliability of the results.

Mapping water quality properties, based on their spectral signatures, represents a robust basis for an innovative water quality monitoring system. The maps derived from the UAS-borne hyperspectral data help to identify patterns and physicochemical conditions of natural waters. Even if the absorption features are not as prominent as the observed between 400–700 nm due to Fe, more subtle absorptions and slope changes in the NIR range provide information about the other chemical compounds and parameters affecting AMD waters. The subtleness of these variations can be detected in the whole VNIR available range, thanks to the multiple and the contiguous number of bands provided by the hyperspectral camera, which offers the possibility to map the unique spectral shape of each pixel with the measured parameters. A map collection for pH (Figure 13), redox, and metal concentration (Fe, Al, and SO_4) makes it possible to detail changes in water quality by combining different criterion in the environmental assessment. Undeniably, the AMD effects in the region are considerable, the contamination of Tintillo even reach the Odiel River before the confluence, probably by groundwater interaction or heavy rainfall events.

It is feasible to estimate the gradient of variables affecting the water quality, discriminating the most affected zones studied. The level of accuracy is remarkable, allowing risk assessors and scientists to take appropriate preventive measures according to the level of contamination by mapping changes in AMD through time, which can occur after sudden rainfall events or stationary changes. The remediation of ecosystems polluted by mining leachates can represent a high-cost investment for mining companies, whereas the proposed approach points out a cost-effective solution during the reclamation or closure planning stage of mining activities by decreasing the time of data acquisition and enhancing the sampling strategy when assessing the most adequate reclamation technologies or treatment methods to apply.

5.3. Innovation

In terms of environmental interpretations and risk assessment, the study has shown that there are several advantages to using UAS-borne hyperspectral data as a complement to traditional environmental monitoring studies. Traditional monitoring of river water quality is mainly based on the chemical analysis of water samples routinely collected over the year, and on the physical parameters of the groundwater measured by instruments located in the flow path. These tasks can be expensive, time-consuming, and constrained by access. In addition, a major advantage of UAS mapping compared to ground surveying is a reduction in the time employed on acquiring data. Furthermore, the application of UAS-surveys enables analysis of locations that may be difficult to access, under protected status, or that involve personal security risks for terrestrial-sampling.

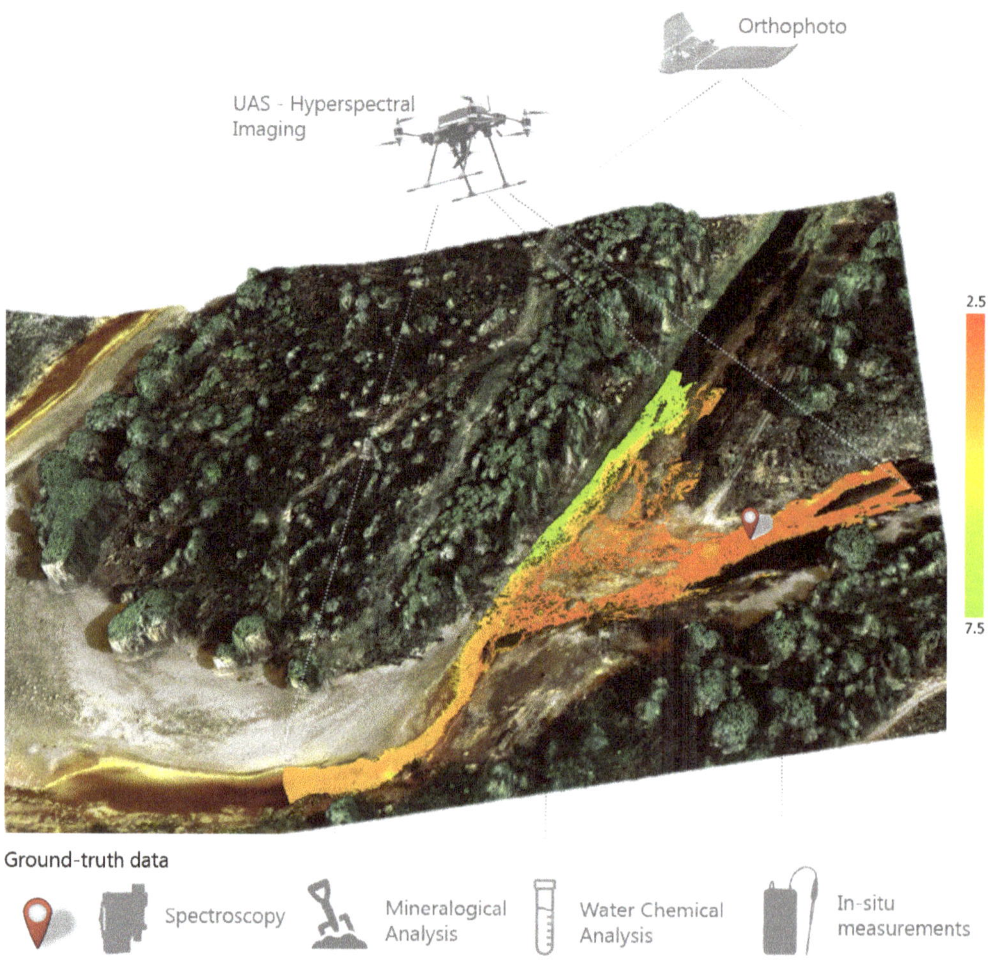

Figure 13. 2.5D surface model representation of the Tintillo–Odiel area. Base layer are the combination of the orthopoto on top of the Digital Surface Model (DSM) and the produced regression–pH map for the river flow path. UAS and further field geochemical techniques used are shown (not drawn to scale).

Environmental monitoring by means of unmanned aerial systems has distinct advantages: (i) low flying heights for high resolution and seamless coverage in fragile areas, (ii) cost-effectiveness after the initial purchase cost, (iii) easy flight preparation after flying permission has been given, and (iv) versatility in the sensors that can be attached to the aircraft/drone. These advantages allow repeatably and recurrent data-acquisition. Therefore, multi-temporal analyses are feasible and may allow constant and repeated monitoring of ecosystems.

Although many instruments with greater spectral resolution and wider wavelength range have been developed, the majority of this equipment tends to be too heavy, fragile, and costly to be mounted on UAS on an operational basis. Renewed efforts have been made on satellite development to increase their spatial resolution by enhancing band acquisition efficiency and making data available in open-source systems but their metric spatial resolution remains a limiting factor.

5.4. Outlook

In order to increase the spatial coverage of this approach, a fixed-wing UAS (e.g., the senseFly eBee [50]) with lightweight sensors may be useful, since longer flight duration (50 min) can be achieved. Although the flight time is longer, the payload is lower, thus a lighter sensor is required to be mounted, for example, a Parrot Sequoia camera [51]. This sensor is multi-spectral and provides only information of four bands (near-infrared [790 nm], red-edge [735 nm], red [660 nm], and green [550 nm]). A band-ratio could be calculated by dividing reflectance-calibrated band 3 (735 nm) by band 4 (790 nm). This ratio may be useful for devising iron-related absorption, as it maps part of the crystal-field absorption produced by the Fe^{2+}/Fe^{3+} [45] around 800 nm. Moreover, the full width at half maximum (FWHM) of the 735 nm band is 10 nm and for the 790 nm band is 40 nm, so some of the reflectance peaks of iron minerals (e.g., from hematite, goethite, jarosite, schwertmannite) might be detected. A second option would be the use of band 2 (660 nm) by band 1 (550 nm) to pick up the differences in part of the charge transfer region of the iron-oxides. Nevertheless, this type of band ratio should be interpreted carefully since the narrow spectral features are combined together or left out using only two single-bands. Hence, multi-spectral data cannot provide as detailed a result as compared to the classification using hyperspectral data can, but overall they could be used for mapping the abundance of the iron-bearing mineral and targeting more acid-affected areas (using AMD minerals as proxy) for subsequent surveys with higher spectral resolution.

The use of the red-edge and NIR bands could be also useful in vegetation health analyses by detecting chlorophyll intensity in the acidic environment. Regardless of the spectral limitations, the multi-spectral sensor has the advantage of being lighter and can be mounted on fixed-wings systems, covering much broader areas than multi-copters.

In summary, the workflow developed by this study for environmental monitoring, based on the integration of remote sensing and geochemical techniques, was successful for the field campaign at the Odiel–Tintillo test site. The acquired mapping data are easy to include in GIS or 3D mining software. This represents an advantage for companies by creating a data catalog, in which initially ignored data parts can be processed to have a complete image of the monitoring system and not having only information from certain points. The high spatial resolution of UAS-borne hyperspectral data and precise orthorectification and georeferencing allow the creation of a three-dimensional, distortion-free framework for intuitive surface visualization and interpretation of the site conditions. Although the topography of the region is basically flat, the Digital Surface Model (DSM) allows us to observe small topographic contrasts and structures such as the hill that exists between the two rivers (Figure 13). Additionally, the 3D data may lead to integrate differences in surface textures and better classification results by using elevation information.

6. Conclusions

UAS-borne hyperspectral imaging represents a fast and non-invasive tool which provides high resolution maps for different hydrogeological applications. We combined the acquired VNIR hyperspectral data cube with constrained hydrogeochemical data using machine learning techniques, aiming to assess surface water quality.

The proposed methodological framework has been tested in the Odiel–Tintillo River confluence. Even though the test-site represents a very variable, confined and mixed environment, it was possible to detect gradual changes in the parameters that control water quality and the spatial distribution of secondary-iron minerals on the borders of the river flow path.

We proposed to up-scale and spread detailed hydrogeochemical information coming from discrete ground-truth points to the whole covered area by discovering the relationship of the water chemistry with the spectral behavior using the supervised machine learning regression model: random forest (RF). Hence, gradual changes in physicochemical properties (pH, Eh) and elemental concentration (Fe, Al, SO_4^{2-}) in surface water have been mapped for assessing the AMD extent.

There are three main outcomes of this study: (1) Data acquisition times have been demonstrated to be fast in comparison with traditional environmental monitoring approaches for producing high-pixel resolution maps that covered 14,000 m^2 of the rivers. Less than one hour was needed for the UAS survey and half of a day for the ground-sampling campaign to acquire reference samples. (2) Laboratory and in-situ analytical data proved fundamental for the validation of the spectral features and to support the construction of robust mineralogical and hydrogeochemical trustworthy training data. (3) Fifteen reference points proved to be sufficient for a good performance of the RF regression model. Initial validation showed a close to 1 R^2 for all the variables and after narrowing the number of samples per classes during the final validation, the performance of the regression still remained good for variables with smaller value ranges (pH $R^2 = 0.72$ and Eh $R^2 = 0.81$) as compared to variables with a wider range of values (chemical composition of Fe, Al, SO_4^{2-} $R^2 = 0.65$).

Overall, the results of this paper emphasize the capabilities of UAS-borne HSI data as a valuable support for environmental monitoring of surface water affected by acid mine drainage. The framework followed may be used for the monitoring of other mining environments potentially affected by acid mine drainage, by targeting sources, and/or contamination in water, promoting continuous supervision or to provide guidance for adequate remediation treatments.

Author Contributions: Conceptualization, H.F., R.J., and R.G.; formal analysis H.F., S.L., and R.J. methodology H.F., S.L., R.Z., I.C.C., and L.T.; software H.F., S.L., I.C.C. and L.T.; validation: H.F. and L.T.; visualization: H.F.; writing—original draft: H.F.; writing—review and editing: H.F., S.L., R.J., I.C.C., L.T., R.Z., and R.G.; supervision: S.L. and R.G. All authors have read and agreed to the published version of the manuscript.

Funding: This research received no external funding.

Institutional Review Board Statement: Not applicable.

Informed Consent Statement: Not applicable.

Data Availability Statement: Not applicable.

Acknowledgments: The Helmholtz Institute Freiberg for Resource Technology is gratefully thanked for supporting and funding this project. H.F. is grateful to Nils Hoth and Marlies Grimmer for their discussions and help with the water analyses at the Mine Water Laboratory of TU Bergakademie Freiberg. Special thanks go to Robert Möckel and Doreen Ebert for their help during XRD and XRF analyses and all the people involved during the field campaign (Margret Fuchs, René Booysen, Roberto De La Rosa, Yuleika Madriz, Erik Herrmann, Joan Blanco, and Kasra Rafiezadeh) and discussions (Louis Andreani) from the Exploration Division.

Conflicts of Interest: The authors declare no conflict of interest.

Appendix A

Table A1. Field parameters of natural stream waters of the Odiel and Tintillo River confluence. Em, redox potential measured with platinum/reference electrode; pe, redox corrected and expressed as electron activity; (**) EC, electrical conductivity measurements laboratory-based; (*) S14 is located in the left border of Odiel River, which shows Tintillo waters already affecting its composition.

River	ID	T (°C)	pH	Em (mV)	pe	EC ** (μS/cm)
Odiel Before Confluence	S11	19.4	6.88	239.4	7.62	409
	S8	19.6	4.73	394.2	10.23	515
	S9	19.5	4.52	369.3	9.81	645
	S10	19.4	7.9	204.2	7.02	373
	S14 *	20	2.63	561.7	13.06	9390
Tintillo	S12	19	2.63	574.5	13.29	9420
	S13	19.6	2.6	574.2	13.28	9460
	S15	19	2.64	575.3	13.30	9430
	S6	19.5	2.6	573.8	13.27	9530
	S5	19.3	2.58	574	13.28	9500
Odiel After Confluence	S2	20.2	3.88	520.6	12.36	1152
	S3	19.7	2.8	577	13.32	5780
	S1	20	3.99	506	12.12	9070
	S4	19.5	2.58	575	13.29	9460
	S7	19.3	2.58	572.3	13.25	9440

Table A2. Elemental composition of water samples from the three analyzed river sections. Photometry for SO_4^{2-} and Cl^- and MP-AES for the rest of the elements.

River	ID	Mg	Cl^-	Zn	Al	Fe	SO_4^{2-}
				mg/L			
Odiel Before Confluence	S11	16.1	12.1	0.17	0.07	<0.02	120
	S8	28.6	12.7	2.53	3.14	0.14	230
	S9	29.3	12.2	2.65	4.09	0.08	330
	S10	17.9	12.1	0.63	0.03	0.02	90
	S14	854	18.7	143	687	300.51	10,000
Tintillo	S12	843	17.7	142	680	306.78	11000
	S13	826	18.0	21.9	672	297.81	9000
	S15	836	17.6	22.6	681	302.02	9400
	S6	824	17.5	23.4	671	301.15	10,000
	S5	884	17.9	21.4	729	325.99	9000
Odiel After Confluence	S2	57.4	13.3	24.6	332.0	157.60	4900
	S3	428	16.0	24.4	339	165.20	5100
	S1	242	14.2	21.6	189	73.80	700
	S4	831	18.3	23.6	674	303.39	9000
	S7	828	21.3	24.1	668	305.46	10,000
	ID	**Cd**	**Cu**	**Mn**	**Co**	**Ni**	**Pb**
				mg/L			
Odiel Before Confluence	S11	<0.03	0.04	0.50	0.05	0.02	<0.05
	S8	<0.03	1.04	2.30	0.05	0.04	<0.05
	S9	<0.03	1.13	2.28	0.05	0.04	<0.05
	S10	<0.03	0.06	0.54	0.05	0.02	<0.05
	S14	0.69	74.1	116	4.36	1.44	<0.05

Table A2. *Cont.*

	ID	Cd	Cu	Mn	Co	Ni	Pb
				mg/L			
Tintillo	S12	0.69	73.1	115	4.29	1.43	<0.05
	S13	0.66	73.5	113	4.48	1.33	<0.05
	S15	0.64	71.1	112	4.26	1.34	<0.05
	S6	0.67	72.1	113	4.25	1.39	<0.05
	S5	0.65	76.2	122	4.42	1.40	<0.05
Odiel After Confluence	S2	0.04	3.67	67.6	0.16	0.10	<0.05
	S3	0.38	32.2	55.6	1.94	0.73	<0.05
	S1	0.21	18.5	31.5	1.04	0.40	<0.05
	S4	0.66	70.8	114	4.26	1.33	<0.05
	S7	0.68	70.4	113	4.47	1.41	<0.05

References

1. Lottermoser, B. *Mine Wastes*; Springer: Berlin/Heidelberg, Germany, 2010; Volume 44, p. 085201. [CrossRef]
2. Ferrier, G. Application of imaging spectrometer data in identifying environmental pollution caused by mining at Rodaquilar, Spain. *Remote Sens. Environ.* **1999**, *68*, 125–137. [CrossRef]
3. Kemper, T.; Sommer, S. Estimate of heavy metal contamination in soils after a mining accident using reflectance spectroscopy. *Environ. Sci. Technol.* **2002**, *36*, 2742–2747. [CrossRef]
4. Swayze, G.A.; Smith, K.S.; Clark, R.N.; Sutley, S.J.; Pearson, R.M.; Vance, J.S.; Hageman, P.L.; Briggs, P.H.; Meier, A.L.; Singleton, M.J.; et al. Using imaging spectroscopy to map acidic mine waste. *Environ. Sci. Technol.* **2000**, *34*, 47–54. [CrossRef]
5. Montero, I.C.; Brimhall, G.H.; Alpers, C.N.; Swayze, G.A. Characterization of waste rock associated with acid drainage at the Penn Mine, California, by ground-based visible to short-wave infrared reflectance spectroscopy assisted by digital mapping. *Chem. Geol.* **2005**, *215*, 453–472. [CrossRef]
6. Sares, M.; Hauff, P.; Peters, D.; Coulter, D. Characterizing Sources of Acid Rock Drainage and Resulting Water Quality Impacts Using Hyperspectral Remote Sensing–Examples from the Upper Arkansas. In Proceedings of the Advanced Integration of Geospatial Technologies in Mining Reclamation, Atlanta, GA, USA, 7–9 December 2004.
7. Richter, N.; Staenz, K.; Kaufmann, H. Spectral unmixing of airborne hyperspectral data for baseline mapping of mine tailings areas. *Int. J. Remote Sens.* **2008**, *29*, 3937–3956. [CrossRef]
8. Choe, E.; van der Meer, F.; van Ruitenbeek, F.; van der Werff, H.; de Smeth, B.; Kim, K.W. Mapping of heavy metal pollution in stream sediments using combined geochemistry, field spectroscopy, and hyperspectral remote sensing: A case study of the Rodalquilar mining area, SE Spain. *Remote Sens. Environ.* **2008**, *112*, 3222–3233. [CrossRef]
9. Shang, J.; Morris, B.; Howarth, P.; Lévesque, J.; Staenz, K.; Neville, B. Mapping mine tailing surface mineralogy using hyperspectral remote sensing. *Can. J. Remote Sens.* **2009**, *35*, S126–S141. [CrossRef]
10. Davies, G.E.; Calvin, W.M. Mapping acidic mine waste with seasonal airborne hyperspectral imagery at varying spatial scales. *Environ. Earth Sci.* **2017**, *76*, 1–14. [CrossRef]
11. Crowley, J.K.; Williams, D.E.; Hammarstrom, J.M.; Piatak, N.; Chou, I.M.; Mars, J.C. Spectral reflectance properties (0.4–2.5 μm) of secondary Fe-oxide, Fe-hydroxide, and Fe-sulphate-hydrate minerals associated with sulphide-bearing mine wastes. *Geochem. Explor. Environ. Anal.* **2003**, *3*, 219–228. [CrossRef]
12. Davies, G.E.; Calvin, W.M. Quantifying Iron Concentration in Local and Synthetic Acid Mine Drainage: A New Technique Using Handheld Field Spectrometers. *Mine Water Environ.* **2016**, *36*, 299–309. [CrossRef]
13. Booysen, R.; Gloaguen, R.; Lorenz, S.; Zimmermann, R.; Nex, P.A. Geological Remote Sensing. In *Reference Module in Earth Systems and Environmental Sciences*, 2nd ed.; Elsevier: Amsterdam, The Netherlands, 2020; Volume 64, pp. 267–274. [CrossRef]
14. Jackisch, R.; Lorenz, S.; Zimmermann, R.; Möckel, R.; Gloaguen, R. Drone-borne hyperspectral monitoring of acid mine drainage: An example from the Sokolov lignite district. *Remote Sens.* **2018**, *10*, 385. [CrossRef]
15. Soriano, C.; Casas, J. Variscan tectonics in the Iberian Pyrite Belt, South Portuguese Zone. *Int. J. Earth Sci.* **2002**, *91*, 882–896. [CrossRef]
16. Abad de los Santos, M.; Olías Álvarez, M.; Carro, B.M. *Geología de Huelva lugares de interés geológico*; Universidad de Huelva: Huelva, Spain, 2009; pp. 68–69.
17. Sánchez, J.; López, E.; Santofimia, E.; Aduvire, O.; Reyes, J.; Barettino, D. Acid mine drainage in the Iberian Pyrite Belt (Odiel river watershed, Huelva, SW Spain): Geochemistry, mineralogy and environmental implications. *Appl. Geochem.* **2005**, *20*, 1320–1356. [CrossRef]
18. Olias Alvarez, M.; Donaire Romero, T.; Mayoral Alfaro, E.; Morales Gonzalez, J.; Alonso Chaves, F.; Ruiz de Almodobar, G. *Geologia de Huelva, Lugares de Interes Geologico*; Aldina (Universidad de Huelva): Huelva, Spain, 2008.
19. Sarmiento, A.M.; Nieto, J.M.; Olías, M.; Cánovas, C.R. Hydrochemical characteristics and seasonal influence on the pollution by acid mine drainage in the Odiel river Basin (SW Spain). *Appl. Geochem.* **2009**, *24*, 697–714. [CrossRef]

20. Davis, R.A.; Welty, A.T.; Borrego, J.; Morales, J.A.; Pendon, J.G.; Ryan, J.G. Rio Tinto estuary (Spain): 5000 Years of pollution. *Environ. Geol.* **2000**, *39*, 1107–1116. [CrossRef]
21. Sánchez, J.; Lopez, E.; Santofimia, E.; Aduvire, O.; Reyes, A.; Martín, J. The Tintillo acidic river (Rio Tinto mines, Huelva, Spain): An example of extreme environmental impact of pyritic mine wastes on the environment or an exceptional site to study acid-sulphate mine drainage systems? *Int. Conf. Min. Environ. Met. Energy Recovery* **2006**, *1*, 278–287.
22. Sánchez España, J.; Pamo, E.L.; Pastor, E.S.; Andrés, J.R.; Rubí, J.A. The impact of acid mine drainage on the water quality of the Odiel river (Huelva, Spain): Evolution of precipitate mineralogy and aqueous geochemistry along the Concepción-Tintillo segment. *Water Air Soil Pollut.* **2006**, *173*, 121–149. [CrossRef]
23. Riaza, A.; Buzzi, J.; García-Meléndez, E.; Carrère, V.; Sarmiento, A.; Müller, A. Monitoring acidic water in a polluted river with hyperspectral remote sensing (HyMap). *Hydrol. Sci. J.* **2015**, *60*, 1064–1077. [CrossRef]
24. Senop Oy. *VIS-VNIR Snapshot Hyperspectral Camera for UAVs*; Technical Report; Senop Oy: Laukaa, Finland, 2020.
25. Jakob, S.; Zimmermann, R.; Gloaguen, R. The Need for Accurate Geometric and Radiometric Corrections of Drone-Borne Hyperspectral Data for Mineral Exploration: MEPHySTo-A Toolbox for Pre-Processing Drone-Borne Hyperspectral Data. *Remote Sens.* **2017**, *9*, 88. [CrossRef]
26. Karpouzli, E.; Malthus, T. The empirical line method for the atmospheric correction of IKONOS imagery. *Int. J. Remote Sens.* **2003**, *24*, 1143–1150. [CrossRef]
27. Savitzky, A.; Golay, M.J. Smoothing and Differentiation of Data by Simplified Least Squares Procedures. *Anal. Chem.* **1964**, *36*, 1627–1639. [CrossRef]
28. Nordstrom, D.K. Thermochemical redox equilibria of ZoBells solution. *Geochim. Cosmochim.* **1977**, *41*, 1835–1841. [CrossRef]
29. Baird, R.; Eaton, A.; Rice, E. *Standard Methods for the Examination of Water and Wastewater*; American Public Health Association: Washington, DC, USA, 2017.
30. Hach. *DR3900 Laboratory Spectrophotometer for Water Analysis User Manual*; Technical Report; Hach: Loveland, CO, USA, 2013.
31. Agilent Technologies. *Microwave Plasma Atomic Emission Spectroscopy (MP-AES) Application eHandbook*; Technical Report; Agilent Technologies: Santa Clara, CA, USA, 2016.
32. Gates-Rector, S.; Blanton, T. The Powder Diffraction File: A quality materials characterization database. *Powder Diffr.* **2019**, *34*, 352–360. [CrossRef]
33. Doebelin, N.; Kleeberg, R. Profex: A graphical user interface for the Rietveld refinement program BGMN. *J. Appl. Crystallogr.* **2015**, *48*, 1573–1580. [CrossRef] [PubMed]
34. Rasti, B.; Ulfarsson, M.O.; Sveinsson, J.R. Hyperspectral Feature Extraction Using Total Variation Component Analysis. *IEEE Trans. Geosci. Remote Sens.* **2016**, *54*, 6976–6985. [CrossRef]
35. Ghamisi, P.; Plaza, J.; Chen, Y.; Li, J.; Plaza, A.J. Advanced Spectral Classifiers for Hyperspectral Images: A review. *IEEE Geosci. Remote Sens. Mag.* **2017**, *5*, 8–32. [CrossRef]
36. Vapnik, V.N. An overview of statistical learning theory. *IEEE Trans. Neural Netw.* **1999**, *10*, 988–999. [CrossRef] [PubMed]
37. Cortes, C.; Vapnik, V.N. Support-Vector Networks. *Mach. Learn.* **1995**, *20*, 273–297. [CrossRef]
38. Waske, B.; Benediktsson, J.A.; Arnason, K.; Sveinsson, J.R. Mapping of hyperspectral AVIRIS data using machine-learning algorithms. *Can. J. Remote Sens.* **2009**, *35*, S106–S116. [CrossRef]
39. Chang, C.C.; Lin, C.J. LIBSVM: A Library for support vector machines. *ACM Trans. Intell. Syst. Technol.* **2011**, *2*, 2–5. [CrossRef]
40. Safavian, S.; Landgrebe, D. A survey of decision tree classifier methodology. *IEEE Trans. Syst. Man Cybern.* **1991**, *21*, 660–674. [CrossRef]
41. Breiman, L. Random Forest. *Mach. Learn.* **2001**, *45*, 5–32. [CrossRef]
42. Rodriguez-Galiano, V.; Sanchez-Castillo, M.; Chica-Olmo, M.; Chica-Rivas, M. Machine learning predictive models for mineral prospectivity: An evaluation of neural networks, random forest, regression trees and support vector machines. *Ore Geol. Rev.* **2015**, *71*, 804–818. [CrossRef]
43. Kokaly, R.; Clark, R.; Swayze, G.; Livo, K.; Hoefen, T.; Pearson, N.; Wise, R.; Benzel, W.; Lowers, H.; Driscoll, R.; et al. *USGS Digital Spectral Library: Version 7*; Technical Report; USGS: Reston, VA, USA, 2007. [CrossRef]
44. Leybourne, M.I.; Pontual, S.; Peter, J.M. Integrating Hyperspectral Mineralogy , Mineral Chemistry, Geochemistry and Geological Data at Different Scales in Iron Ore Mineral Exploration. *Proc. Iron Ore Conf.* **2008**, *3*, 1–10.
45. Hunt, G.R. Spectral Signatures of Particulate Minerals in the Visible and Near Infrared. *Geophysics* **1977**, *42*, 501–513. [CrossRef]
46. Kruse, F.A.; Lefkoff, A.B.; Dietz, J.B. Expert system-based mineral mapping in northern death valley, California/Nevada, using the Airborne Visible/Infrared Imaging Spectrometer (AVIRIS). *Remote Sens. Environ.* **1993**, *44*, 309–336. [CrossRef]
47. Buzzi, J.; Riaza, A.; García-Meléndez, E.; Carrère, V.; Holzwarth, S. Monitoring of River Contamination Derived From Acid Mine Drainage Using Airborne Imaging Spectroscopy (HyMap Data, South-West Spain). *River Res. Appl.* **2016**, *32*, 125–136. [CrossRef]
48. Dozier, J.; Painter, T.H. Multispectral and Hyperspectral Remote Sensing of Alpine Snow Properties. *Annu. Rev. Earth Planet. Sci.* **2004**, *32*, 465–494. [CrossRef]
49. Green, R.O.; Painter, T.H.; Roberts, D.A.; Dozier, J. Measuring the expressed abundance of the three phases of water with an imaging spectrometer over melting snow. *Water Resour. Res.* **2006**, *42*. [CrossRef]
50. SenseFly. *3 Reasons to Choose the eBee Plus*; Technical Report; SenseFly: Cheseaux, Switzerland, 2017.
51. Parrot Group. *Parrot Sequoia Technical Specifications*; Technical Report; Parrot Group: Paris, France, 2016.

MDPI
St. Alban-Anlage 66
4052 Basel
Switzerland
Tel. +41 61 683 77 34
Fax +41 61 302 89 18
www.mdpi.com

Minerals Editorial Office
E-mail: minerals@mdpi.com
www.mdpi.com/journal/minerals

www.ingramcontent.com/pod-product-compliance
Lightning Source LLC
Chambersburg PA
CBHW061602220926
51850CB00005B/406